WILDLIFE SCIENCE

LINKING ECOLOGICAL THEORY AND MANAGEMENT APPLICATIONS

EDITED BY
TIMOTHY E. FULBRIGHT AND DAVID G. HEWITT

T0179104

CRC Press
Taylor & Francis Group
Boca Raton London New York

CRC Press is an imprint of the
Taylor & Francis Group, an **informa** business

CRC Press
Taylor & Francis Group
6000 Broken Sound Parkway NW, Suite 300
Boca Raton, FL 33487-2742

First issued in paperback 2019

ISBN-13: 978-0-367-38895-9

Library of Congress Cataloging-in-Publication Data

Wildlife science : linking ecological theory and management applications / edited by Timothy E. Fulbright and David G. Hewitt.
    p. cm.
    Includes bibliographical references.

    1. Animal ecology. 2. Wildlife management. I. Fulbright, Timothy E. II. Hewitt, David G. III. Title.

QH541.W449 2007
591.7--dc22
                                    2007000601

Visit the Taylor & Francis Web site at
http://www.taylorandfrancis.com

and the CRC Press Web site at
http://www.crcpress.com

WILDLIFE SCIENCE

LINKING ECOLOGICAL THEORY AND MANAGEMENT APPLICATIONS

Contents

Preface

Caesar Kleberg created the Caesar Kleberg Foundation for Wildlife Conservation in his will in 1946. He never knew what would become of it or what direction it would take, but what he believed was true, without question or discussion. His rationale for creating the Caesar Kleberg Foundation for Wildlife Conservation is best described by these words in his Last Will and Testament:

"Because of the importance of wildlife and its beneficial effects on the health, habits, and character of the American people."

The trustees of his foundation, Leroy Denman, Jr., Dr. Duane Leach, and Stephen Justice "Tio" Kleberg, created the Caesar Kleberg Wildlife Research Institute in 1981. In early summer 2004, scientists of the Caesar Kleberg Wildlife Research Institute met to plan a 25th anniversary celebration for the Institute. One of their goals was to select a topic for a 25th Anniversary Symposium that would honor the words of Caesar Kleberg and the wise stewardship of the trustees of the Caesar Kleberg Foundation for Wildlife Conservation. The topic selected was "Linking Ecological Theory and Management Applications"; a topic that we felt emphasized the focus of the Caesar Kleberg Wildlife Research Institute, which is advancing the *science* of wildlife management. It is a topic of fundamental and increasing importance in wildlife science and natural resources conservation.

We invited a group of the best and brightest minds in wildlife science to join us in a symposium held in April 2006. By including foremost international experts in wildlife science, the symposium addressed the critically important theme of linking theory and management applications from the perspective of a number of authors working with diverse wildlife species, in a variety of habitats. This design ensured the symposium had an international scope, but at the same time focused on species important to southern Texas, such as white-winged doves (*Zenaida asiatica*) and ocelots (*Leopardus pardalis*). *Wildlife Science: Linking Ecological Theory and Management Applications* is the compilation of the scientific papers presented at that symposium.

Advancement in wildlife management theory and application is inextricably linked to the evolution of ecological theory. The objective of *Wildlife Science: Linking Ecological Theory and Management Applications* is to elucidate the theoretical underpinnings of wildlife management applications and philosophy and to link evolving ecological concepts to changes in applied wildlife management. Wildlife management is an important part of the field of applied ecology; therefore, the expected results of management practices are predictions of the ecological theory upon which management is based. Developing an understanding of the connection between theory and management is critical for students of wildlife science and wildlife professionals. Advances in wildlife management involve, in part, the ability of wildlife professionals to refine and change management paradigms based on new developments and ideas in ecology. The ability of wildlife professionals to connect management and cutting-edge ecological theory is somewhat constrained because ecological theory and new ideas in wildlife management are published in separate scientific outlets that often have distinctly different readerships.

Wildlife Science: Linking Ecological Theory and Management Applications brings together cutting-edge theory and management in a broad perspective, and attempts to establish the importance of the connection between theory and management. Managers generally have a theory in mind, although sometimes they may not be aware of the details or ramifications of that theory. For example, a manager may recommend a certain level of harvest of an animal based on the assumption that reducing densities will result in improved habitat conditions and greater population productivity. The assumption in the example is based on theory, whether or not the manager is aware of the details of that theory. Sometimes predictions of the practitioners may be more astute than predictions of the theoreticians, as pointed out by Dick Potts in Chapter 2. This incongruence emphasizes that practitioners and theoreticians need to work together, not separately.

The primary audience for *Wildlife Science: Linking Ecological Theory and Management Applications* is wildlife and natural resource professionals; these include university professors; biologists for government agencies; biologists working for state wildlife departments; ecological consultants; and university students. This book may serve as a supplementary text for courses in wildlife ecology, landscape ecology, or conservation biology.

The volume is divided into five parts, reflecting the diverse breadth of wildlife science: birds, mammals, habitat, animal health and genetics, and economic and social issues. Part I focuses on landscape ecology of migratory birds; the increasing need for linking theory and practice in game bird management. Part II deals with the ecology of conserving and managing mammal populations. The emergence of ecosystem management in managing wildlife at the ecosystem scale and increasing understanding of the role of climate in applying ecological theory to habitat management are the topics of Part III. Part IV deals with managing wildlife diseases and, also, the increasing importance and role of genetics in conservation and ecology. Economic and social issues affecting wildlife science are the emphasis of Part V.

Caesar Kleberg recognized that managing and conserving wildlife is important to the welfare and character of society. As theoretical ecologists continually develop new ideas and theories, these new concepts can often serve as the basis for improving and refining approaches to wildlife management. The authors hope that the chapters in this volume will help fulfill the goal of advancing wildlife management by connecting it to relevant ecological theory.

Fred C. Bryant
Leroy G. Denman, Jr. Endowed Director of Wildlife Research
Caesar Kleberg Wildlife Research Institute

Editors

Timothy (Tim) Edward Fulbright is a Regents Professor and is the Meadows Professor in Semiarid Land Ecology at the Caesar Kleberg Wildlife Research Institute at Texas A&M University–Kingsville. He is director of the Center for Semiarid Land Ecology and the Jack R. and Loris J. Welhausen Experimental Station. He graduated magna cum laude from Abilene Christian University in 1976 with a bachelor of science degree in biology with a minor in chemistry. He obtained his master's degree in wildlife biology from Abilene Christian University in 1978. In 1981, he completed his PhD in range ecology at Colorado State University.

Tim became an assistant professor at Texas A&M University–Kingsville (then Texas A&I University) in 1981 and served as chair of the Department of Animal and Wildlife Sciences from 1996 to 2000. His primary research interests are wildlife habitat management, habitat restoration, and rangeland ecology. He has authored or coauthored a book, 63 peer-reviewed scientific publications, and seven book chapters.

Tim served as an associate editor of the *Journal of Range Management* during 1989–1993. He is past president of the Texas Section, Society for Range Management. He received the Regents Professor Service Award in 2000, one of the highest awards given by the Texas A&M University System, and the Vice Chancellor's Award in Excellence in Support of System Academic Partnership Efforts, The Agriculture Program of the Texas A&M University System in 2001. He received the Outstanding Achievement Award from the International Society for Range Management in 2004.

David Glenn Hewitt is the Stuart Stedman Chair for White-tailed Deer Research at the Caesar Kleberg Wildlife Research Institute at Texas A&M University–Kingsville. He graduated with highest distinction and honors from Colorado State University in 1987, with a bachelor of science degree in wildlife biology. He earned a master's degree in wildlife biology from Washington State University and then worked for a year as a research associate at the Texas Agriculture Experiment Station in Uvalde, Texas. In 1994, David completed a PhD in wildlife biology at Virginia Tech.

David taught wildlife courses at Humboldt State University during the 1994–1995 academic year and then spent a year as a postdoctoral scientist at the Jack Berryman Institute at Utah State University. He became an assistant professor at Texas A&M University–Kingsville in 1996. His primary research interests are in wildlife nutrition and white-tailed deer ecology and management. He has authored or coauthored 39 peer-reviewed scientific publications and a book chapter.

David served as associate editor of the *Journal of Wildlife Management* during 1997–1998 and *Rangeland Ecology & Management* during 2004–2006. He was recognized as the Outstanding Young Alumnus from the College of Natural Resources at Virginia Polytechnic Institute and State University in 1999, received the Javelina Alumni Award for Research Excellence in 2004, and the Presidential Award for Excellence in Research and Scholarship from the College of Agriculture and Human Sciences, Texas A&M University–Kingsville, also in 2004.

Contributors

John C. Avise
Department of Ecology and Evolutionary
 Biology
University of California
Irvine, California

Guy A. Baldassarre
College of Environmental Science and
 Forestry
State University of New York
Syracuse, New York

Bart M. Ballard
Caesar Kleberg Wildlife Research Institute
Texas A&M University–Kingsville
Kingsville, Texas

Leonard A. Brennan
Caesar Kleberg Wildlife Research Institute
Texas A&M University–Kingsville
Kingsville, Texas

Tyler A. Campbell
USDA APHIS-Wildlife Services
National Wildlife Research Center Texas Field
 Station
Texas A&M University–Kingsville
Kingsville, Texas

Charles A. DeYoung
Caesar Kleberg Wildlife Research Institute
Texas A&M University–Kingsville
Kingsville, Texas

Randy W. DeYoung
Caesar Kleberg Wildlife Research Institute
Texas A&M University–Kingsville
Kingsville, Texas

Diana Doan-Crider
Caesar Kleberg Wildlife Research Institute
Texas A&M University–Kingsville
Kingsville, Texas

D. Lynn Drawe
Rob and Bessie Welder Wildlife Foundation
Sinton, Texas

Barry H. Dunn
Executive Director of the King Ranch Institute
 for Ranch Management
Texas A&M University–Kingsville
Kingsville, Texas

Alan M. Fedynich
Caesar Kleberg Wildlife Research Institute
Texas A&M University–Kingsville
Kingsville, Texas

Marco Festa-Bianchet
Department of Biology
University of Sherbrooke
Sherbrooke, Québec, Canada

Timothy Edward Fulbright
Caesar Kleberg Wildlife Research Institute
Texas A&M University–Kingsville
Kingsville, Texas

David L. Garshelis
Minnesota Department of Natural Resources
St. Paul, Minnesota

Fred S. Guthery
Department of Natural Resource Ecology and
 Management
Oklahoma State University
Stillwater, Oklahoma

Scott E. Henke
Caesar Kleberg Wildlife Research
 Institute
Texas A&M University–Kingsville
Kingsville, Texas

Fidel Hernández
Caesar Kleberg Wildlife Research Institute
Texas A&M University–Kingsville
Kingsville, Texas

David Glenn Hewitt
Caesar Kleberg Wildlife Research Institute
Texas A&M University–Kingsville
Kingsville, Texas

Maurice G. Hornocker
Director, Selway Institute
Bellevue, Idaho

William P. Kuvlesky Jr.
Caesar Kleberg Wildlife Research Institute
Texas A&M University–Kingsville
Kingsville, Texas

Tom M. Langschied
Caesar Kleberg Wildlife Research Institute
Texas A&M University–Kingsville
Kingsville, Texas

Shane P. Mahoney
Sustainable Development and Strategic Science
Department of Environment and Conservation
Government of Newfoundland and Labrador
St. John's, Newfoundland, Canada

Robert G. McLean
National Wildlife Research Center
WS/APHIS/USDA
Fort Collins, Colorado

Karen V. Noyce
Minnesota Department of Natural Resources
St. Paul, Minnesota

J. Alfonso Ortega-S.
Caesar Kleberg Wildlife Research Institute
Texas A&M University–Kingsville
Kingsville, Texas

Alan S. Pine
Conservation and Research Center
Smithsonian National Zoological Park
Front Royal, Virginia

G. R. (Dick) Potts
The Game Conservancy Trust and the World
  Pheasant Association
Hampshire, United Kingdom

John H. Rappole
Conservation and Research Center
Smithsonian National Zoological Park
Front Royal, Virginia

Allen Rasmussen
Caesar Kleberg Wildlife Research Institute
Texas A&M University–Kingsville
Kingsville, Texas

Eric J. Redeker
Caesar Kleberg Wildlife Research Institute
Texas A&M University–Kingsville
Kingsville, Texas

Stuart W. Stedman
Wesley West Interests, Inc.
Houston, Texas

David A. Swanson
Ohio Division of Wildlife
Athens, Ohio

David R. Synatzske
Texas Parks and Wildlife Department
Artesia Wells, Texas

James G. Teer
Rob and Bessie Welder Wildlife Foundation
Sinton, Texas

Michael E. Tewes
Caesar Kleberg Wildlife Research Institute
Texas A&M University–Kingsville
Kingsville, Texas

Jack Ward Thomas
U.S. Forest Service and
College of Forestry and Conservation
University of Montana
Missoula, Montana

Gary L. Waggerman
Texas Parks & Wildlife Department (Retired)
Austin, Texas

Jackie N. Weir
Sustainable Development and Strategic Science
Department of Environment and Conservation
Government of Newfoundland and Labrador
St. John's, Newfoundland, Canada

Part I

Birds

1 Conservation and Management for Migratory Birds: Insights from Population Data and Theory in the Case of the White-Winged Dove

John H. Rappole, Alan S. Pine, David A. Swanson, and Gary L. Waggerman

CONTENTS

The "landscape" for migratory bird species can involve different continents, with important habitats located hundreds or even thousands of kilometers apart. Factors controlling populations of these species are poorly understood, yet management decisions to conserve both game and nongame migrants must be made. Harvest level often is viewed as the principal management tool for migratory game birds, for example, doves and waterfowl, although populations of many species fluctuate in apparent independence of the number of birds taken by hunting each year (Nichols et al. 1995). The concept of carrying capacity provides some insight into the complexity of migrant population control, where the habitat is in shortest supply, whether on the breeding ground, migration stopover sites, or wintering ground; it can limit populations regardless of the specific causes of mortality (Verhulst 1845, 1847). Understanding the life cycle of migratory birds is the most important starting point for successful management and conservation. Nevertheless, management during some portions of the migrants' life cycle is likely to be beyond the control of managers. In these cases, they must obtain and use the best information available to manage those aspects over which they can have some direct effect, and consider ways in which they can influence those factors currently beyond their control.

The white-winged dove (*Zenaida asiatica*), a game species that breeds in Texas and the southwestern United States, is an example of the kind of population manipulation required for management of a hunted migratory species. In this chapter, we examine the life history, population dynamics, and historical and current management of this species, particularly from the perspective of Texas populations, and discuss what field data and theory can provide in terms of understanding population patterns. We consider how this understanding can be used to develop optimal management practices for the species.

WHITE-WINGED DOVE LIFE HISTORY

The white-winged dove has a broad distribution in dry forest, chaparral, arid shrubland, and savanna of the northern subtropic and tropical regions of the Western Hemisphere (Figure 1.1) (Saunders 1968; George et al. 2000; Schwertner et al. 2002; Pacific Flyway Council 2003). Historically, northern breeding populations have been mostly or entirely migratory, while southern Mexican and Central American populations were composed of resident populations year round that were joined by

FIGURE 1.1 Breeding (gray) and wintering/permanent resident (black) range of the white-winged dove.

northern migrants during the winter (Saunders 1968). Banding data show that breeding populations in Texas, New Mexico, Arizona, Nevada, California, and northern Mexico fall into three major groups that appear to be largely allopatric on both their breeding and wintering areas (Figure 1.2) (George et al. 2000; Pacific Flyway Council 2003).

Up until the late 1980s, most Texas white-winged doves originated from Population #1, *Zenaida asiatica asiatica* (Schwertner et al. 2002) (Figure 1.2), birds whose breeding range covered the Tamaulipan Biotic Province of southern Texas and northeastern Mexico (*sensu* Dice 1943; Blair 1950), and whose winter range covered the Pacific slope of Central America (Saunders 1968; Blankinship et al. 1972; George et al. 2000). These birds arrive on native thorn forest breeding sites in late March or early April, with males arriving first. They depend on fruits of native plants as their principal foods on arrival. Territories initially are "Type A" (Nice 1941), in which the breeding pair uses the territory for mating, nesting, and feeding, excluding other adult conspecifics (Swanson 1989). This social system may be the ancestral type of breeding territory for the species. In Sonoran Desert regions of Arizona (Population #3, Figure 1.2), where large amounts of supplementary foods, for example, agricultural seed crops or bird feeders, are not within easy flying distance for many white-winged populations, pairs still establish and defend Type A territories ranging from 0.1 to 4 ha in size (Viers 1970). The doves in these Arizona desert populations obtain most of their foods from within the territory, largely in the form of native plant fruits and seeds, especially saguaro (*Carnegeia gigantea*) fruits (Arizona Sonora Desert Museum 2003).

At thorn forest or citrus grove nesting sites in south Texas, where seed crops become available later in the season, the territories become "Type B" (i.e., for mating and nesting with feeding areas beyond the territory boundaries) (Swanson and Rappole 1993), and can be very small in size indeed, including little more than the nest site (Blankinship 1970). The nest is built 2–3 m up in a thorn forest tree, for example, Texas ebony (*Pithecellobium ebano*), Texas sugarberry (*Celtis laevigata*),

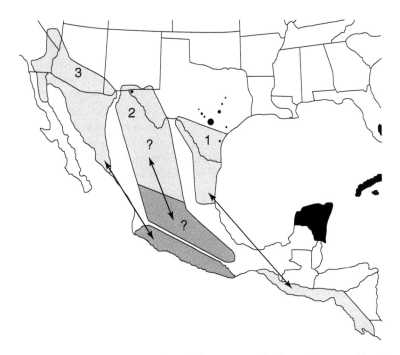

FIGURE 1.2   Breeding (light gray) and wintering (dark gray) range for three migratory whitewing populations: (1) South Texas/Tamaulipan; (2) West Texas/southern New Mexico/north-central Mexico; (3) southwestern United States/northwestern Mexico. Ranges of populations composed mostly or solely of resident birds are shown in black.

and huisache (*Acacia farnesiana*) by both pair members. A clutch, normally of two eggs, is laid, with egg-laying occurring in May. Both parents incubate, and hatching occurs 14 days on average after laying; both parents feed the young using "crop milk" (sloughed esophageal cells) and fledging occurs at 13–18 days post-hatching; parents care for young up to 1 month post-fledging. Second broods are not uncommon. As seed crops become available, use of the breeding territory as a foraging site declines, and individuals and flocks travel back and forth between forest nesting and roosting sites to feeding areas in agricultural fields. Fall migration flights begin in September and continue through early October, flying southward along the Atlantic slope of Mexico, across the Isthmus of Tehuantepec to wintering areas on the Pacific slope (Waggerman and Sorola 1977).

STATISTICAL ANALYSIS OF POPULATION, HABITAT, AND HARVEST DATA

Data on Texas white-winged dove breeding population size, habitat use, and annual harvest size collected by the Texas Parks and Wildlife Department (TPWD) are given in Tables 1.1 and 1.2. Here we present correlation coefficients for the following sets of variables derived from data in Tables 1.1 and 1.2:

1. Size of the breeding population of migratory white-winged doves nesting in thorn forest in the Lower Rio Grande Valley (LRGV) of Texas each year versus amount of thorn forest nesting habitat available.
2. Size of the breeding population of migratory white-winged doves nesting in citrus in the LRGV of Texas each year versus amount of citrus nesting habitat available.
3. Size of the breeding population of migratory white-winged doves nesting in both thorn forest and citrus in the LRGV of Texas each year versus amount of thorn forest plus citrus nesting habitat available.
4. Number of white-winged doves killed by hunters in the LRGV in a given season versus size of the total breeding population of migratory white-winged doves nesting in the LRGV during the following season. In each case, a probability value, p, is also calculated to express the likelihood that the r-value represents a real relationship. We use a value of $p < .05$ to represent probability that the result was significant (i.e., that there was less than a 5% chance that we incorrectly identified a relationship where, in fact, none existed) (Sokal and Rolf 1995; SAS Institute Inc. 2005). These analyses are considered in combination with other whitewing ecological, life history, and population data published in the literature. In addition, we compare actual trends in whitewing populations, illustrated graphically, with models assuming whitewing population control during different seasons of the year.

WHITEWING POPULATION DYNAMICS

It seems likely that by the time ornithologists began recording information on whitewings, planting of seed crops was already a prominent feature for at least the LRGV portion of its Texas range. In the mid-nineteenth century, white-winged doves were reported as "Abundant on the Rio Grande," and it was noted that the species "Finds abundant food from the musquite [sic] and the ebony bean" [McCown (in Lawrence 1858)]. These observations were confirmed by Sennett (1879). The species, however, was limited in its Texas distribution to the Tamaulipan Biotic Province of south Texas (Blair 1950). Whitewings were rare or absent at sites located even a few kilometers north of that region, for example, San Antonio, where whitewings were rare summer visitors and "perhaps" breeding (Attwater 1892). In the early twentieth century, the Texas distribution was, "Southern section of the State. Very abundant summer resident of the Lower Rio Grande counties northwest to Laredo.

TABLE 1.1
Amounts of Breeding Habitat and Population Size by Year for Migratory White-Winged Doves in the LRGV of Texas

Year	Thorn forest (×1000 ha)	Citrus (×1000 ha)	Thorn forest breeding population (×1000)	Citrus breeding population (×1000)	Breeding birds/ha of breeding habitat
1900	400?	—	—	—	—
1923	200?	—	>3 million?	—	—
1939	34	—	500–600	—	15 (thorn forest)
1947	—	—	—	—	—
1948	—	—	—	—	—
1949	—	—	—	—	—
1950	14.5	—	202	839	14 (thorn forest)
1951	—	—	110	—	—
1952	—	—	214	—	—
1953	—	—	137	—	—
1954	—	—	115	—	—
1955	—	—	107	36	—
1956	—	—	115	119	—
1957	—	—	161	173	—
1958	—	—	125	120	—
1959	—	—	167	171	—
1960	—	—	168	273	—
1961	2.5	6.7	209	383	64
1962	3.6	2.7	231	70	48
1963	2.7	2.3	189	88	55
1964	3.6	5.5	302	331	70
1965	—	—	354	250	—
1966	3.9	7.6	426	379	70
1967	4.9	6.7	361	306	58
1968	3.7	10.0	294	227	38
1969	4.3	6.9	219	197	37
1970	4.6	11.6	268	350	38
1971	5.0	11.2	183	342	32
1972	5.0	12.2	173	305	28
1973	4.2	12.4	195	331	32
1974	4.8	12.9	192	337	30
1975	5.8	17.1	290	403	30
1976	4.6	15.3	189	327	26
1977	3.3	21.0	180	276	19
1978	6.0	21.0	200	251	16
1979	8.6	21.0	221	364	20
1980	9.0	21.0	223	285	17
1981	8.2	21.0	250	238	17
1982	8.8	21.0	284	203	16
1983	7.7	21.0	324	253	20
1984	8.0	13.7	227	242	22
1985	7.5	10.1	244	117	21
1986	7.8	11.9	313	159	24
1987	8.6	11.9	314	107	21
1988	8.2	11.9	293	121	21

Continued

TABLE 1.1
Continued

Year	Thorn forest (×1000 ha)	Citrus (×1000 ha)	Thorn forest breeding population (×1000)	Citrus breeding population (×1000)	Breeding birds/ha of breeding habitat
1989	8.8	9.0	296	79	21
1990	8.7	1.0	269	30	31
1991	8.8	0.2	329	9	38
1992	9.7	0.2	364	2	37
1993	—	1.3	430	11	40
1994	11.1	4.1	566	49	40
1995	9.9	2.4	429	25	37
1996	10.3	7.6	356	35	22
1997	9.8	7.1	366	23	23
1998	10.3	5.4	406	18	27
1999	9.8	2.9	410	15	33
2000	9.3	2.5	468	39	43
2001	9.4	2.5	426	39	39
2002	—	2.3	374	40	—
2003	9.9	1.8	363	31	34
2004	10.3	1.9	340	42	31

TABLE 1.2
LRGV Breeding Population Size and Number Killed by Hunters

Year	Total LRGV breeding population size (×1000)	Total kill by hunters (×1000)
1900	—	—
1925	—	—
1939	—	—
1947	—	45
1948	—	144
1949	—	218
1950	202	29
1951	110	28
1952	214	117
1953	137	29
1954	115	—
1955	142	—
1956	234	—
1957	334	375
1958	245	235
1959	338	296
1960	441	60
1961	593	139
1962	301	324
1963	277	—

TABLE 1.2
Continued

Year	Total LRGV breeding population size (×1000)	Total kill by hunters (×1000)
1964	633	675
1965	604	410
1966	805	660
1967	667	797
1968	520	623
1969	416	284
1970	618	241
1971	525	222
1972	475	469
1973	526	386
1974	529	674
1975	693	343
1976	516	483
1977	457	438
1978	451	305
1979	585	498
1980	508	214
1981	488	262
1982	487	391
1983	577	273
1984	469	272
1985	361	—
1986	472	131
1987	421	152
1988	414	124
1989	375	114
1990	303	49
1991	338	46
1992	366	49
1993	441	101
1994	615	113
1995	453	108
1996	391	112
1997	389	267
1998	424	57
1999	425	99
2000	507	212
2001	465	163
2002	414	130
2003	394	193
2004	382	193

Rare summer visitor at San Antonio. Breed at Cotulla, Carrizo Springs, and so forth," according to Strecker (1912). This description of the Texas range was still appropriate as recently as the early 1970s (Oberholser 1974). Before the 1940s, nearly all Texas whitewings bred in Tamaulipan thorn and shrub forest and savanna, variously defined as "Ceniza Shrub" and "Mesquite-Acacia Savanna" (Küchler 1975), "Mesquite Savanna," "Mesquite Chaparral," and "Dry Chaparral" (Rappole and Blacklock 1985) or simply "Brush" (George et al. 2000).

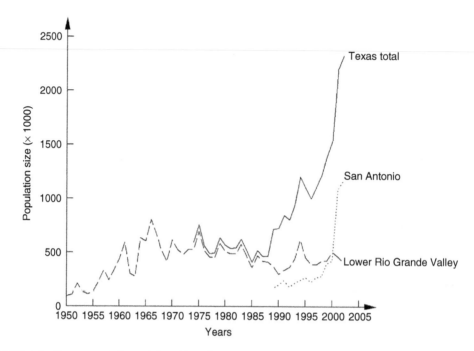

FIGURE 1.3   Texas populations of the white-winged dove, 1951–2002.

Populations in the Texas LRGV, an area that includes Starr, Willacy, Hidalgo, and Cameron counties, were estimated at >3,000,000 in the 1920s (Jones 1945), but had fallen to an estimated 500,000–600,000 by 1939 (Saunders 1940). Annual estimates of breeding population size based on numbers of calling males ("coo counts") were initiated by the TPWD in 1949, and have been carried out until the present. These counts show LRGV breeding populations varying from a high of $\geq$1,000,000 birds in 1950 to a low of 110,000 birds in 1951, subsequent to a severe winter freeze that killed off the citrus in which most of the birds were nesting at that time. The mean population size from 1951 to 2004 is 436,704 (Table 1.1, Figure 1.3).

The four counties of the LRGV comprise 1,099,000 ha. Of these, an estimated 450,000 ha was native thorn forest habitat suitable for whitewing breeding before European colonization. Clearing of native thorn forest habitat in the LRGV began in the early 1800s, and by the 1920s, perhaps half had been altered. By 1942, an additional estimated 200,000 ha of native habitat had been cleared for pasture and agriculture in the LRGV (Marsh and Saunders 1942). By 1961, a low of 2500 ha of native thorn remained in the area (Table 1.1); since then, estimated amounts have increased to roughly 10,000 ha at present (Table 1.1) based on TPWD data.

In the 1940s, citrus orchards began to be established in the LRGV, and by the mid-1950s, some were used extensively for nesting by whitewings. These orchards contain little or no food items for whitewings, and doves using them for nest sites travel to surrounding thorn forest (March–May) or seed crop fields (May–October) to feed. Dependence on seed crop fields distant from the breeding territory for mid- and late-breeding season foods, when available, has been characteristic of thorn forest-breeding doves as well at least since the 1980s, and probably much earlier (Swanson 1989). Breeding populations in citrus and native thorn forest in the LRGV have been monitored separately by TPWD since 1955 (Table 1.1). Amounts of citrus habitat suitable for nesting vary widely from year to year, depending on the frequency of hard winter frosts (George et al. 2000), and have fluctuated from a high of >21,000 ha in 1981 to <200 ha in 1991. Current estimates of citrus available for nesting whitewings in the LRGV are 1800 ha (Table 1.1).

Oberholser (1974) described and pictured the Texas range of the whitewing in a manner comparable to that of Strecker (1912), essentially including the Texas portion of the Tamaulipan Biotic

Province continuing northwest along the Rio Grande to the New Mexico border (Figure 1.1). Documentation of breeding in Bexar County (San Antonio), located just north of the Tamaulipan Biotic Province in the Balconian Biotic Province (Blair 1950), was considered questionable. However, beginning in the 1970s, whitewings began breeding regularly in San Antonio and other urban areas north of their historical range (e.g., Austin and Waco). They also established resident breeding populations in certain urban areas within their historic migratory breeding distribution (e.g., Kingsville and Texas) (Hayslette and Hayslette 1999). Regular censuses by the TPWD were initiated in these urban areas in 1989. These censuses showed sharp increases in urban populations that continue to the present (Figure 1.3). In addition, while south Texas rural populations of whitewings have remained migratory, following much the same annual schedule described by Sennett (1879), Strecker (1912), and Oberholser (1974), the urban populations are mostly or entirely resident year-round.

FACTORS CONTROLLING TEXAS WHITEWING POPULATIONS

White-winged dove breeding populations are estimated based on counts of calling males termed "coo counts," a method that has been applied since 1949 (Uzzell 1949). Coo count estimates of breeding population size can vary significantly from year to year. For instance, in 1963, the estimated LRGV breeding population size was 277,000, while in the following year, the estimated breeding population was 633,000 (Table 1.1). Some variability in breeding estimates of breeding population size likely results from problems concerning assumptions involved in the coo count method. For coo counts to give reliable population estimates, practitioners must assume that there is a precise relationship between the number of calling males and the number of breeding pairs. However, when coo counts are followed by nest survey transects, the results are quite variable (Rappole and Waggerman 1986; Swanson and Rappole 1992; West et al. 1998). The actual relationship between cooing and pair residence is complex, with considerable variability between sites and years. Variability apparently depends on decisions by females as to whether or not to attempt breeding, which may be influenced by rainfall amounts or some other environmental factors. Thus, coo counts and nest survey transects can produce estimates that are nearly identical for some sites, and quite different at other sites or in other years for the same site (Rappole and Waggerman 1986; West et al. 1998).

Most of the annual variation in size of the breeding population is certainly not a function of the variability in the coo count data. This variability likely derives largely from the fact that the LRGV whitewing population is just a portion of the total migratory population in the Tamaulipan Biotic Province. Most of the birds in the population breed in northeastern Mexico. Accurate estimates are difficult to obtain, but Nichols et al. (1986) reported that the total breeding population of migrant whitewings in Tamaulipas numbered in the millions, with at least one colony alone containing >1,000,000 breeding birds.

With the caveat in mind that coo count data contain unknown biases and that we are only looking at the northern tip of the breeding population, the average size of the LRGV breeding population as estimated on this basis has declined from just over 600,000 birds in 1965 to just under 400,000 birds in 1993 (George et al. 1994) (Figure 1.4). Interestingly, Hayslette et al. (1996) found similar results using a completely different methodology, based on transects in which actual numbers of nests with eggs were counted and used to calculate the "average egg density/ha" for each year from 1954 to 1992 for the entire LRGV region in both citrus and thorn forest habitats (Figure 1.5). Over roughly the same time period as that investigated by George et al. (1968–1992), they also found a significant decline in the LRGV breeding population, and they found that the shape of the declines were similar in both citrus and thorn forest, regardless of the amount of each that was available. Taken together, findings reported by the two separate investigations using different methodologies may indicate that the breeding population of migratory whitewings in the LRGV has, in fact, experienced a long-term decline.

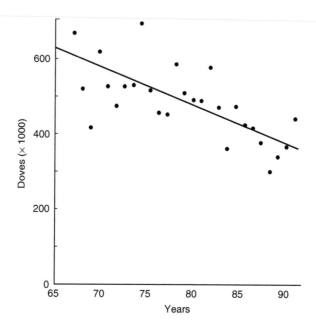

FIGURE 1.4   White-winged dove breeding population estimates (total native thorn forest + citrus) for the Texas LRGV, 1966–1993 [based on coo count data from George, R. R. et al. 1994. In *Migratory Shore and Upland Game Bird Management in North America.* Tacha, T. C. and C. E. Braun (eds). Lawrence, KS: Allen Press, 28.]

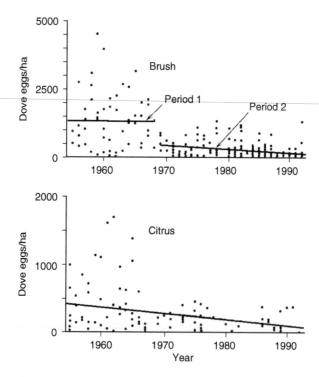

FIGURE 1.5   White-winged dove breeding density in brush (native thorn forest) and citrus habitats in the Texas LRGV, 1954–1993 [based on egg density/ha data from Hayslette, S. E., T. C. Tacha, and G. L. Waggerman. 1996. *J. Wildl. Manage.* 60:298.]. For Citrus, $Y = 18,802 - 9.39X$; $r^2 = .10$; $p = .001$. For Brush, Period 1 $Y = 5489 - 2.09X$; For Brush, Period 2 $Y = 30,522 - 15.25X$; $r^2 = .45$; $p = .0001$.

Many factors have been suggested regarding control of Texas populations of the white-winged dove. Each of these factors is considered below, and evaluated based on available data or theoretical models.

REDUCTION IN BREEDING HABITAT CARRYING CAPACITY

Saunders (1940, 126) thought that whitewing numbers had reached a peak "prior to the beginning of agricultural development and before extensive clearing operations had begun." Others have reached the same conclusion based on the decline in whitewing populations from estimated highs in the early 1920s of >3,000,000 (Jones 1945) to lows of 110,000 at the same time that native breeding habitat declined from >200,000 ha to <12,000 ha (Uzzell 1950; Kiel and Harris 1956; Purdy 1983; George et al. 2000).

The results of the test for correlation between amount of habitat and number of breeding birds (Table 1.1) were as follows:

1. A positive correlation between the size of the LRGV whitewing population using thorn forest and the amount of thorn forest habitat ($r = .55, p = .0001$).
2. A positive correlation between size of the LRGV whitewing population using citrus for nesting and amount of citrus habitat available ($r = .66, p < .0001$).
3. A weak positive correlation between size of the LRGV whitewing breeding population and total amount of breeding habitat (thorn forest + citrus) ($r = .36, p = .0185$).

The strong correlation (.66) between the amount of citrus and the amount of breeding birds results from variation in size of the breeding population in this habitat by several orders of magnitude from 1 year to the next depending on whether or not a freeze occurs. For instance, in 1950, >800,000 birds bred in LRGV citrus, while in 1951 the number was near zero (Table 1.1). During the time period when data were recorded (1950–present) freezes occurred in the winters of 1950–51, 1961–62, 1983–84, 1988–89, 1990–91, and 1991–92; each time marked a drastic reduction in the amount of citrus habitat and the number of birds in this habitat (Table 1.1).

Findings of weak to strong correlations (.36–.66), but very strong evidence of a relationship (p-values all $<.02$) also is not surprising. Breeding populations are calculated by multiplying mean number of birds/ha based on coo counts times the number of ha of each habitat type. This methodology guarantees a very strong positive relationship between the two parameters (i.e., when one increases, the other increases, and vice versa). However, several other factors demonstrate that the amount of breeding habitat is a poor determinant of breeding population size for the migratory breeding population in the LRGV (except on a very gross scale for citrus), and certainly does not function as a limiting factor:

1. Total amounts of breeding habitat in the LRGV differing by a factor of 3 can support roughly the same number of breeding birds. Saunders (1940) estimated a total of 500,000–600,000 birds breeding in an estimated 34,000 ha of habitat (Marion 1974). Presently, 400,000–500,000 birds breed in 10,000 ha.
2. There is no evidence of density-dependent interaction in which some individuals are denied access to breeding territories (Swanson and Rappole 1993).
3. The number of breeding birds in a given piece of apparently suitable habitat can vary from 0 to 400 birds/ha (George et al. 2000).
4. There is no obvious relationship between the total amount of habitat and breeding bird density in any given year. Reference to Table 1.1 shows breeding birds/ha varying from a low of 13 in 1950 to a high of 70 in 1964 and 1966. The chief factor that appears to dictate

    density is amount of habitat, that is, when there are fewer habitats, the birds nest in greater density. This behavior is not indicative of a situation in which density is controlled by the amount of breeding habitat.

5. Hayslette et al. (1996) found whitewings in low density or even absent from apparently suitable breeding habitat, a behavior that is not indicative of situations in which breeding habitat is limiting (Fretwell 1972; Rappole and McDonald 1994).

6. In some years during the period of long-term breeding population decline, for example, 1989, citrus amounted to half or more of the total amount of LRGV breeding habitat, while in others (e.g., 1951) it was near zero due to severe freezes. Despite the extraordinary year-to-year variability in relative total amounts of citrus versus thorn forest breeding habitat, the slope value of the long-term decline in LRGV breeding whitewing density in both thorn forest and citrus was similar (Hayslette et al. 1996). This finding indicates that both values are responding in a way similar to some third variable, which is why there is evidence of a correlation.

FRAGMENTATION OF BREEDING HABITAT

Several researchers have suggested that the size of breeding habitat sites can impact suitability and affect productivity for a number of migratory bird species (Robbins et al. 1989). Hayslette et al. (2000) assessed the impact of woodlot size on whitewing nest density and productivity, and found no apparent relationship.

BREEDING SEASON FOOD AVAILABILITY

Dolton (1975) suggested that lack of sufficient food resources could limit whitewing breeding populations in South Texas. However, the number of grain fields surrounding whitewing nesting sites are unrelated to dove productivity, indicating that food does not limit breeding population size (Hayslette et al. 2000). Similarly, food availability does not explain patterns of whitewing reproduction in the LRGV from 1954 to 1993 (Hayslette et al. 1996).

NEST FAILURE AND PREDATION

The great-tailed grackle (*Quiscalus mexicanus*) is an important predator under certain circumstances on whitewing eggs and young. Removal of grackles from a woodlot of native thorn forest in which both grackles and whitewings occur in high nesting densities causes a sharp increase in whitewing productivity (Blankinship 1966). However, grackles apparently do not affect overall productivity of LRGV whitewings (Hayslette et al. 2000). In addition, urban populations of whitewings share nesting habitat with large numbers of grackles, and these populations are presently increasing at a logarithmic rate (Figure 1.3).

REDUCTION IN WINTERING HABITAT CARRYING CAPACITY

South Texas populations of white-winged doves winter (October–March) in dry forest and shrub habitats from southwestern Mexico to northwestern Costa Rica. George et al. (2000, 7) state:

> Cottam and Trefethen (1968) concluded that winter range in the 1960s was probably not a limiting factor for Texas-reared whitewings. Political unrest and guerilla warfare in Central America during the 1970s and early 1980s may have actually benefited whitewings since agricultural development and deforestation of roosting habitat were probably curtailed during the wartime period.

Despite this sanguine assessment, there is evidence that dry forest and shrub habitats in the whitewing winter range have been severely reduced in amount (Dinnerstein et al. 1995). Nevertheless, destruction of winter habitat might have little effect on whitewing populations if abundant seed crops were available during this period. At present, there are no data we are aware of on whitewing winter ecology, making a data-based assessment of the likelihood of winter population limitation impossible.

FALL HUNTING MORTALITY

The number of birds killed during the fall hunting season in South Texas is quite large, at times equaling or even exceeding South Texas breeding population size (as in 1957; see Table 1.2). Hunting could be a major factor controlling whitewing breeding population size (Marsh and Saunders 1942; Kiel and Harris 1956). In Table 1.2, we present the estimated hunter kill size and South Texas breeding population size for 1951–1997. We correlated the size of the LRGV whitewing kill in a given year with the size of the breeding population the following year, and found that the size of the LRGV whitewing harvest in a given year was positively correlated to the size of the breeding population the following year ($r = .51, p = .0002$). We take this to mean that both the size of the fall kill and the size of the next year's breeding population are controlled by the same factor, namely breeding productivity during the season previous to the hunt. Whether this interpretation is correct or not, the fact that there is a positive correlation between fall harvest and the next year's breeding population does not support the argument that harvest rates have a negative effect on breeding population size.

ADDITIONAL FACTORS UNDER INVESTIGATION

Researchers are currently exploring other factors that could exercise control over LRGV whitewing populations. For instance, Bautch (2004) and Pruitt (2005) suggest that it is possible that food quality (rather than quantity) could limit dove production, whereas Glass et al. (2001) have proposed that the parasite *Trichomonas gallinae* may affect LRGV whitewing population dynamics.

DISCUSSION

Some factor is exerting control over breeding populations of migratory white-winged doves in the LRGV, and presumably the rest of the breeding population in Tamaulipas, holding population levels well below those dictated by either food or nesting habitat availability. The chief evidence of this fact is twofold: Long-term decline in thorn forest breeding density despite stable or increasing amounts of thorn forest for the past 20 years; and the extraordinary range in breeding bird density (0–200 birds/ha) that can occur in different places in the same habitat.

One of the main reasons it is extraordinarily difficult to determine what that factor is derives from the fact that the LRGV breeding population is just a small part of the total breeding population, probably only 5–10%. It is hard to overestimate the significance of this point because, while adults in a migrant breeding population generally return each year to the locality where they have nested in previous years, the young of many migrant species behave as though the entire breeding range was available for them as a source of breeding sites (Graves 1997). Thus, when these young birds return, they act as though they were free to select breeding territories anywhere within the breeding range. This behavior can readily explain large, rapid swings in breeding bird numbers from year to year if only a portion of the breeding range is examined. The LRGV population of migratory whitewings is like the tip of an iceberg.

Nevertheless, the fact that two different measures of breeding population size have shown long-term declines during the past several years that are not related to breeding habitat loss in any obvious way (George et al. 1994; Hayslette et al. 1996) is clearly significant. Interestingly, whatever factor is controlling migrant whitewing populations is not having the same effect on resident breeding populations of the species, which provides further confirmation of the fact that amounts of breeding habitat (in this case, trees along residential streets) do not control the size of the breeding population in the region. Resident populations, which were virtually nonexistent before the 1970s, behave entirely different from migratory populations, almost to the degree of different species. Consider, for instance, the San Antonio population: It is new; it is nonmigratory; it is mostly not hunted. This resident, urban population is currently undergoing rapid logarithmic growth in contrast to the LRGV migratory population, which appears to have undergone a significant, long-term decline dating from at least the 1960s (Figure 1.3) (George et al. 1994; Hayslette et al. 1996). Perhaps even more revealing is the fact that, while migratory populations in rural thorn forest and citrus have declined, resident urban populations in the same region (e.g., Brownsville, Kingsville, and Corpus Christi) are increasing in a manner similar to that seen in the resident San Antonio population. Texas Parks and Wildlife Department data show that in Kleberg County, where the town of Kingsville is located, whitewing breeding populations increased from 2,100 in 1990 to 21,450 in 2003, most of which were resident birds in Kingsville. In addition, while the migrant whitewing breeding season begins in May and runs through August, resident populations begin the breeding season in March, before most migratory whitewings have returned from their Mexican or Central American wintering grounds, and continue into September, after migratory whitewings have commenced preparations for southward migration (Hayslette and Hayslette 1999; Small et al. 2005).

We propose that the migratory white-winged doves may be controlled by wintering ground carrying capacity in southwestern Mexico and western Central America. In a migratory population, for example, the rural LRGV white-winged doves, at least two separate carrying capacities potentially exert control over the population, that is, breeding and wintering ground carrying capacities. In this situation, we propose that the habitat with the lower carrying capacity dictates population size (using an equation developed by A. R. Pine in Pine and Rappole [2007]). This possibility is illustrated in Figure 1.6 in which a hypothetical whitewing population is subject to control by (1) a breeding ground carrying capacity ("Habitat 1") during the summer; and (2) a wintering ground carrying capacity ("Habitat 2") in winter. In this figure, a graph is shown for a model population of migratory white-winged doves in the LRGV for summer and winter habitats

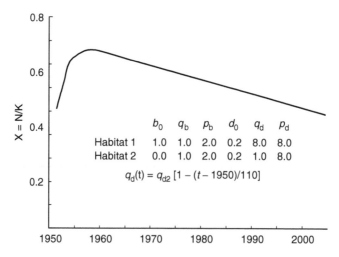

FIGURE 1.6 Hypothetical population growth curve for the migratory white-winged dove population of the LRGV predicted by the equation $dx/dt = x\{b_0 \exp[-(x/q_b)^{P_b}] - (d_0 - 1) \exp[-(x/q_d)^{P_d}] - 1\}$, as described in the text for the time period 1950–2005. The seasonal variations given by this sequential habitat model have been averaged for the purposes of graphic depiction.

according to the following equation:

$$\frac{dx}{dt} = x \left\{ b_0 \exp\left[-\left(\frac{x}{q_b}\right)^{p_b} \right] - (d_0 - 1) \exp\left[-\left(\frac{x}{q_d}\right)^{p_d} \right] - 1 \right\},$$

where x is the population N relative to the carrying capacity K over time t; b_0 and d_0 are the birth and death rates in a given habitat; q_b and q_d are the carrying capacities for the birth and death rates relative to K; p_b and p_d are the saturation powers for the birth and death rates representing the abruptness of the response at the respective carrying capacities. All parameters can vary with the habitat; initial carrying capacity in 1950 for both Habitat 1 and Habitat 2 are set as equivalent; carrying capacity for Habitat 1 is considered to have remained the same from 1950 to the present, while that for Habitat 2 is considered to have declined linearly to roughly half its value in 1950 by 2005; the initial population size (x_0) in 1951 is set at 50% of the carrying capacity in habitats 1 and 2 as of that date after the freeze of 1950; other parameters used are shown in Figure 1.6. Annual birth and death rates for the population are based on a combination of the birth and death rates for Habitats 1 and 2 (note that birth rate in Habitat 2, the wintering ground, is assumed to be zero). The shape of the curve in Figure 1.6 will vary somewhat, depending on the specific parameters used. However, so long as the average birth rate for the two habitats exceeds the death rate, the population will decline in accordance with whichever carrying capacity is smaller. The "carrying capacity" function in the equation essentially acts as a flexible death rate. The population goes to extinction regardless of carrying capacity if average death rate exceeds birth rate, unless offset by immigration, a factor not taken into account in this equation.

Following the logic of this model, we propose (1) that actual breeding habitat carrying capacity, although unknown, exceeds current population levels; (2) that actual winter habitat carrying capacity for migratory whitewings, although also unknown, is less than the number of birds migrating southward after years of good production; and (3) that this winter habitat carrying capacity has been lower than breeding habitat carrying capacity, and has been in decline since at least the early 1960s. These factors result in a curve showing an average long-term decline similar to that reported by Hayslette et al. (1996).

Testing validity of our conclusions concerning the effects of winter habitat on migratory whitewing populations requires investigation of the wintering ground to determine how much habitat is available, and how the birds use it. Regardless of whether or not our hypothesis is correct, it should be apparent that, for a species that spends half of its annual cycle (October–March) in a habitat >1000 km away from its breeding habitat, this portion of its life cycle should have significant relevance to its ecology, population dynamics, and conservation.

ACKNOWLEDGMENTS

This chapter is dedicated to the memory of Dr. Sam Beasom (1945–1995), director of the Caesar Kleberg Wildlife Research Institute from 1984 to 1995. Data for the analyses included in this paper were gathered by personnel from the TPWD. We thank TPWD for use of these excellent data gathered over many years by a cadre of dedicated personnel, and for their efforts in managing this important species. Steven Hayslette was helpful in providing literature and information on the key points raised in the paper. Loren Smith and Michael Small also provided literature and information on urban dove populations.

REFERENCES

Arizona Sonora Desert Museum. 2003. Migratory pollinators program: Spring 2003 update; 2003. White-winged doves (*Zenaida asiatica*), http://www.desertmuseum.org/pollination/doves.html.

Attwater, H. P. 1892. List of birds observed in the vicinity of San Antonio, Bexar County, Texas. *Auk* 9:229.

Bautch, K. A. 2004. Historic and current forage area locations and food abundance in relation to nesting sites for white-winged doves in the lower Rio Grande Valley of Texas. Masters Thesis, Texas A&M University — Kingsville, Kingsville, TX.

Blair, W. F. 1950. The biotic provinces of Texas. *Texas J. Sci.* 2:93.

Blankinship, D. R. 1966. The relationship of white-winged dove production to control of great-tailed grackles in the Lower Rio Grande Valley of Texas. *Trans. N. Am. Wildl. Nat. Resourc. Conf.* 31:45.

Blankinship, D. R. 1970. White-winged dove nesting colonies in northeastern Mexico. *Trans. N. Am. Wildl. Nat. Resourc. Conf.* 35:171.

Blankinship, D. R., J. G. Teer, and W. H. Kiel, Jr. 1972. Movements and mortality of white-winged doves banded in Tamaulipas, Mexico. *Trans. N. Am. Wildl. Nat. Resour. Conf.* 37:312.

Cottam, C., and J. B. Trefethen. 1968. *White-Wings: The Life History, Status, and Management of the White-Winged Dove.* Princeton, NJ: D. Van Nostrand.

Dice, L. R. 1943. *The Biotic Provinces of North America.* Ann Arbor, MI: University of Michigan Press.

Dinnerstein, E., D. M. Olson, D. J. Graham, A. L. Webster, S. A. Primm, M. P. Bookbinder, and G. Ledec. 1995. *A Conservation Assessment of the Terrestrial Ecoregions of Latin America and the Caribbean.* Washington, DC: World Wildlife Fund.

Dolton, D. D. 1975. Patterns and influencing factors of white-winged dove feeding activity in the Lower Rio Grande Valley of Texas and Mexico. Masters thesis, Texas A&M University, College Station, TX.

Fretwell, S. 1972. *Populations in a Seasonal Environment.* Princeton, NJ: Princeton University Press.

George, R. R., R. E. Tomlinson, R. W. Engel–Wilson, G. L. Waggerman, and A. G. Sprah. 1994. White-winged dove. In *Migratory Shore and Upland Game Bird Management in North America.* Tacha, T. C. and C. E. Braun (eds). Lawrence, KS: Allen Press, 28.

George, R. R., G. L. Waggerman, D. M. McCardy, R. E. Tomlinson, D. Blankinship, and J. H. Dunks. 2000. *Migration, Harvest, and Population Dynamics of White-Winged Doves Banded in Texas and Northeastern Mexico, 1950–1978.* Austin, TX: Texas Parks Wildlife Department.

Glass, J. W., A. M. Fedyrich, M. F. Small, and S. J. Benn. 2001. *Trichomonas gallinae* in an expanding population of white-winged doves from Texas. *Southw. Nat.* 46:234.

Graves, G. 1997. Geographic clines of age ratios of black-throated blue warblers (*Dendroica caerulescens*). *Ecology* 78:2524.

Hayslette, S. E., and B. A. Hayslette. 1999. Late and early season reproduction of urban white-winged doves in southern Texas. *Texas J. Sci.* 51:173.

Hayslette, S. E., T. C. Tacha, and G. L. Waggerman. 1996. Changes in white-winged dove reproduction in southern Texas, 1954–93. *J. Wildl. Manage.* 60:298.

Hayslette, S. E., T. C. Tacha, and. G. L. Waggerman. 2000. Factors affecting white-winged, white-tipped, and mourning dove reproduction in lower Rio Grande valley. *J. Wildl. Manage.* 64:286.

Jones, C. G. 1945. Past and present whitewings. *Texas Game and Fish* 3:13.

Kiel, W. H., Jr., and. J. T. Harris. 1956. Status of the white-winged dove in Texas. *Trans. N. Am. Wildl. Nat. Resourc. Conf.* 21:376.

Küchler, A. W. 1975. *Potential Natural Vegetation of the Conterminous United States.* New York: American Geographical Society.

Lawrence, G. W. 1858. Ornithological notes No. 2. *Ann. Lyc. Nat. Hist. New York* 6:7.

Marion, W. R. 1974. Status of the plain chachalaca in south Texas. *Wilson Bull.* 86:200.

Marsh, E. G., and G. B. Saunders. 1942. The status of the white-winged dove in Texas. *Wilson Bull.* 54:145.

Nice, M. M. 1941. The role of territory in bird life. *Am. Midl. Nat.* 26:441.

Nichols, J. D., F. A. Johnson, and B. K. Williams. 1995. Managing North American waterfowl in the face of uncertainty. *Ann. Rev. Ecol. Syst.* 26:177.

Nichols, J. D., R. E. Tomlinson, and G. L. Waggerman. 1986. Estimating nest detection probabilities for white-winged dove nest transects in Tamaulipas, Mexico. *Auk* 103:825.

Oberholser, H. C. 1974. *The Bird Life of Texas.* Austin: University of Texas Press.

Pacific Flyway Council. 2003. *Pacific Flyway Management Plan for Western White-Winged Doves.* Portland, OR: U.S. Fish Wildl. Serv.

Pine, A. S., and J. H. Rappole. 2007. Population limits with multiple habitats and mechanisms: I. age-independent periodic breeders. Ecology (submitted).

Pruitt, K. D. 2005. Bioenergetics and foraging of white-winged doves in the lower Rio Grande Valley of southern Texas. Masters Thesis, Texas A&M University — Kingsville, Kingsville, TX.

Purdy, P. C. 1983. Agricultural, industrial, and urban development in relation the white-winged dove. Masters Thesis, Colorado State University, Fort Collins, CO.

Rappole, J. H., and G. W. Blacklock. 1985. *Birds of the Texas Coastal Bend: Abundance and Distribution.* College Station, TX: Texas A&M University Press.

Rappole, J. H., and M. V. McDonald. 1994. Cause and effect in migratory bird population changes. *Auk* 111:652.

Rappole, J. H., and G. L. Waggerman. 1986. Calling males as an index of density for breeding white-winged Doves. *Wildl. Soc. Bull.* 14:141.

Robbins, C. S., D. K. Dawson, and B. A. Dowell. 1989. Habitat area requirements of breeding birds of the middle Atlantic states. *Wildl. Monogr.* 103:1.

SAS Institute, Inc. 2005. *SAS/STAT: Guide for Personal Computers. Version 9.13.* Cary, NC: SAS Institute, Incorporated.

Saunders, G. B. 1940. *Eastern White-Winged Dove (Melopelia asiatica asiatica) in Southeastern Texas.* Washington, DC: U.S. Biological Survey.

Schwertner, T. W., H. A. Mathewson, J. A. Roberson, M. F. Small, and G. L. Waggerman. 2002. White-winged dove. In *The Birds of North America.* Poole, A. and F. Gill (eds), Philadelphia: National Academy of Sciences.

Sennett, G. B. 1879. Further notes on the ornithology of the lower Rio Grande of Texas, from observations made during the spring of 1878. *Bull. U.S. Geol. Geog. Surv. Terr.* 5(3):371.

Sokal, R. R., and F. J. Rolf. 1995. *Introduction to Biostatistics*, 2nd edn. New York: Freeman and Co.

Small, M. F., C. L. Schaefer, J. T. Baccus, and J. A. Roberson. 2005. Breeding ecology of white-winged doves in a recently colonized urban environment. *Wilson Bull.* 117:172.

Strecker, J. K., Jr. 1912. The birds of Texas: An annotated check-list. *Baylor Univ. Bull.* 25(1):1–70.

Swanson, D. A. 1989. Breeding biology of the white-winged dove (*Zenaida asiatica*) in south Texas. Master's thesis, Texas A&I University — Kingsville, Kingsville, TX.

Swanson, D. A., and J. H. Rappole. 1992. Status of the white-winged dove in southern Texas. *Southw. Nat.* 37:93.

Swanson, D. A., and J. H. Rappole. 1993. Breeding biology of the eastern white-winged dove in southern Texas. *Southw. Nat.* 38:68.

Uzzell, P. B. 1949. Status of the white-winged dove in Texas. Texas Parks Wildl. Dep., Fed. Aid Perf., Proj. W-30-R.

Uzzell, P. B. 1950. Status of the white-winged dove in Texas. Texas Parks Wildl. Dep., Fed. Aid Perf., Proj. W-30-R.

Verhulst, P. F. 1845. Recherches mathématiques sur la loi d'accroissement de la population. *Nouv. mém. de l'Academie Royale des Sci. et Belles-Lettres de Bruxelles* 18:1.

Verhulst, P. F. 1847. Deuxième mémoire sur la loi d'accroissement de la population. *Mém. de l'Academie Royale des Sci., des Lettres et des Beaux-Arts de Belgique* 20:1.

Viers, C. E., Jr. 1970. The relationship of calling behavior of white-winged doves to population and production in southern Arizona. PhD Dissertation, University of Arizona, Tucson, AZ.

Waggerman, G. L., and S. Sorola. 1977. Whitewings in the winter, *Texas Parks Wildl.* 35:6.

West, L. M., L. M. Smith, and R. R. George. 1998. The relationship between white-winged dove call-count surveys and nest densities in an urban environment. *Wildl. Soc. Bull.* 26:259.

2 Avian Ecology at the Landscape Scale in South Texas: Applying Metapopulation Theory to Grassland Bird Conservation

William P. Kuvlesky, Jr., Leonard A. Brennan,
Bart M. Ballard, and Tom M. Langschied

CONTENTS

South Texas is among the most ecologically diverse regions in North America. Indeed, the number of plant and vertebrate species on the 365-km^2 Lower Rio Grande Valley National Wildlife Refuge is comparable to or exceeds the number of plant and vertebrate species found in both the 6102-km^2 Everglades National Park and the 9034-km^2 Yellowstone National Park (Fulbright and Bryant 2001).

The tremendous diversity of wildlife in South Texas reflects the extraordinary diversity of vegetation communities that results from the spatially and temporally variable environmental conditions characterizing the region. Moreover, the unique land ownership patterns of South Texas ranchers also contribute to the continued preservation of South Texas biodiversity. Unlike numerous rural landowners in Texas, many South Texas landowners are third- and fourth-generation ranchers who have resisted commercial development of their family properties and have thereby minimized the landscape fragmentation that is so prevalent elsewhere in Texas. Moreover, many of these landowners have managed extensive areas of their properties in a manner that has minimally disturbed native vegetation communities. Consequently, as Fulbright and Bryant (2001) so aptly indicate, South Texas truly represents one of the last great habitats in North America.

Bird communities that inhabit the region are uniquely diverse and abundant because of the unique climate and vegetation community diversity that characterizes South Texas, and the fact that at least three bird migration corridors converge immediately north of Corpus Christi, which funnel millions of migratory birds into the Lower Texas Gulf Coast. According to the most recent edition of the *Texas State Bird Checklist*, more than 620 species of birds have been documented in Texas. There are over 200 resident species in South Texas alone, and an additional 200 migratory species are commonly observed during spring and fall months. Many of these migratory species are transients that stop and use specific habitats for short periods to rest and replenish energy reserves before continuing south to wintering areas in Mexico or Central and South America. However, numerous migratory grassland bird species remain in South Texas for the winter, and many of the 200 resident species are grassland birds that complete their annual life cycles on the Rio Grande Plains. Clearly, the South Texas region is critical to both migratory and resident grassland bird species. However, similar to grassland habitats elsewhere throughout North America, some of the land management activities utilized by landowners on South Texas grasslands during the past 150 years have degraded the utility of these habitats for grassland bird species. Some of these harmful land management activities continue on thousands of acres of grasslands today. Important grassland bird habitat is being lost to agricultural and livestock production, exotic plant invasions, and the explosive development of the Lower Rio Grande Valley. Yet, few avian ecologists understand how these activities are impacting grassland bird populations that depend on South Texas landscapes to complete their life cycles successfully.

GRASSLAND BIRDS

Grassland birds have been important components of grassland ecosystems in North America for thousands of years. The pre-Columbian abundance and diversity of grassland bird species is unknown. Nevertheless, it is probably safe to assume that the abundance of most endemic grassland bird species was greater before Europeans settled grassland ecosystems and disrupted their natural functional dynamics. Grassland bird communities inhabiting the short-, mid-, and tallgrass prairies of North America partition habitat according to specialized life history requirements that vary on a seasonal basis. Natural fires generated by lightening, seasonal flooding of moist prairie soil types, and the grazing patterns of bison and prairie dogs created a mosaic of habitat types that comprised prairie landscapes (Askins 1999).

During the mid-nineteenth century to the early twentieth century, European settlers began to systematically disrupt North American prairie ecosystems. First, commercial hide hunters and the U.S. Army virtually exterminated bison, and then stockmen overgrazed much of the dominant herbaceous prairie vegetation. The incidence and impacts of fires were simultaneously reduced, which, in turn, promoted growth of woody vegetation. Furthermore, farmers converted millions of hectares of grasslands to agricultural fields. Ranchers, along with state and federal agencies, introduced invasive exotic grasses, which have rapidly replaced native prairies species. It has been estimated that less than 1% of the tallgrass prairie that existed before Europeans arrived remains today (Samson and

Knopf 1994). Reduction of the continental mid- and short-grass prairies has not been quite as extreme as it has been for tallgrass prairie (Knopf 1994). It should, therefore, be no surprise that grassland bird populations have declined simultaneously with the widespread loss of grassland habitat in North America (Brennan and Kuvlesky 2005).

Grassland bird populations have likely been declining for some time; however, continental-scale documentation of these declines was impossible until the National Breeding Bird Survey (BBS) was implemented in the mid-1960s (Robbins et al. 1986). Consequently, the serious nature of grassland bird population declines were not apparent until relatively recently when scientists analyzed long-term trends in BBS data. Peterjohn and Sauer (1999) analyzed BBS data collected between 1966 and 1996, and found that as a group, grassland birds displayed a lower proportional increase than any other avian guild in North America. Furthermore, they reported that populations of 13 species of grassland birds had declined, while only two species increased over the 30-year period. Moreover, the grassland bird decline has been remarkably consistent across geographic areas, compared with recent forest-obligate Neotropical migrant declines, which can vary considerably on a geographic basis (James et al. 1992; Knopf 1994; Walters 1998). Indeed, grassland bird population declines in North America appear to be occurring in conjunction with a worldwide decline in grassland bird populations (Goriup 1988).

The continental decline in grassland bird populations in North America prompted a flurry of research activity beginning in the mid-1980s among scientists working for numerous state and federal agencies, as well as non-governmental organizations (NGOs) interested in bird conservation. Declines of most species were linked to the degradation and loss of grasslands in North America (Knopf 1994; Noss et al. 1995). Most avian ecologists and bird conservationists agreed with this assessment; however, many recognized that the ecological mechanisms driving the bird declines, which were associated with habitat loss, need to be identified. Consequently, numerous studies were conducted to determine potential factors responsible for the population declines of specific species or groups of North American grassland birds (Bollinger et al. 1990; Frawley and Best 1991; Bollinger and Gavin 1992; Bock et al. 1993; Bowen and Kruse 1993; Saab et al. 1995; Zimmerman 1996; Best et al. 1997; Dale et al. 1997; Griebel et al. 1998; Herkert and Glass 1999; Madden et al. 1999; Temple et al. 1999; Winter 1999; Gordon 2000).

Much has been learned about grassland bird ecology during the past 10–15 years (Vickery and Herkert 2001). In the recent flurry of research, factors contributing to the negative demographic trends of grassland birds have been identified, and this information has enhanced grassland bird conservation. Most of these recent studies also highlight critical research needs. Unfortunately, as noted earlier, most of the recent grassland bird work has been conducted on breeding grounds in the Upper Midwest and Northern Great Plains. Little work has been conducted on breeding and wintering grassland bird populations in South Texas. This oversight is lamentable, because it will be very difficult to reverse population declines if researchers continue to ignore regions where grassland birds complete about one-half of their annual life cycle. Furthermore, several declining grassland bird species appear to breed and fledge young successfully, indicating that their declines may be associated with problems occurring during migration or on the wintering grounds (Vickery and Herkert 2001). Expanded efforts to quantify the wintering ecology of grassland birds are clearly justifiable.

SOUTH TEXAS GRASSLAND BIRDS

Most of the grassland bird research conducted in South Texas has been devoted to understanding the ecology of the northern bobwhite (*Colinus virginanus*) because of their economic value to South Texas landowners. In contrast, few studies relevant to the ecology of grassland nongame birds that inhabit South Texas landscapes are published in the scientific literature. Most published studies included both grassland and shrubland birds (Roth 1977; Grzybowski 1982, 1983; Baker and Guthery 1990; Chavez-Ramirez and Prieto 1994; Vega and Rappole 1994; Fulbright and Guthery 1996; Nolte and Fulbright 1997; Van't Hul et al. 1997; Reynolds and Krausman 1998; Igl and Ballard 1999; Mix 2004; Flanders et al. 2006).

Clearly, little research has been accomplished for grassland birds in South Texas, and most of the research that has been carried out on bird communities focused on bird ecology, the impacts of range management on habitat, or had narrow scales of resolution. To our knowledge, no one has attempted to determine how avian communities interact with South Texas landscapes, which is remarkable given the large number of studies that have been carried out recently and that have addressed the landscape ecology of birds elsewhere in North America. Therefore, avian ecologists do not know how fragmentation of important habitats are impacting grassland birds in South Texas, and it is quite possible that fragmentation of the region's landscapes are contributing to the population declines currently exhibited by some grassland bird species. Opportunities certainly exist to determine how birds interact with a gradient of South Texas landscapes that range from significantly fragmented to minimally disrupted. The purpose of this chapter is to review the research that has been accomplished regarding how birds interact with the landscapes they occupy, and then use the predictions of metapopulation theory to better manage the landscapes of South Texas for grassland birds.

HABITAT FRAGMENTATION

Much of the research relevant to the landscape ecology of bird communities has focused on the impacts of habitat fragmentation on individual bird species or bird communities occupying forested, grassland, or agricultural landscapes. Avian ecologists believed that anthropogenic activities such as farming, timber management, and urbanization had fragmented large, continuous landscapes into islands too small and isolated to support self-sustaining bird communities. Consequently, researchers began to focus on the response of bird populations to the fragmentation of broad landscapes. Based on the results of these studies, population declines that are ongoing for numerous forest and grassland bird species seem to be related to habitat fragmentation, though certainly not all grassland and forest bird species exhibit area sensitivity (Coppedge et al. 2001; Johnson and Igl 2001; Bakker et al. 2002; Coppedge et al. 2004).

BREEDING GROUNDS

A majority of the bird-habitat fragmentation studies in North America have been conducted in breeding habitats. The majority of these studies indicated that bird populations in fragmented landscapes are declining due to poor survival and poor reproductive success. Askins et al. (1987), for example, conducted one of the first studies in which researchers evaluated the impact of habitat fragmentation on breeding forest bird communities in the United States. They found that abundance and species diversity of forest interior birds were lower in smaller forests, and in isolated forest tracts. Forest area was the best predictor of density and species richness for forest interior bird species inhabiting small forest tracts, and isolation was the best predictor for large forests. Two of the primary threats to Neotropical migratory forest and grassland birds are habitat fragmentation and habitat loss, which are closely associated with fragmentation (Robinson 1995). Moreover, habitat fragmentation is associated with reduced North American breeding forest bird numbers and increased temporal variability in the number of species, which is associated with higher local extinction and turnover rates (Boulinier et al. 2001). Similarly, Rosenberg et al. (1999) assumed a continental perspective in examining the impacts of forest fragmentation on breeding tanagers (*Piranga* spp.) in North America and found that three widespread tanager species exhibited negative responses to fragmentation. They stated that effects of fragmentation were similar for breeding tanagers across most of North America, although sensitivity to fragmentation was local and seemed lower in regions with greater forest cover.

Nesting success was too low for Neotropical migrants inhabiting woodlots in fragmented landscapes in Illinois for these bird populations to be self-sustaining because of the small size and isolation of woodlots (Brawn and Robinson 1996). In another Illinois study, Bollinger et al. (1997) reported poorer productivity among Neotropical migrants on fragmented landscapes. In addition, Donovan et al. (1995a) reported that habitat fragmentation negatively impacted forest-nesting breeding birds in Missouri, Wisconsin, and Minnesota. Total nest failure for three migrant species was significantly higher in forest fragments than in continuous forest. In addition, forest-breeding birds in southern Ontario benefit from maintenance of large forest fragments in eastern deciduous forest landscapes (Nol et al. 2005). In the boreal forests of central Saskatchewan, Bayne and Hobson (2001) documented fewer paired ovenbird (*Seiurus aurocapillus*) males/site in fragments created by forestry and agriculture compared with study sites composed of contiguous forest. Wood thrush (*Hylocichla mustelina*) experience lower reproductive success and higher nest losses and rates of nest parasitism in small forest tracts than large tracts (Weinberg and Roth 1998). Fragmentation of pine barrens in northwestern Wisconsin was one of the primary reasons for the decline of sharp-tailed grouse (*Tympanuchus phasianellus*) populations, and maintenance of important grouse habitat fragments of sufficient size and spatial arrangement is critical to maintaining viable grouse populations (Akcakaya et al. 2003). Likewise, fragmentation of managed pine woodlands may be a potential reason for recent population declines experienced by Bachman's sparrow (*Aimophila aestivalis*) (Dunning et al. 1995).

In addition to forested landscapes, numerous studies on the impacts of habitat fragmentation on breeding bird populations and communities inhabiting grassland landscapes have been accomplished. Vickery et al. (1994) documented area sensitivity for six of the ten grassland bird species on grassland-barren sites in Maine and reported that grassland fragments <200 ha in size are unlikely to support a diverse bird community. Almost half of the 15 bird species that Johnson and Igl (2001) examined in four northern Great Plains states exhibited area sensitivity, because their occurrence and abundance were higher in large grassland fragments than in smaller ones. In South Dakota, Bakker et al. (2002) also documented area sensitivity for grasshopper sparrows and dickcissels (*Spiza americana*) in mixed grass prairie and for savanna sparrows (*Passerculus sandwichensis*) in tallgrass prairie. Greater prairie chickens (*Tympanuchus cupido*), Henslow sparrows (*Ammodramus henslowii*), and dickcissels were extremely area sensitive or negatively impacted by fragmentation in native tallgrass prairie in southwestern Missouri (Winter and Faaborg 1999). Prairie chickens were absent from small prairie fragments, Henslow sparrow densities were low in small fragments, and dickcissels' nesting success was reduced on small prairie fragments.

Grassland bird densities in southern Wisconsin are lower in more diverse landscapes that are more fragmented than landscapes that are less diverse and consist of primarily grassland, pasture, and hay fields (Ribic and Sample 2001). Most grassland bird species exhibited consistent declines on Oklahoma grasslands that were in the process of fragmentation by invading juniper (*Juniperus* spp.), although open habitat generalists and woodland and successional scrub species increased on these invaded grassland landscapes (Coppedge et al. 2001). Indeed, Coppedge et al. (2004) projected a decline in some obligate and facultative grassland bird species as juniper continues to invade these remnant grasslands in Oklahoma. Conserving large tracts of mixed grass prairies in southern Saskatchewan is important to grassland bird conservation (Davis 2004). Abundance of numerous species of grassland birds studied by Davis (2004) was lower in small prairie fragments than larger fragments or in more continuous tracts of prairie. Implementation of the Conserve Reserve Program (CRP) provides additional evidence that the fragmentation of grasslands has negatively impacted numerous grassland bird species, because several grassland specialist populations that had been declining in the Northern Plains states have increased since establishment of CRP in 1985 (Reynolds et al. 1994; Johnson and Igl 1995). Perennial grass fields established via CRP are particularly important as breeding habitats (Igl and Johnson 1997).

WINTERING GROUNDS

Much of the research devoted to quantifying effects of habitat fragmentation on Neotropical migratory birds has been conducted on grassland and forest-breeding grounds. These studies demonstrate that fragmentation of breeding landscapes negatively impacts a variety of forest and grassland species. As a consequence of these studies, remedial practices have been identified that could be implemented to reverse or stabilize declines of bird populations that are negatively impacted by fragmentation. Nevertheless, fewer studies have been devoted to quantifying the impacts of habitat fragmentation on wintering grounds, which is inexplicable, because most migrants spend more than half of each year on wintering grounds. The primary objective of wintering birds is surviving and maintaining body condition (Faaborg et al. 1995), which requires habitats that provide sufficient amounts of food and adequate amounts of cover for birds to successfully survive winter. This is why many species migrate thousands of miles between temperate and tropical climates. Loss or degradation of wintering habitats would force Neotropical migrants into suboptimal habitats, where obtaining sufficient food and evading predators would be more difficult, thereby reducing winter survival and forcing birds to begin spring migrations to breeding grounds in poorer body condition. Neotropical migrating birds occupy an array of habitats on wintering grounds. Determining which habitats are optimal and suboptimal requires long-term studies that measure survival and fitness (Conway et al. 1995). Comprehensive habitat studies have not been conducted on wintering grounds, possibly because of the difficulty associated with collecting these data (Rappole and McDonald 1994; Faaborg et al. 1995).

Another reason why fewer fragmentation studies have been conducted on the wintering grounds of Neotropical migrants may be the belief that habitat fragmentation is not particularly relevant on wintering landscapes. Habitat fragmentation did not appear to detrimentally impact winter residents in the Caribbean and Central America (Robbins et al. 1987; Hagan and Johnston 1992). Many grassland bird species show less habitat specificity on wintering areas relative to breeding areas (Confer and Holmes 1995; Igl and Ballard 1999). Owing to this, these species possibly use suboptimal habitats when loss of optimal habitat occurs on wintering grounds, because they are primarily concerned with survival (Faaborg et al. 1995). However, differing spring migration strategies among species may dictate how important wintering habitats are to migrant birds and, thus, their ability to accumulate energy in a timely manner to optimize migration.

Many Neotropical migratory bird species appear to be controlled by wintering ground events (Rappole and McDonald 1994). Declines of several populations of Neotropical migratory bird species were not correlated with loss of breeding habitat but were instead strongly correlated with loss of wintering habitat in the tropics (Rappole and McDonald 1994). In addition, a loggerhead shrike in Minnesota declined because of winter habitat loss (conversion of pasturelands and hayfields to row crops) on wintering grounds in Gulf Coastal regions of the United States (Brooks and Temple 1990). Destruction of important wintering habitat in South America is contributing to the declines of upland sandpipers (*Bartramia longicauda*) and bobolinks (*Dolichonyx oryzivorus*) (Bucher and Nores 1988; Herkert 1997). Golden-cheeked warblers (*Dendroica chrysoparia*) tolerate moderate levels of logging and grazing of their wintering habitats in Mexico and Central America, but understory clearing of forests significantly threatens winter habitat availability (Rappole et al. 1999).

Declines of some Neotropical migrant species could be related to the loss of optimal wintering habitats; however, data regarding the wintering habitat preferences of numerous species are limited (Faaborg et al. 1995). Total Neotropical migrant abundance is lower in wet or dry limestone forest than in mangrove and coastal forests, which are presumably better habitat, because the specific habitat requirements of some bird species are not met in the more disturbed wet or dry limestone forests (Confer and Holmes 1995). Similarly, conversion of tropical forest habitat (optimal) to second-growth habitat (suboptimal) possibly degrades the quantity and quality of wintering habitat of American redstarts (*Setophaga ruticilla*), which may contribute to population declines on breeding grounds in New Hampshire (Holmes and Sherry 2001). Holmes and Sherry (2001) concluded that winter habitat conditions have a carryover effect into the breeding season.

Suboptimal second-growth habitats reduce winter body condition of American redstarts, thereby reducing survival of subordinate individuals relegated to using lower-quality habitats (Marra et al. 1998a). In addition, poor winter habitat quality delays departure of American redstarts to breeding grounds and, thus, potentially reduces reproductive output (Marra et al. 1998b). Similarly, winter habitat quality influences reproductive success of American redstarts on their breeding grounds, and continued loss of winter tropical forest habitat may have negative effects on Neotropical migrants the following breeding season (Norris et al. 2003). In addition, quality of wintering habitat not only influences physical condition and survival of black-tailed godwits (*Limosa limosa icelandica*) on wintering locales in Great Britain, but also influences the birds' breeding success in Iceland (Gill et al. 2001). A decline in habitat quality, however, may have profound effects on wintering bird species, regardless of whether it occurs in optimal or suboptimal habitats. Second-growth wintering habitats in the Neotropics may not be of the highest quality and, therefore, may not be suitable for maintaining body condition and survival over winter (Holmes and Sherry 2001). One should not assume, consequently, that because many Neotropical migrants are habitat generalists and can occupy a variety of winter habitats (Faaborg et al. 1995), all suboptimal winter habitats provide resources necessary to prevent winter population declines and reductions in body conditions that impact breeding success. Loss of preferred tropical forest winter habitat continues unabated; hence, it is likely that improving suboptimal winter habitat will become increasingly important to maintaining Neotropical bird populations.

SOUTH TEXAS

Clearly, most of the research relevant to the impacts of habitat fragmentation on Neotropical birds has been conducted on North American breeding grounds and to a lesser extent, on wintering grounds in tropical Mexico and Central and South America. However, with the exception of Roth's (1977) research relating the composition of breeding bird communities to vegetation complexity at four locations in South Texas, and Igl and Ballard's (1999) study of the habitat associations of 21 species of wintering grassland birds conducted on several hundred thousand hectares of rangeland, few studies have focused on landscape ecology of grassland birds in South Texas. In fact, impacts of habitat fragmentation on South Texas grassland birds have been ignored by avian ecologists. This represents a major research opportunity, because South Texas represents an important breeding habitat for numerous species of resident birds; also, the region is of much importance to Neotropical migrants. However, because of the agricultural and urban development occurring in the lower Rio Grande Valley Corpus Christi, and Laredo areas, and brush and livestock management that is applied to South Texas rangelands, habitat fragmentation probably impacts at least some of the breeding, winter resident, and transient migrants in South Texas. Consequently, avian ecologists have a tremendous opportunity to use what has been learned from all of the other published studies that examined impacts of habitat fragmentation on North American Neotropical migrants, particularly on breeding grounds, to identify specific bird species and perhaps guilds that exhibit area sensitivity on fragmented South Texas landscapes.

APPLYING METAPOPULATION THEORY TO BIRDS
OCCUPYING FRAGMENTED LANDSCAPES

One of the most practical ways to explore impacts of habitat fragmentation on South Texas rangeland bird communities is to use metapopulation theory because of its applicability to populations occupying fragmented landscapes (Askins et al. 1987; Opdam 1991; Donovan et al. 1995b; Doncaster et al. 1997; With and King 2001). Metapopulations are spatially structured populations composed of discrete subpopulations separated by space or barriers and connected by dispersal (Opdam 1991). Metapopulations are characterized by turnover of subpopulations that experience extinction and reestablishment, which results in a spatial distribution pattern of subpopulations that shifts temporally. Extinction rates of bird subpopulations seem to be related to the size of habitat fragments, whereas recolonization rates depend on the degree of isolation of habitat patches. Thus, the probability of occurrence depends on the size of habitat fragments, their position in the landscape, and number of corridors that effectively lower landscape resistance. Numerous bird studies demonstrate that metapopulation models can be applied at multiple scales encompassing counties, states, regions, or continents (Donovan et al. 1995b; Bollinger et al. 1997; Esler 2000; Boulinier et al. 2001; Johnson and Igl 2001; Nol et al. 2005).

THE SOURCE–SINK PARADIGM

Metapopulation theory was originally conceived by Levins (1968) to develop better strategies to control insect pests occupying patchy habitats. He thought that understanding how extinction and recolonization rates of discrete small populations (subpopulations) interact to maintain a larger population (metapopulation) would lead to potential strategies for pest control. Later, Pulliam (1988) used metapopulation theory to develop his source–sink model and apply it to birds. Connected subpopulations experience extinction only to be recolonized by individuals from another subpopulation, and metapopulation models assume equal probabilities of extinction and recolonization for each subpopulation (Donovan et al. 1995b). Metapopulation models explain and predict the distribution of occupied and unoccupied patches and factors affecting dispersal between patches and, therefore, persistence of the metapopulation. Metapopulation theory, however, is concerned primarily with migration from one subpopulation to other subpopulations, regardless of the size of the subpopulations (Hanski and Gilpin 1997) (Figure 2.1), whereas source–sink models are concerned with both the distribution and size of connected populations, where birth and death rates vary for each discrete subpopulation (Figure 2.2). Sink subpopulations occupy poor quality habitat patches so they are not viable and require immigration to sustain themselves over time (Donovan et al. 1995b). Conversely, source subpopulations occupy higher quality habitat patches relative to productivity and

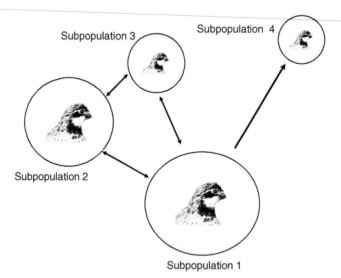

FIGURE 2.1 A theoretical representation of a classical bird metapopulation model consisting of four subpopulations. Each bobwhite symbol represents the presence of a bird subpopulation within a habitat patch. The arrows represent migration from one subpopulation to another. Movement of birds is unidirectional from subpopulation 1 to subpopulation 4, indicating that the persistence of subpopulation 4 is dependent on dispersal of birds from subpopulation 1.

survival so they are viable in the absence of immigration from other subpopulations. Source populations can supply immigrants to sinks, because birth rates exceed death rates. Dispersal is therefore unidirectional or fixed, because source–sink models predict that bird movement from sources to sinks is constrained by either density-independent or density-dependent mechanism (Diffendorfer 1998). Unlike metapopulation models concerned with only presence or absence of species in subpopulations, determining the size of the entire population (metapopulation) of source–sink models requires estimating density of subpopulations, their birth and death rates, and dispersal rates between subpopulations (Donovan et al. 1995b). Therefore, estimating vital demographic rates are essential requirements of source–sink models.

Numerous studies have provided evidence that source–sink dynamics operate for bird metapopulations that occupy fragmented landscapes. A variety of factors promote source–sink dynamics, depending on landscape characteristics. For instance, size, degree of isolation, number of patches, and the distribution of patches throughout the landscape influence the ability of subpopulations to successfully occupy a habitat patch and the ability for recolonization from source patches (van Dorp and Opdam 1987; Breininger et al. 1995; Donovan et al. 1995b; Foppen et al. 2000; Bojima et al. 2001; Nol et al. 2005). Further, subpopulations occupying rich habitat patches often function as sources for sink subpopulations occupying poor quality patches (Blondel et al. 1992). Habitat sinks also occur seasonally, such as when habitat quality is degraded during drought, resulting in reproductive success below that required to maintain subpopulations inhabiting those patches (Verner and Purcell 1999). Murphy (2001) modeled a similar relationship that he termed a *source–"leaky" sink system*, which operates during certain years when sources depend on immigration from sinks, because adult survival is too low to sustain source populations.

More complex source–sink dynamics also exist. For example, research on breeding brown jay (*Cyanocorax morio*) populations in the forests of Costa Rica indicated that a source–sink system may exist for this cooperatively breeding bird (Williams and Rabenold 2005). Dispersal was male-biased and most common during nest-building and egg-laying. It appeared that males made occasional monitoring forays to other breeding groups to determine if breeding vacancies were available and eventually dispersed to groups where vacancies became apparent. Moreover, males primarily bred

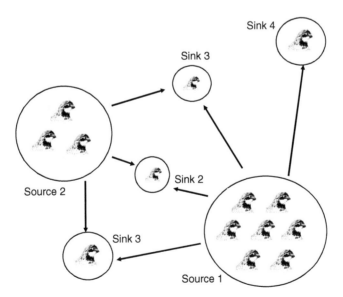

FIGURE 2.2   A theoretical representation of a source–sink model consisting of two source subpopulations and four sink subpopulations. Sources consist of multiple bobwhites indicating that these subpopulations have densities and vital rates that enable these subpopulations to sustain themselves and provide dispersing individuals that maintain sink subpopulations. The arrows represent unidirectional dispersal from sources to sinks.

outside their natal groups. Females did not disperse as frequently as males, but when they did disperse, they emigrated to populations where reproduction was poor. Brown jay males, therefore, emigrated from sources with excess males to sinks with male vacancies, and females emigrated from sources where reproduction was good to sinks with poor reproduction.

THE BALANCED DISPERSAL PARADIGM

Despite the evident popularity of the source–sink paradigm, source–sink models might not accurately reflect the mechanisms of bird metapopulation dynamics in heterogeneous (fragmented) landscapes (Diffendorfer 1998). Balanced dispersal models are possibly more relevant to bird metapopulation dynamics in fragmented landscapes. Instead of the unidirectional or fixed dispersal from sources to sinks predicted by source–sink models, the balanced dispersal model predicts that an equal number of individuals move between patches so that dispersal is balanced (Figure 2.3) (Diffendorfer 1998). Dispersal operates under the Ideal Free Distribution described by Fretwell and Lucas (1970), where birds would distribute themselves among patches in a landscape according to their perceptions of the habitat quality of patches. Bird dispersal depends on an individual's assessment of fitness gains to be obtained by occupying a patch relative to the patch's current carrying capacity. Dispersal rates, therefore, are higher from habitats with low carrying capacities and lower from habitats with high carrying capacities. As patch quality varies over space and time, immigration and migration rates are balanced between patches in a fragmented landscape.

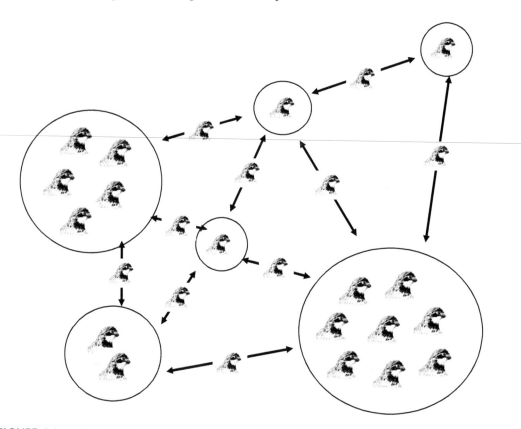

FIGURE 2.3 A theoretical representation of balanced dispersal models consisting of six subpopulations occupying habitat patches in a fragmented landscape. The number of quail heads represents the size of quail subpopulations and indicate that density is a reflection of a bird's perception of habitat quality. Quail heads superimposed on arrows indicate that dispersal is balanced between large and small subpopulations.

Diffendorfer (1998) compared and contrasted source–sink and balanced dispersal models, noting both models share some assumptions, while other assumptions differ. The first assumption shared by the two models deals with patch numbers. Source–sink dynamics are generally modeled in systems consisting of two or more patches, whereas the balanced dispersal model has been restricted to two patches. Adding additional patches, however, would not change the outcome of balanced dispersal models. The second shared assumption is that dispersal is density independent. Density-independent dispersal is assumed for balanced dispersal models, whereas source–sink models permit either density-independent or density-dependent dispersal. The third and final assumption the two models share is that density-dependent regulation occurs in nonsink habitats. The major difference between source–sink and balanced dispersal models is that source–sink models assume that dispersal is constrained and that sinks exist, whereas balanced dispersal models do not. Rather, balanced dispersal models permit both fixed and conditional dispersal rates based on patch occupancy. Also, instead of sinks, they assume that carrying capacities of habitat varies between patches, so that habitat quality in some patches is good while it is poor in other patches.

A continuum may exist between an organism's vagility and the dispersal patterns in fragmented landscapes (Diffendorfer 1998). Organisms with high vagility and the ability to assess habitat quality exhibit balanced dispersal, because they have sufficient mobility to explore landscapes and assess quality of patches. Therefore, since most birds display high vagility and are able to assess the habitat quality of a patch based on structure and composition of vegetation, and the occupancy of the patch, birds might be expected to exhibit balanced dispersal. In contrast, source–sink dynamics are likely more common for organisms with low vagility and organisms dependent on abiotic dispersal (wind or water) mechanisms, such as the seeds of plants. Several bird studies lend support to the balanced dispersal paradigm. Balanced dispersal appeared to exist among juvenile black-capped chickadee (*Parus atricapillus*) in a southeastern Wisconsin population, because juveniles emigrating from an area were replaced by immigrants so that little net change in numbers occurred (Weise and Meyer 1979). Similarly, a balanced dispersal system operated among collared flycatchers (*Ficedula albicollis*) that occupied fragmented landscapes in Sweden (Doncaster et al. 1997). Subpopulations near saturation level occupied discrete habitat patches existing as simultaneous sources and sinks to neighboring patches. Equal numbers of immigrations and emigrations occurred between patches. There was no evidence of source and sink subpopulations.

OTHER METAPOPULATION SYSTEMS IN FRAGMENTED LANDSCAPES

Whether source–sink or balanced dispersal dynamics better represent a species' response to fragmented landscapes likely varies by the species and the characteristics of the landscape. Responses of birds to habitat fragmentation, for example, appear highly individualistic based on the number, size, and configuration of patches within a landscape (Watson et al. 2005). Therefore, the same species could exhibit source–sink dynamics in one fragmented landscape and balanced dispersal in another fragmented landscape. A significant problem that may contribute to misclassifying how birds interact with fragmented landscapes is that most studies wherein researchers have identified source–sink dynamics for bird metapopulations inhabiting fragmented landscapes have not been conducted over a long enough period of time to ascertain that sources and sinks actually existed (Opdam 1991). Moreover, Diffendorfer (1998) emphasized that few studies have provided adequate empirical demographic data to support assumptions of source–sink or balanced dispersal models, which has hindered testing model predictions. He suggested that in real systems, metapopulations occupying fragmented landscapes may be described by a combination of source–sink and balanced dispersal models, or other alternative models may better describe how organisms are distributed across fragmented landscapes.

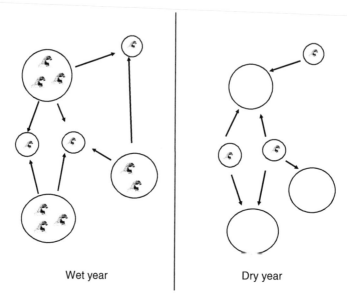

Wet year   Dry year

FIGURE 2.4   A theoretical representation of a reciprocating dispersal model consisting of six subpopulations during a wet year when subpopulations are increasing and during a dry year when populations are declining. During the wet year, dispersal is from high-density subpopulations to low-density subpopulations. During the dry year, habitat conditions have deteriorated such that high-density subpopulations are dependent of reciprocal dispersal from low-density populations to persist.

In fact, evidence exists that neither source–sink nor balanced dispersal dynamics explains how some bird species interact with fragmented landscapes. Bird communities may exhibit a dispersal strategy similar to the reciprocal dispersal strategy described for rodent populations, where asymmetric dispersal occurs as populations increase and decline. In this strategy, individuals disperse from high-density patches to low-density patches during periods of population increases, whereas the reverse is true during population declines (Morris and Diffendorfer 2004, Morris et al. 2004, Shenbrot 2004) (Figure 2.4).

Recent research has shown that a reciprocal dispersal strategy may also operate in birds. For example, asymmetric exchange of individuals occurs between populations of citril finches (*Serinus citrinella*) occupying habitats of differing quality (Senar et al. 2002). Instead of emigration from high-quality habitats to low-quality habitats, citril finches in poorer condition emigrated from low-quality habitats where survival was poor, to high-quality habitats where survival was higher.

APPLYING METAPOPULATION THEORY TO SOUTH TEXAS GRASSLAND BIRDS

Most studies that have quantified how birds interact with fragmented landscapes have been conducted in forest ecosystems, and to a lesser extent in agricultural landscapes (Opdam 1991). Moreover, few studies have involved ecosystems characterized by natural fragmentation. Thus, it is not clear if natural fragmentation impacts bird metapopulations different from fragmentation caused by human activities. Landscapes of South Texas provide an opportunity to not only evaluate how bird metapopulations interact with fragmented rangeland landscapes, but also to study landscapes that have not been substantially altered by anthropogenic activities, such as coastal prairies. Opportunities also exist to test predictions of a variety of metapopulation models because (1) of the diversity of upland vegetation communities in South Texas; (2) fragmentation patterns vary spatially and temporally according to the manner in which individual ranches are managed and urban areas have developed; and (3) there is rich diversity of birds that are annual residents, winter residents, or migratory transients that utilize the landscapes.

Numerous metapopulation models or combinations of models, may, for example, be in effect simultaneously on a single landscape that supports a diversity of species, because bird interactions with their landscapes may be species specific (Watson et al. 2005). Source–sink dynamics may apply to certain species of the bird community using a landscape, whereas the balanced dispersal model, the reciprocating dispersal model, or combinations of various models may be apparent for other members of the bird community. Moreover, it is possible that the manner in which species interact with landscapes changes seasonally. Breeding residents may exhibit source–sink dynamics during the breeding season and then switch to balanced dispersal dynamics during winter when migratory species arrive to share vegetation communities with resident species. In addition, a species interaction with a landscape could be landscape-specific, depending on the degree of fragmentation. An individual bird species may exhibit balanced dispersal dynamics on one landscape and source–sink dynamics on another. Also, because populations of numerous South Texas birds fluctuate in response to impacts that fluctuating precipitation patterns have on habitat quality, the reciprocating dispersal hypothesis may explain how these species interact with South Texas landscapes. Unraveling the metapopulation dynamics of the diversity of breeding and wintering bird species that occupy South Texas landscapes is an intriguing challenge. Nevertheless, it is a challenge that, if approached in a scientifically rigorous manner, could yield insights into how a diverse bird community interacts with fragmented landscapes. It will also allow avian ecologists to test the predictions of various metapopulation models and determine what models are applicable for specific species on South Texas landscapes.

NORTHERN BOBWHITE RESEARCH: A MODEL FOR GRASSLAND BIRDS

A model currently used by quail scientists at the Caesar Kleberg Wildlife Research Institute (CKWRI) to quantify the landscape ecology of northern bobwhites in South Texas could help determine how grassland birds interact with fragmented landscapes (Chapter 4, this volume). Scientists at CKWRI initiated a long-term bobwhite research project during the late 1990s and have accumulated almost 10 years of data relevant to bobwhite demography and habitat use. Annual age and sex ratios, age and sex-specific survival rates, and density data have been estimated during this period (Hernandez et al. 2005; Rusk et al. in press). Moreover, radio-telemetry enabled quail researchers to obtain habitat preference information for almost a decade. Multivariate statistical models were used in a recent study to determine how nest success is influenced by habitat and environmental variables (Rader 2006). Ongoing molecular genetics research is also providing results relevant to minimum habitat area, effective population size, and dispersal rates of a bobwhite metapopulation (Whelan 2006). Furthermore, CKWRI quail researchers are using GIS/GPS technology to quantify how bobwhites use various vegetation communities on South Texas landscapes.

Therefore, CKWRI quail researchers have obtained, or are currently obtaining, vital demographic rate and habitat-specific data that will enable them to link metapopulation models with landscape models to illuminate how northern bobwhites interact with fragmented South Texas landscapes. This will enable them to test the predictions of source–sink, balanced dispersal, or reciprocating dispersal metapopulation theories. Results of this research will enable landowners to manage landscapes to benefit quail metapopulations in South Texas. Because northern bobwhites are grassland birds, a research approach similar to what CKWRI quail scientists have initiated could be applied successfully to the grassland bird community to determine how grassland birds interact with fragmented landscapes in South Texas.

LANDSCAPE MANAGEMENT OPPORTUNITIES

TRADITIONAL LAND MANAGEMENT ACTIVITIES

Results of scientifically rigorous studies that quantify how grassland bird metapopulations interact with fragmented South Texas landscapes would clarify further grassland bird conservation and management. Landscape patterns and processes that either positively or negatively impact grassland bird population demography could be identified. It would be possible to learn how livestock and crop production and urbanization impact grassland birds that use South Texas rangelands. In addition, it would be possible to learn how widespread white-tailed deer (*Odocoileus virginianus*) and bobwhite habitat management are impacting grassland birds. For example, habitat management techniques used to improve rangelands for quail and deer could benefit grassland bird populations, because habitat management activities create early successional habitats by reducing brush cover. Moreover, quail habitat management might particularly benefit grassland birds, because, in addition to reducing brush cover, quail management also emphasizes restoring grasslands by reducing the impacts of livestock on herbaceous vegetation. Landowners interested in maximizing their land for quail either significantly reduce cattle stocking rates or remove cattle completely and use prescribed burning to manipulate herbaceous vegetation. Consequently, the current deer and quail landscape management that prevails over a significant portion of South Texas could represent a boon to grassland bird populations.

ECOLOGICAL TRAPS

Certain game management practices, however, and traditional rangeland management practices could negatively impact grassland bird populations in South Texas. For example, continued planting of exotic invasive grasses for cattle production and nutritional supplementation of deer on a broad scale, as is done on many South Texas ranches, may limit grassland bird populations by creating ecological traps, which could represent a demographic drain on grassland bird populations. Ecological traps are habitats that retain cues or specific characteristics that attract birds for specific purposes such as foraging, breeding, roosting, or loafing but have changed in a manner that affects fitness, resulting in population declines (Stochat et al. 2005). For example, an area could possess vegetation that a bird species perceives as suitable nesting habitat, but if this habitat is also attractive to ground nest predators, birds nesting in the area may experience high mortality and low reproductive success, resulting in a population decline. Therefore, ecological traps represent habitats where habitat selection becomes decoupled from habitat quality (Kristan 2003). Ecological traps may be particularly relevant to bird communities in South Texas landscapes, because extensive portions of the region's landscapes are dominated by exotic grass species that could be perceived as attractive nesting habitats, but in fact provide insufficient resources to sustain breeding bird populations. Breeding bird abundance, for instance, was lower on areas dominated by exotic grasses compared with areas dominated by native grasses in the western Rio Grande Plains (Flanders et al. 2006). Moreover, the ground-foraging guild of birds was less abundant on exotic grass sites, possibly due to the lower abundance of invertebrates on exotic grass sites compared with native grass sites. Therefore, patches of grasslands dominated by exotic grasses may represent ecological traps, because they are perceived as grasslands and thus attract grassland birds, but do not supply sufficient food resources and nesting substrates to maintain stable populations.

SUPPLEMENTAL FEEDING

In addition to exotic grass infestations, supplemental feeding is a popular management technique used by South Texas landowners to improve white-tailed deer antler quality and attract deer and bobwhite quail to areas where they are more accessible to hunters. Concentrations of feeders and food plots may attract birds to areas that they would not ordinarily frequent, and these may also support abundant predator populations. Consequently, areas where supplemental feeding is concentrated may also represent ecological traps to South Texas grassland birds (Haines et al. 2004).

LINKING METAPOPULATION THEORY TO MANAGEMENT OF GRASSLAND BIRDS

A greater understanding of how habitat fragmentation impacts the demography of bird populations, and how grassland birds use landscapes with differing fragmentation patterns, could help South Texas landowners improve their property for grassland birds. It is likely that appropriate scales of landscape resolution could be identified for individual species and entire bird communities, which would not only further our knowledge about landscape ecology of grassland birds, but also indicate the size of landscapes that viable grassland bird populations require. Moreover, ecological traps could be identified and remedial measures could be instituted to transform at least some ecological traps into source subpopulations. Furthermore, improving South Texas landscapes for grassland birds may contribute to at least stabilizing or perhaps reducing the current declines exhibited by a number of grassland bird species by improving wintering landscapes. Grassland bird habitat could be improved using metapopulation theory on large South Texas numerous ranches that encompass tens of thousands of acres of largely continuous dense mixed thorn scrub. Herbaceous openings on these ranches often exist, but in many cases these openings are islands that are either too small to be of much use to grassland birds or are located far enough from similar opens to impede bird dispersal between islands. From metapopulation theory, we know that sufficient habitat must be available to sustain several subpopulations of birds that comprise a metapopulation. Sufficient habitat could be composed of numerous islands of herbaceous habitat created through brush management that are large enough to sustain a viable breeding population of grassland birds. In addition, these islands of herbaceous habitat would need to be close enough to one another to permit interaction of bird subpopulations via dispersal. Designing and applying habitat management that would accomplish the objective of establishing and maintaining a metapopulation of a specific grassland bird species, such as lark buntings (*Calamospiza melanocorys*), could be accomplished via roller-chopping and prescribed burning. On a ranch composed of several thousands of acres of dense thorn scrub, roller-chopping could create openings that would be large enough to maintain several viable breeding subpopulations of lark buntings (Figure 2.5). Metapopulation/landscape models developed via research would provide information about what constitutes sufficient habitat for a viable breeding subpopulation of lark buntings in South Texas. In the absence of this information, published information regarding home-range sizes, sex and age ratios, and survival rates of lark buntings using similar prairie habitats elsewhere could be used to determine how large these openings in the thorn scrub would need to be. Likewise, South Texas-specific research or information from published studies could be used to determine how many of these openings would need to be created to maintain a metapopulation of lark buntings. Similarly, dispersal rates and distances from original research or from published studies would be used to determine the spatial distribution of openings. Herbaceous openings must be distributed in a manner that permits lark bunting dispersal between subpopulations. However, because stochastic events such as drought are common in South Texas and because rainfall events are patchy, numerous herbaceous openings of sufficient size and spaced at appropriate distances would need to be created to ensure that enough viable subpopulations are maintained during drought to sustain the lark bunting metapopulation. Ideally, these openings would be distributed throughout the landscape in a manner that facilitates lark bunting dispersal between openings.

Post-mechanical treatment of herbaceous openings would be necessary to maintain viable populations of lark buntings, because in the absence of follow-up treatment, woody plants would begin to invade herbaceous openings, thereby reducing their quality for lark buntings. Consequently, a rotational 4- to 5-year prescribed fire schedule could be developed to suppress invasion of woody plants

FIGURE 2.5   A roller-chopping operation being conducted on the western Rio Grande Plains of South Texas to improve bobwhite quail and white-tailed deer habitat. Roller-chopping is removing thorn scrub to restore herbaceous habitat conditions that will benefit grassland bird species (Photograph Copyright David G. Hewitt).

in herbaceous openings. It would be undesirable to burn all of the openings in one year, because the immediate post-burn shock effect might prevent herbaceous openings from supporting viable populations of lark buntings for perhaps a year or longer if drought occurs after prescribed burning has been conducted. Instead, perhaps only one-third of the herbaceous openings would be burned during any year so that all of the openings are burned over a 3-year period beginning 4 to 5 years after the roller-chopping event was completed. Follow-up prescribed burns would then occur on a 4- to 5-year schedule. The roller-chopping/prescribed burning program described in the preceding paragraphs represents an example of how wildlife habitat management that is commonly utilized in South Texas can be applied using metapopulation theory to improve grassland bird management. Other habitat management techniques could also be used to accomplish the same results. What is important is to understand that to maintain a metapopulation of a grassland bird species, one must provide sufficient herbaceous habitat to sustain numerous subpopulations during stochastic events that can depress subpopulations and to distribute improved habitats throughout the landscape in a manner that enables members of subpopulations to disperse to other subpopulations.

CONCLUSION

Our focus in this chapter was on theories explaining the dynamics of bird interactions within the landscapes they occupy; in particular, the impacts that fragmented landscapes have on bird communities. Bird species respond to habitat fragmentation differently, depending on their area sensitivity and characteristics of the fragmented landscape. Size and isolation of habitat patches influence the abundance and species diversity of birds inhabiting them (Askins et al. 1987; Vickery et al. 1994;

FIGURE 2.6 Metapopulation theory used to improve grassland bird conservation in South Texas by applying contemporary wildlife habitat management practices aimed at improving bobwhite quail and white-tailed deer habitat could result in improved habitat for species such as painted buntings (*Passerina ciris*) (Photograph Copyright Timothy E. Fulbright).

Boulinier et al. 2001; Johnson and Igl 2001). Metapopulation theory provides a template for investigating impacts of fragmentation on bird communities. Metapopulation models have been successfully applied to landscapes at highly varying spatial scales to assess the impacts of fragmentation on bird communities (Donovan et al. 1995b; Bollinger et al. 1997; Boulinier et al. 2001; Nol et al. 2005). In a similar sense, investigation of source–sink dynamics can also provide insight into the impacts of fragmentation by adding spatially explicit information on vital rates throughout the landscape. In general, the quality of a habitat patch is important in determining birth and death rates, and dispersal. Patches of low quality may produce sink populations that require immigration from high-quality patches (sources) to be maintained (Donovan et al. 1995b). Source–sink dynamics are driven by features of the landscape such as size and number of patches, their degree of isolation, and the distribution of patches throughout the landscape (van Dorp and Opdam 1987; Breininger et al. 1995; Donovan et al. 1995b; Nol et al. 2005). Balanced dispersal models have also been used to explain responses of birds to habitat fragmentation (Weise and Meyer 1979; Doncaster et al. 1997). These models are thought by some to be more representative of bird community dynamics in fragmented landscapes than source–sink models (Diffendorfer 1998). Dispersal is predicted to be somewhat balanced under these models with greater dispersal from low-quality patches and less dispersal from high-quality patches. Regardless of which model better explains bird community responses to heterogeneous landscapes, it is clear that one model is insufficient for explaining the diverse responses by different species to landscape features (Watson et al. 2005). More specifically, different species may have markedly different responses within the same landscape, and the same species respond differently among different landscapes.

We suggest that wintering grassland birds provide an appropriate model to study the impacts of landscape fragmentation on bird communities. Most research on habitat associations within this guild have been on breeding areas; however, factors on wintering areas may be critical in explaining continent-wide declines in many Neotropical migratory bird species (Rappole and McDonald 1994).

Further, population declines in this guild have been linked to declines in quantity and quality of grassland habitats throughout North America (Knopf 1994; Noss et al. 1995). South Texas supports landscape characteristics and avian assemblages that provide unique opportunities for investigating interactions between birds and their landscape.

Conservation and management of grassland birds and other avian groups will be strengthened by scientifically rigorous studies that quantify how these bird communities interact with fragmented landscapes. In addition, the metapopulation theory could be used to improve grassland bird conservation in South Texas by applying contemporary wildlife habitat management practices used by landowners to improve bobwhite quail and white-tailed deer habitat. In landscapes dominated by thorn scrub, roller-chopping could be used to create numerous herbaceous openings in sizes that will support viable grassland bird populations, and these openings could be distributed throughout the landscape in a manner that permits dispersal of birds from the various subpopulations that occupy each opening.

ACKNOWLEDGMENTS

We are grateful to T. Fulbright and D. Hewitt for the editorial assistance they provided during the preparation of this manuscript. The Caesar Kleberg Foundation for Wildlife Conservation supported Kuvlesky, Ballard, and Langschied and the Richard M. Kleberg, Jr., Center for Quail Research supported Brennan.

REFERENCES

Akcakaya, H. R., V. C. Radeloff, D. J. Mladenoff, and H. S. He. 2003. Integrating landscape and metapopulation modeling approach: Viability of the sharp-tailed grouse in a dynamic landscape. *Cons. Biol.* 18:526.

Askins, R. A. 1999. *Restoring North America's Birds: Lessons from Landscape Ecology.* New Haven: Yale University Press.

Askins, R. A., M. J. Philbrick, and D. S. Sugeno. 1987. Relationship between the regional abundance of forest and the composition of forest bur communities. *Biol. Cons.* 39:129.

Baker, D. L., and F. S. Guthery. 1990. Effects of continuous grazing on habitat and density of ground-foraging birds in South Texas. *J. Range Manage.* 43:2.

Bakker, K. K., D. E. Naugle, and K. F. Higgins. 2002. Incorporating landscape attributes into models for migratory grassland bird conservation. *Cons. Biol.* 16:1638.

Bayne, E. M., and K. A. Hobson. 2001. Effect of habitat fragmentation on pairing success of ovenbirds: Importance of male age and floater behavior. *Auk* 118:380.

Best, L. B., H. Campa III, K. E. Kemp, R. J. Robel, M. R. Ryan, J. A. Savidge, H. P. Weeks, and S. R. Winterstein. 1997. Bird abundance in CRP and nesting in CRP fields and cropland in the Midwest: A regional approach. *Wildl. Soc. Bull.* 25:64.

Blondel, J., P. Perret, M. Maistre, and P. C. Dias. 1992. Do harlequin Mediterranean environments function as source sink for blue tits (*Parus caeruleus* L.)? *Land. Ecol.* 3:213.

Bock, C. E., V. A. Saab, T. D. Rich, and D. S. Dobkin. 1993. Effects of livestock grazing on Neotropical migratory land birds in western North America. In *Status and Management of Neotropical Migratory Birds.* Finch, D. M., and P. W. Stangel (eds),. Fort Collins: USDA Forest Service, RM-GTR-229, p. 269.

Bojima, R. A., T. L. DeVault, P. E. Scott, and S. L. Lima. 2001. Reclaimed coal mine grasslands and their significance for Henslow's sparrows in the American Midwest. *Auk* 118:422.

Bollinger, E. K., P. B. Bollinger, and T. A. Gavin. 1990. Effects of hay-cropping on eastern populations of the bobolink. *Wildl. Soc. Bull.* 18:142.

Bollinger, E. K., and T. A. Gavin. 1992. Eastern bobolink populations: Ecology and conservation in an agricultural landscape. In *Ecology and Conservation of Neotropical Migrant Land Birds.* Hagan, J. M., III, and D. W. Johnson (eds). Washington, DC: Smithsonian Institute Press, pp. 497–506.

Bollinger, E. K., B. D. Peer, and R. W. Jansen. 1997. Status of Neotropical migrants in three forest fragments in Illinois. *Wilson Bull.* 109:521.

Boulinier, T., J. D. Nichols, J. E. Hines, J. R. Sauer, C. H. Flather, and K. H. Pollock. 2001. Forest fragmentation and bird community dynamics: Inference at regional scales. *Ecology* 82:1159.

Bowen, B. S., and A. D. Kruse. 1993. Effects of grazing on nesting by upland sandpipers in south-central North Dakota. *J. Wildl. Manage.* 57:301.

Brawn, J. D., and S. K. Robinson. 1996. Source–sink population dynamics may complicate the interpretation of long-term census data. *Ecology* 77:3.

Breininger, D. R., V. L. Larson, B. W. Duncan, D. M. Oddy, and M. F. Goodchild. 1995. Landscape patterns of Florida scrub jay habitat use and demographic success. *Cons. Biol.* 9:1442.

Brennan, L. A., and W. P. Kuvlesky, Jr. 2005. North American grassland birds: An unfolding conservation crisis? *J. Wildl. Manage.* 69:1.

Brooks, B. L., and S. A. Temple. 1990. Dynamics of a loggerhead shrike population in Minnesota. *Wilson Bull.* 102:441.

Bucher, E. H., and M. Nores. 1988. Present status of birds in steppes and savannas of northern and central Argentina. In *Ecology and Conservation of Grassland Birds.* Goriup, P. D. (ed.). *ICBP Tech. Publ.* 7:71.

Chavez-Ramirez, F., and F. G. Prieto. 1994. Effects of prescribed fires on habitat use by wintering raptors on a Texas barrier island grassland. *J. Raptor Res.* 28:262.

Confer, J. L., and R. T. Holmes. 1995. Neotropical migrants in undisturbed and human-altered forests of Jamaica. *Wilson Bull.* 107:577.

Conway, C. J., G. V. N. Powell, and J. D. Nichols. 1995. Overwinter survival of Neotropical migratory birds in early successional and mature tropical forests. *Cons. Biol.* 9:855.

Coppedge, B. R., D. M. Engle, R. E. Masters, and M. S. Gregory. 2001. Avian response to landscape change in fragmented southern Great Plains Grasslands. *Ecol. Appl.* 11:47.

Coppedge, B. R., D. M. Engle, R. E. Masters, and M. S. Gregory. 2004. Predicting juniper encroachment and CRP effects on avian community dynamics in southern mixed-grass prairie, USA. *Biol. Cons.* 115:431.

Dale, B. C., P. A. Martin, and P. S. Taylor. 1997. Effects of hay management on grassland songbirds in Saskatchewan. *Wildl. Soc. Bull.* 25:616.

Davis, S. K. 2004. Area sensitivity in grassland passerines: Effect of patch size, patch shape, and vegetation structure on bird abundance and occurrence in southern Saskatchewan. *Auk* 121:1130.

Diffendorfer, J. E. 1998. Testing models of source–sink dynamics and balanced dispersal. *Oikos* 81:417.

Doncaster, C. P., J. Clobert, B. Doligez, L. Gustafsson, and E. Danchin. 1997. Balanced dispersal between spatially varying local populations: An alternative to the source–sink model. *Am. Midl. Nat.* 150:425.

Donovan, T. M., F. R. Thompson III, J. Faaborg and J. R. Prob St. 1995a. Reproductive success of migratory birds in habitat sources and sinks. *Cons. Biol.* 9:1380.

Donovan, T. M., D. A. Clarke, R. W. Howe, and B. J. Danielson. 1995b. Metapopulations, sources and sinks, and the conservation of neotropical migratory birds in the Midwest. In *Management of Midwestern Landscapes for the Conservation of Neotropical Migratory Birds.* Thompson, F. R., III (ed.). St. Paul: U.S. Forest Service, North Central Forest Experiment Station, GTR-NC-187, p. 41.

Dunning, J. B., Jr. R. Borgella, Jr. K. Clements, and G. K. Meffe. 1995. Patch isolation, corridor effects and colonization by a resident sparrow in a managed pine woodland. *Cons. Biol.* 9:542.

Esler, D. 2000. Applying metapopulation theory to conservation of migratory birds. *Cons. Biol.* 14:366.

Faaborg, J., A. D. Anders, M. E. Baltz, and W. K. Gram. 1995. Non-breeding season considerations for the conservation of migratory birds in the Midwest; post-breeding and wintering grounds. In *Management of Midwestern Landscapes for the Conservation of Neotropical Migratory Birds.* Thompson, F. R., III (ed.). St. Paul: U.S. Forest Service, North Central Forest Experiment Station, GTR-NC-187, p. 189.

Flanders, A. A., W. P. Kuvlesky, Jr, D. C. Ruthven, III, R. E. Zaiglin, R. L. Bingham, T. E. Fulbright, F. Hernandez, and L. A. Brennan. 2006. Effects of invasive exotic grasses on South Texas rangeland breeding birds. *Auk* 123:171.

Foppen, R. P. B., J. P. Chardon, and W. Liefveld. 2000. Understanding the role of sink patches in source–sink metapopulations: Reed warbler in an agricultural landscape. *Cons. Biol.* 14:1523.

Frawley, B. J., and L. B. Best. 1991. Effects of mowing on breeding bird abundance and species composition in alfalfa fields. *Wildl. Soc. Bull.* 19:135.

Fretwell, S. D., and H. L. Lucas. 1970. On territorial behavior and other factors influencing habitat distribution in birds. I. Theoretical development. *Acta Biotheor.* 19:16.

Fulbright, T. E., and F. R. Bryant. 2001. The last great habitat. Special Publication CKWRI 1.

Fulbright, T. E., and F. S. Guthery. 1996. Mechanical manipulation of plants. In *Rangeland Wildlife*. Krausman, P. R. (ed.), Vol. 1. Denver: The Society for Range Management, 1, p. 39.

Gill, J. A., K. Norris, P. M. Potts, T. G. Gunnarson, P. W. ActKinson, and W. J. Sutherland. 2001. The buffer effect and large-scale population regulation in migratory birds. *Nature* 412:436.

Gordon, C. E. 2000. Fire and cattle grazing on wintering sparrows in Arizona grasslands. *J. Range Manage.* 53:384.

Goriup, P. D. 1988. Ecology and conservation of grassland bird communities. International Council for Bird Preservation, Tech. Publ. No. 7.

Griebel, R. L., S. L. Winter, and A. A. Steuter. 1998. Grassland birds and habitat structure in sandhills prairie managed using bison and cattle plus fire. *Great Plains Res.* 8:255.

Grzybowski, J. A. 1982. Population structure in grassland bird communities during winter. *Condor* 84:137.

Grzybowski, J. A. 1983. Patterns of space use in grassland bird communities during winter. *Wilson Bull.* 95:591.

Hagan, J. M., III, and D. W. Johnston (eds) 1992. *Ecology and Conservation of Neotropical Migrant Landbirds.* Washington, DC: Smithsonian Institution Press, p. 609.

Haines, A. M., F. Hernandez, S. E. Henke, and R. L. Bingham. 2004. Effect of road baiting on home range and survival of northern bobwhites in southern Texas. *Wildl. Soc. Bull.* 32:401.

Hanski, I., and M. E. Gilpin (eds) 1997. *Metapopulation Biology: Ecology, Genetics and Evolution.* San Diego: Academic Press.

Herkert, J. M. 1997. Bobolink *Dolichonyx oryzivorous* population decline in agricultural landscapes in the Midwestern USA. *Biol. Cons.* 80:107.

Herkert, J. R., and W. D. Glass. 1999. Henslow's sparrow response to prescribed fire in an Illinois prairie remnant. *Studies Avian Biol.* 19:160.

Hernandez, F., J. A. Arredondo, F. C. Bryant, L. A. Brennan, and R. L. Bingham. 2005. Influence of precipitation on demographics of northern bobwhites in southern Texas. *Wildl. Soc. Bull.* 33:1071.

Holmes, R. T., and T. W. Sherry. 2001. Thirty-year bird population trends in an unfragmented temperate deciduous forest: Importance of habitat change. *Auk* 118:589.

Igl, L. D., and B. M. Ballard. 1999. Habitat associations of migrating and overwintering grassland birds in southern Texas. *Condor* 101:771.

Igl, L. D., and D. H. Johnson. 1997. Changes in breeding bird populations in North Dakota: 1967 to 1992–93. *Auk* 114:74.

James, F. C., D. A. Wiedenfield, and C. E. McCullough. 1992. Trends in breeding populations of warblers: Declines in the southern highlands and increase in the lowlands. In *Ecology and Conservation of Neotropical Migrant Landbirds*. Hagen, J. M., III, and D. W. Johnston (eds). Washington, DC: Smithsonian Institution Press, 1, p. 43.

Johnson, D. H., and L. D. Igl. 1995. Contributions of the conservation reserve program to populations of breeding birds in North Dakota. *Wilson Bull.* 107:709.

Johnson, D. H., and L. D. Igl. 2001. Area requirements of grassland birds: A regional perspective. *Auk* 118:24.

Knopf, F. L. 1994. Avian assemblages on altered grasslands. *Studies Avian Biol.* 15:247.

Kristan, W. B. 2003. The role of habitat selection behavior in population dynamics: Source–sink systems and ecological traps. *Oikos* 103:457.

Levins, R. 1968. Evolution in changing environments: Some theoretical explorations. *Monogr. Pop. Biol.*, Princeton University Press.

Madden, E. M., A. J. Hansen, and K. R. Murphy. 1999. Influence of prescribed fire history on habitat and abundance of passerine in northern mixed-grass prairie. *Canadian Field Natur.* 113:627.

Marra, P. P., K. A. Hobson, and R. T. Holmes. 1998a. Linking winter and summer events in a migratory bird using stable-carbon isotopes. *Science* 282:1884.

Marra, P. P., K. A. Hobson, and H. S. Horn. 1998b. Corticosterone levels as an indicator of habitat quality: effects of habitat segregation in a migratory bird during the non-breeding period. *Oecologica* 116:284.

Mix, K. D. 2004. The impacts of summer prescribed fires on the vegetation, avian and invertebrate communities of the Welder Wildlife Refuge. M.S. Thesis, Texas A&M University, Kingsville, p. 84.

Morris, D. W., and J. E. Diffendorfer. 2004. Reciprocating dispersal by habitat-selecting white-footed mice. *Oikos* 107:549.

Morris, D. W., J. E. Diffendorfer, and P. Lundberg. 2004. Dispersal among habitats varying in fitness: Reciprocating migration through ideal habitat selection. *Oikos* 107:559.

Murphy, M. T. 2001. Source–sink dynamics of a declining eastern kingbird population and the value of sink habitats. *Cons. Biol.* 15:737.

Nol, E., C. M. Francis, and D. M. Burke. 2005. Using distance from putative source woodlots to predict occurrence of forest birds in putative sinks. *Cons. Biol.* 19:836.

Nolte, K. R., and T. E. Fulbright. 1997. Plant, small mammal and avian diversity following control of honey mesquite. *J. Range Manage.* 50:205.

Norris, D. R., P. B. Marra, T. K. Kyser, T. W. Sherry, and L. M. Ratcliff. 2003. Tropical winter habitat limits reproductive success on the temperate breeding grounds in a migratory bird. *Proc. R. Soc. Lond. B.* 3:683.

Noss, R. F., E. T. Laroe, and J. M. Scott. 1995. Endangered ecosystems of the United States: A preliminary assessment of loss and degradation. Report No. 0611-R-01 (MF), National Biological Service, U.S. Department of the Interior, Washington, DC.

Opdam, P. 1991. Metapopulation theory and habitat fragmentation: A review of holarctic breeding bird studies. *Land. Ecol.* 5:93.

Peterjohn, B. G., and J. R. Sauer. 1999. Population status of North American grassland birds. *Studies Avian Biol.* 19:27.

Pulliam, H. R. 1988. Source, sinks and population regulation. *Am. Nat.* 137:50.

Rader, M. J. 2005. 1996 Factors influencing nest success of northern bobwhites in south Texas. Ph.D. Dissertation, Texas A&M University, Kingsville.

Rappole, J. H., D. I. King, and W. C. Barrow, Jr. 1999. Winter ecology of the endangered golden-cheeked warbler. *Condor* 101:762.

Rappole, J. H., and M. V. McDonald. 1994. Cause and effect in population declines of migratory birds. *Auk* 111:652.

Reynolds, R. E., T. L. Schaffer, J. R. Squer, and B. G. Peterjohn. 1994. Conservation Reserve Program: Benefit for grassland birds in the northern plains. *Trans. North Am. Wildl. Nat. Resour. Conf.* 59:328.

Reynolds, M. C., and P. R. Krausman. 1998. Effects of wintering burning on birds in mesquite grassland. *Wildl. Soc. Bull.* 26:867.

Ribic, C. A., and D. W. Sample. 2001. Associations of grassland birds with landscape factors in southern Wisconsin. *Am. Midl. Nat.* 146:105.

Robbins, C. S., D. Bystrak, and P. H. Geissler. 1986. The breeding bird survey: Its first fifteen years, 1965–1979. *U.S. Fish Wildl. Serv. Resource Publ.*, 157.

Robbins, C. S., B. A. Dowell, D. K. Dawson, J. Colon, F. Espinosa, J. Rodriguzz, R. Sutton, and T. Vargas. 1987. Comparison of Neotropical winter bird populations in isolated patches vs. extensive forests. *Acta Oecol.* 8:285.

Robinson, S. K. 1995. Threats to breeding neotropical migratory birds in the Midwest. In Management of Midwestern landscapes for the conservation of neotropical migratory birds. Thompson, F. R., III (ed.). St. Paul: U.S. Forest Service, North Central Forest Experiment Station, GTR-NC-187, p. 1.

Rosenberg, K. V., J. D. Lowe, and A. A. Dhondt. 1999. Effects of forest fragmentation on breeding tanagers: A continental perspective. *Cons. Biol.* 13:586.

Roth, R. R. 1977. The composition of four bird communities in south Texas brush grasslands. *Condor* 79:417.

Rusk, J. P., F. Hernandez, J. A. Arredando, F. C. Bryan, D. G. Hewitt, E. J. Redeker, and L. A. Brennan. In press. Refining the morning covey-call survey to estimate northern bobwhite (*Colinus virginanus*) abundance. *J. Wildl. Manage.*

Saab, V. A., C. E. Book, T. D. Rich, and D. S. Dobkin. 1995. Livestock grazing effects in western North America. In *Ecology and Management of Neotropical Migratory Birds*. Martin, T. E., and D. M. Finch (eds). New York: Oxford University Press, p. 311.

Samson, F., and F. Knopf. 1994. Prairie conservation in North America. *Bioscience* 44:418.

Senar, J. C., M. J. Conroy, and A. Borras. 2002. Asymmetric exchange between populations differing in habitat quality: A metapopulation study on the citril finch. *J. Appl. Stat.* 29:425.

Shenbrot, G. 2004. Habitat selection in a seasonally variable environment: Test of the isodar theory with the fat sand rat, *Psammomys obesus*, in the Negev Desert, Israel. *Oikos* 106:359.

Stochat, E. et al. 2005. Ecological traps in isodars: Effects of tallgrass prairie management on bird nest success. *Oikos* 111:159.

Temple, S. A. et al. 1999. Nesting birds and grazing cattle: Accommodating both on Midwestern pastures. *Studies Avian Biol.* 19:196.

van Dorp, D., and P. F. M. Opdam. 1987. Effects of patch size, isolation and regional abundance of forest bird communities. *Land. Ecol.* 1:59.

Van't Hul, J. T., R. S. Lutz, and N. E. Mathews. 1997. Impact of prescribed burning on vegetation and bird abundance at Matagorda Island, Texas. *J. Range Manage.* 50:346.

Vega, J. H., and J. H. Rappole. 1994. Effects of scrub mechanical management on the nongame bird community in the Rio Grande Plains of Texas. *Wildl. Soc. Bull.* 22:165.

Verner, J., and K. L. Purcell. 1999. Fluctuating populations of house wrens and Bewick's wrens in foothills of the western Sierra Nevada of California. *Condor* 101:219.

Vickery, P. D., and J. R. Herkert. 2001. Recent advances in grassland bird research: Where do we go from here? *Auk* 118:11.

Vickery, P. D., M. L. Hunter, and S. M. Melvin. 1994. Effects of habitat area on the distribution of grassland birds in Maine. *Cons. Biol.* 8:1087.

Walters, J. R. 1998. The ecological basis of avian sensitivity to habitat fragmentation. In *Avian Conservation Research and Management*. Marzluff, J. M., and R. Sallabanks (eds). Washington DC: Island Press, p. 181.

Watson, J. E. M., R. J. Whittaker, and D. Freudenberger. 2005. Bird community responses to habitat fragmentation: How consistent are they across landscapes? *J. Biogeogr.* 32:1353.

Weinberg, H. J., and R. R. Roth. 1998. Forest area and habitat quality for nesting wood thrushes. *Auk* 115:879.

Weise, C. M., and R. J. Meyer. 1979. Juvenile dispersal and development of site-fidelity in the black-capped chickadee. *Auk* 96:40.

Whelan, E. M. 2006. A landscape-scale assessment of genetic diversity and population structure in northern bobwhite (*Colinus virginianus*). M.S. Thesis, Texas A&M University, Kingsville.

Williams, D. A., and K. N. Rabenold. 2005. Male-biased dispersal, female philopatry, and routes to fitness in a social corvid. *J. Animal Ecol.* 74:150.

Winter, M. 1999. Relationships of fire history to territory size, breeding density and habitat of Baird's sparrow in North Dakota. *Studies Avian Biol.* 19:178.

Winter, M., and J. Faaborg. 1999. Patterns of area sensitivity in grassland-nesting papers. *Cons. Biol.* 13:1424.

With, K. A., and A. W. King. 2001. Analysis of landscape sources and sinks: The effect of spatial pattern on avian demography. *Biol. Cons.* 100:75.

Zimmerman, J. L. 1996. Avian community response to fire, grazing and drought in tallgrass prairie. In *Ecology and Conservation of Great Plains Vertebrates*. Knopf, F., and F. B. Samson (eds). New York: Springer-Verlag, p. 167.

3 Global Biodiversity Conservation: We Need More Managers and Better Theorists

G. R. (Dick) Potts

CONTENTS

This review aims to investigate some links between theory and practice in conservation ecology and then to assess whether the linkages have been useful. From the outset, let us be clear; management can be successful in the complete absence of a link with theory. Just think about the farmers and fanciers who quickly bred and genetically fixed the many spectacular domesticated varieties without any knowledge of natural selection. In this case, there was a link, but it was in the opposite direction; Darwin developed his theory of natural selection in part from the practice of selection by the stockbreeders. So, given that theory can evolve from practice, is the reverse possible — can practice evolve from theory? Put another way, do we have the equivalent of a reference library of useful theory to help in management, or is it better to rely on the experience of a manager? A question asked (Maynard-Smith 1978) but not yet answered. Suppose, for example, that we are modeling energy flow through an ecosystem, we may need the metabolic rate of a large number of species, a laborious and expensive task. Fortunately, because the theory is well established in this case, all it is necessary to know is the weight of the individual animals concerned and whether they are single-celled or warm-blooded, or not, as the case may be. The equations of Kleiber's Law will then provide us the precise information we need. A similar approach is possible in plants (Reich et al. 2006). How often can we parallel such downloading in biodiversity management?

The aim here is to answer this question by weighing the usefulness of the views of managers, relative to the views of theorists. The approach is presented as a series of case studies built around studies of the grey partridge (*Perdix perdix*) that will suggest a general way forward.

STUDIES ON THE PARTRIDGE

For present purposes, these are centered on the 75 years of research on the grey partridge based at the Game Conservancy Trust, near the New Forest in southern England. Most studies of game bird population dynamics fall into one of two categories: linear (through years) or spatial (surveys). From the start, my grey partridge (henceforth partridge) research aimed to be both linear (now 39 years) and spatial, 62 km^2 reviewing, plus all relevant work in >30 countries, almost all at sites that I have visited, with fieldwork carried out in France, Germany, Austria, Spain (the red-legged partridge *Alectoris rufa*), and, most recently, Russia.

Throughout these studies, partridge fecundity was measured spatially and temporally from counts of birds on cereal stubbles in August and September using standard techniques. Adult survival was measured from the number of old females present in August of year t_{+1} divided by the number of all females present in August of year t (Potts 1980, 1986). In Sussex, the supply of insects for partridge chicks, obtained in cereal crops, was monitored annually using the Dietrick Vacuum Insect Net. Weeds, cropping, and all pesticide use were monitored annually on 12 farms (Ewald and Aebischer 2000), as was game management (Potts 1986).

The data on partridges were used to construct a computer simulation model (Potts 1980, 1986) that determined the most important factors limiting population size. The factors are as follows: (1) use of pesticides lowering the chick survival rate by reducing the insect food supply; (2) nest predation, particularly by the red fox (*Vulpes vulpes*) and Corvidae; (3) the supply and quality of nesting cover; and (4) the impact of shooting. Effects of pesticides on chick survival (Rands 1985, 1986a) and of predation at nests (Tapper et al. 1996) were measured in controlled and replicated experiments that verified the model. The effects of hedgerow management and of shooting were verified by comparing results of varying estate management regimes (Rands 1986b; Aebischer and Ewald 2004). As indicated above, the model was also given additional validation by reference to the independent studies overseas. Variation in equilibrium densities was explained over a range from 0.5 to 56 pairs km^{-2}. The model could not, however, explain the recent very high densities of up to 100 pairs km^{-2} in northern France until a new factor, provision of grain throughout the year, was incorporated (Potts 2003) — a relatively unexplored subject to which we will return later.

Despite this work, I had not had an opportunity to submit my accumulated knowledge to the acid test of actual hands-on management until 2003. In addition, a new factor had been proposed, parasitism by a cecal nematode *Heterakis gallinarum*, spreading as a result of an increasing number of reared pheasants (Tompkins et al. 2000).

By the chance retirement of a farmer in early 2003, one of the twelve farms in the Sussex Downs study area became available for experimental studies, a triangular-shaped arable area of 155 ha with steep wooded valleys descending on two sides. In spring 2003, a program for improving nesting and brood-rearing habitat was initiated by subdividing each of the six fields with a total of 4.6 km of beetle banks. These banks are low ridges planted with tall perennial grasses that in winter provide hibernating cover for beneficial predatory Carabidae, beetles that help contain the numbers of cereal aphids and that provide nesting cover for the partridge in spring. Kale (*Brassica oleracea* var.) was sown along one side of each beetle bank in strips 4 m wide (covering a total area of 2.2 ha). Next to this was a natural regeneration stubble set aside strip 6 m wide. All cereal headlands were managed under the U.K. government's DEFRA: CH2 option (no fertilizer, herbicides, or insecticides in outer 10 m of crop), amounting to 12.3% of the area. From October 1 to July 1, wheat grain was provided along each beetle bank in hoppers placed at intervals of 75 m. Predation control followed the methods of Tapper et al. (1996).

Several partridges radio tracked in the area in 2000–2001 moved freely over the nonwooded boundary (Watson 2004). By winter 2003–2004, however, only one female partridge remained on the 155 ha, and it too spent some time over the boundary. There was, therefore, the high probability of having a much-improved habitat but with no partridges. Accordingly, during the night of March 3, 2004, nine pairs were translocated from a distant estate to the arable part of the farm. In February 2005,

TABLE 3.1

Comparison of Observed and Predicted Parameter Values for Experimental Management of Grey Partridge (*P. perdix*) on Sussex Downs England: 2003–2005

	Before experiment	During experiment	
	Observed	Observed	Predicted
Fecundity	1.91 ± 0.332	5.91 ± 0.20[a]	5.35
Adult survival	0.62 ± 0.023	0.66 ± 0.065	0.69

[a] The total number of partridges on the experimental farm and on surrounding fields was 23 in 2003, 93 in 2004, 198 in 2005, and 253 in 2006. The numbers on the control area were 121, 78, 141, and 194, respectively.

some of their progeny moved as pairs to adjacent fields, returning with young in autumn 2005. Adult survival was, therefore, calculated to cover both the experimental area of 155 ha and the adjacent fields.

To concentrate management on the partridge, the three pheasant release pens on the experimental farm were not used from autumn 2003. Despite this, the grain provided in hoppers for the partridges attracted many pheasants from adjacent areas, where releasing continued, and these birds and their forbears represented a potential source of parasites. Replicated controlled experiments had shown that, in the absence of new fecal material, the infectivity of soil with the cecal nematode *H. gallinarum* declines slowly (Farr 1961). Her data predict that infectivity in 2004 and 2005 would be on the order of 90% and 75%, respectively, of that in 2003. With 1782 pheasants km^{-2} released on the experimental farm each year before 2003, seven times the U.K. average and not treated with anthelmintics, the partridges were predicted to be challenged by a level of *H. gallinarum* infection much greater than the U.K. average, even if no pheasants had been present in 2004–2005. In fact, autumn numbers of pheasants fell only 30%, and spring numbers increased.

The high numbers of pheasants did not have any of the adverse effects predicted by Tompkins et al. (2000) (Table 3.1). The explanation comes from new investigations of postmortem results for the partridge, carried out at the Game Conservancy over the period 1926–2005 and also from 31 short-term independent studies. Analysis shows that, far from increasing with the numbers of pheasants, *H. gallinarum* has undergone a global decline in prevalence to levels far below those capable of adversely affecting partridge population density (Potts, unpublished data).

The model is now successfully guiding the restoration of partridge numbers on many estates throughout England.

THE CASE STUDIES

THE CHARACTER OF THE LANDSCAPE BEFORE FARMING

So, what was the countryside like before farming? The answer is very important, because it gives us the natural habitat of the partridge that in turn will provide a context for the bird's current behavior. Until very recently, the theory was that a closed-canopy forest dominated by large tall trees had been the natural state of the European countryside. The forest would differ geographically in species composition, but everywhere there would have been a cover of "climax vegetation," a theory originating with Clements (1916).

Where would the partridges live in such a tall closed-canopy continuous forest? Some textbooks mentioned marram (*Ammophila arenaria*) ecosystems on coastal sand dunes as natural habitats, but

TABLE 3.2
Presence of the Partridge (*P. perdix*) in Europe during the Pleistocene and Holocene

Period	1000 Years BP	Sites/layers with *Perdix*[a]	
		Britain	Mainland Europe
Earlier inter-glacial	100–1000	1	24
Later inter-glacial[b]	10–100	15	225
After beginning of cropping	<10	11	14

[a] Most of the early and middle Pleistocene records are ascribed to *P. palaeoperdix* or the sub-species *P. perdix jurcsaki*. I have combined these, because there is no DNA evidence, because *P. perdix* has been reported as early as the Pliocene (Tyrberg 1998) and especially because the distinguishing features are sizes of bones that fall within, or very near to, the size ranges of the current subspecies (Roberts and Parfitt 1999; Stewart 2003). I have also included *Alectoris sutcliffei*, which has been reassigned to *P. perdix* (Stewart 2003).
[b] Last ended about 12,000 years BP (years before 1950).

most authorities concluded that the partridge did not arrive in Europe until agriculture had cleared the forests. That, however, does not fit the subfossil record, as set out by Tommy Tyrberg of the University of Linköping, Sweden, who awakened me to the wider relevance of his meticulous compilations. In the following section, I have extracted the *Perdix* records from his 1998 book (and web updates to August 2, 2001) (Table 3.2). Tyrberg's work draws together data from archaeological "digs" at hundreds of sites, many of them caves. Also included were the results of special studies, particularly those by Cecile Mourer-Chauviré at the Musée d'Histoire Naturelle de Lyon, who measured the bones of hundreds of subfossil partridges.

It is interesting that early humans were so keen on eating such a relatively small quarry as the partridge and so evidently adept at catching them, presumably using simple snares of the kind still used to exploit smaller game birds in many tropical rain forests and mountainous regions. Most interesting has been the finding of bones indistinguishable from *P. perdix* at the Boxgrove site very near the Sussex Downs study area and dated 475,000 years BP (Roberts and Parfitt 1999; Stewart 2003). This site is noted as the first in Britain where early humans used a variety of stone tools to butcher horse and other meat. When Mark Roberts took me to see the horizons in the quarry at one of the trenches, I was amazed to see that the layer from which the bones had come was only 3–4 cm thick. There were no signs that tree roots had penetrated the layer below. The climate for a relatively brief period was similar to that of the present day. However, in the absence of agriculture, the area was clearly grassland, kept that way by elephants (*Elephantidae*), wild horses (*Equus ferus*), an extinct rhinoceros (*Stephanorhinus hundsheimensis*), bison (*Bison* sp.), five kinds of deer including the giant deer (*Megaloceros* cf. *verticornis*) (similar to the better-known Irish elk), and a species of goat/sheep (Roberts and Parfitt 1999).

Thus, partridges were in the United Kingdom and on the European mainland long before agriculture. The partridges were not living in climax forests, but on savannas, and the evidence about how these can develop in moist climates has been put together by Dr. Frans Vera of the Netherlands Ministry of Agriculture, Nature Management and Fisheries in The Hague (Vera 2000). Given that no large area of pristine forest remains in lowland Europe, it is not surprising that the subject is controversial (e.g., Bobiec 2002). However, a new level of understanding has emerged from Dr. Vera's work, and the key is the interaction of large herbivores and vegetation; especially, the grazing and browsing of oak (*Quercus* spp.) and hazel (*Corylus avellana*). These species are unusual in having seeds (acorns and nuts) with large stores of food that are dispersed and buried by jays (*Garrulus glandarius*) and red squirrel (*Sciurus vulgaris*). When protected by grazing-resistant spiny vegetation, these trees thus have a good start in life; this is part of the story.

The most important evidence is that about the effects of the large grazers on trees in general. Even after the mega-herbivores had gone, grazers, such as aurochs (*Bos taurus*), moose (*Alces alces*), tarpan (*Equus caballus*), wild pig (*Sus scrofa*), deer (Cervidae), and bison (*Bison bonasus*) had, Vera concluded, lived in a long-term dynamic relationship with trees, and it was this that had produced the mosaic of grassland and trees. He argued that analyses of the fossil pollen record failed to allow for the fact that grazed grass produces relatively little pollen. Thus, the frequency of grasses was underestimated, indeed, contrary to what some workers had concluded; the pollen record indicated a park-like landscape. Vera then dispelled the notion that forests had disappeared through overgrazing by livestock and drew attention to the unnatural assemblage of species now found in forest reserves from which large grazing animals have been excluded. The key observation was that oak and hazel were characteristic features of pristine West European floras. Oak seedlings are easily grazed, browsed, and outcompeted by forest trees, but, similar to hazel, they regenerate well under the sort of spiny blackthorn/hawthorn scrub that is the product of grazing.

Vera summed it up: "The natural vegetation consists of a mosaic of large and small grasslands, scrub, solitary trees, and groups of trees, in which the indigenous fauna of large herbivores is essential for the regeneration of the characteristic trees and shrubs of Europe." Our final evidence comes from a consideration of the factors that determine the abundance of the largest herbivores.

The maximum size of prey that can be killed by predators in the Serengeti Ecosystem can be high relative to their own mass, especially when they hunt in packs; nevertheless, these predators are most efficient when preying on animals about 1.5 times their own body mass. Because the lion (*Panthera leo*) is the heaviest predator at about 170 kg, species such as buffalo (*Syncerus caffer*) and hippopotamus (*Hippopotamus amphibius*) (~3–4 times a lion's weight), and the heavier giraffe (*Giraffa* sp.) are rarely affected by predation, while the still heavier rhinoceroses (Rhinocerotidae) and elephant (*Loxodonta africana*) almost never suffer predation as adults and only rarely as juveniles (Sinclair et al. 2003). Thus, the large herbivores are not predator limited. They are mainly food limited and can have a major impact on the flora. The effects can be catastrophic when numbers are unnaturally high (Cumming et al. 1997). For example, elephant populations lived in equilibrium with Baobab trees for centuries though now, in some areas, they are destroying them, and no natural regeneration can take place, for example, in Tarangire Reserve, Tanzania (Potts, pers. observ., 1988). It follows that, in the example of the Sussex savannas of 475,000 BP, the numerous elephant, rhinoceros, bison, horse, and giant deer would most probably not have been controlled by the cave bear (*Ursus deningeri*), lion, wolf (*Canis lupus*), or spotted hyena (*Crocuta crocuta*) then present, leaving these large herbivores able to determine the flora. In summary of this case, much of the original theory about climax vegetation was weak; a wide range of field-workers and naturalists are correcting the theory.

THE VALUE OF PREDATOR CONTROL IN RELATION TO "COMPENSATORY MORTALITY"

In his *The Origin of Species*, Darwin (1859), who found some time to hunt partridges, wrote:

> The amount of food for each species of course gives the extreme limit to which each can increase; but very frequently it is not the obtaining of food, but the serving as prey to other animals, which determines the average number of a species. Thus, there seems to be little doubt that the stock of partridges, grouse and hares on any large estate depends chiefly on the destruction of vermin. If not one head of game were shot during the next twenty years in England, and, at the same time if no vermin were destroyed, there would, in all probability, be less game than at present, although hundreds of thousands of game animals are now annually shot.

TABLE 3.3

Effects of Removing Predators (but Not Raptors) on Various Game Birds, Using the Method of Bump et al. (1947)

Species, area, and period	Stock	Production	Bag	Authority
Partridge, Germany, 1959–69	+	+	>×1.91	Frank (1970)
Pheasant, South Dakota, 1964–71	+	×2.16	+	Trautman et al. (1974)
Grouse, Sweden[a], 1976–84	×2.70	×2.19	n/a	Marcström et al. (1988)
Partridge, Salisbury Plain, 1984–91	×2.60	×3.50	×3.65	Tapper et al. (1996)
Partridge, Otterburn, 2000–5	×5.00	×6.0	None	Baines (unpublished data, 2005)

[a] See also Summers et al. (2004) for similar cases in Scotland.

Unfortunately, by the time that I began to work on partridges such views were considered, at best, anachronistic, and a series of studies had reduced the relevant predators to the role of virtual scavengers. The key work was on northern bobwhite quail (*Colinus virginianus*) (Errington and Hamerstrom 1936), which gave rise to Errington's theory of compensatory mortality (Errington 1945, 1946). This excellent work was somewhat misinterpreted by followers who stressed the idea of the doomed surplus in winter but ignored conclusions on the importance of summer predation. Studies on the effects of shooting on wood pigeons (*Columba palumbus*) (Murton et al. 1966), of parasites and predation on red grouse (*Lagopus lagopus*) (Jenkins et al. 1964), and of predation by spotted hyenas (*C. crocuta*) on wildebeest (*Connochaetes taurinus*) and gazelles (*Gazella* spp.) (Kruuk 1972), all suggested that much mortality, attributable to predators, parasites, or shooting, could be compensated by lower mortality elsewhere in the life cycle. In the case of the woodpigeons, for example, fewer shot meant more starved and vice versa. One strong conclusion from the grouse work was that predators need not be killed; they too were only removing a "doomed surplus." Maybe this was arguable in an ethical framework, but what was the biological reality?

The managers, the gamekeepers, could see predation was important, but the theorists held the day. Some early work on northern bobwhite (Stoddard 1931) and on ruffed grouse (Bump et al. 1947) was virtually ignored, and the position only began to change in the mid-1980s when Marcström et al. (1988) published their results. At the Game Conservancy, we had noted Vidar Marcström's work in Sweden and in 1984 began a replicate study on the partridge. The basic approach was to select two similar areas, then to control the predators on one for a few years, after which the treatment and control were switched. It is interesting that the effects of predator removal were broadly similar in each experiment of this type (Table 3.3).

No raptors were removed in any of the five experiments. The key reason was the belief that raptor predation was not important, but in any event, it would not have been possible to obtain the necessary licenses. Whether biologists really believed raptor predation to be unimportant or simply wished to stop raptor persecution is a moot point. This was clarified with the experience at Langholm; at one time the most prolific grouse moor in Scotland, where predation by the peregrine falcon (*Falco peregrinus*) and especially hen harrier (*Circus cyaneus*) was shown to reduce grouse numbers greatly. The basic story is that, when the persecution on raptors at Langholm was stopped in 1992, the formerly persecuted hen harriers rapidly increased reducing the grouse to a very low density (Thirgood et al. 2000a) (Table 3.4); their predation at Langholm clearly limited the population of grouse. Today, foxes are no longer controlled at Langholm, and the densities of both grouse and hen harrier are at very low levels. Four studies of raptor predation on partridges show smaller (Table 3.5) but not insignificant (Watson 2004) predation rates with the decline in partridges in the Beauce also considered partly due to the increasing hen harrier population (Aebischer and Reitz 2000; Bro et al. 2004).

The key turning point in these investigations came with the realization that, although decreases in particular survival rates may not reduce overall survival rates, in the long term, through density

TABLE 3.4
Model Prediction of Long-Term Effects of Raptor Predation on Red Grouse (*Lagopus scoticus*) (Density km^{-2}), Given the Same Starting Point

		Density km^{-2}	
Year	Month	Raptors	No raptors
1	April	30	30
	October	45	90
2	April	31	49
	October	47	147
3	April	32	62
	October	48	188

Source: From Thirgood, S. J., et al. 2000a. *J. Anim. Ecol.* 69: 504.

TABLE 3.5
Red Grouse Annual Losses to Raptors at Langholm (3.) Presented in Comparison to Four Studies on Grey Partridge

S. no.	Raptor	Adult males (%)	Adult females (%)	Chicks (%)	Authority
1	Goshawk	24	13	?	Brüll (1964)
2	Great horned (*Bubo virginianus*) and Snowy owls (*Nyctea scandiaca*)	>16	<16	?	Carroll (1989)
3	Hen harrier and Peregrine falcon	>30	>30	37	Thirgood et al. (2000a)
4	Hen harrier	29	11	?	Aebischer and Reitz (2000)
5	Sparrowhawk (*Accipiter nisus*)	14	10	?	Watson (2004) and Watson et al. (in press)

dependence, they can still depress population density. It is this depression that releases the survival or fecundity needed to sustain the population (Potts 1986; Potts and Aebischer 1989). Almost 90 years ago, a Russian fisheries scientist clearly understood the true position — "a fishery, by thinning out a fish population, itself creates the production by which it is maintained" (Baranov 1918). Similarly, in partridges, predator control produces the surplus that can be harvested while still maintaining the population at equilibrium where mortality rates and fecundity balance (Table 3.6).

Whether or not the population is at equilibrium is important, obviously fishing and shooting can cause populations to drift down towards zero, if intensive enough. A case of excess shooting has been described for the partridge in one part of the Sussex Downs study area (Aebischer and Ewald 2004). Baranov was writing about sustainable fisheries and would have been appalled at the disastrous state of fishing today, only sustainable by continually switching to smaller species further down the food chain (Pauly and Maclean 2003).

But, does a new mortality, even though it may not increase total mortality, always reduce population density? Most relevant here are some valuable data on the demography of the red grouse (*L. lagopus*) gathered through a huge amount of diligent fieldwork by Adam Watson and colleagues

TABLE 3.6

Results of Computer Simulations of Partridge Population Density Showing Adverse Effect of Annual Shooting on Equilibrium Density Despite Compensatory Survival

Management	Survive shooting (%)	Survive winter and spring (%)	Survive shooting, winter, and spring (%)	Equilibrium density
Nest predation	100	62	62	12
Nest predation controlled	100	36	36	43
Predation control and shooting	59 (MSY)	58	34	20

Source: From Potts, G. R. 1986. *The Partridge: Pesticides, Predation and Conservation.* London: Collins.

TABLE 3.7

Percentage of Tabbed Territorial and Nonterritorial Red Grouse That Were Found Dead or Disappeared During November to May: Glen Esk (1957–61) and Kerloch (1961–67)

Grouse	Sex	Number	Found dead (%)	Disappeared (%)
Territorial	Male	385	4	3
	Female	281	1	1
Nonterritorial	Male	240	30	69
	Female	346	39	61

Source: From Watson, A. 1985. *Oecologia* 67:493.

(Watson 1985). In this species, the coveys break up in October, territories are established on the moor, and pairs are formed. When deep snow intervenes, the grouse form packs but quickly reestablish territories, if the weather ameliorates. The grouse Watson studied were individually back-tabbed and of known sex. His observations show that whether or not grouse died was determined by whether they had a territory and a mate and not by the force of mortality (Table 3.7). In its essentials, this is similar to the situation Paul Errington found in the bobwhite quail.

The subject is very controversial (Bergerud 1987), but these observations show that a new source of winter and spring predation would not reduce spring numbers, if it were to fall on the nonterritorial birds. It must be pointed out, however, that the moors Watson studied were managed to provide maximum density of territories and there was intensive predation control with only light shooting, an unusual combination. Watson explained — "Other, extrinsic kinds of population limitation may occur in less favorable regions, or even locally in northeast Scotland, such as heavy predation, severe disease, or catastrophic damage to food, and these might be sufficient to depress spring numbers appreciably" (Watson 1985). Heavy predation of territorial grouse was obviously the case at Langholm. Similarly, predation and shooting outside the period of territoriality would impact all birds, not just nonterritory holders and may, in consequence, depress spring numbers and predation on incubating females would obviously depress the productivity of the population. Whether or not birds took up territories, may also have been determined by the level of disease caused by the parasitic nematode *Trichostrongylus tenuis*, which seriously reduced the body condition of the grouse. At low densities on Speyside, where disease was far less important than at Glen Esk, or Kerloch, Hudson and coworkers found that overwinter predation was high, that it reduced the survival rate of territory holders, added to mortality rates, and reduced breeding densities. In consequence,

40% of nonterritorial birds survived and bred (Hudson 1992). In summary, the theorist's claims of a doomed surplus and compensatory mortality have not been helpful to the manager in this case.

THE TROPHIC PYRAMID AND INTRA-GUILD PREDATION

Lindeman (1942) developed the theory that available food decreases by 80–90% at each level of consumption. Often, this was represented as a trophic pyramid depicting four or more levels: plants, herbivores, carnivores, and top-carnivores. However, it has long been established that this arrangement is unsatisfactory, because many species feed from a pool of species rather than from one layer of the pyramid (Odum 1971; Pimm and Lawton 1978). For an early example, in Shakespeare, Macbeth Act II Scene IV we have:

> Tis unnatural,
> Even like the deed that's done. On Tuesday last,
> A falcon, towering in her pride of place,
> Was by a mousing owl hawk'd at and killed.

I have seen such a falcon (kestrel, *Falco tinnunculus*) intensely harass and irritate a mousing owl (short-eared owl, *Asio flammeus*), and maybe one got its own back? Such predation may not be unusual; reviewing causes of death in radio-tracked birds, Newton (1998) summarized estimates of the predation rate on adult raptors and owls, mostly by raptors and owls, and these are, on average, half the overall predation rate on most adult game birds! This mortality can be substantial, based on new research with radio-marked birds. In one study, raptors and foxes predated 36% of tawny owls (*Strix aluco*) during the post-fledging period (Sunde 2005).

This kind of behavior is termed intra-guild predation, a subject that was greatly popularized through work on spiders and scorpions by Polis et al. (1989). For the most part, spiders catch other predators inadvertently in their webs; scorpions are, however, true intra-guild predators, as are many vertebrates.

The subject of intra-guild predation became of general interest to conservationists following the work of Soulé et al. (1988), and it has grown as the top-predators have recovered from the losses they endured in the past, whether in Europe (e.g., for birds of prey; Bijleveld 1974) or in North America (e.g., Rudd 1964). The abundance of small predators such as the rat (*Rattus norvegicus*) has for centuries been blamed partly on the control of its predators, but the earliest clear recognition of the management value of intra-guild predation was when Brüll (1964) wrote:

> Basically, in the countryside of today, the hunter must replace the controls of the past–the wolf, lynx, eagle owl…to bring home a bag, however modest.

Similarly, but much more recently, Korpimäki and Nordström (2004) concluded:

> the return of the golden eagle (*Aquila chrysaetos*) and wolves in central and southern Finland could have beneficial effects on small game …. This might reduce the recent need for gamekeepers to control medium-sized carnivore populations.

As far as the United Kingdom is concerned, Tapper (1999), thinking particularly of wolves, is right, "resurrecting the primeval food web can only be a fireside dream." Even so, there may be many opportunities to reduce the need to control meso-predators through the encouragement of top-predators. Interestingly, approaching the issue from the opposite viewpoint, encouraging meso-predators in the biocontrol of insect pests, top-predators can be a hindrance (Prasad and Snyder 2006).

One of the best examples of intra-guild effects in a mammal is that which results from the competition between the coyote (*Canis latrans*) and the red fox (Table 3.8). Sargeant and Allen (1989) summarized 42 accounts of interaction between the two predators. In 79% (33) of the cases,

TABLE 3.8

The Differential Effects of Coyotes and Red Foxes on Duck Nests in the Dakotas, United States

Principal canid	Study areas	% Nesting success (95% confidence limits)	% Losses due to red fox
Red fox	13	17 (11–25)	27
Coyote	17	32 (25–40)	4
Both	6	25 (13–47)	?

Source: From Sovada, M.A. et al. 1995. *J. Wildl. Manage.* 59:1.

TABLE 3.9

Studies Reporting Intra-guild Effects of Top-predators on Meso-predator Densities

Top-predator	Meso-predator	Authority
Badger (*Meles meles*)	Hedgehog (*Erinaceus europaeus*)	Doncaster (1992)
Red fox	Pine marten (*Martes martes*)	Lindström et al. (1995)
Ural owl (*Strix uralensis*)	Tengmalm's owl (*Aegolius funereus*)	Hakkarainen and Korpimäki (1996)
Red fox	Hen harrier	Potts (1998)
Coyote	Raccoon (*Procyon lotor*), Skunk (*Mephitis mephitis*), and Badger (*Taxidea taxus*)	Henke and Bryant (1999)
Goshawk	Buzzard	Krüger (2002)
Eagle owl	Black kite (*Milvus migrans*)	Sergio et al. (2003)
Goshawk	Kestrel	Petty et al.(2003)
Eagle owl	Goshawk	Busche et al. (2004)
Golden eagle	Red fox	Korpimäki and Nordström (2004)
Goshawk	Buzzard	Hakkarainen et al. (2004)

the coyote was dominant, whereas no case of the reverse was reported. Nevertheless, they concluded that "avoidance of coyotes by red foxes is believed to be the principal cause of spatial separation of the two species." It follows that the removal of top-predators (e.g., coyote) can benefit meso-predators (e.g., red fox) and vice versa. Since these studies were made, an epidemic of sarcoptic mange affected both species for a few years and a comprehensive update of the current position is awaited. In South Texas, coyote removal was followed by an increase in five species of subdominant predator (Henke and Bryant 1999) and Henke (2002) considered "the benefit of coyotes to bobwhites may outweigh the occasional loss of birds to coyotes." This is a rapidly growing area of research and of eleven comparable published studies of the effects of one predator on another predator's density, six have appeared since 2000 (Table 3.9).

Further evidence of the intricate way in which intra-guild predation might benefit small game is to be seen in the results of Uttendörffer's incredibly detailed work on the food of raptors in northwest Germany in the 1930s. Based on the results summarized in Table 3.10, the bottom-predator, the sparrowhawk (*Accipiter nisus*) (European equivalent of the sharp-shin *Accipiter striatus*) is not itself capable of killing other raptors, but it is preyed upon by the goshawk (*Accipiter gentilis*), which in turn was included in the prey of the eagle owl (*Bubo bubo*). In a study in Schleswig Holstein in the early 1950s, the goshawk killed 13% of adult female partridges per annum (Brüll 1964) (Table 3.5). Since these studies were made, the eagle owl has reoccupied the area. The impact of this on partridges has not been measured, but we can estimate this given that the eagle owl is apparently capable of

TABLE 3.10
Summary of the Studies on Food of Birds of Prey in Germany

Predator	Killed		
	Birds	**Partridges**	**Partridge predators (mammals excluded)**
Sparrowhawk	58,077	87	0
Goshawk	8,309	880	376
Eagle owl	1,611	261	86

Source: Uttendörfer, O. 1939. *Die Ernährung der deutschen Raubvögel und Eulen und ihre Bedeutung in der heimischen Natur.* Neudamm: Neumann.

halving the breeding density of the goshawk (Busche et al. 2004), and thus presumably halving the percentage of partridges killed by the goshawk. At the point of halving, the density of goshawks and eagle owls is similar. The occurrence of all game birds averaged 10.4% in 28 studies of goshawk diet compared with 4.2% in 15 studies of the eagle owl (Valkama et al. 2004). Using the cube root adjustment for the size and food consumption of the birds brings the eagle owl estimate to 5.6%. From this, we might conclude that the probability of an eagle owl eating a game bird is (5.6/10.4) or 54% that of a goshawk. Thus, the goshawk take of 13% of partridges becomes, with the addition of eagle owls, 6.5% + 54% of 6.5% = 10%. This is, however, only part of the story, because the eagle owl also kills far more potential game bird egg predators than the goshawk. In one particularly detailed study of eagle owl diet, partridges accounted for 8.4% of items and pheasants for 2% with their potential egg predators accounting for over 31% (Bezzel et al. 1976). The most important game bird egg predator the hooded/carrion crow (*Corvus corone*) features in eagle owl diets more than twice as often as in goshawk diets (Uttendörfer 1939; Brüll 1964; Bezzel et al. 1976). In conclusion of this part, it is apparent that more predators will not mean more partridges killed, if there are intra-guild effects.

Modeling these relationships using a similar approach to that of Courchamp et al. (1999), but incorporating far more field data than they did, would be interesting.

Establishing the true position in the field in a combined study of predators and prey would be so complicated that it would need the discipline of a modeling approach. Buzzards (*Buteo buteo*), for example, are significantly adversely affected by predation and competition from goshawks (Hakkarainen et al. 2004), but the goshawks also provide the buzzards some protection from egg predation by Corvids (Krüger 2002). Another complication is the degree of specialization among the predators, a subject raised as a result of studies of the great skua (*Catharacta skua*) by Bayes et al. (1964) but virtually ignored in other species. Some eagle owls may even specialize in killing raptors; in 1948, Hagen reported that a pair in Norway killed 13 raptors and owls in one nesting season (Bannerman and Lodge 1955). Similarly, in France a pair of peregrine falcons killed 17 other raptors in one season (Kayser 1999).

In all, there are seven studies where it is considered that benefits might arise through top-predators controlling meso-predators and thereby benefiting small game and essentially similar species. These are listed in the order of publication in Table 3.11.

Because many top-predators are currently missing from the United Kingdom, the findings are potentially very significant, especially for those who can see the evidence about the need for predator control but who oppose any direct intervention by humans.

In some parts of Africa, a full complement of predators is still present and, maybe because of this, the resident game birds often thrive without management. Several wildlife managers I have

TABLE 3.11

Some Studies on Birds and Mammals Raising the Possibility of Conservation Benefits Arising from Intra-guild Predation

Top-predator	Meso-predator (s)	Benefiting species	Authority
Goshawk	Hooded crow	Willow grouse	Milonoff (1994)
Coyote	Red fox	Ducks	Sovada et al. (1995)
Pardel lynx (*Lynx pardina*)	Red fox and mongoose	Rabbit	Palomares et al. (1995)
Coyote	Feral cat	California quail	Crooks and Soulé (1999)
Feral cat	Rat	Island birds in burrows	Courchamp et al. (1999)
Golden eagle	Hen harrier	Red grouse?	Thirgood et al. (2000b)
Dingo (*Canis lupus d.*)	Red fox	Marsupials	Johnson et al. (2007)*

* this was published online early in 2006.

met in South Africa consider that reintroducing such top-predators such as the leopard (*Panthera pardus*) and martial eagle (*Polemaetus bellicosus*) can help to control numbers of meso-predators such as the mongooses (especially *Herpestes ichneumon*) and baboon (*Papio hamadryas*), which in turn help increase the numbers of guinea fowl (*Numida meleagris*) and francolin. In East Africa, four species of francolin can be abundant, <80 km^{-2} in the case of *Francolinus leucoscepus* (del Hoyo et al. 1994) despite a natural abundance of predators. In Spain, in the best areas for the red-legged partridge, four species of eagle and the eagle owl can be seen in a single partridge counting session and at least three of the species are important predators of partridge predators (Borralho et al. 1992).

How important, generally, are top-predator/meso-predator relationships? The picture is very unclear. For example, it was originally claimed that the increase in density of meso-predators, such as coatimundi (*Nasua narica*), on Barro Colorado Island, Panama, caused high predation rates on ground nesting birds, because the island was too small to hold top carnivores (Sieving 1992). This conclusion, however, may be unsafe, because humans may have previously kept down the meso-predators before the island was created (Karr 1982). As is now increasingly recognized, Native Americans were for a long time "the ultimate keystone predator" (Kay 1998). Summarizing this case, the pursuers of both theory and practice have been too timid and appropriate research whether in theory or practice is sadly lacking.

INTERACTIONS OF FOOD SUPPLY AND PREDATION

The first formal game management appeared in the thirteenth century in China, combining hunting regulations with the scattering of grain during the winter (Leopold 1933). Considering this long history, the importance of such provision of grain has been neglected although 35 years ago the late Kaj Westerskov, a Danish biologist then working in France, drew attention to the fact that partridges suffered from being fed in ways that were more appropriate to the pheasant (Westerskov 1977). There was no problem with the kind of food provided; rather, it was the poor dispersion of large food hoppers through the landscape, mostly in woodland. A decade later, Jean Grala, a farmer from near Lille in northeast France, began experimenting with smaller hoppers evenly dispersed in and around fields, one or two for each pair, and kept supplied through the spring and early summer. Further southwest, Jacques Hicter adopted the same system known as l'agrainage. Jacques has two farms — the system was introduced on Bellicourt by 1990 and on Savy in 1997, both in the Province of Aisne. Integrated into a program of predation control and reducing field size (the opposite to what other farmers were doing at the time), the two farms achieved all-time world record pair densities in 2006.

At the end of the century, the Office National de la Chasse (now ONCFS) found that l'agrainage or hopper feeding of wheat was the most widely used method of partridge management. Despite the fact that there are several thousand scientific papers on the partridge, no scientist has drawn attention to the importance of providing grain for partridges beyond the winter. Spring feeding has been recommended for pheasants (Hoodless et al. 1999), but not for the partridge.

At this point, we should review the context of these studies. First, it had become clear that there had been a substantial decline in the availability of grain on stubble fields and in several species of weed that provided relatively large seeds, for example, *Polygonum* spp. The importance of relatively large seeds of these species and of the even larger stubble-grain was well known. The amount of *Polygonum* seed found in partridges declined from 31% of the diet in the 1930s to 2% by 1977 (Potts 1986). What is more, it was calculated that one grain of wheat, obtained by one peck, provides the energy value of at least 20 pecks at cereal leaves (Potts 1986). Thus, by eating grain rather than cereal leaves, a partridge can greatly reduce the amount of effort and time necessary for obtaining food.

Some research had been carried out on the artificial feeding of partridges, but it was limited in scope to the choice of food by the partridge. In 1926, a questionnaire survey revealed that supplementary feeding was common and normally carried out from October to March. The choice of cereal grains among the gamekeepers was wheat > millet > barley > oats = sorghum (Parker 1927). In Wisconsin, a series of field trials, mostly with multicompartment self-feeding hoppers and troughs with the aim of "holding" partridges found that the partridges preferred grain in the order — maize > wheat > oats > sorghum > barley (Hawkins 1937). In Finland, food selection boxes were used, first in pens and then with the results checked in field trials, giving the following preferences — wheat > rye > barley > oats (Pulliainen 1965).

Thus, it is clear that the hopper system was using the right food with partridges well able to obtain such essentials as vitamin A from green food, weeds, growing crops, and arthropods. In many countries, forage legumes such as clover (*Trifolium* spp.) and lucerne (*Medicago sativa*) were preferred foods of the partridge, and these crops were abundant until the 1970s. The demise of mixed farming, explored in the next section of this paper, explains how the supply of these easily digested and nutritious plants was reduced on arable farms.

As the supply of good-quality partridge food declined, several species of raptor that at least occasionally preyed on partridges were rapidly increasing. It was known that partridges are most vulnerable to raptor predation during that part of the spring when they are seeking mates and nesting sites; it explains why one bird in the pair remains vigilant while the other feeds (Potts 1986). Only very recently has it been established that raptor predation can depress partridge densities in the UK (Watson 2004). Adult losses, however, remain very low on Jacques Hicter's farms where the hopper system is deployed but where raptors are common.

The clue to the importance of the hopper system comes from the way in which they are used. The fact that partridges are active up to 45 min after sunset has been well established for some years (Rotella and Ratti 1988). However, it was only in 2004, after we started hopper feeding in spring, that I found feeding after sunset; in spring, the pairs are usually found near hoppers that they use in a brief but intensive feeding bout when it is almost dark.

This may be important because the diurnal raptors that prey on the partridge, the accipiters, buzzards, and harriers, do not seek prey during this crepuscular period and thus do not threaten partridges. The briefness of the period of feeding may be to minimize losses from nocturnal owls and mammals. Francois Reitz of the Office National de la Chasse (now ONCFS) has reviewed what is known in France concluding that, although much of the research is inconclusive, it is possible to improve the status of partridge populations by giving some extra food during the prebreeding phase and during nesting (Reitz 2004). The main problem at present is that the research has not sufficiently involved the behavioral aspects such as vigilance against raptors. One feature of Jacques Hicter's partridges is their very low emigration rate in spring and it seems to me that this should be the starting point for the research on the attractiveness of the hopper feeding system. A large-scale experiment began in 2001 at Tall timbers Research Station, Tallahassee, FL, United States, and it has already shown significantly higher northern bobwhite densities on the fed area compared with the control area, perhaps by reducing predation (Palmer, pers. comm., 2006). As in Europe, however, many game managers had already adopted the techniques, enabling the co-existence of game-birds and clean farming. In summary of this part, it is clear that practical managers are ahead of research. More generally, it is interesting that the interaction of predation, vigilance, and feeding has been much neglected, with research only now under way (e.g., Zanette et al. 2003).

THE FUNCTIONAL RELATIONSHIP BETWEEN BIODIVERSITY AND SUSTAINABLE PRODUCTIVITY

At the outbreak of Civil War in England, Sir Richard Weston, a Roman Catholic and Royalist, was forced into exile in Flanders. There, he found a system of growing clover and cereals together in a rotation that, adapted to England, revolutionized its agriculture and ultimately that of much of North America. His key work was published in 1645 (Prothero 1917). The part of the revolution that concerns us here is the system of using cereals as a nurse crop for clover and grasses known as under sowing. Where the system originated is unclear, but the Romans grew cereals and grasses together to give better grazing after harvest (Fussell 1972). How the practice became adopted in England is also lost in the mists of time, but it had certainly reached the Sussex Downs Study Area by 1689 (Passmore 1992, 1997).

In the early nineteenth century, George Sinclair, head gardener to the Duke of Bedford, planted 242 plots with mixtures of herbs and grasses with results that led Charles Darwin (once again!) to write:

> It has been experimentally proved, if a plot of ground be sown with one species of grass, and a similar plot be sown with several distinct genera of grasses, a greater number of plants and dry herbage can be raised in the latter than in the former case.
>
> (Hector and Hooper 2002)

This basic principle, established in the early nineteenth century or earlier, was followed on all farms in the Sussex Downs study area until 1966, when one of the main farms ceased the practice to maximise the government subsidies from systems that are more modern. Others followed and only one farm now continues the system; Applesham, with 344 ha. On this farm, a spring cereal is sown, together with a mixture of six other species, each one with a purpose (Table 3.12). Because there is no cultivation of the soil between the cereal crop and the following spring when insects that hibernate in the soil emerge, the system is the most beneficial known for insect-eating farmland birds. The geographical polarization of pastures, stock, and cereal crops that followed the use of fertilizers instead of under sown legumes has been considered the most inimical event affecting farmland wildlife during the twentieth Century (Potts 1986).

TABLE 3.12
Composition of Christopher Passmore's Species Mixture Under Sown in Cereal Crops at Applesham, Sussex Downs

Species	kg ha^{-1}	Reason
Ryegrass (*Lolium perenne*)	7	Rapid growth easily digested
Cocksfoot (*Dactylis glomerata*)	3	Deep rooted: drought resistant
Timothy (*Phleum pratense*)	2	Grows well in cool weather
White clover (*Trifolium repens*)	2.75	Legume: slow growth but persistent
Alsike clover (*Trifolium hybridum*)	3	Legume: rapid bulking biennial
Black medick (*Trifolium lupulina*)	3	Legume: rapid bulking biennial

Our focus here should be on the reasons why so many ecological and agricultural theorists supported the monocultures that the above changes brought about. Notwithstanding the trials reported by Charles Darwin, this was certainly the case until the unique work of David Tilman and colleagues at the University of Minnesota who showed with a series of now classic experiments the benefits to overall productivity of mixtures of legumes and grasses including drought-resistant species (Tilman et al. 1996). In their plots, the more diverse assemblage of species used and retained nutrients more efficiently than did less diverse plots. The consequences of the increased uptake of nitrate and reduced leaching could be clearly seen in the form of increased productivity in the diverse plots. The Applesham mixture is at least as sophisticated, yet it owes nothing at all to theorists or modern experiments. In fact, an understanding of the way legumes combine symbiotically with *Rhizobium* spp. (soil bacteria) to divert available soil nitrogen to nonlegumes, has only recently emerged (Hall 1995).

To retain a representative global biodiversity, it will be necessary to further intensify farm production, encourage biodiversity on farmland, and thus minimize clearing of forest for food production (Potts 2003). Incorporation of diversity into cropping has been sorely neglected. Sometimes the consequences have been severe, such as, when advised by Lysenko, Stalin ordered that clover and cereals should not be grown together. This made the 1935 famine in the Ukraine much worse than it otherwise would have been, and millions starved (Volin 1970). It could be argued that Russian agriculture had not fully recovered from such policies by 1991 when the collapse of the Soviet Union made the picture worse still though with the future prospect of a strong and competitive agri-food sector (Kwiecinski 1998; Potts, pers. observ., 2004).

Today, many comparisons are being made between organic and conventional farming systems as a whole, but there is very little research on multispecies cropping. This extends to livestock, with no research on the optimum mixtures of plants in the diet of grazers and none on optimum mixtures of species grazing. Some of these issues in relation to agriculture were explored decades ago, but with results that were not followed up (Voisin 1959). André Voisin obtained clear evidence that cows were very selective in their choice of food, if given the opportunity, something that had been virtually ignored apart from a classic study on the palatability of different species of plants in Sweden in 1740! Voisin thought it odd that farmers, who were very choosy about their wine, expected their cows to eat anything they were given. As he put it "It is essential to seek the cow's opinion; that of the scientist is not enough." The basic issue extends to management of grazing in nature reserves where single species continuous grazing or browsing has too often been employed only for managers to express disappointment at the results. Yet, it is axiomatic that any natural system has a suite of seminomadic specialized grazers. Fortunately, in the Netherlands (in particular), there is a growing realization that management should aim for a variety of grazers that reflect the reality in nature that "the interdependence between large herbivores and the vegetation will have to be restored" (Vera 2000). This is the aim of a research project on the Nature Reserve at Ooostvaardersplassen.

Returning to agriculture, it is a tragedy that so little holistic research has been carried out on the merits of mixed (i.e., high biodiversity) farming. The highly influential post-war review by Astor et al. (1946) lists five advantages of mixed farming with only one of possible benefit to the environment and five disadvantages of mixed farming, all of them economic. Hindsight is a wonderful thing, but the fact that farmers had for thousands of years countered the loss of soil fertility by manuring fields, fixing nitrogen with the use of legumes, and by rotating cropping and stocking was ignored (Tilman et al. 2002). Only now has the evidence emerged that manures, legumes, and leys increase soil carbon and reduce leaching. What is more, the same experiments reveal that the yield depression compared with conventional systems is small (Drinkwater et al. 1998).

There is a fundamental problem here; agriculturalists have increasingly aimed for the supposedly efficient production of a single product such as wheat in a monoculture. In this process, the environmental costs of the monoculture are externalized. Ecologists, on the other hand, have emphasized biodiversity and sustainability of ecosystems. Whether in theory or practice, there are two different worlds here, and some bridging of the gap is long overdue. Globally, there are serious issues at the interface between biodiversity and agriculture. It is clear that modern agriculture reduces biodiversity on farmed land, but it is equally clear that the process reduces the need to convert pristine habitats to agricultural land (Potts 2003). Obtaining the correct balance is difficult, because agricultural economics externalize costs at all scales. For example, free trade, as advocated by WTO (Doha Round), will concentrate crops where they can be grown most cheaply, not where they cause least damage to the environment. There are serious consequences for global biodiversity; crops will become even more geographically concentrated as monocultures and thus will eventually be exposed to disease or drought. The vast area given over to soybeans in Brazil has recently been hit by both. Many crops can be grown in the already developed but declining agricultures of North America and Europe, especially Russia (not in the WTO), but are not simply because of cost or, in the case of Russia, problems of infrastructure already mentioned. The problem is severe; at present, Europe (the EC12) has set-aside 4 million ha of arable land, devotes another 4 million ha to cereals for export, while importing the equivalent of 15 million ha of vegetable proteins and oils (Herman and Kuper 2003). Much of the area from where the imports originate has until recently been pristine habitat. Examples are palm oil (for bio diesel) in Malaysia and Indonesia that could be replaced by oilseed rape, soybeans in the Cerrado (pronounced Seyhado) of Brazil that could be replaced by lupins and sugarcane (for bio-ethanol) that could be replaced by sugar beet. Joint thought is as scarce as hen's teeth here, and there remains a fundamental disharmony between the ecological and agricultural approaches. In conclusion of this case, the theorists, whether in agriculture or ecology, have until recently misled the policy makers.

ASSESSMENT

First we need to establish a few working definitions and relationships; whom we mean by managers and theorists (Table 3.13).

The manager is typified by the pre-Darwinian breeders, the theorist by those who considered predation to be unimportant to game birds. I have assessed each of the case studies on this basis and scored them according to the relative success of managers and theorists (Table 3.14).

In summary, these managers score 3–4 out of 5; the theorists 0 out of 4. As a result, I see the situation today much as it is described by Watt (1962) all those years ago; ecology probably has too much theory rather than too little.

In the future, progress must depend on evolving out of the current polarization of cultures, with managers in the field and theorists at their computers, not talking to one another. As Botkin (1990) points out, it is a real problem that many field ecologists cannot relate to the continuous time equations of the theorists. In such equations, mortality and fecundity are often dealt with simultaneously in a continuous time not easy to relate to the real world. Only with models can we disentangle the complexity of ecological systems, yet the innovative mathematical approach modeling top-predator

TABLE 3.13
Temporal Progression and Expansion of the Conceptual Terms Used in This Paper

Manager	Theorist
Practical challenge	Heurism
Diagnosis	Evidence
Trials	Hypothesis
Remedy	Opinion

TABLE 3.14
Summary of the Conclusions from the Five Case Studies

S. no.	Case study	Manager	Theorist
1	Landscape character	?	Incorrect
2	Predation control	Correct	Incorrect
3	Intra-guild predation	Incorrect	?
4	Spring food supply	Correct	Incorrect
5	Biodiversity function	Correct	Incorrect

effects in the cat, rat, and bird systems (Fan et al. 2005) is not accessible to field ecologists. A different approach would be the modeling of the relationships between top-predators, meso-predators, and small game, with a focus on field data and experimental management. A considerable amount of data exists (e.g., Craighead and Craighead 1969; Jedrzegewska and Jedrzegewska 1998; Krebs et al. 2001). Even so, the main problem will often be a lack of basic fieldwork.

The approach here is bound to be colored by personal experience, but I hope this is not too egocentric, given that it is the product of 44 years as a professional ecologist at the research and policy interface between agriculture and environment internationally. I started from the perspective of the partridge, but this has produced what I hope will be a new and refreshing approach to some serious concerns about global agriculture and practical wildlife management.

REFERENCES

Aebischer, N. J., and E. J. A. Ewald. 2004. Managing the UK Grey Partridge *Perdix perdix* recovery: Population change, reproduction, habitat and shooting. *Ibis* 146 (Suppl.): 181.

Aebischer, N. J., and F. Reitz. 2000. Estimating brood production and chick survival rates of grey partridges: an evaluation. In *Perdix VIII: Proceedings of an International Symposium on partridges, quails and pheasants in the western Palaearctic and Nearctic*, 26–29 October 1998. S. Farago (ed.), Sopron: Ministry of Agr. Dept. Game Manage. & Fishery, p. 191.

Astor, V., B. S. Rowntree, and F. W. Bateson. 1946. *Mixed Farming and Muddled Thinking: An Analysis of Current Agricultural Policy*. London: Macdonald.

Bannerman, D. A., and G. Lodge. 1955. *The Birds of the British Isles*, vol. 4. Edinburg: Oliver & Boyd.

Baranov, F. I. 1918. On the question of the biological basis of fisheries. Nauchnyi issledovatelskii iktiologicheskii Institute, Iszvestiia, 1:18.

Bayes, J. C., M. J. Dawson, and G. R. Potts. 1964. The food and feeding behaviour of the great skua (*Catharacta skua*) in the Faroes. *Bird Study* 11:272.

Bergerud, A. T. 1987. Reply to Watson and Moss. *Can. J. Zool.* 65:1048.

Bezzel, E., J. Obst, and K.-H. Wickl. 1976. Zur Ernährung und Nahrungswahl des Uhus (*Bubo bubo*). *J. für Ornithologie* 117:S.210.

Bijleveld, M. 1974. *Birds of Prey in Europe*. London: Macmillan.

Bobiec, A. 2002. "Grazing Ecology" from the Bialowieza primal forest perspective — The book review. *Acta Theriol.* 47:509.

Borralho, R., F. Rego, and N. Onofre. 1992. Raptors and game: The assessment of a net predation rate. *Gibier Faune Sauvage* 10:155.

Botkin, D. B. 1990. *Discordant Harmonies*. New York: Oxford University Press.

Bro, E., P. Mayot, E. Corda, and F. Reitz. 2004. Impact of habitat management on grey partridge populations: Assessing wildlife cover using a multisite BACI experiment. *J. Appl. Ecol.* 41:846.

Brüll, H. 1964. A study of the importance of the goshawk (*Accipiter gentilis*) and sparrowhawk (*Accipiter nisus*) in their ecosystem. In *Proceedings of the International Congress for Bird Protection: Birds of Prey and Owls*, Caen, p. 24.

Bump, G., R. W. Darrow, F. G. Edminster, and W. F. Crissey. 1947. *The Ruffed Grouse, Its Life Story, Ecology and Management.* New York: New York State Legislature.

Busche, G., H., J. Raddatz, and A. Kostrzewa. 2004. Nistplatz-Konkurrenz und Prädation zwischen Uhu (*Bubo bubo*) und Habicht (*Accipiter gentilis*): erste Ergebnisse aus Norddeutschland. *Vogelwarte* 42:169.

Carroll, J. P. 1989. Ecology of gray partridge in North Dakota. PhD thesis, University of North Dakota, Grand Forks.

Clements, F. E. 1916. *Plant Succession: An Analysis of the Development of Vegetation.* Washington, DC: Carnegie Institute of Washington.

Courchamp, F., M. Langlais, and G. Sugihara. 1999. Cats protecting birds: Modelling the mesopredator release effect. *J. Anim. Ecol.* 68:282.

Craighead, J. J., and F. C. Craighead, Jr. 1969. *Hawks, Owls and Wildlife.* London: Constable.

Crooks, K. R., and M. E. Soulé. 1999. Top dogs maintain diversity. *Nature* 400:563.

Cumming, D. H. M., M. B. Fenton, I. L. Rautenbach, R. D. Taylor, G. S. Cumming, M. S. Cumming, J. M. Dunlop, A. G. Ford, M. D. Havorka, D. S. Johnston, M. Kalcounis, Z. Mahlangu, and C. V. R. Portfors. 1997. Elephants, woodlands and biodiversity in southern Africa. *South African J. Sci.* 93:231.

Darwin, C. 1859. *The Origin of Species.* Mentor Edition 1958, New York.

del Hoyo, J., A. Elliott, and J. Sargatal. 1994. *Handbook of the Birds of the World*, vol. 2. Barcelona: Lynx Edicions.

Doncaster, C. P. 1992. Testing the role of intraguild predators in regulating hedgehog populations. *Proc. R. Soc. Ser. B* 249:113.

Drinkwater, L. E., P. Wagoner, and M. Sarrantonio. 1998. Legume-based cropping systems have reduced carbon and nitrogen losses. *Nature* 396:262.

Errington, P., and F. N. Hamerstrom. 1936. The Northern Bob-white's winter territory. *Agr. Exp. Sta., Iowa State College of Agr. Res. Bull.* 201:301.

Errington, P. L. 1945. Some contributions of a fifteen year local study of the Northern Bobwhite to knowledge of population phenomena. *Ecol. Monogr.* 15:1.

Errington, P. L. 1946. Predation and vertebrate populations. *Quart. Rev. Biol.* 21:144.

Ewald, J. A., and N. J. Aebischer. 2000. Trends in pesticide use and efficacy during 26 years of changing agriculture in Southern England. *Environ. Monit. Assess.* 64:493.

Fan, M., Y. Kuang, and Z. Feng. 2005. Cats protecting birds re-visited. *Bull. Math. Biol.* 67:1081.

Farr, M. M. 1961. Further observations on survival of the protozoan parasite, *Histomonas meleagridis* and eggs of poultry nematodes in faeces of infected birds. *Cornell Veterinary* 51:3.

Frank, H. 1970. Die Auswirkung von Raubvild und Raubzeugminderung auf die Strecken von Hase, Fasan,und Rebhuhn in eimen revier mit intensivister land wirtschaftlicher Nutzung. *Int. Congr. Game Biologists* 9:472.

Fussell, G. E. 1972. *The Classical Tradition in Western European Farming.* Newton Abbott: David and Charles.

Hakkarainen, H., and E. Korpimäki. 1996. Competitive and predatory interactions among raptors: An observational and experimental study. *Ecology* 77:1143.

Hakkarainen, H., S. Mykrä, S. Kurki, R. Tornberg, and S. Jungell. 2004. Competitive interactions amongst raptors in boreal forests. *Oecologia* 141:420.

Hall, R. L. 1995. Plant diversity in arable ecosystems. In *Ecology and Integrated Farming Systems.* D. M. Glen, M. P. Greaves, and H. M. Anderson (eds). Chichester: Wiley.

Hawkins, A. S. 1937. Winter feeding at Faville Grove, 1935, 1937. *J. Wildl. Manage.* 1:3.

Hector, A., and R. Hooper. 2002. Darwin and the first ecological experiment. *Science* 295:639.

Henke, S. E. 2002. Coyotes: Friend or foe of northern bobwhite in Southern Texas? In *Proceedings of the Fifth National Quail Symposium*, P. 57.

Henke, S. E., and F. C. Bryant. 1999. Effect of coyote removal on the faunal community in western Texas. *J. Wildl. Manage.* 63:1066.

Herman, P., and R. Kuper. 2003. *Food for Thought.* London: Pluto Press.

Hoodless, A. N., R. A. Draycott, M. N. Ludiman, and P. A. Robertson. 1999. Effects of supplementary feeding on the territoriality, breeding success and survival of pheasants. *J. Appl. Ecol.* 36:53.

Hudson, P. J. 1992. *Grouse in Space and Time.* Fordingbridge: The Game Conservancy.

Jedrzegewska, B., and W. Jedrzegewska. 1998. *Predation in Vertebrate Communities: The Bialowieza Primeval Forest as a Case Study.* Frankfurt: Springer.

Jenkins, D., A. Watson, and G. R. Miller. 1964. Predation and red grouse populations. *J. Appl. Ecol.* 1:183.

Johnson, N. J., J. L. Isaac, and D. O. Fisher. 2007. Rarity of a top predator triggers continent wide collapse of mammal prey: dingoes and marsupials in Australia. *Proc. R. Soc B*, 274:341–346.

Karieva, P. 1996. Diversity and sustainability on the prairie. *Nature* 379:673.

Karr, J. R. 1982. Avian extinction on Barro Colorado Island, Panama, since the island was formed in 1914. *Amer. Nat.* 119:220.

Kay, C. E. 1998. Are ecosystems structured from the top–down or bottom–up: A new look at an old debate. *Wildl. Soc. Bull.* 26:484.

Kayser, Y. 1999. Forte prédation sur des oiseaux de proies par un couple de Faucons pèlerins *Falco peregrinus* dans le Parc National du Mercantour, France. *Nos Oiseaux* 46:205.

Korpimäki, E., and M. Nordström. 2004. Native predators, alien predators and the return of native top-predators: Beneficial and detrimental effects on small game? *Suomen Riista* 50:33.

Krebs, C. J., S. Boutin, and R. Boonstra. 2001. *Ecosystem Dynamics of the Boreal Forest: The Kluane Project.* Oxford: Oxford University Press.

Krüger, O. 2002. Interactions between common buzzard *Buteo buteo* and goshawk *Accipiter gentilis*: Trade-offs revealed by a field experiment. *Oikos* 96:441.

Kruuk, H. 1972. *The Spotted Hyena.* Chicago: University of Chicago Press.

Kwiecinski, A. 1998. The slow transformation of Russian agriculture. *OECD Observer* 214:1.

Leopold, A. 1933. *Game Management.* Madison: University of Wisconsin Press.

Lindeman, R. L. 1942. The trophic-dynamic aspect of ecology. *Ecology* 23:399.

Lindström, E. R., S. M. Brainerd, J. O. Helldin, and K. Overskaug. 1995. Pine marten-red fox interactions: A case of intraguild predation? *Annales Zoologica Fennicae* 32:123.

Marcström, V., R. E. Kenward, and E. Engren. 1988. The impact of predation on boreal Tetraonids during vole cycles: An experimental study: 1976–1984. *J. Anim. Ecol.* 57:859.

Maynard-Smith, J. 1978. *Models in Ecology.* Cambridge: Cambridge University Press.

Milonoff, M. 1994. An overlooked connection between goshawk and tetraonids-corvids! *Suomen Riista* 40:91.

Murton, R. K., A. J. Isaacson, and N. J. Westwood. 1966. The relationship between wood-pigeons and their clover food supply and the mechanism of population control. *J. Appl. Ecol.* 3:55.

Newton, I. 1998. *Population Limitation in Birds.* London: Academic Press Ltd.

Odum, H. T. 1971. *Environment, Power and Society.* New York: Wiley-Interscience.

Palmer, B. 2006. Personal communication.

Palomares, F., P. Gaona, P. Ferreras, and M. Delibes. 1995. Positive effects on game species of top predators by controlling smaller predator populations -an example with lynx, mongooses and rabbits. *Cons. Biol.*, 9:295.

Parker, E. 1927. *Partridges: Yesterday and Today.* London: The Field.

Passmore, C. W. 1992. Farming in an environmentally sensitive area. *J. R. Agricult. Soc. England*, 153:45.

Passmore, C. W. 1997. Science into practice: Farming on the South Downs. *British Grassland Society Occasional Symposium*, 32:128.

Pauly, D., and J. Maclean. 2003. *In a Perfect Ocean.* Washington, DC: Island Press.

Petty, S. J., D. I. K. Anderson, M. Davison, B. Little, T. N. Sherratt, C. J. Thomas, and X. Lambin. 2003. The decline of the common kestrel *Falco tinnunculus* in a forested area of Northern England: The role of the Northern Goshawk, *Accipiter gentilis*. *Ibis* 145:472.

Pimm, S. L., and J. H. Lawton. 1978. On feeding in more than one trophic level. *Nature* 275:542.

Polis, G. A., C. A. Myers, and R. D. Holt. 1989. The ecology and evolution of intra-guild predation: Potential competitors that eat each other. *Ann. Rev. Ecol. Syst.* 20:297.

Potts, G. R. 1980. The effects of modern agriculture, nest predation and game management on the population ecology of the partridges *Perdix perdix* and *Alectoris rufa*. *Adv. Ecol. Res.* 11:2.

Potts, G. R. 1986. *The Partridge: Pesticides, Predation and Conservation.* London: Collins.

Potts, G. R. 1998. Global dispersion of nesting hen harriers *Circus cyaneus*. *Ibis* 140:76.

Potts, G. R. 2003. Balancing biodiversity and agriculture. In *Proceedings of the BCPC International Congress on Crop Science & Technology 2003.* Alton: The British Crop Protection Council, 35.

Potts, G. R., and N. J. Aebischer. 1989. Control of population size in birds: The grey partridge as a case study. In *Proceedings of the 30th Symposium of the British Ecological Society*. P. J. Grubb and J. B. Whittaker (eds). Oxford: Blackwell.

Prasad, R. P., and W. E. Snyder. 2006. Polyphagy complicates conservation biological control that targets generalist predators. *J. Appl. Ecol.* 43:343.

Prothero, R. E. 1917. *English Farming: Past and Present*. London: Green & Co.

Pulliainen, E. 1965. Studies on the weight, food and feeding behaviour of the partridge (*Perdix perdix* L.) in Finland. *Annales Academeiae Scientiarum Fennicae: Series A IV Biologica* 93:1.

Rands, M. R. W. 1985. Pesticide use on cereals and the survival of partridge chicks: A field experiment. *J. Appl. Ecol.* 22:49.

Rands, M. R. W. 1986a. The survival of game bird chicks in relation to pesticide use on cereals. *Ibis* 128:57.

Rands, M. R. W. 1986b. The effect of hedgerow characteristics on partridge breeding densities. *J. Appl. Ecol.* 23:479.

Reich, P. B., M. G. Tjoelker, J. -L. Machado, and J. Oleksyn. 2006. Universal scaling of respiratory metabolism, size and nitrogen in plants. *Nature* 439:457.

Reitz, F. 2004. Internal ONCFS Report — Grey partridge feeding: What impact does it have on populations? Saint Benoist: Office National de la Chasse.

Roberts, M. B., and S. A. Parfitt. 1999. *Boxgrove: A Middle Pleistocene Hominid site at Eartham Quarry, West Sussex*. London: English Heritage.

Rotella, J. J., and J. T. Ratti. 1988. Seasonal vocalisation in gray partridge vocal behavior. *Condor* 90:304.

Rudd, R. L. 1964. *Pesticides and the Living Landscape*. New York: University of Wisconsin Press.

Sargeant, A. B., and S. H. Allen. 1989. Observed interactions between coyotes and red foxes. *J. Mammal.* 70:631.

Sergio, F., L. Marchesi, and P. Pedrini. 2003. Spatial refugia and the co-existence of a diurnal raptor with its intra-guild owl predator. *J. Anim. Ecol.* 72:232.

Sieving, K. 1992. Nest predation and differential insular extinction among selected forest birds of Central Panama. *Ecology* 73:2310.

Sinclair, A. R. E., S. Mduma, and J. S. Brashares. 2003. Patterns of predation in a diverse predator–prey system. *Nature* 425:288.

Soulé, M. E., D. T. Bolger, A. C. Alberts, J. Wright, M. Sorice, and S. Hill. 1988. Reconstructed dynamics of rapid extinctions of chaparral-requiring birds in urban habitat islands. *Cons. Biol.* 2:75.

Sovada, M. A., A. B. Sargeant, and J. W. Greir. 1995. Differential Effects of Coyotes and Red Foxes on Duck Nest Success. *J. Wildl. Manage.* 59:1.

Stewart, J. R. 2003. Report: The bird remains from Boxgrove Project D (the Hominid area). Department of Anthropology and AHRB Centre for the Evolutionary Analysis of Cultural Behaviour, University College, London.

Stoddard, H. L. 1931. *The Bobwhite Quail, its Habits, Preservation and Increase*. Charles Scribner's Sons, New York.

Summers, R. W., R. E. Green, R. Proctor, D. Dugan, D. M. R. Lambie, R. Moss, and D. Baines. 2004. An experimental study of the effects of predation on the breeding productivity of capercaillie and black grouse. *J. Appl. Ecol.* 41:513.

Sunde, P. 2005. Predators control post-fledging mortality in tawny owl *Strix aluco*. *Oikos* 110:461.

Tapper, S. C. 1999. *A Question of Balance: Game Animals and Their Role in the British Countryside*. Fordingbridge: The Game Conservancy Trust.

Tapper, S. C., G. R. Potts, and M. H. Brockless. 1996. The effect of an experimental reduction in predation pressure on the breeding success and population density of grey partridges *Perdix perdix*. *J. Appl. Ecol.* 33:965.

Thirgood, S. J., S. M. Redpath, P. Rothery, and N. J. Aebischer. 2000a. Raptor predation and population limitation in red grouse. *J. Animal Ecol.* 69:504.

Thirgood, S. J., S. M. Redpath, I. Newton, and P. J. Hudson. 2000b. Raptors and red grouse: Conservation conflicts and management solutions, *Cons. Biol.*, 14:95.

Tilman, D., D. Wedin, and J. Knops. 1996. Productivity and sustainability influenced by biodiversity in grassland ecosystems. *Nature* 379:718.

Tilman, D., K. G. Cassman, P. A. Matson, R. Naylor, and S. Polasky. 2002. Agricultural sustainability and intensive production practices. *Nature* 418, 671.

Tompkins, D. M., J. V. Greenman, P. A. Robertson, and P. J. Hudson. 2000. The role of shared parasites in the exclusion of wildlife hosts: *Heterakis gallinarum* in the ring necked pheasant and the grey partridge. *J. Anim. Ecol.* 69:829.

Trautman, C. G., L. F. Frederickson, and A. V. Carter. 1974. Relationship of red foxes and other predators to populations of ring-necked pheasants and other prey in South Dakota. *Trans. 39th N. Am. Wildl. Nat. Res. Conf.* 241.

Tyrberg, T. 1998. Pleistocene birds of the Palaearctic: A catalogue. *Bull. Nuttall Onith. Club* 27:1.

Uttendörfer, O. 1939. *Die Ernährung der deutschen Raubvögel und Eulen und ihre Bedeutung in der heimischen Natur.* Neudamm: Neumann.

Valkama, J., E. Korpimäki, B. Arroyo, P. Beja, V. Bretagnolle, E. Bro, R. Kenward, S. Mañosa, S. M. Redpath, S. Thirgood, and J. Viñuela. 2005. Birds of prey as limiting factors of gamebird populations in Europe: A review. *Biol. Rev.* 80:171.

Vera, F. W. M. 2000. *Grazing Ecology and Forest History.* Wallingford: CABI Publishing.

Volin, L. 1970. *A Century of Russian Agriculture: From Alexander II to Khrushchev.* Cambridge, MA: University of Harvard Press.

Watson, A. 1985. Social class, socially-induced loss, recruitment, and breeding of red grouse. *Oecologia* 67:493.

Watson, M. 2004. The effects of raptor predation on grey partridges *Perdix perdix.* DPhil, University of Oxford, Oxford.

Watson, M., N. J. Aebischer, G. R. Potts, and J. E. Ewald. 2007. The relative effects of raptor predation and shooting on over-winter mortality of grey partridges in the UK. *J. Appl. Ecol.* 44 (in press).

Watt, K. E. F. 1962. Use of mathematics in population ecology. *Annu. Rev. Entom.* 7:243.

Westerskov, K. E. 1977. Covey-orientated partridge management in France. *Biol. Cons.* 11:185.

Zanette, L., J. N. M. Smith, H. van Oort, and M. Clinchy. 2003. Synergistic effects of food and predators on annual reproductive success in song sparrows. *Proc. R. Soc. Lond. B* 270:799.

4 Upland Game Bird Management: Linking Theory and Practice in South Texas

Leonard A. Brennan, Fidel Hernández,
William P. Kuvlesky, Jr., and Fred S. Guthery

CONTENTS

During the past 50 years, the South Texas landscape has provided a rich laboratory for developing a scientific basis for game bird management, primarily based on studies of northern bobwhites (*Colinus virginianus*) (Brennan 1999; Hernandez et al. 2002). Beginning with landmark natural history investigations (Lehmann 1984) and continuing with a series of rigorous empirical studies and development of models by Guthery and his associates (Guthery 2002), the emphasis on quail research innovation in the United States has clearly shifted from the southeastern states to Texas and Oklahoma.

Few, if any, wildlife species other than perhaps white-tailed deer (*Odocoileus virginianus*) have received research attention comparable to the northern bobwhite. For example, the recent *Technology of Bobwhite Management* (Guthery 2002) contains conceptual, quantitative, and theoretical models of all aspects of bobwhite life history: Energetics and energy-based carrying capacity, the physiological need for water, population ecology, viability and production, harvest theory, hunter–covey interface (HCI) theory, and theory of habitat management. Linking such theoretical models with management practices remains a major challenge, but has the potential to address and clarify issues that have important implications for game bird management policy and regulation.

EMERGING LINKS BETWEEN THEORY AND MANAGEMENT

Presently, there are at least four emerging examples of links between theoretical models and game bird management practices in South Texas:

1. Various scenarios using harvest theory management models seem to be useful in clarifying the implications of an agency proposal to double the daily quail harvest in Texas.
2. Empirical data suggest that theoretical HCI models provide meaningful results that can help manage harvest pressure while optimizing hunting opportunities.
3. Role of heat in regulating bobwhite populations has been given a new prominence in quail management, both from a climactic perspective and from understanding how operative temperatures can make vast areas potentially lethal to bobwhites during significant periods of the day.
4. Simulation modeling to assess interactions among predation dynamics, usable space for nesting, along with precipitation and heat, indicate that while modest increase in bobwhite population might be gained from nest predator control, annual production can be devastated by only moderate decrease in nesting cover.

A unifying theme emerging from these four examples is that concepts related to habitat theory and usable space remain the cornerstone of successful bobwhite management applications. The objective of this chapter is to briefly review and describe how these four examples form a linkage between theoretical ideas and applied management actions for sustaining wild quail populations and quail hunting in South Texas. Given the space constraints of this chapter, it is impossible to address each topic in depth. Rather, we focus on these four issues and the ways they serve as examples of how theory and management can be linked for wildlife conservation.

HARVEST MANAGEMENT

The Texas Quail Conservation Initiative (TQCI) has been developed to stabilize, increase, and restore wild quail populations (Brennan et al. 2005). The foundation on which the TQCI is based is habitat. The initiative recognizes that a great deal of usable space for quail has been lost across the Texas landscape and that successful efforts to reverse the quail decline will mean delivery of programs that result in net gains of habitat for quail.

Despite the appropriate focus of the TQCI on habitat, administrators and politicians in Texas attempted to promote doubling the daily quail bag limit from 15 to 30 birds as an incentive to inspire private landowners to implement habitat management. Rather than inspiring people to implement habitat management, the proposal to double the daily bag limit for private property owners who implement quail management backfired on its proponents. It sparked a vociferous public debate about the impact of hunting, including bag limits, season length, and associated issues. Although the intensity of opposition to the 30-bird daily bag limit caused the proposal to be withdrawn, it pointed to a number of issues pertaining to quail harvest in general and shifted the debate to a new direction.

Opponents of the 30-bird daily quail bag limit argued that there was no biological basis for such a proposition, which seems like an appealing argument until one considers that the present 15-bird per day limit in Texas also has no biological basis, as do most other statewide daily bag limits for quail

(Williams et al. 2004). Fifteen birds per day is simply a number that is deemed an intuitive, happy medium somewhere between zero (no hunting allowed) and infinity (no limit required) that provides opportunities to maximize recreational quail hunting opportunities in a manner assumed consistent with wise use and sustainability. In Texas, the 30-bird bag limit debate, as it became known, veered into conceptual territory that began to address the question: What, exactly, were the biological bases for setting a daily quail bag limit, or even a season length?

Although quail harvest management remains a contentious issue, recent syntheses of past work, along with refinement of associated additive harvest models, appear to have added a bit of clarity and new direction (Guthery 2002) to the debate. In its simplest form, the additive model appears as

$$Q_a = V_o + S_o H_o,$$

where Q_a is the total mortality from start to end of the hunting period, H_o is the hunting mortality rate for a given hunting effort in a population not subject to natural mortality, V_o is the natural mortality rate in a population not subject to hunting mortality, and S_o is the survival rate in a population not subject to harvest.

Despite being imperfect, the additive model can be used as a basis for approximating harvest rates on specific areas. Managers will need data on fall density, spring density, and fall-to-spring survival (an approximate guess based on the previous year or a range of historical data) to make meaningful harvest prescriptions that result in a predetermined spring density. Some simple assumptions about percent summer population gain from breeding also need to be made.

EMPIRICAL EXAMPLE

Using a modification of the additive model, harvest prescriptions for specific areas can be calculated based on

$$H = (T - N)(1 - N),$$

where H is the harvest rate, T is the total mortality over winter $=$ (fall density $-$ spring density goal)/(fall density), and N is the nonhunting winter mortality.

Assuming a fall density of 1 bird/acre (2.4 birds/ha) and a spring density goal of 0.5 bird/acre (or about 1.2 birds/ha)

$$T = (1 - 0.5)/1 = 0.5$$
$$N = 0.4.$$

Therefore,

$$H = (0.5 - 0.4)/(1 - 0.4) = 0.17 \text{ or a harvest rate of } 17\%.$$

If fall density estimates showed 1000 birds present on a 1000-acre pasture, the recommended harvest for that given hunting season would be 170 quail. Theoretically, it should not matter much when those 170 birds were harvested during the hunting season, although it would probably be preferable to do so during the early to middle part of the season to avoid potentially additive effects of late-season hunting mortality. In contrast, given the current 15 birds per day bag limit, and a 120-day quail season in Texas, it would take only 33 hunter-days at 15 birds/day to legally inflict local extinction on the quail population in the 1000-acre pasture. Thus, implementing harvest management prescriptions, that is, harvesting a given number of birds during the season based on calculations such as those shown above, on specific areas indicates that not only would losses to hunting be potentially lower than under the present policy, but hunting losses could also be tailored to year to year variation in rainfall and associated habitat conditions that influence summer gain (Table 4.1).

TABLE 4.1

Quail Harvest Prescriptions on a Management Unit under Various Annual Environmental Conditions

Recall from the text example the 17% harvest rate prescription to achieve a breeding population density of 0.5 bird/acre. Now, consider the following three different potential production scenarios for year 2:

Example 1: Good production

    Breeding season gain of 100% = +500 birds

    Start of next hunting season = 1000 birds

    $H = (0.5 - 0.4)/(1 - 0.4)$ for a harvest rate of 17% or 170 birds out of 1000

Example 2: Excellent production

    Breeding season gain of 150% = +750 birds

    Start of next hunting season = 1200 birds

    Assuming a spring density management objective of 0.5 bird/acre, and a fall density of 1.2 birds/acre:

    $T = (1.2 - 0.5)/(1.2) = 0.58$, and $N = 0.4$ (as noted in text); therefore,

    $H = (0.58 - 0.4)/(1 - 0.4) = 0.8/0.6 = 0.3$ for a harvest rate of 30% or 360 birds out of 1200

Example 3: Low production

    Breeding season gain of 50% = +250 birds

    Start of next hunting season = 750 birds

    Assuming a spring density management objective of 0.5 bird/acre, and a fall density of 0.75 birds/acre:

    $T = (0.75 - 0.5)/(0.75) = 0.33$, and $N = 0.4$ (as noted in text); therefore,

    $H = (0.33 - 0.4)/(1 - 0.4) = -0.07/0.6 = -0.12$ or no harvest

Note that under the three scenarios presented, both the total annual bag and harvest rate, covary as fall density changes in relation to variation in breeding season gain.

RESEARCH AND DATA NEEDS

The data needed to implement prescriptions for quail harvest management are at once both simple and complex. They are simple in the sense that specific harvest quotas to achieve a breeding population density are based on estimates of population density and fall-to-spring survival and some measure of summer gain in relation to environmental conditions. They are complex in the dual sense that (1) many factors can influence and bias estimates of population density and fall-to-spring survival, and (2) most field managers find collecting such data challenging at best, and intimidating at worst. Of course, there is also the potential for resentment among managers who would consider collecting such data unnecessary, because they are entrenched in *status quo*.

CULTURE AND PERCEPTION

Application of harvest management prescriptions has not gained widespread acceptance as a means of regulating quail hunting. First, it is a relatively new idea that is not well known by managers even though the basic ideas behind it are based on landmark quail research (Errington 1945; Roseberry and Klimstra 1984; Guthery 2002). Second, the incumbent data and associated work required for collecting such data are a significant obstacle for many managers. Third, the theoretical lack of a daily bag limit gives lay people the impression of unlimited harvest opportunities when in fact annual harvest will usually be more conservative than what is currently allowed. Some see a seasonal quota based on harvest management prescriptions as an opportunity for hunters to be legally sanctioned game hogs, when actually just the opposite is the case.

Despite the obstacles noted above, harvest management prescriptions will need to be considered as we move towards new definitions of quail management in the twenty-first century. For example,

Williams et al. (2004) presented a compelling case for such a change in perspective. They argued that quail habitat management efforts need to be scaled up from individual efforts on single properties to coordinated efforts on regional cooperatives, and that quail hunting pressure needs to be scaled down from statewide bag limits that have no biological basis to localized harvest prescriptions based on meaningful and conservative seasonal quail hunting quotas designed to sustain populations.

HUNTER–COVEY INTERFACE MODELS

Quail hunting is a complex behavioral process that until recently did not seem amenable to quantification. However, application of static and dynamic approaches that identified data needed to describe the HCI have appeared during the past decade (Radomsky and Guthery 2000; Guthery 2002) and produce meaningful results and predictions when used with empirical data (Hardin et al. 2005).

The models on which HCI theory is based contain some of the same variables as those of operations research in warfare. Fortunately, for hunters and pointing dogs, quail are quarry that do not return fire. However, hunting is hunting whether you are after a submarine or a bobwhite. The potential impacts of "friendly fire" (e.g., politicians accidentally shooting attorneys while quail hunting) are not part of HCI theory.

STATIC MODELS

Application of static models based HCI theory has the potential to complement harvest management prescriptions in that such models provide a mechanism to optimize hunting opportunities in a spatial context through time. That is, by increasing or decreasing the rate at which the hunt takes place — and by extension the rate at which coveys are encountered and birds are killed — harvest, and harvest opportunities, can be modulated to potentially achieve a management outcome (Figure 4.1). However, HCI is only about a decade old, and has been tested only on one area in South Texas (Hardin et al. 2005), although a less formalized application of spatial data has been used to quantify willow ptarmigan (*Lagopus lagopus*) hunting in Norway (Brøseth and Pederson 2000).

HCI theory has yet to become assimilated into the mainstream of quail management, despite having potential as a tool for managing harvest pressure on intensively hunted areas. The quantitative nature of HCI theory further contributes to maintaining its obscurity. Perhaps such issues will change over time as wildlife managers become more comfortable with quantitative techniques or software becomes available that will put a user-friendly face on implementing HCI theory for management.

DYNAMIC MODELS

In HCI theory, dynamic models are used to assess the role of how quails learn to avoid predators and thus influence the outcome of the hunt. Any seasoned quail hunter appreciates that quail tend to "flush wild" or "run wild" or otherwise seemingly increase their ability to avoid being shot at as the quail hunting season progresses. It has been noted that there are at least four different scenarios (Figure 4.2) that are related to the rate at which quail learn and the intensity at which hunting occurs (Radomsky and Guthery 2000; Guthery 2002). Empirical data indicate that bobwhites in South Texas most likely fit a "low hunting intensity–low learning rate" scenario (Hardin et al. 2005) (Figure 4.3).

ASSUMPTIONS

One of the most fascinating aspects of HCI theory is that it produces meaningful predictions and results when tested with empirical data, despite failure to meet underlying assumptions. The key assumptions upon which HCI theory is based (Table 4.2) are seldom met under field conditions for a variety of reasons. For example, the assumption that coveys are randomly distributed across the landscape can be met during some years but not others (Figure 4.4). Additionally, the assumption that

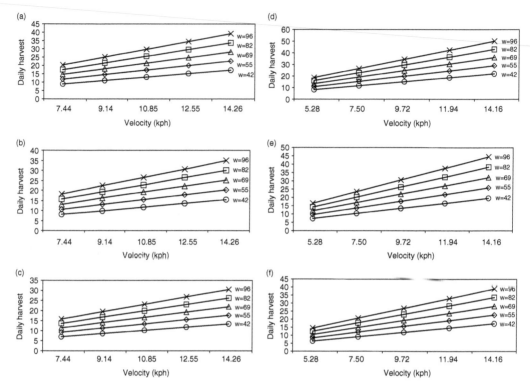

FIGURE 4.1 Static model output from HCI analyses of quail hunting data. (From Hardin, J. B., et al. 2005. *J. Wildl. Manage.* 69: 498. With permission.)

hunting patterns and areas covered during a hunt are not redundant is easily violated (Figure 4.5). If nonredundancy is more or less constant, this would not affect correlations between empirical and theoretical kill rates. If hunts are applied randomly to space, or some semblance thereof, HCI ought to work, as it seems to work in line transect sampling.

Despite the difficulty in meeting assumptions, it seems that HCI theory is "robust" to violation of the assumptions upon which it is based. This is a seemingly remarkable phenomenon that needs to be tested with sensitivity analyses and additional simulations.

ROLE OF HEAT

Poultry scientists have long-recognized the deleterious effects of excess heat on laying chickens such as declines in feed intake, egg production, eggshell thickness, and quality of yolk and albumin inside the egg (Card and Nesheim 1972; North 1972). This was also noted in bobwhites more than three decades ago (Case and Robel 1974). Thus, it is curious that, until recently, quail researchers ignored heat as a factor that limits quail production (Guthery et al. 2000; Guthery 2002). In this section, we explore two scale-dependent issues that relate to how heat affects quail populations, the first on a continental scale in the context of climate change and global warming, and the second related to how heat impacts usable habitat space on a local, fine-grained scale.

CONTINENTAL SCALE

By disrupting breeding and habitat use on a local scale, global warming could potentially contribute to quail population declines on a continental scale. Although largely overlooked by most quail

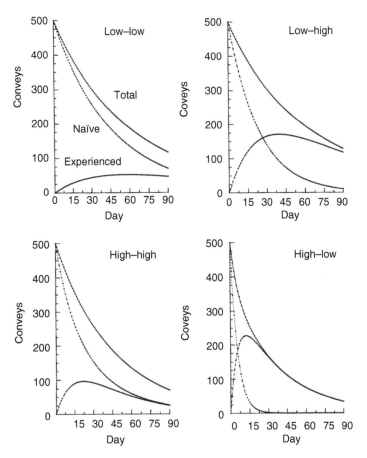

FIGURE 4.2 Dynamic model predictions of HCI theory. (From Radomsky, A. A., and F. S. Guthery. 2000. In *National Quail Symposium Proceedings*. Brennan, L. A., et al. (eds), vol. 4. Tallahassee, FL: Tall Timbers Research Station, p. 78. With permission.)

researchers, global warming has been hypothesized to be potentially responsible for long-term quail declines (Guthery et al. 2000). Habitat loss and fragmentation are most frequently invoked as being the cause of quail population declines, but some have noted (Guthery et al. 2000) that there are other places where declines have occurred where there is no apparent loss of habitat. Intellectually, long-term warming trends could potentially be responsible for quail population declines, at least from the perspective of climate change as one in a series of multiple working hypotheses (Chamberlain 1890).

LOCAL SCALE

Few people seem to appreciate that quail are sensitive to operative temperatures >38.7°C, that such temperatures occur regularly at southern latitudes, and that operative temperatures greater than this threshold can be lethal to quail (Figure 4.6). In the south, operative temperatures >45°C regularly occur at ground level during summer (Figure 4.7). At such temperatures, the heat intake exceeds heat loss in the quail body and death can occur in <1 h.

What this means from a management point of view is that shade from woody or robust herbaceous cover is critical for providing thermal refugia for quail during daily periods of peak temperature. Thus, if a pasture has, for example, <20% woody cover, then >80% of the habitat at ground level

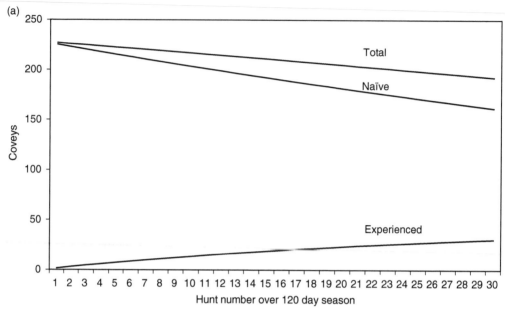

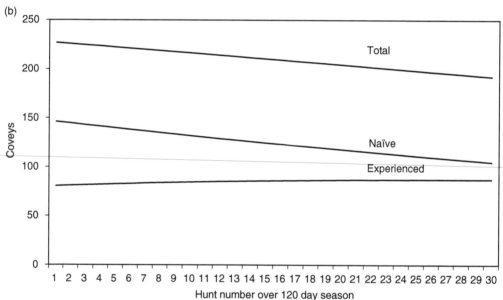

FIGURE 4.3 Dynamic model output from HCI analyses of quail hunting data. (From Hardin, J. B., et al. 2005. *J. Wildl. Manage.* 69:498. With permission.)

may not only be unusable, it may also be lethal for quail during most of a given summer day. Empirical data support this theoretical assertion for quail in South Texas (Forrester et al. 1998). This simple linkage among heat, lethality, and lack of usable habitat space is remarkably underappreciated within many quail management circles, especially as related to how relatively short times of excess heat exposure can be potentially lethal during extended periods of the day.

Along with such potential lethality, excess heat can also reduce the laying season for quail by as much as 60 days, which of course will lead to drastic declines or even complete failure of reproduction. However, we cannot at this time say this is a causal relation. There are certainly

TABLE 4.2
Assumptions of HCI Theory and Factors That Do and Do Not Allow Them to Be Met

Assumptions	Factors
1. Hunting is not redundant (static and dynamic models)	GPS data from hunting events document that hunting dogs and quail-hunting trucks frequently cover the same areas repeatedly
2. Coveys are randomly distributed over the hunting area (static and dynamic models)	Data indicate that coveys can be randomly distributed during years of high-population density (habitat saturation) but not during years of low density
3. All coveys are naive at the beginning of the hunting season (dynamic models)	Adult quail that have survived from previous years may retain learned avoidance behaviors
4. There is a greater probability of flushing a naive covey than an experienced on given an encounter (dynamic models)	An unknown portion of the population that was considered naive may have actually been experienced, based on encounter rates

correlative relationships between excess heat and poor reproduction from Illinois, the Southeast, and South Texas. All we know for sure is that hot summers are associated with low production and short laying seasons.

NEST PREDATION DYNAMICS AND USABLE SPACE

Despite an omnibus lack of evidence, predators continue to bear more than their share of the blame for ongoing quail declines (Rollins and Carroll 2001). Few people, including many wildlife managers, understand that correlation does not imply causation. During recent decades, we have seen a dramatic recovery of raptor populations and concomitant decline in quail numbers, yet there is no empirical evidence that raptors are suppressing quail populations, despite what many quail enthusiasts might think.

Nevertheless, predation remains an important life history component of quail, especially during the nesting season. As ground nesters, quail are vulnerable to predation losses, and have evolved important mechanisms (large clutch sizes, indeterminate laying, and multiple nesting attempts) to cope with these losses. Furthermore, predation is a process that influences quail populations, but people tend to perceive predation as an event, or perhaps as an unrelated series of events, which leads to shortsighted and oversimplified perspectives.

To understand how predation as a process influences quail populations, it is necessary to accumulate a large series of records of predation events. This is now possible using infrared video camera technology (Staller et al. 2005; Rader 2006), which, among other things, has documented that the bobwhite nest predator context is quite different between the southeastern United States and South Texas. One promising development for advancing our understanding of predation as a process that influences quail populations is the combination of inventory data on predation events from the infrared cameras with simulation modeling using systems analyses (Rader 2006). A systems model, based on difference equations that produce accurate empirical results, allows researchers to conduct "what if?" thought experiments — or simulations — to explore various scenarios to examine trade-offs among heat, precipitation, habitat structure, and nest predator reduction.

PREDICTIONS FROM SIMULATIONS

Based on simulation analyses, Rader (2006) observed that reduced nest-clump availability had the greatest negative effect on quail populations by lowering median bobwhite densities 1.8 quail/ha

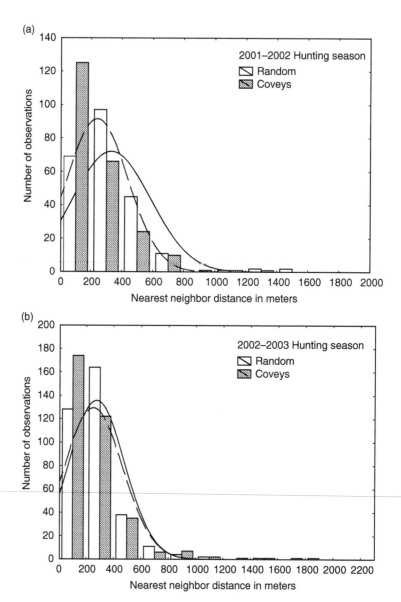

FIGURE 4.4 Evidence for annual variation in random and nonrandom distribution of quail coveys. (From Hardin, J. B., et al. 2005. *J. Wildl. Manage.* 69:498. With permission.)

(75%) from the baseline scenario. Simulation of eliminating losses to the top three predators increased density by 1.3 quail/ha (54%) from baseline conditions. Predator control also indicated some potential to mitigate the effects of low precipitation and extreme heat, but effects were not significant for the reduced nest-clump availability scenario. Over time, bobwhite median densities for all scenarios, except predator control, showed gradual declines from the initial density of 3.7 quail/ha.

Simulation analyses conducted by Rader (2006) provide insight into dynamics of the predation process that are not directly amenable to field conditions or empirical manipulation. Rader's (2006) simulation results pointed to the importance of nest clump density as the primary factor that influences quail population density. Furthermore, these results showed that predator reduction was secondary in importance to nest clump density when it comes to producing quail, thus providing support for the importance of habitat. From a practical standpoint, the simulation results provided by Rader (2006)

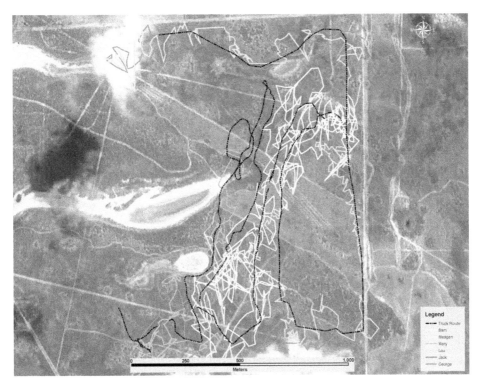

FIGURE 4.5 Example of how spatial redundancies in quail hunting can violate the assumption of no spatial redundancies in HCI theory. (From Hardin, J. B., et al. 2005. *J. Wildl. Manage.* 69:498. With permission.)

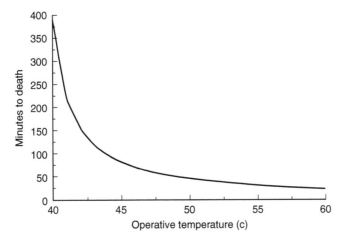

FIGURE 4.6 Approximate time to death from hyperthermia in 180-g northern bobwhites. (From Radomsky, A. A., and F. S. Guthery. 2000. In *National Quail Symposium Proceedings*. Brennan, L. A., et al. (eds), vol. 4. Tallahassee, FL: Tall Timbers Research Station, p. 78. With permission.)

support the philosophy that it is far easier and more economical for a manager interested in stabilizing and increasing a quail population in South Texas to provide abundant, high-quality nesting cover than it is to administer a pogrom against nest predators. From a metaphysical standpoint, Rader's (2006) results support conclusions made nearly three decades earlier by Beasom (1974) and Guthery and Beasom (1977).

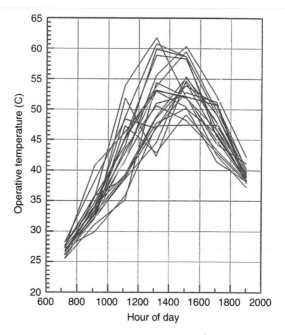

FIGURE 4.7   Hourly trends in operative temperature at ground level at a location in Webb County, Texas. (From Radomsky, A. A., and F. S. Guthery. 2000. In *National Quail Symposium Proceedings*. Brennan, L. A., et al. (eds), vol. 4. Tallahassee, FL: Tall Timbers Research Station, p. 78. With permission.)

HABITAT: THE CONCEPTUAL LINK BETWEEN THEORY AND APPLICATION

Habitat is the common thread among the four quail examples used in this chapter to illustrate linkages between theory and game bird management applications. For example, no amount of quail harvest management of any kind can be possible in the absence of the habitat required to produce quail in the first place. Habitat structure is the key element that can mitigate the potentially lethal effects of heat, especially on a local scale — which is the scale at which management actions are conducted. Habitat saturation, or the circumstances that result in a random distribution of coveys across a landscape, is important to meeting a key assumption of HCI theory. Finally, simulation analyses indicate that the reduction of predators — something that is extremely difficult and expensive — results in only modest population increases. In contrast, the same series of simulation analyses indicate that providing adequate nesting cover will result in a much better population response that can mitigate the effects of predators, heat, and drought.

Although the importance of habitat is clearly not new (Guthery 1997), an appreciation of the role habitat plays in the context of management is fundamental to understanding linkages between theory and application (Block and Brennan 1994). Hutchinson (1965) used the title *The Ecological Theater and the Evolutionary Play* to demonstrate linkages between theoretical ideas and real-world observations in ecology. He used a wide, metaphorical array of stage and set props provided by habitat as factors that introduce countless twists on a classic plot: The struggle for existence of organisms in the natural world through space and time.

ACKNOWLEDGMENTS

We appreciate the editorial guidance from T. Fulbright and D. Hewitt that helped improve this chapter. Brennan and Hernandez were supported by the Richard M. Kleberg, Jr., Center for Quail Research, and Guthery was supported by the Bollenbach Endowment for Quail Research at Oklahoma State

University. We especially thank M. J. Rader for sharing results from his unpublished dissertation, and S. J. DeMaso for discussions regarding harvest management. This manuscript has been approved for publication by Oklahoma State University.

REFERENCES

Beasom, S. L. 1974. Intensive short-term predator removal as a game management tool. *Trans. N. Am. Wildl. Nat. Res. Conf.* 39:230.

Block, W. M., and L. A. Brennan. 1994. The habitat concept in ornithology: Theory and application. *Curr. Ornith.* 10:63.

Brennan, L. A. 1999. Northern bobwhite. In *Birds of North America*. Poole, A. and Gill, F. (eds). Philadelphia: The Birds of North America, Inc., No. 397.

Brennan, L. A., S. DeMaso, F. Guthery, J. Hardin, C. Kowaleski, S. Lerich, R. Perez, M. Porter, D. Rollins, M. Sams, T. Trail, and D. Wilhelm. 2005. *Where Have All the Quail Gone? A Proactive Approach for Restoring Quail Populations by Improving Wildlife Habitat.* Austin, TX: Texas Parks and Wildlife Department.

Brøseth, H., and H. C. Pederson. 2000. Hunting effort and game vulnerability studies on a small scale: A new technique combining radio-telemetry, GPS and GIS. *J. Appl. Ecol.* 37:182.

Card, L. E., and M. C. Nesheim. 1972. *Poultry Production*. Philadelphia, PA: Lea and Febiger.

Case, R. M., and R. J. Robel. 1974. Bioenergetics of the bobwhite. *J. Wildl. Manage.* 38:638.

Chamberlain, T. C. 1890. The method of multiple working hypotheses. *Science* 15:92.

Errington, P. L. 1945. Some contributions of a fifteen-year local study of the northern bobwhite to a knowledge of population phenomena. *Ecol. Monogr.* 15:1.

Forrester, N. D., F. S. Guthery, S. D. Kopp, and W. E. Cohen. 1998. Operative temperature reduces habitat space for northern bobwhite. *J. Wildl. Manage.* 62:1505.

Guthery, F. S. 1997. A philosophy of habitat management for northern bobwhites. *J. Wildl. Manage.* 61:291.

Guthery, F. S. 2002. *The Technology of Bobwhite Management: The Theory behind the Practice.* Ames, IA: Iowa State Press.

Guthery, F. S., and S. L. Beasom. 1977. Responses of game and nongame wildlife to predator control in South Texas. *J. Range Manage.* 30:404.

Guthery, F. S., N. D. Forrester, K. R. Nolte, W. E. Cohen, and W. P. Kuvlesky, Jr. 2000. Potential effects of global warming on quail populations. In *National Quail Symposium Proceedings*. Brennan, L. A., W. E. Palmer, L. W. Burger, Jr, and T. L. Pruden, (eds). 2000. National Quail Symposium Proceedings. vol. 5. Tallahassee, FL: Tall Timbers Research Station, p. 198.

Hardin, J. B., L. A. Brennan, F. Hernandez, E. J. Redeker, and W. P. Kuvlesky, Jr. 2005. Empirical tests of hunter–covey interface models. *J. Wildl. Manage.* 69:498.

Hernandez, F., F. S. Guthery, and W. P. Kuvlesky, Jr. 2002. The legacy of bobwhite research in South Texas. *J. Wildl. Manage.* 66:1.

Hutchinson, G. E. 1965. *The Ecological Theater and the Evolutionary Play.* New Haven, CT: Yale University Press.

Lehmann, V. W. 1984. *Bobwhites on the Rio Grande Plain of Texas.* Texas: Texas A&M University Press.

North, M. O. 1972. *Commercial Chicken Production Manual.* Westport, CT: AVI Publishing Co.

Rader, M. J. 2006. Factors influencing nest success of northern bobwhites in southern Texas. PhD Dissertation, Texas A&M University, Kingsville, TX.

Radomsky, A. A., and F. S. Guthery. 2000. Theory of the hunter–covey interface. In *National Quail Symposium Proceedings*. Brennan, L. A., et al. (eds), vol. 4. Tallahassee, FL: Tall Timbers Research Station, 78.

Rollins, D., and J. P. Carroll. 2001. Impacts of predation on northern bobwhite and scaled quail. *Wildl. Soc. Bull.* 29:39.

Roseberry, J. L., and W. D. Klimstra. 1984. *Population Ecology of the Bobwhite.* Southern Illinois, IL: Illinois University Press.

Staller, E. L., W. E. Palmer, J. P. Carroll, R. P. Thornton, and D. C. Sisson. 2005. Identifying predators at northern bobwhite nests. *J. Wildl. Manage.* 69:124.

Williams, C. K., F. S. Guthery, R. D. Applegate, and M. J. Peterson. 2004. The northern bobwhite decline: Scaling our management for the twenty-first century. *Wildl. Soc. Bull.* 32:861.

5 An Ecological Basis for Management of Wetland Birds

Guy A. Baldassarre

CONTENTS

Wetland birds are species dependent on fresh, salt, or brackish-water wetlands to satisfy most, if not all, life history requirements. They are a subset of the larger waterbirds group, which comprises about 800 species dependent on any aquatic habitat, but not necessarily wetlands (Reid 1993). Hence, the wetland birds group excludes all seabirds (e.g., albatrosses, auks, and boobies) and most coastal waterbirds (e.g., gulls, terns, pelicans, and cormorants), but the group is nonetheless still large, diverse, and widely distributed.

Globally, the wetland birds group comprises about 620 species, of which 197 (31.8%) occur in North America (Table 5.1). The largest groups are the waterfowl (Anatidae — 157 species), followed by the rails, gallinules, and coots (Rallidae — 134 species). Other significant groups are the sandpipers (Scolopacidae — 87 species), plovers (Charadriidae — 66 species), egrets, herons, and bitterns (Ardeidae — 63 species), ibises and spoonbills (Threskiornithidae — 33 species), grebes (Podicipedidae — 19 species), storks (Ciconiidae — 19 species), and cranes (Gruidae — 15 species). Several smaller families of wetland birds contain 1–10 species.

Unfortunately for wetland birds, wetland habitats of all types have undergone massive loss and alteration. In the United States, for example, the "best estimate" is that 89.5 million ha of wetlands existed in the lower 48 states at the time of colonial America, with another 69 million ha in Alaska and 24,000 ha in Hawaii. By the mid-1980s, however, only about 42 million ha remained in the lower 48 states, which represented a loss of 53% (Dahl 1990). Some 22 states have lost 50% or more of their original wetlands; 11 have lost more than 70%. By 1997, 42.7 million ha remained in the lower 48 states, of which 95% were inland freshwater wetlands (Dahl 2000).

Globally, wetland loss may approach 50% (Dugan 1993), although loss is more severe in some regions than others. Based on a summary appearing in Mitsch and Gosselink (2000, 38), losses exceed 90% in Europe and New Zealand, 60% in China, and more than 50% in Australia. Within nations, certain regions or types of wetlands may be particularly affected. For example, about 67%

TABLE 5.1
Global and North American Diversity of Wetland Birds[a]

Family	Common name	Global species[a]	North American species[b]	Percentage of North American species
Anatidae	Ducks, geese, and swans	157	62	−39.5
Rallidae	Rails, gallinules, and coots	134	17	12.7
Scolopacidae	Sandpipers and allies	87	64	73.6
Charadriidae	Plovers and allies	66	16	24.2
Ardeidae	Herons, egrets, and bitterns	63	16	−25.4
Threskiornithidae	Ibises and spoonbills	33	5	15.2
Podocipedidae	Grebes	19	7	−36.8
Ciconiidae	Storks	19	2	10.5
Gruidae	Cranes	15	3	20.0
Recurvirostridae	Avocets and stilts	10	3	30.0
Burhinidae	Thick-knees	9	1	11.1
Jacanidae	Jacanas	8	1	12.5
Total species		620	197	31.8

[a] Data are from Clements, J. F. 2000. *Birds of the World: A Checklist*, 5th edn. Temecula, CA: Ibis Publishing.
[b] Data are from American Ornithologists' Union (AOU). 1998. *Check list of North American Birds*, 7th edn. Washington, DC: American Ornithologists' Union.

of all mangrove swamps in the Philippines are gone, as are an estimated 80% of the Pacific coastal estuarine wetlands and 71% of the prairie potholes in Canada (Whigham et al. 1993).

This extensive loss of habitat is exacerbated, because wetland destruction has differentially involved the "best" wetlands for wildlife. Such wetlands often occur on the most productive soils for agriculture (e.g., prairie and riparian wetlands), wherein those wetlands strongly compete with humans for space. Productive coastal wetlands have been differentially targeted for expansion of coastal cities, shipping channels, and agricultural development, especially rice and various forms of aquaculture (e.g., shrimp). Hence, many of the remaining wetlands in the United States and elsewhere often are poor-quality wildlife habitat. For example, in an early assessment of wetlands in the United States and their importance as waterfowl habitat, 70% were ranked as low or of negligible value (Shaw and Fredine 1956). Arctic wetlands are the major exception to this general pattern, because soils are poor and growing seasons short; hence, agricultural activities are virtually nonexistent in the Arctic, and arctic wetlands are critical breeding habitat for some waterfowl and shorebirds.

Nonetheless, the extensive quantitative and qualitative loss of wetlands has severely affected wetland birds of all taxonomic groups. Indeed, in comparison to species numbers summarized in Table 5.1, some 175 (28.2%) are listed by International Union for the Conservation of Nature and Natural Resources (IUCN) in the *Red List of Threatened Species* (IUCN 2006; Table 5.2); 30 (4.8%) are listed extinct, 21 (3.4%) as critically endangered, 37 (6.0%) as endangered, 46 (7.4%) as vulnerable, and 41 (6.6%) as near threatened (Table 5.2). The most affected group is the cranes (66.6% listed), followed by the rails (35.1%), sandpipers (26.4%), waterfowl (26.1%), herons (20.6%), and plovers (19.7%).

NEED AND BASIS FOR WETLAND MANAGEMENT

Acquisition, easement, or legal designations cannot adequately protect wetlands, because they are especially subject to rapid changes in structure and plant composition. Hence, active management

TABLE 5.2
Total Number of Species of Wetland Birds Listed in Four Major Categories by the International Union for the Conservation of Nature and Natural Resources (2006)

Family	Common name	Extinct[a]	Critically endangered	Endangered	Vulnerable	Near threatened
Anatidae	Ducks, geese, and swans	6	6	9	12	8
Rallidae	Rails, gallinules, and coots	15	3	7	14	8
Scolopacidae	Sandpipers and allies	2	2	4	4	11
Charadriidae	Plovers and allies	0	2	2	4	5
Ardeidae	Herons, egrets, and bitterns	4	0	6	2	1
Threskiornithidae	Ibises and spoonbills	1	4	3	1	2
Podocipedidae	Grebes	2	2	1	1	1
Ciconiidae	Storks	0	0	3	2	2
Gruidae	Cranes	0	1	2	6	1
Recurvirostridae	Avocets and allies	0	1	0	0	0
Burhinidae	Thick-knees	0	0	0	0	2
Jacanidae	Jacanas	0	0	0	0	0
Total	All species	30	21	37	46	41

[a] Species extinct since 1500.

is usually critical to maintain the functional values that led to protection in the first place! Active management of remaining wetlands is especially essential, because a lesser amount of habitat must now maintain population levels of wetland birds and other wetland wildlife once supported by a wetland base nearly twice as large as that in North America today.

Wetlands management basically synchronizes availability of habitat and habitat components (e.g., food) to coincide temporally and spatially with life history events affecting survival and reproduction of populations of target species or species groups. Most wetland birds are migratory, and species-specific migratory patterns, habitat use, and other life history requirements are fairly well known, especially in North America. Hence, the spatial and temporal considerations of wetland management (i.e., "where and when") are well known, and techniques (i.e., "how") for actual management are very well documented (Payne 1992). However, managers must understand *why* protection and manipulation of certain habitats benefits some species and species groups but not others, and then reconcile issues of size, juxtaposition, connectivity, and habitat diversity. Managers must understand *why* certain factors are important to issues of biodiversity, because they are now called upon to address an array of wetland-dependent biota in addition to a focus species or group. Finally, managers must understand why a particular management technique is warranted in one situation but not another. Wetland managers may wish away this level of understanding in the decision-making process, but such complexity is only accelerating with the new millennium, coincident with perhaps the most urgent need ever to protect and manage wetland habitats and associated biota such as wetland birds. Further, despite increasing attention to wetland conservation issues everywhere, wetland loss will continue wherein management of remaining wetlands will be of paramount concern. Thus, increasingly complex issues will characterize the landscape for wetland managers in the future, but that difficulty is not insurmountable, and certainly not an *a priori* recipe for failure.

Wetland managers must understand the ecological underpinnings of certain management approaches (i.e., an ecological approach) to understand why they are successful. A "how-to" approach dooms managers to a "hit-or-miss" strategy that is often ineffective, unrepeatable, nontransferable among managers, and costly in terms of both time and money. The purpose of this chapter is to demonstrate the value of an ecological approach to the management of wetland birds. To achieve that objective, I review the ecological basis for four major approaches to wetland management that I believe are most effective in the decision-making process associated with wetland bird conservation and management. Two approaches focus on habitat considerations (landscape ecology and wetland plant succession), whereas the other two focus on wetland birds themselves (demographic models and phylogenetics).

LANDSCAPE ECOLOGY AND WETLAND MANAGEMENT

All wetlands occur as individual entities, but wetlands also aggregate into groups or complexes that spatially occur at scales ranging from local to global, the array of which leads management into the realm of landscape ecology. Along the path, managers also confront issues such as wetland size and juxtaposition, as well as assessment of species richness and habitat functions. Habitat acquisition and legal protection of wetlands are especially concerned with all of the above issues, and landscape ecology has figured prominently in these deliberations even before its formal emergence in ecology. The decision-making tools available to managers have transformed from crude, black-and-white aerial photographs to high-resolution digital satellite images accurate to within a few meters, and the emergence of geographic information systems (GIS) and associated techniques has facilitated detailed quantitative analysis where earlier efforts involved much guesswork. Nonetheless, improvement of data quality has not negated application of the principles of landscape ecology to wetland protection efforts.

Historically, waterfowl managers were perhaps first to use concepts of landscape ecology in association with acquisition of national wildlife refuges. Early managers recognized that nearly all species of waterfowl in North American were highly migratory; hence, management efforts were needed to establish refuges on breeding, wintering, and migration areas, if the annual cycle needs of waterfowl were to be satisfied for the array of species involved. For migratory birds in general, these issues were later recognized as "connectivity," which realizes that individuals move back and forth between specific breeding and nonbreeding areas (Webster et al. 2002). Regardless, early wetland acquisition efforts often focused on large tracts of wetlands that formed the basis of many national wildlife refuges. Small wetland areas received direct focus from the U.S. Fish and Wildlife Service with creation of the Waterfowl Production Areas program in 1959, which recognized the importance of small wetlands to breeding waterfowl.

More detailed relationships about habitat size and species richness of wetland birds stemmed from early studies that spawned the initial theory of "island biogeography," which established a relationship between the size and isolation of islands and subsequent species richness of birds, spawning the species-area equation now so familiar in conservation biology (MacArthur and Wilson 1967). Subsequent studies examined these issues in an array of insular habitats such as prairies, forests, cemeteries, and, of course, wetlands.

The first major study occurred in 1983–84 when Brown and Dinsmore (1986) addressed the influence of size and isolation on the diversity of breeding birds in 30 Iowa wetlands ranging in size from 0.2 to 182.0 ha. Data from their study yielded a significant correlation between species richness and wetland area ($r = .82$). Gibbs et al. (1991) later examined use of 87 wetlands in Maine by 15 breeding waterbirds and also found a significant correlation between area and species richness ($r = .66$). Similarly, Grover and Baldassarre (1995) found wetland area correlated ($r = .65 - .66$) with richness of wetland birds using active and inactive beaver (*Castor canadensis*) ponds in south-central New York.

Relative to isolation of habitats, Brown and Dinsmore (1986) found that total area of wetlands within 5 km of a given wetland explained the most variation in species richness ($r = .42$), and was the only significant factor among 12 isolation variables measured. Indeed, wetlands in complexes had more species (11) but were half as large (14 ha) as isolated wetlands (30 ha, nine species).

Gibbs et al. (1991) found that isolation was weakly correlated ($r = .24$) with the species richness of birds. In a two-variable model, however, wetland size and isolation explained 74% of the variation associated with species richness in Iowa and 43% in Maine.

These and other studies certainly provide managers with the guidance that protection of large wetlands and wetland complexes will protect species richness. Weller (1981) was perhaps the first to promote the idea that protection of a wetland complex was the best way to protect regional wetland flora and fauna. Within such complexes, however, protection of small wetlands may be especially significant in designing programs to protect species diversity. For example, Gibbs (1993) modeled the effect of removing small wetlands (i.e., 4.05 ha) within a 600-km^2 area in Maine that contained 354 wetlands ranging in size from 0.05 to 105.3 ha. Loss of small wetlands reduced total wetland area by only 19% but decreased the total wetland number by 62%, and increased the inter-wetland distance by 67%. The simulation predicted that local populations of turtles, small birds, and small mammals faced significant risk of extinction with loss of small wetlands. This study thereby underscored the value of small wetlands as a means of maintaining species richness of wetland wildlife, including waterbirds. Unfortunately, small wetlands are the easiest wetlands to drain or fill. Further, because of their ephemeral nature, small wetlands are often not afforded legal protection and do not garner significant attention from managers. Hence, despite their importance, small wetlands are often the first to disappear from wetland complexes.

Large wetlands are, of course, critically important within wetland complexes for many reasons, but particularly because their absence affects area-dependent species. For example, in their Iowa study, Brown and Dinsmore (1986) found that 10 of the 25 species detected did not occur in wetlands <5 ha. Pied-billed grebes (*Podilymbus podiceps*) and yellow-headed blackbirds (*Xanthocephalus xanthocephalus*) are area-dependent species, rarely exploiting habitat outside the vicinity of nesting areas (Naugle et al. 1999).

Managers of waterbirds at a landscape level must also recognize that features other than wetlands can profoundly influence some species. For example, western willets (*Catoptrophorus semipalmatus inornatus*), breeding in the Great Basin, moved daily from upland nesting sites to wetland foraging sites (Haig et al. 2002). Black terns (*Chlidonias niger*), breeding in South Dakota, were more likely to occur in landscapes that contained grasslands instead of agricultural fields (Naugle et al. 1999).

Although many studies have connected issues of species occurrence and richness to landscape-level consideration of wetlands, significantly fewer studies, however, have demonstrated causal relationships. Such data largely come from studies of waterfowl, but they are extremely significant, because they can explain why certain patterns occur, which provides a powerful approach to management decisions. Seasonal and temporary wetlands, for example, are heavily used by ducks breeding in the Prairie Pothole Region of both the United States and Canada, because such wetlands provide habitat for the aquatic macroinvertebrates that provide protein essential for egg production (Krapu 1974; Swanson et al. 1979). However, temporary wetlands are substantially less available in drought years. This reduced wetland availability relegates mallards (*Anas platyrhynchos*) and other dabbling ducks to foraging in deeper, more permanent wetlands where invertebrates are less abundant. Such a situation then affects mallard fitness, which is reflected in a shorter nesting period and production of fewer nests. During a drought year in North Dakota, for instance, mallards produced dramatically fewer nests than during a wet year, and females remained on study area wetlands for 44 days in the wet year compared with only 16 in the drought year (Krapu et al. 1983).

Other studies also have concluded that wetland complexes increase habitat heterogeneity, which is important to waterfowl and other waterbirds, because life history requirements usually require various types of wetlands (Leitch and Kaminski 1985; Murkin et al. 1997). Wetland complexes in North Dakota, for example, contain seasonal, semipermanent, and permanent wetlands that provide optimal brood-rearing habitat for mallards, and subsequent production was greatly enhanced where such complexes were protected by managers (Talent et al. 1982). Indeed, mallard broods used up to 11 different wetlands, which again emphasizes the importance of maintaining a complex of wetlands as a prerequisite for good waterfowl production.

Upland landscape variables measured at 10.4 and 41.4 km^2 best explained nest survival for ducks nesting in North Dakota (Stephens et al. 2005). Not surprisingly, nest success was positively related to the amount of grassland habitat, but success was quadratically related to the amount of grassland edge. Stephens et al. (2005) speculated that the latter relationship likely occurred because predator communities became most diverse at levels of intermediate fragmentation. In contrast, landscapes with more contiguous grasslands favored endemic predators such as coyotes (*Canis latrans*) and badgers (*Taxidea taxus*) over red fox (*Vulpes vulpes*), raccoons (*Procyon lotor*), and striped skunk (*Mephitis mephitis*); the latter three species are significant predators of waterfowl nests. Hence, nest success may be higher where grassland landscapes are more intact, because such sites will favor coyotes and badgers, which have larger home ranges and occur at lower densities than other nest predators such as the raccoon, skunk, and red fox (Sargeant et al. 1993). Additionally, coyotes can be especially effective predators on other mammalian nest predators like the red fox. For example, Sovada et al. (1995) found that duck nest success in North Dakota was merely 2% at a study area predominately occupied by red foxes compared with 32% in an area where coyotes (*C. latrans*) were the dominate predator — the two areas were 5 km apart. Sovada et al. (2000) also reported that daily survival rates of duck nests were significantly greater for nests in large patches of grassland habitat in one of six possible year × patch size comparisons, and the trend was similar for the remaining five.

Overall, the underpinnings of landscape ecology are demonstrating that the best strategy for waterbird conservation, in terms of both species richness and fitness, is to protect entire wetland complexes, where possible, and to enhance existing complexes by acquiring or restoring those wetlands missing from the complex. Furthermore, managers must also recognize that nonwetland habitats such as grasslands and other upland nesting sites are critically important landscape components that dramatically affect species fitness. Protection of the most heterogeneous array of habitats possible, in terms of both size and types, is a clear guiding principle for waterfowl communities, because pairs, broods, and fledged juveniles select different habitats within a diverse array of wetland types to complete their life-cycle requirements (Patterson 1976; Nelson and Wishart 1988). Hence, the strategy of protecting wetland complexes is likely a sound approach for protecting diversity and enhancing fitness of other species of wetland birds. An interesting approach to restoration of wetland complexes was detailed by Taft and Haig (2003) who used historical accounts from early explorers and travelers and contemporary knowledge to develop a profile of the wetlands in the Willamette Valley of Oregon and their use by nonbreeding waterbirds.

In general, landscape ecology has figured prominently in management of wetland birds from the beginnings of the national wildlife refuge system to virtually every aspect of habitat protection from local to national scales and beyond. In the United States, for example, the humble beginnings of the national wildlife refuge system have expanded from the first tiny refuge set aside in Florida in 1903 (Pelican Island) to a landscape-oriented refuge system unrivaled in the world — as of 2004, there were 632 units in the system, including 545 refuges, spread across all 50 states and most territories. The system protects nearly 39 million ha of habitat, of which some 18.5 million ha are wetlands. Such protection and management of wetlands continue today, powered by an understanding of why a landscape-level approach to wetland conservation is a powerful means to protect wetland birds.

WETLAND PLANT SUCCESSION

Active habitat management is especially essential for wetlands, which are prone to rapid plant succession or colonization by exotic plants wherein wetland function for wildlife can be completely altered within only one or two growing seasons. Manipulation of wetland vegetation is an especially critical focus of waterfowl management, because plants are differentially desirable for food and cover (Baldassarre and Bolen 2006). Other wetland plants can form undesirable monotypes (e.g., cattails); whereas, still others [e.g., purple loosestrife (*Lythrum salicaria*)] are invasive exotics that can seriously diminish the value of a given wetland as wildlife habitat (Thompson et al. 1987;

Blossey et al. 2001). Hence, the basic concepts of wetland plant succession form the essential foundation for habitat management in wetlands. Successful management must, therefore, be grounded in an ecologically sound conceptual basis if managers are to understand, test, and hence predictably repeat management initiatives.

van der Valk (1981) advanced a Gleasonian view of plant succession in wetlands in that changes are powered by allogenic or external forces, especially the frequency and duration of flooding (i.e., hydroperiod). The effect of these allogenic factors especially depends on the seed bank, which is the amount of viable seed in the upper few centimeters of the substrate (van der Valk and Davis 1976). However, the propagules of each species in the seed bank germinate differentially in response to allogenic factors, particularly the extremes of wet and dry conditions. Hence, wetland managers must understand the ecology of seed banks, because each management approach (e.g., deep or shallow flooding) will differentially affect the emergence and subsequent growth and persistence of the various species in the seed bank.

In a paper especially important toward understanding plant succession in wetlands, van der Valk and Davis (1978) noted that variability in water levels differentially affected germination from the seed bank of prairie wetlands in a manner that led to a description of four basic phases in the cycle of marsh vegetation: (1) dry marsh, (2) regenerating marsh, (3) degenerating marsh, and (4) lake marsh. Dry marsh develops during drought and is a time when seeds requiring exposed mudflats (i.e., annuals) will germinate from the seed bank. Regenerating marsh occurs when drought ends. During the regenerating phase, mudflat species from the dry marsh stage are eliminated, as is germination of new emergents from the seed bank; free-floating and submergent plants begin to appear wherein the marsh becomes a mix of emergents and submergents. The marsh next enters a degenerating phase, which is characterized by a rapid decline in emergents leading to the final phase where the marsh is now largely open water (lake marsh) and dominated by submergent and floating plants. Overall, the vegetative community of each marsh phase was a function of water level, but species composition was a function of the seed bank.

van der Valk (1981) next used the allogenic approach to develop a model of vegetation succession in freshwater wetlands. Hence, following manipulation of the hydroperiod via control of water levels within a given wetland, managers could reasonably predict the outcome by knowing only three life history attributes of the species in the seed bank or colonizing the site: (1) life span, (2) propagule longevity, and (3) establishment requirements for seedlings. The requirements for seedling establishment are especially important but rather simple: (1) species that establish only when and where there is no water, and (2) species that establish when and where there is water. van der Valk (1981) also combined the three general life history attributes into a model demonstrating that allogenic influences act as an environmental "sieve" that determines when and where each species will occur in a given wetland. Indeed, using the three life history attributes, van der Valk (1981) identified only 12 "types" of wetland plants for his model. These models of vegetation dynamics in prairie wetlands were further refined during the Marsh Ecology Research Program conducted during the 1980s at the Delta Waterfowl and Wetlands Research Station in Manitoba, Canada, which generated significant new information that ultimately appeared in a comprehensive book, *Prairie Wetland Ecology, The Contribution of the Marsh Ecology Research Program* (Murkin et al. 2000).

The Marsh Ecology Research Program led to new and more refined models of wetland vegetation dynamics, again using the idea of environmental filters, especially changes in water levels. As with earlier models, hydroperiod was the major factor affecting wetland vegetation, but the effect of hydroperiod is modified by subtleties of soil moisture, temperature, and salinity. van der Valk (2000) thus refined earlier models by adding tolerance of a given species to water depth. The new model thus recognized four types of plants in relation to water depth: (1) annuals, (2) wet-meadow species, (3) emergents, and (4) submergents. Mudflat species only occur when sediments are exposed; whereas, wet-meadow species can tolerate short intervals of standing water but not long-term inundation. Emergents, in contrast, can tolerate permanent flooding and periods without standing water; whereas, submergents generally require permanent water.

Obviously, these changes in wetland plant communities affect marsh physiognomy, which in turn affects use of wetlands by wildlife. For example, shorebirds generally use shallow water (0–18 cm), but about 70% of all species prefer depths <10 cm and areas where vegetation cover is <25% (Helmers 1992, 1993; Collazo et al. 2002). Small species of shorebirds (i.e., *Calidris* spp.) require especially shallow water (0–4 cm); hence, accessible habitat for these species might only be a small proportion of a given wetland. For example, only 21–22 ha (13%) of 161-ha managed impoundment at Pea Island National Wildlife Refuge in North Carolina contained habitat in the 0–8 cm range deemed important for foraging dunlins (*Calidris alpina*) and semipalmated sandpipers (*Calidris pusilla*; Collazo et al. 2002). In contrast, ducks generally prefer deeper water (Fredrickson and Taylor 1982). Further still, drawdowns scheduled during the breeding season of wading birds can concentrate their prey sources, which increases the foraging opportunities for adult birds feeding their nestlings (Parsons 2002). In addition, various types and densities of emergent vegetation provide highly suitable nest sites for an array of wetland birds such as bitterns, rails, coots, gallinules, waterfowl, and others.

THE ROLE OF MODELS

In comparison to the overall field of ecology, "modelers" are relative newcomers. Modelers, however, are now quite advanced in creating models that describe patterns in nature, whether they focus on habitat, populations, or both. In other words, models have come of age. For example, models have driven the development of two important concepts familiar to almost all wildlife managers and conservation biologists: population viability analysis and metapopulation dynamics. Indeed, the importance of models in the decision-making process of conservation in general has been stated as follows: "In conservation, predicting the future behavior of a population or system under different management programs is of paramount interest, and no credible prediction is possible without a formal or informal model of the system" (Beissinger et al. 2006, 3). The habitat models of van der Valk (1981, 2000) were reviewed above and demonstrated the utility of models in predicting plant succession in wetlands. Hence, this section focuses on some of the relevant demographic models that relate specifically to waterbirds and their habitats. For birds in particular, the 2006 Ornithological Monograph *Modeling Approaches in Avian Conservation and the Role of Field Biologists* is especially relevant and notes "Use of models in avian conservation may not be a panacea, but neither is it a passing fancy. Thus, modelers and field biologists increasingly depend on one another to achieve effective conservation" (Beissinger et al. 2006, 40). This excellent publication forms the basis for much of the following discussion.

To begin, Beissinger et al. (2006) discuss the general form of ecological models, highlighting the approach and application of six types of models especially useful in avian conservation: (1) deterministic, single-population matrix models; (2) stochastic population viability analysis (PVA) for single populations; (3) metapopulation models; (4) spatially explicit models; (5) genetic models; and (6) species distribution models. Overall, their publication is perhaps the best available in reviewing models and their applicability in guiding conservation decisions on birds in general, and space does not allow such a review here. Hence, in this section, I provide examples where models have proven especially useful in guiding conservation actions associated with wetland birds.

A radiotelemetry study of mallards nesting in North Dakota, for example, generated an important model that identified a 15% nest success rate as necessary to maintain mallard populations in agricultural regions (Cowardin et al. 1985). Hoekman et al. (2002) later expanded this and other efforts by using a variety of both published and unpublished data on female mallards to generate stage-based matrix models that were subjected to sensitivity analyses to examine the relative importance of vital population parameters on population growth rates. Nest success explained most (43%) of the population growth, followed by adult survival during the breeding season (19%), and survival of ducklings (14%). Nonbreeding survival, in contrast, only accounted for 9% of the variation in population growth. Overall, predation on the breeding grounds was deemed the proximate factor most limiting mallard population growth. Such sensitivity analyses of population data are especially significant, because they can identify key life cycle stages most influential on population growth, which can then act as a focus for management (Beissinger et al. 2006).

Flint et al. (1998) used demographic models to assess the effects of survival and reproduction on the population dynamics of northern pintails (*Anas acuta*) breeding on the Yukon-Kuskokwim Delta in Alaska. Population parameters indicated that 13% of the females produced all of the young in the population, and that each female produced an average of 0.16 young females per nesting season. Adult female survival had the greatest effect on population growth rate (0.8825), followed by 0.1175 for both reproductive success and first-year survival. However, the resultant population projection model suggested that the population was declining rapidly, and that nest success and duckling survival each needed to increase about 40% for the population to stabilize.

Stochastic, single population models especially see use in predicting extinction likelihood via population viability analysis or PVA (Beissinger and Westphal 1998). Ryan et al. (1993) used this approach to model viability of piping plover (*Charadrius melodus*) populations on the Great Plains and noted a strong probability for extinction in the next 100 years, barring significant conservation action. Similarly, Reed et al. (1998) used PVA to assess the endangered Hawaiian stilt (*Himantopus mexicanus knudseni*). Their analysis identified nest failure and adult survival as important factors affecting population growth, noting that the population was likely to go extinct, again, without significant conservation efforts.

These and other models that focused on waterfowl have strong ramifications for conservation. For example, managers could affect the two important variables influencing Hawaiian stilt populations by controlling predators and stabilizing water levels (Reed et al. 1998). Flint et al. (1998) recommended that managers seeking to increase populations of northern pintails should focus their efforts on increasing adult survival in general, which could be achieved by focusing harvest on nonreproductive or reproductively unsuccessful segments of the population (Clark et al. 1988). The results from Hoekman et al. (2002) also support the idea that management should focus on increasing survival of females during the breeding season, because population growth is highly sensitive to adult survival, and, because mortality during the breeding season was >65% of total annual mortality. Such a finding was supported by Blums and Clark (2004), who studied the effect of female age and other factors on the lifetime reproductive success of three species of European ducks. Their study reported that the number of breeding attempts was most strongly correlated with lifetime reproductive success. Thus, adult survival is especially important to overall population growth, because surviving females have more opportunities to attempt breeding efforts and therefore encounter conditions suitable for duckling survival.

A second major use of models in waterbird conservation is to generate predictors of habitat quality that can affect reproductive parameters of interest. Such an approach can be an effective guide for management actions and efficiently target scarce conservation dollars. For example, models developed by Bancroft et al. (2002) showed that abundance of four species of wading birds in the Florida Everglades was related to water level and vegetation community, and there was a threshold of water depth above which wading bird abundance would decline. Such findings have major ramifications for restoration efforts in the Everglades, which is of national concern. Accordingly, Bancroft et al. (2002) used their results to recommend that restoration efforts result in more natural hydrologic cycles and slough habitat, which together would improve foraging habitat for wading birds. An earlier study by Frederick and Collopy (1989) used logistic regression models to assess water levels in relation to the nesting success of five species of wading birds in the Everglades, noting that ibises were especially affected by water-level fluctuations. Overall, general habitat models for the Everglades have long demonstrated that the pristine system was characterized by longer periods of inundation over larger areas than occurs today under a managed and altered ecosystem (Fennema et al. 1994). However, Curnutt et al. (2000) later used spatially explicit species index models to compare the response of long-legged waders and other wetland birds to different management scenarios, concluding that no one management scenario was best for all species. Nonetheless, their work demonstrates the utility and role of models in restoration of one of the world's most famous wetlands.

Royle et al. (2002) generated a two-state model containing wet and dry states of prairie pothole wetlands, the results of which predict the wet probability of a given basin and the amount of water in that basin. The model is especially useful in estimating the number of wet basins and the subsequent amount of water likely to occur over a given spatial region. Such predications are significant, because wetland conditions are correlated with important demographic parameters such as population size and age structure of fall waterfowl populations, which in turn affect harvest management strategies (Johnson et al. 1997). Among other uses, the model could guide management actions effectively, because it could characterize habitat structure at any given point on the landscape.

Reynolds et al. (2001) used models to evaluate the effect of the Conservation Reserve Program (CRP) on duck production in the Prairie Pothole Region of the United States. Their models were used to compare nest success and recruitment for five ducks species during peak years of CRP (1990–94) in comparison to a simulated scenario where cropland replaced CRP cover. Results were dramatic: Nest success was 46% higher and recruitment 30% higher with CRP versus cropland cover on the landscape, resulting in an estimated 12.4 million additional recruits from their study areas. Such findings are critical to agricultural policy, because they demonstrate the effectiveness of a major federal agriculture program, which has significant ramifications outside the Prairie Pothole Region: Ducks produced in the region migrate to $\geq$44 states (Reynolds et al. 2001).

Relative to shorebirds, Collazo et al. (2002) used models to identify water level targets that maximized accessible habitat for species wintering at two national wildlife refuges on the Atlantic Coast. They further noted how estimates of habitat accessibility, along with turnover rates of prey bases, are essential to establish and then implement management goals for shorebirds.

TAXONOMY AND PHYLOGENETIC SYSTEMATICS

Every wildlife biologist is familiar with the concepts of taxonomy, having spent countless hours as an undergraduate or graduate student memorizing genus and species names, as well as families, orders, and more in association with typical vertebrate ecology courses such as ornithology and mammalogy and a plant-oriented course such as systematic botany. Although these classification approaches produced scenarios of ancestry and evolutionary relationships among taxa, the resultant taxonomic schemes could not be subjected to critical analysis. In contrast, the science of modern phylogenetic systematics seeks to empirically capture the orderly relatedness among similar taxa, which has resulted from patterns of phylogenetic ancestry and descent (Eldredge and Cracraft 1980; Cracraft 1981). In essence, a phylogenetic systematist maps the path of evolution by looking at the intrinsic features of organisms, with the goal of developing a classification that represents the true ancestry (i.e., historical "traits") and relatedness of those organisms. The resultant classification reflects relationships among organisms that are familiar to all biologists: Species within a genus are more closely related than species assigned to another genus, family, or order.

Such an approach to taxonomy is significant, because true phylogenies can be used to analyze additional evolutionary patterns exhibited by ecomorphological characteristics such as body mass, clutch size, mating systems, propensity for hybridization, sexual dimorphism, diet, and nest-site selection, among others. In other words, if groups of species are related by a set of morphological or molecular characteristics, then relatedness should also be reflected ecologically. Hence, phylogeny is seeing new and exciting potential in conservation detailed in books such as *Phylogeny and Conservation* (Purvis et al. 2005).

The use of phylogenetic diversity is gaining currency as a metric for conservation of evolutionary history (Mooers et al. 2005), which is beyond our scope here. However, other aspects associated with phylogenetic concept of systematics have wide applicability at the management level. For example, it is well known that preferred water depths for foraging vary for shorebirds, ducks, and wading birds, and that such habitat differences are strongly correlated with morphologies (Pöysa 1983; Davis and Smith 1998; DuBowy 1988), which reflect phylogenies. All waterfowl biologists know that dabbling ducks in the genus *Anas* typically forage in shallow water by "tipping-up," as opposed to diving ducks in the genus *Aythya*, which typically forage in much deeper water and dive for food. The ramifications for management are obvious: Shallow water is required for foraging dabbling ducks versus deep, more open water for diving ducks. Species of shorebirds in the genus *Calidris* are commonly referred to as "peeps." The short-legs characteristic of the group mandate they feed on mudflats or very shallow water. More generally, shorebirds in the family Charadriidae (plovers) peck at food on the surface, whereas those in the Scolopacidae (sandpipers) probe beneath the surface. In general, shallowly flooded wetlands in the San Joaquin Valley increased use by shorebirds, dabbling ducks, and the black-necked stilt (*H. mexicanus*), but use and density of diving ducks was greater on deeper-water sites (Colwell and Taft 2000). Researchers conducting another study of ten waterbird taxa in the San Joaquin Valley (six shorebirds and four dabbling ducks) found that water depth explained 86% of the difference in habitat use among taxa, with four groups identified based on habitat use by water depth: small shorebirds (*Calidris* spp.), large shorebirds, teal, and large dabbling ducks (Isola et al. 2000). Wetland managers have recognized all the above relationships in their management approaches, perhaps not realizing that the underlying science was reflected by phylogenetic systematics.

Waterfowl and the issue of hybridization are well known to managers, but the underlying mechanisms often are not. Hybridization is generally rare for birds, but very high in waterfowl. Indeed, interspecific and intergeneric hybridization within waterfowl is among the highest observed among all orders of birds (Johnsgard 1960; Grant and Grant 1992). Furthermore, a significant proportion of the hybrids may be fertile; wherein, introgressive gene flow leads to backcrossing with the parent populations (Rhymer and Simberloff 1996).

Effects of such introgression are especially well known between mallards and American black ducks (*Anas rubripes*), where hybridization has been implicated in the decline of the latter species (Ankney et al. 1987), and the Mexican duck (*Anas diazi*; Hubbard 1977). Indeed, genetic distance between mallards and American black ducks has decreased from 0.146 before 1940 to 0.008 within birds collected in 1998, a breakdown in genetic differentiation that represents a breakdown in species integrity (Mank et al. 2004). Mallard × Mexican duck hybridization is so pervasive that the Mexican duck was eventually declared conspecific with the mallard (American Ornithologists' Union 1998). In New Zealand, mallards introduced for hunting have hybridized extensively with the native gray duck (*Anas superciliosa*) leading to the conclusion that speciation is undergoing reversal (Rhymer et al. 1994). Mallards are also established in Australia, where they are encountering gray ducks. In Hawaii, mallards introduced for hunting are hybridizing with the endemic Hawaiian duck (*Anas wyvilliana*; Engilis et al. 2002). These and similar issues emphasize the importance of a well-grounded understanding of genetic relationships, including ancestry, as a tool in waterfowl and other waterbird management. No waterfowl manager familiar with phylogenetics and its ramifications in management would ever have allowed mallards to become established in New Zealand, Australia, or Hawaii.

Phylogenetics also can be useful in guiding research, which is of course the prerequisite to the entire topic of ecologically based management. Consider, for example, that a database search of Biblioline's "Wildlife and Ecology Studies Worldwide" (1950–2005) yields 11,800 "hits" for the keyword "waterfowl," but only 2,700 for shorebirds, and 500 for wading birds. Clearly, there is need for research emphasis in the latter two groups.

Within these three major groups, there are other revealing patterns from searching this same database. Among waterfowl, for example, there were 8000 "hits" for *Anas* (dabbling ducks) versus 2500 for *Aythya* (diving ducks). Within the wading-bird group, the least bittern (*Lxobrychus exilis*) and American bittern (*Botaurus lentiginosus*) received 60 to 70 hits each, compared with 600 for the great blue heron (*Ardea herodias*), and only 150 for the reddish egret (*Egretta rufescens*). Such a simple analysis, if expanded to compare and contrast all species involved, could provide a meaningful guide for research gaps and strengthen interpretations of management predications. For example, the least bittern and American bittern are not in the same genus, so it is reasonable to assume that management for one might not similarly benefit the other, even though both are "bitterns."

To sum this section, an appreciation and understanding of phylogenetics is mandatory for managers. For example, managers have long understood that management for one group of species will not benefit another group (e.g., dabbling versus diving ducks), but they have perhaps not appreciated that phylogenetic systematics is the reason why it is so.

AN UPSHOT

The beginning of the twenty-first century is witness to the greatest human concern for wetlands and associated wildlife. At the global level, the conservation community has responded to wetland loss in general and wetland birds in particular. For example, the Ramsar Treaty (1971), despite an initial focus on waterfowl as evidenced in the full name of the treaty — *Convention on Wetlands of International Importance, Especially as Waterfowl Habitat* — addresses habitat protection for all birds ecologically dependent on wetlands. Indeed, formal criteria for listing wetlands within signatory nations were expanded in the years after the 1971 conference and now include several criteria specifically targeting the array of wetland-dependent birds (e.g., support 20,000 or more waterbirds, or regularly support 1% of a population of a species or subspecies of waterbird). As of 2006, some 150 contracting parties to the Convention have listed 1592 areas exceeding 134 million ha in six administrative regions: Africa, Asia, the Neotropics, Europe, North America, and Oceana. In the Western Hemisphere alone, Mexico listed 34 new Ramsar sites — the largest number ever declared at one time — on World Wetlands Day 2004.

The North American Waterfowl Management Plan is a continental undertaking by Canada, the United States, and Mexico to protect and enhance wetland habitat for the benefit of waterfowl and other wetland-dependent birds. Originally signed by the United States and Canada in 1986, by 2004 the plan reported expenditure of more than $3 billion to protect and manage some 5.3 million ha of wetland habitat spread across the three signatory nations.

The Western Hemisphere Shorebird Reserve Network emerged in 1985 and expanded the landscape idea to both continents in the Western Hemisphere, recognizing that many species of shorebirds were using such an extensive landscape to complete their life cycle. The program now involves 63 sites in eight countries.

In the United States, the Wetlands Reserve Program as administered by the Natural Resources Conservation Service had enrolled 7831 projects affecting some 0.6 million ha by the close of 2004. The Partners for Wildlife Program administered by the U.S. Fish and Wildlife Service had entered into 35,000 agreements with private landowners from 1987 to 2004, which has led to the restoration of 293,000 ha of wetlands and 637,000 ha of prairie, native grasslands, and other uplands. The CRP, also administered by the Natural Resources Conservation Service, had enrolled 14 million ha by 2003, which greatly benefited species of waterfowl requiring upland nesting habitat (Reynolds et al. 2001).

These achievements in wetland protection are the product of decades of hard and dedicated work by the conservation community. The result is that the legacy provided by wetlands will be passed to the next generation as it was passed to this generation. Active, ecologically based management must go forward with that legacy to ensure the future of the birds and other wildlife that depend on these habitats.

ACKNOWLEDGMENTS

This paper benefited from Mike Erwin, Jaime Collazo, Sue Haig, and Rick Kaminski, who directed me to key research and associated publications. Jim Goetz facilitated library research.

REFERENCES

American Ornithologists' Union (AOU). 1998. *Check List of North American Birds*, 7th edn. Washington, DC: American Ornithologists' Union.

Ankney, C. D., D. G. Dennis, and R. C. Bailey. 1987. Increasing mallards, decreasing black ducks: Coincidence or cause and effect? *J. Wildl. Manage.* 51:523.

Baldassarre, G. A., and E. G. Bolen. 2006. *Waterfowl Ecology and Management.* Malabar, FL: Krieger Publishing.

Bancroft, G. T., D. E. Gwalik, and K. Rutchey. 2002. Distribution of wading birds relative to vegetation and water depths in the northern Everglades of Florida. *Waterbirds* 25:265.

Beissinger, S. R., and M. L. Westphal. 1998. On the use of demographic models of population viability in endangered species management. *J. Wildl. Manage.* 62:821.

Beissinger, S. R., J. R. Walters, D. G. Catanzaro, K. G. Smith, J. B. Dunning, Jr., S. M. Haig, B. R. Noon, and B. M. Stith. 2006. Modeling approaches in avian conservation and the role of field biologists. *Ornith. Monogr.* 59.

Blossey, B., L. C. Skinner, and J. Taylor. 2001. Impact and management of purple loosestrife (*Lythrum salicaria*) in North America. *Biodiver. Conserv.* 10:1787.

Blums, P., and R. G. Clark. 2004. Correlates of lifetime reproductive success in three species of European ducks. *Oecologia* 140:61.

Brown, M., and J. J. Dinsmore. 1986. Implications of marsh size and isolation for marsh bird management. *J. Wildl. Manage.* 50:392.

Clark, R. G., et al. 1988. The relationship between nest chronology and vulnerability to hunting of dabbling ducks. *Wildfowl* 39:137.

Clements, J. F. 2000. *Birds of the World: A Checklist*, 5th edn. Temecula, CA: Ibis Publishing.

Collazo, J. A., D. A. O'Harra, and C. A. Kelly. 2002. Accessible habitat for shorebirds: Factors influencing its availability and conservation implications. *Waterbirds* 25 (Spec. Publ. 2):13.

Colwell, M. A., and O. W. Taft. 2000. Waterbird communities in managed wetlands of varying water depth. *Waterbirds* 23:45.

Cowardin, L. M., D. S. Gilmer, and C.W. Shaiffer. 1985. Mallard recruitment in the agricultural environment of North Dakota. *Wildl. Monogr.* 92.

Cracraft, J. 1981. Toward a phylogenetic classification of the Recent birds of the world (Class Aves). *Auk* 98:681.

Curnutt, J. L., et al. 2000. Landscape-based spatially explicit species index models for Everglades restoration. *Ecol. Appl.* 10:1849.

Dahl, T. E. 1990. *Wetlands Losses in the United States, 1780s to 1980s.* Washington, DC: U.S. Fish and Wildlife Service.

Dahl, T. E. 2000. *Status and Trends of Wetlands in the Conterminous United States, 1986 to 1997.* Washington, DC: U.S. Fish and Wildlife Service.

Davis, C. A., and L. M. Smith. 1998. Ecology and management of migrant shorebirds in the Playa Lakes Region of Texas. *Wildl. Monogr.* 140.

DuBowy, P. J. 1988. Waterfowl communities and seasonal environments: Temporal variation in interspecific competition. *Ecology* 69:1439.

Dugan, P. 1993. *Wetlands in Danger.* London: Reed International Books.

Eldredge, N., and J. Cracraft. 1980. *Phylogenetic Patterns and the Evolutionary Process.* New York: Columbia University Press.

Engilis, A., Jr., K. J. Uyehara, and J. G. Griffin. 2002. Hawaiian duck (*Anas wyvilliana*). In *The Birds of North America*. Washington, DC: The American Ornithologists' Union and Philadelphia: The Academy of Natural Sciences, p. 694.

Fennema, R. J., et al. 1994. A computer model to simulate natural Everglades hydrology. In *Everglades: The Ecosystem and Its Restoration*, S. M. Davis, and J. C. Ogden (eds). Delray Beach, FL: St Lucie Press.

Flint, P. L, J. B. Grand, and R. F. Rockwell. 1998. A model of northern pintail productivity and population growth rate. *J. Wildl. Manage.* 62:1110.

Frederick, P. C., and M. W. Collopy. 1989. Nesting success of five ciconiform species in relation to water in the Florida Everglades. *Auk* 106:625.

Fredrickson, L. H., and T. S. Taylor. 1982. *Management of Seasonally Flooded Impoundments for Wildlife.* U.S. Fish Wildl. Serv. Resour. Publ., p. 148.

Gibbs, J. P. 1993. Importance of small wetlands for the persistence of local populations of wetland-associated animals. *Wetlands* 13:25.

Gibbs, J. P., et al. 1991. *Use of Wetland Habitats by Selected Nongame Water Birds in Maine.* U.S. Fish and Wildl. Serv., Fish and Wildl. Res., p. 9.

Grant, P. R., and B. R. Grant. 1992. Hybridization of bird species. *Science* 256:193.

Grover, A. M., and G. A. Baldassarre. 1995. Bird species richness within beaver ponds in south-central New York. *Wetlands* 15:108.

Haig, S. M., et al. 2002. Space use, migratory connectivity, and population segregation among willets breeding in the western Great Basin. *Condor* 104:620.

Helmers, D. L. 1992. *Shorebird Management Manual.* Manomet, MA: Western Hemisphere Shorebird Reserve Network.

Helmers, D. L. 1993. Enhancing the management of wetlands for migrant shorebirds. *Trans. N. Am. Wildl. Nat. Resour. Conf.* 58:335.

Hoekman, S. T., et al. 2002. Sensitivity analyses of the life cycle of midcontinent mallards. *J. Wildl. Manage.* 66:883.

Hubbard, J. P. 1977. *The biological and taxonomic status of the Mexican duck.* New Mexico Dept. Game Fish Bull. 16.

International Union for the Conservation of Nature and Natural Resources (IUCN). 2006. *IUCN Red List of Threatened Species.* Gland, Switzerland.

Isola, C. R., et al. 2000. Interspecific differences in habitat use of shorebirds and waterfowl foraging in managed wetlands of California's San Joaquin Valley. *Waterbirds* 23:196.

Johnsgard, P. A. 1960. Hybridization in the Anatidae and its taxonomic implications. *Condor* 62:25.

Johnson, F. A., et al. 1997. Uncertainty and the management of mallard harvests. *J. Wildl. Manage.* 61:202.

Krapu, G. L. 1974. Feeding ecology of pintail hens during reproduction. *Auk* 91:278.

Krapu, G. L., A. T. Klett, and D. G. Jorde. 1983. The effect of variable spring water conditions on mallard reproduction. *Auk* 100:689.

Leitch, W. G., and R. M. Kaminski. 1985. Long-term wetland–waterfowl trends in Saskatchewan grassland. *J. Wildl. Manage.* 49:212.

MacArthur, R. H., and E. O. Wilson. 1967. *The Theory of Island Biogeography.* Princeton, NJ: Princeton University Press.

Mank, J. E., J. E. Carlson, and M. C. Brittingham. 2004. A century of hybridization: Decreasing genetic distance between American black ducks and mallards. *Conser. Genet.* 5:395.

Mitsch, W. J., and J. G. Gosselink. 2000. *Wetlands*, 3rd edn. New York: John Wiley & Sons.

Mooers, A. Ø., S. B. Heard, and E. Chrostowski. 2005. Evolutionary heritage as a metric for conservation. In *Phylogeny and Conservation*, A. Purvis, T. L. Brooks, and J. L. Gittleman (eds). Oxford: Oxford University Press.

Murkin, H. R., E. J. Murkin, and J. P. Ball. 1997. Avian habitat selection and prairie wetland dynamics: A 10-year experiment. *Ecol. Appl.* 7:1144.

Murkin, H. R., A. G. van der Valk, and W. R. Clark (eds). 2000. *Prairie Wetland Ecology: The Contribution of the Marsh Ecology Research Program.* Ames, IA: Iowa State University Press.

Naugle, D. E., et al. 1999. Scale-dependent habitat use in three species of prairie wetland birds. *Landscape Ecol.* 14:267.

Nelson, J. W., and R. A. Wishart. 1988. Management of wetland complexes for waterfowl production: Planning for the prairie habitat joint venture. *Trans. N. Am. Wildl. Nat. Resour. Conf.* 53:444.

Parsons, K. C. 2002. Integrated management of waterbird habitats at impounded wetlands in Delaware Bay, U.S.A. *Waterbirds* 25 (Spec. Publ. 2):25.

Patterson, J. H. 1976. The role of environmental heterogeneity in the regulation of duck populations. *J. Wildl. Manage.* 40:22.

Payne, N. F. 1992. *Techniques for Wildlife Habitat Management of Wetlands.* New York: McGraw-Hill.

Pöysa, H. 1983. Morphology-mediated niche organization in a guild of dabbling ducks. *Ornis Scand.* 14:317.

Purvis, A., J. L. Gittleman, and T. Brooks (eds). 2005. *Phylogeny and Conservation. Conserv. Biol.* 8.

Reed, J. M., C. S. Elphick, and L. W. Oring. 1998. Life history and viability analysis of the endangered Hawaiian stilt. *Biol. Conserv.* 84:35.

Reid, F. A. 1993. Managing wetlands for waterbirds. *Trans. N. Am. Wildl. Nat. Resour. Conf.* 58:345.

Reynolds, R. E., et al. 2001. Impact of the Conservation Reserve Program on duck recruitment in the U.S. Prairie Pothole Region. *J. Wildl. Manage.* 65:765.

Rhymer, J. M., and D. Simberloff. 1996. Extinction by hybridization and introgression. *Ann. Rev. Ecol. Syst.* 27:83.

Rhymer, J. M., M. J. Williams, and M. J. Braun. 1994. Mitochondrial analyses of gene flow between New Zealand mallards (*Anas platyrhynchos*) and grey ducks (*A. superciliosa*). *Auk* 111:970.

Royle, J. A., M. D. Koneff, and R. E. Reynolds. 2002. Spatial modeling of wetland condition in the U.S. Prairie Pothole Region. *Biometrics* 58:270.

Ryan, M. R., B. G. Root, and P. M. Mayer. 1993. Status of piping plovers in the Great Plains of North America: A demographic simulation model. *Conserv. Biol.* 7:581.

Sargeant, A. B., et al. 1993. Distribution and abundance of predators that affect duck production — Prairie Pothole Region. U.S. Fish Wildl. Serv. Resour. Publ. 194.

Shaw, S. P., and C. G. Fredine. 1956. Wetlands of the United States: Their extent and their value to waterfowl and other wildlife. U.S. Fish Wildl. Serv. Circ. 39.

Sovada, M. A., A. B. Sargeant, and J. W. Grier. 1995. Differential effects of coyotes and red foxes on duck nest success. *J. Wildl. Manage.* 59:1.

Sovada, M. A., M. C. Zicus, R. J. Greenwood, D. P. Rave, W. E. Newton, R. O. Woodward, and J. A. Beiser. 2000. Relationships of habitat patch size to predator community and survival of duck nests. *J. Wildl. Manage.* 64:820.

Stephens, S. E., et al. 2005. Duck nest survival in the Missouri Coteau of North Dakota: Landscape effects at multiple spatial scales. *Ecol. Appl.* 15:2137.

Swanson, G. A., G. L. Krapu, and J. R. Serie. 1979. Foods of laying female dabbling ducks on the breeding grounds. In *Waterfowl and Wetlands — An Integrated Review*, T. A. Bookhout (ed.). La Crosse, WI: La Crosse Printing.

Taft, O. W., and S. M. Haig. 2003. Historical wetlands in Oregon's Willamette Valley: Implications for restoration of winter waterbird habitat. *Wetlands* 23:51.

Talent, L. G., G. L. Krapu, and R. L. Jarvis. 1982. Habitat use by mallard broods in south central North Dakota. *J. Wildl. Manage.* 46:629.

Thompson, D. Q., R. L. Stuckey, and E. B. Thompson. 1987. Spread, impact, and control of purple loosestrife (*Lythrum salicaria*) in North American wetlands. U.S. Fish and Wildl. Serv., Fish and Wildl. Res. 2.

van der Valk, A. G. 1981. Succession in wetlands: A Gleasonian approach. *Ecology* 62:688.

van der Valk, A. G. 2000. Vegetation dynamics and models. In *Prairie Wetland Ecology: The Contribution of the Marsh Ecology Research Program*, H. R. Murkin, A. G. van der Valk, and W. R. Clark (eds). Ames, IA: Iowa State University Press.

van der Valk, A. G., and C. B. Davis. 1976. The seed banks of prairie glacial marshes. *Can. J. Bot.* 54:1832.

van der Valk, A. G., and C. B. Davis. 1978. The role of seed banks in the vegetation dynamics of prairie glacial marshes. *Ecology* 59:322.

Webster, M. S., et al. 2002. Links between worlds: Unraveling migratory connectivity. *Trends Ecol. Evol.* 17:76.

Weller, M. W. 1981. *Freshwater Marshes, Ecology and Wildlife Management*. Minneapolis, MN: University Minnesota Press.

Whigham, D., D. Dykyjová, and S. Hejný. 1993. *Wetlands of the World: Inventory, Ecology and Management*, vol. I. Dordrecht, The Netherlands: Kluwer Academic.

6 Linking Waterfowl Ecology and Management: A Texas Coast Perspective

Bart M. Ballard

CONTENTS

The Texas coast ranks as one of the highest priority areas for bird conservation in North America because of its great abundance and diversity of bird life. This region provides breeding, wintering, or migratory stopover habitat for about 400 species of birds (Rappole and Blacklock 1994; DeGraaf and Rappole 1995). Further, potentially more than 100 million birds migrate through this region each fall and spring. Some species rely heavily on the Texas coast for part or all of the annual cycle. For example, about 75% of all redhead ducks (*Aythya americana*) spend winter in the Laguna Madres of Texas and Tamaulipas (Weller 1964; U.S. Fish and Wildlife Service, unpublished data, 2004); however, all these birds most likely use Texas coastal areas during migration and portions of the winter. Similarly, most reddish egrets (*Egretta rufescens*) breed in Texas estuaries, whereas a large proportion of western Gulf coast mottled ducks (*Anas fulvigula*) rely on coastal habitats in Texas throughout the annual cycle. Many species of neotropical migrant birds breed in temperate regions of North America and migrate to and from wintering grounds in Central and South America. A narrowing of the North American continent along with an east–west restriction of habitats suitable to many species causes large-scale convergence of migratory corridors along the Texas coast (Lincoln et al. 1998).

The most direct route from breeding to wintering areas in Central and South America for species that breed in eastern portions of North America is across the Gulf of Mexico. However, many species have limited flight range capabilities and are unable to fly the long segments of nonstop flight over open water or other segments with limited access to stopover habitats. These species typically follow a circum-Gulf route along the coastline of Texas to more southerly breeding areas (Rappole et al. 1979) and rely on stopover habitats en route to rest and refuel during migration. Species that are strong fliers and able to store adequate energy reserves to traverse the Gulf of Mexico benefit from a more direct route from breeding areas in eastern North America to wintering areas in South and Central America. However, Texas coastal habitats often become important to these species when adverse weather in the Gulf of Mexico diverts them to the mainland during migration. During these events, known as "fall outs," immense abundance and diversity of birds can appear along the Texas coast to rest and refuel before continuing on their journey.

The importance of the Texas coast to migrating and wintering waterfowl is well established (Bellrose 1980; Stutzenbaker and Weller 1989). This region supports an estimated 2–4 million waterfowl each winter (U.S. Fish and Wildlife Service 1999) comprising over 25 species. Four species of geese and 16 species of ducks are common winter residents along the Texas coast, while five other species of waterfowl are observed regularly but are uncommon (Rappole and Blacklock 1994). This great abundance of waterfowl relies on the diverse assemblage of habitats throughout the Texas Coastal Plain for a large portion of the annual cycle. Most species arrive by mid-to-late October and remain through mid-March (Stutzenbaker and Weller 1989).

THE TEXAS COAST

COASTAL HABITATS

The Texas Gulf Coast extends almost 600 km from the Sabine River on the Texas–Louisiana border to the Rio Grande River along the Texas–Mexico border (Figure 6.1). Several bays and estuaries of varying size provide a convoluted nature to the 2300 km of shoreline (Brown et al. 1977). The extensive geographic range of the Texas coast provides variation in climate with generally warmer and drier conditions progressing southwest along the coast. A diverse array of habitats occurs as a result of this variation in climate.

The Coastal Plain of Texas contains fertile alluvial soils that are appealing for agricultural uses, and the port cities provide natural centers for industry. As a result, 6% of the state's land area encompassed by the coastal plain contains over 33% of the state's human population (Brown et al. 1977). Thus, within much of the coastal plain, there is direct conflict between land use interests and coastal habitat. Several state wildlife management areas and national wildlife refuges protect small portions of the upper and central portions of the coast, while large cattle ranches and Laguna Atascosa National Wildlife Refuge provide a buffer to development and access to much of the lower coast (Fulbright and Bryant 2002).

The upper Texas coast includes more extensive coastal marsh relative to central and lower coastal areas (Tacha et al. 1993; Muehl 1994). Ancient beach ridges, or cheniers, form east–west levees that create linear wetlands paralleling the coast and a salinity gradient within the coastal marsh, ranging from saline near the coast to fresh further inland. The coastal prairie that lies inland from the chenier marsh is dominated by agricultural land uses. Farmed rice fields, vegetated freshwater wetlands, and vegetated estuarine wetlands are most abundant and comprise about 82% of the 226,887 ha of wetland area (Tacha et al. 1993). Greatest abundances of waterfowl are found in fresh and saline marsh habitats proximal to the coast (Tacha et al. 1993).

Waterfowl habitats along the central Texas coast are characterized by a thin fringe of coastal marsh and a greater extent of wet prairie and depressional wetlands inland relative to the upper coast as the coastal plain extends inland the furthest here (Hobaugh et al. 1989; Stutzenbaker and Weller 1989). The central coast is the largest section of the Texas coast and contains 62% more wetland area than the upper coast (Tacha et al. 1993; Muehl 1994). Much of the wet prairie zone had been converted to rice production by the mid-1900s, resulting in large-scale landscape changes in both habitat and waterfowl distribution (Hobaugh et al. 1989). Vegetated freshwater wetlands are most abundant, comprising over half of the estimated 594,776 ha of wetland habitat in this region (Muehl 1994).

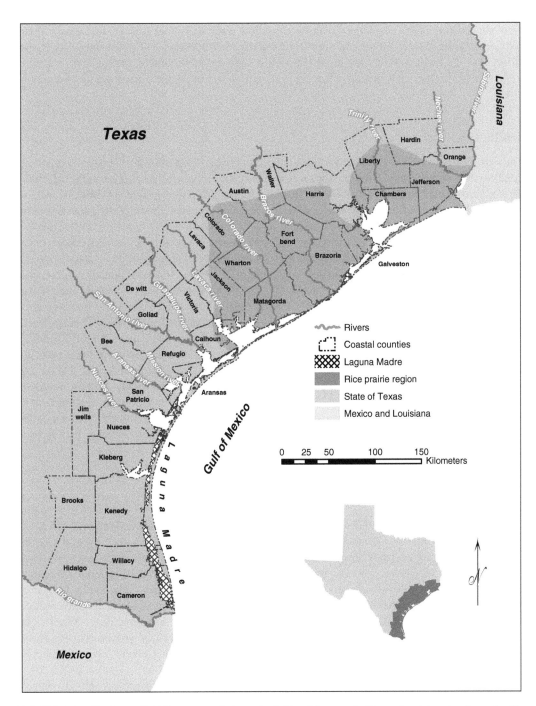

FIGURE 6.1 The Texas Gulf Coast extends from the Sabine River on the Texas–Louisiana border to the Rio Grande River along the Texas–Mexico border.

Along the lower coast, freshwater inflow from mainland drainages is limited, and evaporation typically exceeds precipitation often resulting in hypersaline conditions in the Laguna Madre (McMahan 1968; Tunnel 2002). Coastal marsh in this region is less extensive than other areas along the Texas coast because of the narrower shoreline gradient (Stutzenbaker and Weller 1989). The Laguna Madre has vast meadows of sea grasses with shoal grass (*Halodule wrightii*) dominating in most areas (Onuf 1996). The large expanses of shallow water and shoal grass in the Laguna Madre provide foraging habitat for several species of waterfowl (McMahan 1970). Freshwater wetlands adjacent to the Laguna Madre are important in this semiarid environment, particularly as sources of dietary freshwater for waterfowl foraging in the saline waters of the Laguna Madre (Adair et al. 1996).

THE RICE PRAIRIES

The rice prairies in Texas extend across 16 counties comprising about 9000 km^2 and are an important component to the waterfowl habitat along the upper and central coasts of Texas (Hobaugh et al. 1989). Before the advent of rice agriculture on the Texas landscape, this region provided freshwater wetland habitat for large numbers of ducks; however, most geese were found in the brackish marshes near the coast (Bateman et al. 1988). A large proportion of geese that historically relied on coastal marshes during winter have altered their distribution and moved into rice-producing areas (Bellrose 1980). Many other species of waterfowl have shifted their winter distribution along the Texas coast to include the rice prairie region to a greater degree (Hobaugh et al. 1989). Northern pintails (*Anas acuta*) and other waterfowl use rice habitats extensively throughout winter where rice seeds comprise an important part of the diet (Miller 1987). Rice fields provide abundant, high-energy foods in areas where native wetlands have declined because of changes in land use. Rice fields have the potential to produce more energy per unit area than native wetlands (Fredrickson and Taylor 1982; Miller 1987) and may have partially mitigated the loss of natural wetland habitats along the Texas coast due to their ability to support large numbers of waterfowl in relatively concentrated areas. Fallow rice fields also provide a diverse assemblage of seed-bearing plants that are widely used by waterfowl during winter (Miller 1987). Rice seeds are mostly depleted by mid-winter from consumption by waterfowl, deterioration, or sprouting (Manley 1999). During spring, ducks forage on invertebrates in rice fields to acquire protein that is important for egg production and to increase lean mass in preparation for migration (Miller 1987).

DECLINES IN WETLAND HABITATS

Moulton et al. (1997) assessed changes in wetland area that occurred between the mid-1950s and early 1990s throughout the Coastal Plain of Texas and found that wetland area declined by 85,229 ha. Freshwater emergent wetlands declined more than any wetland type in their study. These wetlands were reduced by 95,144 ha; however, many of these wetlands were converted to farmed wetlands. Fifty-two percent of all wetlands in the coastal plain (687,981 ha) were classified as farmed by Moulton et al. (1997). Alteration to natural hydrology caused by cultivating basins often reduces vegetation diversity and results in lower wildlife value (Mitsch and Gosselink 2000). Much of the loss in wetland area in the upper coast has been at the cost of wetland types important to waterfowl, and these wetlands are often replaced with poorer-quality, open-water wetland types (Tacha et al. 1993).

Compounding loss of native waterfowl habitat along the Texas coast, rice acreage has also largely declined in the state (Alston et al. 2000). Rice acreage in Texas declined 56% (Hobaugh et al. 1989) between 1980 and 1987, and further declines have occurred since then (Alston et al. 2000). Because of reduced price supports and urban encroachment, much acreage planted in rice has been developed or converted to row crops or rangeland (Alston et al. 2000).

Natural wetlands provide important habitat for over 20% of North America's bird species (Weller 1999). Protection provided to wetland habitats is more limited now than in recent years, and conflicts between wildlife habitat and human land use needs will increase over time. Efforts to mitigate wetland habitat loss through creation or restoration of wetlands have had varying success. Bird species response to created wetlands has been limited in some cases compared with the more diverse and productive natural wetland habitats (Snell-Rood and Cristol 2003). Shallow wetlands with shorter hydroperiods are typically more prone to destruction or degradation compared to deeper wetlands with more permanent inundation (Fredrickson and Laubhan 1994). These more ephemeral wetland types typically provide higher-quality habitat for waterfowl and other water birds than deeper wetland types (Ball et al. 1989). The large-scale reduction in wetland habitat along the Texas coast over the past several decades has likely reduced carrying capacity of water birds in this region and potentially reduced the size of some bird populations (Galarza and Telleria 2003). This is especially of concern in regions such as the Texas coast where habitats have the potential to maintain the majority of continental migratory populations during much of the nonbreeding period (Galarza and Telleria 2003).

THE NORTHERN PINTAIL

The northern pintail is a dabbling duck that winters throughout much of North America. Areas of greatest concentration during winter include the Central Valley of California, and the Gulf Coast of Texas and Louisiana (Bellrose 1980). Rice production is a land use common to each of these areas and appears to concentrate pintails during winter. Within the Central Flyway, up to 78% of northern pintails tallied on winter surveys are in Texas, and largely within the Coastal Plain (Bellrose 1980; Texas Parks and Wildlife Department, unpublished data). The largest proportion of northern pintails along the Texas coast occurs in rice fields (Anderson 1994). Additionally, northern pintails and other ducks wintering along the Texas coast extensively use estuarine sea grass meadows and inland, vegetated, freshwater wetlands (Briggs 1982; McAdams 1987; Anderson 1994). I will use the northern pintail to illustrate how an understanding of a species' ecology can provide insight to further our management effectiveness.

TRENDS IN NORTHERN PINTAIL ABUNDANCE

The northern pintail has declined in North America over the past three decades. Spring breeding population estimates during 2000–2005 ranged from 20 to 56% below long-term averages (U.S. Fish and Wildlife Service 2005) and have remained well below goals established by the North American Waterfowl Management Plan (NAWMP) for the past 30 years (NAWMP 2004). Historically, northern pintails have been the second most abundant duck in North America and an important component of the waterfowl harvest in several flyways (Bellrose 1980).

Northern pintails breed across a relatively broad range in North America; however, the largest breeding concentrations occur in the Prairie Pothole region of the Northern Great Plains and in Alaska and western Canada (Bellrose 1980; Miller and Duncan 1999). Spring wetland availability within the Prairie Pothole region varies greatly depending on annual variation in precipitation, primarily snow accumulation. Breeding distributions of many species of waterfowl are positively correlated with spring wetland availability in this region (Johnson and Grier 1988). Further, because breeding conditions improve with increased moisture, annual recruitment of prairie breeding ducks is also directly influenced by spring wetland availability in this region. The greater availability of nesting and brood-rearing habitat in wet years results in increased nesting success and a greater proportion of hens nesting. In contrast, recruitment suffers in dry years when wetland availability is low, and a greater proportion of hens and nests are depredated. In addition, during dry years, pintails tend to over-fly the Prairie Pothole region and travel further north to nest (Smith 1970). As a consequence, they arrive on breeding areas in poorer condition and experience relatively low reproductive success, further impacting annual recruitment (Smith 1970). Hens presumably utilize endogenous stores of energy to make these movements further north that would otherwise be allocated to the production of the clutch when settling in the Prairie Pothole region, thus clutch sizes tend to be smaller (Smith 1970; Krapu 1974). Fall populations of northern pintail have historically tracked wetland conditions in the Prairie Pothole region because of this relationship (Johnson and Grier 1988; Austin and Miller 1995). In recent years, however, northern pintails have not responded with increased populations when the Prairie Pothole region has had favorable wetland conditions (Miller and Duncan 1999). Although habitat conditions were excellent throughout most of the Prairie Pothole region during the early and mid-1990s, northern pintail populations did not recover from low numbers in the late 1980s and early 1990s. Almost all other species of prairie-nesting dabbling ducks responded to the excellent habitat conditions in the 1990s with populations above goals set by the NAWMP and several reaching record high numbers (U.S. Fish and Wildlife Service 2005). Thus, there appeared to be new factors influencing northern pintail populations that reduced their ability to respond to wet conditions in their primary breeding area.

Habitat conditions in Alaska do not show the great annual variability that is characteristic of the Prairie Pothole region, and, as a result, northern pintail populations there remain relatively stable (Miller and Duncan 1999). Thus, the decline in continental pintail numbers appears to be driven primarily by northern pintails nesting in the Prairie Pothole region.

FACTORS POTENTIALLY LIMITING POPULATION

Identification of factors limiting the northern pintail's ability to recover to historic levels has been a primary focus of many waterfowl researchers (i.e., implementation of a Northern Pintail Recovery Group). Most biologists working with pintails agree that factors are likely related to low adult survival or inadequate recruitment (Miller and Duncan 1999). Most research investigating the long-term decline has been on breeding areas because of its direct relation to recruitment and because female mortality is typically high during nesting in dabbling ducks (Sargent and Raveling 1992). Recruitment is decreasing in traditional nesting areas in the Canadian Prairies, and it appears that a greater proportion of northern pintails are nesting in areas further north as a result of large-scale habitat loss (Herbert and Wassenaar 2005; Runge and Boomer 2005).

Research on breeding areas provides compelling evidence regarding potential factors limiting growth of northern pintail populations. Because most of the important vital rates relate directly to recruitment, it is intuitive that breeding ground factors can be important in population regulation. What is less intuitive is how factors outside the breeding season influence these vital rates. Because waterfowl spend much of their annual cycle in nonbreeding locations, investigation of nonbreeding ecology is also necessary to gain a complete understanding of regulating factors throughout their annual cycle.

Northern pintails nest early relative to other dabbling ducks (Higgins 1977; Grand et al. 1997) and, as a consequence, are believed to rely more on stored reserves obtained from wintering areas and spring migration areas for initial clutch formation (Krapu 1974; Mann and Sedinger 1993; Esler and Grand 1994); thus, there is great potential for cross-seasonal effects on both survival and recruitment. Cross-seasonal effects are effects manifested in one portion of the annual cycle that are realized in subsequent portions. The suggestion that cross-seasonal effects influence populations of dabbling ducks has largely been based on a pure correlative nature of relationships. However, the ability to track the origin of energy used on breeding areas or to understand what proportion of energy used for reproductive activities originates on nonbreeding areas has only recently been available. Recent research using stable isotope techniques provides strong evidence that energy accumulated during the nonbreeding period has an influence on reproductive success in some species of waterfowl (Hobson et al. 2005).

Raveling and Heitmeyer (1989) reported evidence for cross-seasonal effects in northern pintails when they found a direct relationship between winter habitat conditions and recruitment. It is well substantiated that winter habitat conditions play a large role in the ability of wintering pintails and other dabbling ducks to build and maintain endogenous reserves (Delnicki and Reinecke 1986; Miller 1986; Smith and Sheeley 1993; Ballard et al. 2006). Winter body condition of waterfowl can influence survival (Haramis et al. 1986; Conroy et al. 1989), reproductive potential (Heitmeyer and Fredrickson 1981; Krapu 1981; Raveling and Heitmeyer 1989), and timing of annual cycle events such as prebasic molt (Richardson and Kaminski 1992) and pair bond formation (Hepp 1986). Female mallards carrying more reserves during winter initiate prebasic molt earlier than leaner females, which allows them to form pair bonds earlier (Heitmeyer 1985). Heitmeyer (1985) suggested that individuals that were able to complete these events earlier experienced greater reproductive success.

Large differences exist in endogenous reserve dynamics between pintails wintering in the mid-continent region and those wintering in California. Birds along the lower Texas coast, in the Playa Lake Region of Texas, and in Yucatan, Mexico, depart wintering areas with greatly reduced carcass fat and protein levels compared with birds in California (Miller 1986; Thompson and Baldassarre 1990; Smith and Sheeley 1993; Ballard et al. 2006). If in fact nutrient stores at this time have any influence on reproductive potential, pintails departing these areas may be at a disadvantage well before arriving on breeding areas.

Pintails along the lower Texas coast and possibly elsewhere (e.g., Playa Lakes region and Mexico) appear to be under an endogenous control of body mass during winter (Ballard et al. 2006). According to this hypothesis, winter survival is optimized through lower energy requirements because of the bird's reduced body mass (Reinecke et al. 1982; Williams and Kendeigh 1982). Considering that the primary functions of lipid reserves are to provide insulation and emergency energy stores (King 1972; Raveling 1979), the utility of carrying excess reserves may be limited in environments with mild winter temperatures and dependable availability of resources. However, having resources readily available during critical periods would be important for the success of this strategy. For instance, access to resources before spring migration when energy requirements increase may be important to optimize migration. Management for late winter food availability would likely be more important for birds utilizing this strategy than having it available early and depleted by late winter.

An additional advantage of maintaining minimal endogenous reserves is to allow better flight maneuverability to evade avian predators. Coastal areas are known to concentrate raptors during migration as migration routes of many species of potential prey are focused here as well (Aborn 1994). This is particularly true for the Texas coast as annual "hawk watches" can tally over a million raptors in a single location and during a single migratory period. Large falcons are particularly capable of preying on ducks, and peregrine falcons (*Falco peregrinus*), prairie falcons (*F. mexicanus*), and aplomado falcons (*F. femoralis*) are common members of the raptor guild during migration and winter along the coast (Rappole and Blacklock 1994).

WINTER SURVIVAL

Winter survival may be especially important for northern pintails that spend a larger proportion of their annual cycle on wintering areas than other duck species. Population dynamics of pintails in Alaska are highly sensitive to variation in survival of females (Flint et al. 1998). Additionally, pintails display high fidelity to coastal wintering areas such as in Texas, and low winter survival in these areas can have significant impacts on local populations (Hestbeck 1993). Therefore, female survival on wintering areas is an important factor to consider when assessing the long-term decline in continental pintail numbers.

Winter survival estimates for female northern pintails using conventional radio-telemetry techniques show geographic variation throughout North America. Adult female northern pintails wintering in coastal areas of Louisiana (Cox et al. 1998) and Texas (Ballard, unpublished data, 2004) appear to experience lower winter survival ($\leq$0.71) than those wintering along the west coast of Mexico (0.91; Migoya and Baldassarre 1993), Central California (0.76–0.88; Miller et al. 1995; Fleskes et al. 2002), and the Playa Lakes region of Texas (0.93; Moon and Haukos 2006), which experience relatively high winter survival. Differences in harvest mortality among these areas appear to explain variation among most estimates. Hunting mortality rates were over twice as high in Louisiana and Texas compared with the other areas. Natural mortality was quite low in all areas except the Texas coast, where mortality due to natural causes was estimated to be greater than harvest mortality. From these studies, it appears that pintails wintering along the Gulf Coast, and particularly Texas, experience lower survival from greater hunting pressure and from non-harvest-related factors.

Survival of northern pintails wintering along the Gulf Coast appears to decrease in late winter following the hunting season and before departure (Ballard, unpublished data, 2004). This decline coincides with annual large-scale habitat loss in this region. A high proportion of wetlands along the central Texas coast are under hydrologic control, primarily for waterfowl hunting. Water levels in these wetlands are manipulated to optimize food availability that is timed for the waterfowl season. Drainage of many of these wetlands is initiated during the final week of the hunting season in late January and most are dry soon after the close of duck season (Ballard, unpublished data, 2004). This management approach significantly reduces the amount of habitat available for waterfowl in a very short amount of time in late winter and occurs at a time when pintails and other migrating species are typically building nutrient reserves before migration. Habitat loss that occurs immediately and on a large-scale before migration may influence the ability of wintering pintails to increase nutrient reserves in preparation for migration and result in later departure or a more protracted migration (Alerstam and Lindstrom 1990).

Optimal migration theory suggests that birds face many decisions during migration, such as (1) timing departure from wintering grounds or staging areas to optimize arrival on breeding areas, (2) deciding how much energy to accumulate before departing on the next leg of migration, and (3) choosing stopover sites that optimize fuel deposition rates and predation risk (Alerstam and Lindstrom 1990). Optimal migration ties all these decisions together for migration to be as fast, energy conserving, and as safe from predation as possible. If early arrival on breeding areas is beneficial, as in most dabbling duck species, then migration that minimizes time may be important. However, if suitable stopover options along migration routes are not limited, conserving energy during travel may be a better strategy. For species that incur significant predation risk during migration, fat accumulation rates, habitat selection, and travel rate may be governed by strategies that minimize the associated mortality risk. Flight maneuverability and speed are reduced with increasing fat reserves. Thus, larger fuel loads reduce predator-evading capabilities and increase predation risk. Birds trying to reduce predation risk should depart wintering and stopover sites with smaller fuel loads that would be optimal with energy-minimizing migration. Some species may realize lower predation risk by migrating at night when most aerial predators are inactive. Nocturnal migrants may also increase the efficiency of migration by traveling during periods when fat deposition is less efficient.

Miller et al. (2005) and Haukos et al. (2006) have recently provided information on migration routes of pintails from wintering areas in California and mid-continent region, respectively, with the aid of satellite telemetry. Although their findings are based on a small sample of individuals (e.g., $n = 5$ from Texas coast), it appears that pintails in these wintering populations pursue different migration strategies that may at least partially account for differences in stored fuel loads. Pintails that depart California wintering sites typically make a short flight to Oregon before many individuals make long-distance flights either across the Pacific Ocean to breeding areas in Alaska, or across the Rocky Mountains to breeding areas in Alberta. Because of the presumably limited choices in stopover sites en route, these birds need to depart southern Oregon with sufficient fuel loads to make the long journey without refueling. Late snow accumulation on breeding grounds can result in pintails having to wait until snowmelt to nest. During these periods, foraging opportunities can be limited as well because snow and ice prevent foods from being available. Therefore, an overloading strategy may be invoked to provide energy stores that would be available over and above that required to migrate to breeding areas (Wilson 1981).

Pintails migrating from mid-continent wintering areas to the Prairie Pothole region exhibit a more staggered migration (Haukos et al. 2006). These birds appear to take advantage of few barriers and presumably plentiful stopover options through the mid-continent region (Pederson et al. 1989). Thus, large departure fuel loads are probably not as critical to these birds as to those leaving California wintering areas, because the migration can be divided into shorter segments that include more foraging opportunities to keep up with energy demands. It is costly to carry large fuel loads, thus being able to migrate with reduced fuel loads can be advantageous. Also, as mentioned previously, predation risk may be reduced because of the maneuverability advantages and increase in flight performance over carrying heavy fuel loads.

Pintails wintering in California appear to depart wintering areas much earlier than pintails wintering in mid-continent regions (Miller et al. 2005; Haukos et al. 2006). Recent information from a large sample of female pintails in Texas equipped with conventional VHF transmitters also shows pintails wintering along the Texas coast depart later than pintails wintering in the Playa Lakes region (Moon and Haukos 2006; Ballard, unpublished data, 2004). Early arrival on breeding grounds is beneficial, because reproductive success is inversely related to date of egg laying in many migratory species (Daan et al. 1990; Rowe et al. 1994). Waterfowl that initiate nests early experience greater nest success (Flint and Grand 1996) with larger clutches (Duncan 1987; Blums et al. 1997). Further, brood survival tends to be greater for earlier hatched nests due to declines in seasonal wetland availability and food resources (Rotella and Ratti 1992; Mauser et al. 1994; Cox et al. 1998; Guyn and Clark 1999), or higher predation rates on ducklings due to lower availability of alternate prey later in the breeding season (Grand and Flint 1996). Reproductive success in migratory barnacle geese (*Branta leucopsis*) is related to arrival dates on breeding areas and to the amount of fat accumulated before and during migration (Prop et al. 2003). Although barnacle geese forage extensively on breeding grounds, individuals that arrived with larger fat stores benefited by earlier nest initiations due to less time required to build endogenous reserves for nesting. This provided a head start for reproduction and presumably increased reproductive success for those individuals with early fledging young (Prop et al. 2003). These studies emphasize the importance relative to fitness of a timely arrival on breeding areas and of arriving with adequate energy stores. Northern pintails have an affinity to nest early and in association with ephemeral wetland habitats (Stewart and Kantrud 1973; Duncan 1987). They are one of the earliest nesting species in the Prairie Pothole region (Bellrose 1980), and often arrive on breeding areas before snowmelt, necessitating energy stores large enough to wait out snowmelt and still provide nutrients and energy to produce a clutch.

The Rainwater Basin in Nebraska is thought to provide the most important stopover habitat for spring migrating pintails in the Northern Great Plains (Bellrose 1980; Pederson et al. 1989), and many management resources have been diverted to this region because of its importance. Wetland plant and animal foods and waste grain in adjacent agricultural land provide a diverse and energy-rich diet to migrating waterfowl staging in this region. These wetlands are also important for pairing before reaching breeding areas (LaGrange and Dinsmore 1988). Increased energy acquisition to meet demands of migration, courtship, and gonadal development occur during late winter and spring migration. Consequently, habitat quality plays a large role on where and how easily ducks meet these demands.

The Rainwater Basin is considered to be the primary spring staging area for northern pintails in the mid-continent region (Bellrose 1980). Based on recent information from 634 pintails marked with conventional VHF radio transmitters, most pintails radio marked in the Playa Lakes region of Texas passed through the Rainwater Basins during spring migration (Cox, unpublished data, 2004). However, few female pintails radio marked along the Gulf Coast of Texas were detected in the Rainwater Basins during spring (Cox, unpublished data, 2004). These birds appeared to take a more easterly migration route presumably through the Missouri River drainage. Similarly, northern pintails radio marked in southern Louisiana primarily followed the Des Moines River corridor through Iowa en route to northern breeding areas (Cox 1996). Thus, it appears that northern pintail migration routes have changed since the 1970s, or new technology has provided better insight as to the importance of spring staging habitat east of the Rainwater Basins for northern pintails.

Whether depressed fuel loads in late winter can be easily mitigated on spring stopover sites is unclear. In any case, winter survival may be an important regulator in northern pintail population dynamics (Flint et al. 1998). Considerably low winter survival in a wintering area that holds the majority of pintails in the Central flyway could have impacts noticeable at the continental level. Whether a late departure has consequences on fitness is less clear; however, given pintails' affinity for ephemeral habitats that are available early, their greater reliance on endogenous reserves relative to other species, and their finicky nature toward suboptimal habitat conditions and tendency to forgo nesting in favor of survival, a late departure may prove costly. This is particularly relevant given the abundance of migratory bird studies that show reproductive success to be negatively impacted by late nest initiations. Habitat quantity and quality on wintering areas may play a larger role than expected in allowing pintails to optimize migration and timing of arrival on breeding grounds.

LINKING OPTIMAL MIGRATION THEORY AND MANAGEMENT

A better understanding of the relationship between the dynamics of body stores and the consequences that decisions during winter and migration have on the fitness of pintails needs to occur before we can make clear the most cost-effective strategy to allocate management resources within the annual cycle of mid-continent northern pintails. Wetland management objectives, particularly with respect to timing of food availability, may need to be reevaluated to better suit mid-continent pintail ecology.

Large-scale and immediate loss of habitat concurrent with the close of duck season in late January along the Texas Coast is of concern. Because this time of the year is important for waterfowl to increase fat deposition to meet energy demands for migration, molt, and reproduction, resource availability is especially important. Educating landowners or hunting lease managers as to the importance of late-season habitat for waterfowl may be the first logical step to acknowledge this issue. A staggered drainage regime where certain impoundments are drained while others remain inundated through the wintering period may provide a compromise beneficial to both land managers and waterfowl. Slowing the rate of drainage would also help alleviate large-scale habitat loss. A slower rate of drainage will allow a proportion of habitat to remain available to waterfowl later and concentrate aquatic invertebrates in the receding water. Landowner assistance programs for wetland projects may need to place higher priority on late season water and establish agreements where this is a requirement.

Many resources, in terms of research and management, have been focused on the Rainwater Basins for pintail conservation. However, new information from recent and ongoing research suggests that managers may need to reevaluate management strategies to address a larger component of northern pintails within the mid-continent. Improved habitat quality on wintering areas may mitigate loss and seasonal variability in wetland habitats along migration routes.

REFERENCES

Aborn, D. A. 1994. Correlation between raptor and songbird numbers at a migratory stopover site. *Wilson Bull.* 106:150.

Adair, S. E., J. L. Moore, and W. H. Keil, Jr. 1996. Winter diving duck use of coastal ponds: An analyses of alternative hypotheses. *J. Wildl. Manage.* 60:83.

Alerstam, T., and A. Lindstrom. 1990. Optimal bird migration: The relative importance of time, energy, and safety. In *Bird Migration: Physiology and Ecophysiology*, E. Gwinner (ed.), Chap. 5. New York: Springer-Verlag.

Alston, L. T., et al. 2000. Ecological, economic, and policy alternatives for Texas rice agriculture. TR-181, Institute for Science, Technology, and Public Policy, George Bush School of Government and Public Service, Texas Water Resources Institute.

Anderson, J. T. 1994. Wetland use and selection by waterfowl wintering in coastal Texas. MS thesis, Texas A&M University–Kingsville, Kingsville, TX.

Austin, J. E., and M. R. Miller. 1995. Northern pintail (*Anas acuta*), No. 163. In *The Birds of North America*, A. Poole, and F. Gill (eds). Philadephia: The Academy of Natural Sciences and Washington, DC: The American Ornithologists' Union.

Ball, I. J., et al. 1989. Northwest Riverine and Pacific Coast. In *Habitat Management for Migrating and Wintering Waterfowl in North America*, L. M. Smith, R. L. Pederson, and R. M. Kaminski (eds.), 429. Lubbock, TX: Texas Tech University Press.

Ballard, B. M., J. E. Thompson, and M. J. Petrie. 2006. Carcass composition and digestive tract dynamics of northern pintails wintering along the lower Texas coast. *J. Wildl. Manage.* 70:1316.

Bateman, H. A., T. Joanen, and C. D. Stutzenbaker. 1988. History and status of mid-continent snow geese on their Gulf Coast winter range. In *Waterfowl in Winter*, M. W. Weller (ed.). Minneapolis, MN: University of Minnesota Press.

Bellrose, F. C. 1980. *Ducks, Geese, and Swans of North America*. Harrisburg: Stackpole Books.

Blums, P., G. R. Hepp, and A. Mednis. 1997. Age-specific reproduction in three species of European ducks. *Auk* 114:737.

Briggs, R. 1982. Avian use of small aquatic habitats in South Texas. MS thesis, Texas A&M University — Kingsville, Kingsville, TX.

Brown, L. F., et al. 1977. Environmental geological atlas of the Texas coastal zone–Kingsville area. Bureau of Economic Geology, University of Texas, Austin, TX.

Conroy, M. J., G. R. Costanzo, and D. B. Stotts. 1989. Winter survival of female American black ducks on the Atlantic Coast. *J. Wildl. Manage.* 53:99.

Cox, R. R., Jr. 1996. Movements, habitat use, and survival of female northern pintails in southwestern Louisiana. PhD thesis, Louisiana State University, Baton Rouge, LA.

Cox, R. R., Jr., A. D. Afton, and R. M. Pace, III. 1998. Survival of female northern pintails wintering in southwestern Louisiana. *J. Wildl. Manage.* 62:1512.

Cox, R. R., Jr., M. A. Hanson, C. C. Roy, N. H. Euliss, Jr., D. H. Johnson, and M. G. Butler. 1998. Mallard duckling growth and survival in relation to aquatic invertebrates. *J. Wildl. Manage.* 62:124.

Daan, S., C. Dijkstra, and J. M. Tinbergen. 1990. Family planning in the kestrel (*Falco tinnunculus*): The ultimate control of covariation of laying date and clutch size. *Behaviour* 114:83.

DeGraaf, R. M., and J. H. Rappole. 1995. *Neotropical Migratory Birds: Natural History, Distribution, and Population Change*. Ithaca, NY: Comstock Publishing Associates.

Delnicki, D., and K. J. Reinecke. 1986. Mid-winter food use and body weights of mallards and wood ducks in Mississippi. *J. Wildl. Manage.* 50:43.

Duncan, D. C. 1987. Nest-site distribution and overland brood movements of northern pintails in Alberta. *J. Wildl. Manage.* 51:716.

Esler, D., and J. B. Grand. 1994. The role of nutrient reserves for clutch formation by northern pintails in Alaska. *Condor* 96:422.

Fleskes, J. P., R. L. Jarvis, and D. S. Gilmer. 2002. September–March survival of female northern pintails radio-tagged in San Joaquin Valley, California. *J. Wildl. Manage.* 66:901.

Flint, P. L., and J. B. Grand. 1996. Nesting success of northern pintails on the coastal Yukon-Kuskokwim Delta, Alaska. *Condor* 98:54.

Flint, P. L., J. B. Grand, and R. F. Rockwell. 1998. A model of northern pintail productivity and population growth rate. *J. Wildl. Manage.* 62:1110.

Fredrickson, L. H., and M. K. Laubhan. 1994. Intensive wetland management: A key to biodiversity. *Trans. North Am. Wildl. Nat. Resour. Conf.*, 59:555.

Fredrickson, L. H., and Taylor, T. S. 1982. Management of seasonally flooded impoundments for wildlife. *U.S. Fish Wildl. Resour. Publ.* 148.

Fulbright, T. E., and F. C. Bryant. 2002. The last great habitat. Special Publication No. 1, Caesar Kleberg Wildlife Research Institute, Kingsville, TX.

Galarza, A., and J. L. Telleria. 2003. Linking processes: Effects of migratory routes on the distribution of abundance of wintering passerines. *Anim. Biodivers. Conserv.* 26:19.

Grand, J. B., and P. L. Flint. 1996. Survival of northern pintail ducklings on the Yukon-Kuskokwim Delta, Alaska. *Condor* 98:48.

Grand, J. B., P. L. Flint, and P. J. Heglund. 1997. Habitat use by nesting and brood rearing northern pintails on the Yukon-Kuskokwim Delta, Alaska. *J. Wildl. Manage.* 61:1199.

Guyn, K. L., and R. G. Clark. 1999. Factors affecting survival of northern pintail ducklings in Alberta. *Condor* 101:369.

Haramis, G. M., et al. 1986. The relationship between body mass and survival of wintering canvasbacks. *Auk* 103:506.

Haukos, D. A., et al. 2006. Spring migration of northern pintails from Texas and New Mexico, USA. *Waterbirds* 29:127.

Heitmeyer, M. E. 1985. Wintering strategies of female mallards related to dynamics of lowland hardwood wetlands in the upper Mississippi delta. PhD thesis, University of Missouri, Columbia, MO.

Heitmeyer, M. E., and L. H. Fredrickson. 1981. Do wetland conditions in the Mississippi Delta hardwoods influence mallard recruitment? *Trans. North Am. Wildl. Nat. Resourc. Conf.* 46:44.

Hepp, G. R. 1986. Effects of body weight and age on the time of pairing of American black ducks. *Auk* 103:477.

Herbert, C. E., and L. I. Wassenaar. 2005. Stable isotopes provide evidence for poor northern pintail production on the Canadian prairies. *J. Wildl. Manage.* 69:101.

Hestbeck, J. B. 1993. Overwinter distribution of northern pintail populations in North America. *J. Wildl. Manage.* 57:582.

Higgins, K. F. 1977. Duck nesting in intensively farmed areas of North Dakota. *J. Wildl. Manage.* 41:232.

Hobaugh, W. C., C. D. Stutzenbaker, and E. L. Flickinger. 1989. The rice prairies. In *Habitat Management for Migrating and Wintering Waterfowl in North America*, L. M. Smith, R. L. Pederson, and R. M. Kaminski. Lubbock: Texas Tech University Press.

Hobson, K. A., et al. 2005. Tracing nutrient allocation to reproduction in Barrow's goleneye. *J. Wildl. Manage.* 69:1221.

Johnson, D. H., and J. W. Grier. 1988. Determinants of breeding distributions of ducks. *Wildl. Monogr.* 100:1.

King, J. R. 1972. Adaptive periodic fat storage by birds. In *Proceedings of the International Ornithological Congress*, 1970:200.

Krapu, G. L. 1974. Foods of breeding pintails in North Dakota. *J. Wildl. Manage.* 38:408.

Krapu, G. L. 1981. The role of nutrient reserves in mallard reproduction. *Auk* 98.

LaGrange, T. G., and J. J. Dinsmore. 1988. Nutrient reserve dynamics of female mallards during spring migration through central Iowa. In *Waterfowl in Winter*, M. W. Weller (ed.), Chap. 20. Minneapolis, MN: University of Minnesota Press.

Lincoln, F. C., S. R. Peterson, and J. L. Zimmerman. 1998. Migration of birds. U.S. Department of the Interior, U.S. Fish and Wildlife Service, Washington, DC, Circular 16, Northern Prairie Wildlife Research Center, Jamestown, ND. http://www.npwrc.usgs.gov/resource/birds/migratio/migratio.htm (Version 02APR2002).

Manley, S. W. 1999. Ecological and agricultural values of winter-flooded rice fields in Mississippi. PhD thesis, Mississippi State University, Starkville.

Mann, R. E., and J. S. Sedinger. 1993. Nutrient-reserve dynamics and control of clutch size in northern pintails breeding in Alaska. *Auk* 110:264.

Mauser, D. M., R. L. Jarvis, and D. S. Gilmer. 1994. Survival of radio-marked mallard ducklings in northeastern California. *J. Wildl. Manage.* 58:82.

McAdams, M. S. 1987. Classification and waterfowl use of ponds in South Texas. MS thesis, Texas A&M University, Collage Station, TX.

McMahan, C. A. 1968. Biomass and salinity tolerances of shoalgrass and manateegrass in lower Laguna Madre, Texas. *J. Wildl. Manage.* 32:501.

McMahan, C. A. 1970. Food habits of ducks wintering on the Laguna Madre, Texas. *J. Wildl. Manage.* 34:946.

Migoya, R. G., and G. A. Baldassarre. 1993. Harvest and food habits of waterfowl wintering in Sinaloa, Mexico. *Southwestern Nat.*, 38:168.

Miller, M. R. 1986. Northern pintail body condition during wet and dry winters in the Sacramento Valley, California. *J. Wildl. Manage.* 50:189.

Miller, M. R. 1987. Fall and winter foods of northern pintails in the Sacramento Valley, California. *J. Wildl. Manage.* 51:405.

Miller, M. R., and D. C. Duncan. 1999. The northern pintail in North America: Status and conservation needs of a struggling population. *Wildl. Soc. Bull.* 27:788.

Miller, M. R., J. P. Fleskes, D. L. Orthmeyer, W. E. Newton, and D. S. Gilmer. 1995. Survival of adult female northern pintails in Sacramento Valley, California. *J. Wildl. Manage.* 59:478.

Miller, M. R., et al. 2005. Spring migration of northern pintails from California's Central Valley wintering area tracked with satellite telemetry: Routes, timing, and destinations. *Canadian J. Zool.* 83:1314.

Mitsch, W. J., and J. G. Gosselink. 2000. *Wetlands*, 3rd edn., Chap. 17. New York: John Wiley & Sons, Inc.

Moon, J. A., and D. A. Haukos. 2006. Survival of female northern pintails wintering in the Playa Lakes region of northwestern Texas. *J. Wildl. Manage.* 70:777.

Moulton, D. W., T. E. Dahl, and D. M Dall. 1997. Texas coastal wetlands: Status and trends, mid-1950s to early 1990s. U.S. Department of the Interior, Fish and Wildlife Service, Albuquerque, NM.

Muehl, G. T. 1994. Distribution and abundance of water birds and wetlands in coastal Texas. MS thesis, Texas A&M University — Kingsville, Kingsville, TX.

North American Waterfowl Management Plan (NAWMP). 2004. North American Waterfowl Management Plan strategic guidance: Strengthening the biological foundation. Plan Committee, Canadian Wildlife Service, U.S. Fish and Wildlife Service, Secretaria de Medio Ambiente y Recursos Naturales.

Onuf, C. P. 1996. Biomass patterns in sea grass meadows of the Laguna Madre, Texas. *Bull. Mar. Sci.* 58:404.

Pederson, R. L., Jorde, D. G., and Simpson, S. G. 1989. Northern Great Plains. In *Habitat Management for Migrating and Wintering Waterfowl in North America*, L. M. Smith, R. L. Pederson, and R. M. Kaminski (eds). TX: Texas Tech University Press.

Prop, J., J. M. Black, and P. Shimmings. 2003. Travel schedules to the high arctic: Barnacle geese trade-off the timing of migration with accumulation of fat deposits. *Oikos* 103:403.

Rappole, J. H., and G. W. Blacklock. 1994. *A Field Guide to the Birds of Texas*. College Station, TX: Texas A&M University Press.

Rappole, J. H., et al. 1979. Timing of migration and route selection in North American songbirds. In *Proceedings of the First Welder Wildl. Found. Symposium*, D. L. Drawe (ed.). Sinton, 199.

Raveling, D. G. 1979. The annual cycle of body composition of Canada geese with special reference to control of reproduction. *Auk* 96:234.

Raveling, D. G., and M. E. Heitmeyer. 1989. Relationships of population size and recruitment of pintails to habitat conditions and harvest. *J. Wildl. Manage.* 53:1088.

Reinecke, K. J., T. L. Stone, and R. B. Owen, Jr. 1982. Seasonal carcass composition and energy balance of female black ducks in Maine. *Condor* 84:420.

Richardson, D. M., and R. M. Kaminski. 1992. Diet restriction, diet quality, and prebasic molt in female mallards. *J. Wildl. Manage.* 56:531.

Rotella, J. J., and J. T. Ratti. 1992. Mallard brood survival and wetland habitat conditions in southwestern Manitoba. *J. Wildl. Manage.* 56:499.

Rowe, L., D. Ludwig, and D. Schluter. 1994. Time, condition and the seasonal decline of avian clutch size. *Am. Nat.* 143:698.

Runge, M. C., and G. S. Boomer. 2005. Population dynamics and harvest management of the continental northern pintail population. Division of Migratory Bird Management, United States Fish and Wildlife Service, Washington, DC.

Sargent, A. B., and D. G. Raveling. 1992. Mortality during the breeding season. In *Ecology and Management of Breeding Waterfowl*, B. D. J. Batt, et al. (eds), Chap. 12. Minneapolis, MN: University of Minnesota Press.

Smith, R. I. 1970. Response of pintail breeding populations to drought. *J. Wildl. Manage.* 34:934.

Smith, L. M., and D. G. Sheeley. 1993. Factors affecting condition of northern pintails wintering in the southern High Plains. *J. Wildl. Manage.* 57:62.

Snell-Rood, E. C., and D. A. Cristol. 2003. Avian communities of created and natural wetlands: Bottomland forests in Virginia. *Condor* 105:303.

Stewart, R. E., and H. A. Kantrud. 1973. Ecological distribution of breeding waterfowl populations in North Dakota. *J. Wildl. Manage.* 37:39.

Stutzenbaker, C. D., and M. W. Weller. 1989. The Texas coast. In *Habitat Management for Migrating and Wintering Waterfowl in North America*, L. M. Smith, R. L. Pederson, and R. M. Kaminski (eds). TX: Texas Tech University Press.

Tacha, T. C., A. M. Holzem, and D. W. Bauer. 1993. Changes in waterfowl and wetland abundance in the Chenier Plain of Texas 1970s–1990s. *Texas J. Agric. Nat. Resour.* 6:31.

Thompson, J. D., and G. A. Baldassarre. 1990. Carcass composition of nonbreeding blue winged teal and northern pintails in Yucatan, Mexico. *Condor* 92:1057.

Tunnel, J. W., Jr. 2002. Geography, climate, and hydrography. In *The Laguna Madre of Texas and Tamaulipas*, J. W. Tunnel, Jr., and F. W. Judd (eds), Chap. 2. College Station, TX: Texas A&M University Press.

U.S. Fish and Wildlife Service. 1999. Analyses of selected mid-winter waterfowl survey data (1955–1999), Region 2 (Central Flyway portion). U.S. Fish and Wildlife Service, Albuquerque, NM.

U.S. Fish and Wildlife Service. 2005. Waterfowl population status. U.S. Department of Interior, Washington, DC.

Weller, M. W. 1999. *Wetland Birds*, Chap. 3. Cambridge: Cambridge University Press.

Williams, J. E., and S. C. Kendeigh. 1982. Energetics of the Canada goose. *J. Wildl. Manage.* 46:588.

Wilson, J. R. 1981. The migration of High Arctic shorebirds through Iceland. *Bird Study* 28:21.

Part II

Mammals

7 Conserving the Cats, Cougar Management as a Model: A Review

Maurice G. Hornocker

CONTENTS

Cats have been part of the culture, mythology, and environment of human beings since the beginning of recorded history. The tiger has been important in the culture and art of the great East Asian civilizations. The lion has been used as a symbol of royalty throughout the ages. The jaguar figured prominently in pre-Columbian civilizations in Mexico and Central America, as did the mountain lion. Cats were revered in ancient Egypt, where they were first domesticated some 3600 years ago.

At the same time, most species of wild cats have declined drastically because of human activities — direct killing and habitat destruction. Realistically, it may not be possible to save some populations of cats — they may simply be overwhelmed by human beings. However, it behooves us to make every effort to reverse this decline in cats because of their cultural significance, and because they play an important role in ecological balance. The big cats, in particular, are at the pinnacle of the food chain and, because of the large areas they require, can literally define an ecosystem.

In 1987, the Cat Specialist Group Species Survival Commission of the International Union for Conservation of Nature and Natural Resources (IUCN) prepared a manifesto for IUCN's World Conservation Strategy. That statement is as timely today as it was then, and portions of it are appropriate here.

WHY CATS SHOULD BE CONSERVED

The decline of a carnivore generally alters the ecological balance of its biological community. Cats are linked through predation to herbivores, which are, in turn, linked to each other through competition and to plant communities by their foraging. They are particularly sensitive to environmental disturbance, and decline or disappearance of these vulnerable cat species serves as an indicator of changes in their ecosystem, which may be the result of natural phenomena or, as is increasingly the case in present times, of the impact of human activities. These changes frequently involve deterioration in the human environment, such as the loss of forests and grasslands and their valuable animal and plant products, or impairment of water supplies essential to human life and agriculture. Furthermore large cats, being at the pinnacle of the food chain, need considerable space and are, therefore, key species in determining the area required to define an appropriate ecosystem. In addition to the ecological consequences of the disappearance of these carnivores, many people feel a sense of inner loss when such magnificent and mysterious animals are gone from the wild.

PROBLEMS FACED BY CATS

Accelerating loss of habitat has now reached a critical stage as the human population continues to soar. In many cat ranges, remaining habitat is but a small percentage of what existed in the past, and what remains could be wiped out in the near future. Cats have long been hunted. They are killed because they have been viewed as competitors for prey. They are killed because they have taken livestock. They are killed for sport, and their body parts are used in some places as medicine. Young cats are captured for adoption as pets. Moreover, some, especially spotted cats, are killed for the fashion trade, which has often led to overexploitation. At the same time, the disappearance of natural prey has frequently deprived cats of their normal sustenance and contributed to conflict with humans and their livestock, leading inevitably to reprisal killing of cats, often including those not actually involved.

Where cat populations have been reduced to small numbers, they are increasingly vulnerable to extinction due to fortuitous local events, such as epidemics, fires, and floods. Some scientists also fear the possibility of deterioration through inbreeding depression and loss of genetic diversity in the long term, which might reduce the ability of small populations to adapt to changes in their environment.

PROBLEMS OF CAT CONSERVATION

There is still only limited knowledge of the distribution, numbers, biology, and behavior of many species of cats. Research to increase understanding of these factors is essential to planning and implementation of effective conservation measures.

Economic planners and decision-makers often fail to recognize the importance of wild lands for human welfare, including ecosystems of which cats are part. Consequently, development programs are carried out with little or no consideration of the longer-term impact, which may result in the decline and extinction of many species, including cats, and impoverishing the human environment.

As a result of increasing fragmentation of habitat and the pressure of human activities in their vicinity, large cats may become problem animals, particularly through livestock predation, and in rare cases taking human life. Demands may then arise for elimination, not only of the offending animals, but also of all the large cats in the area. Insufficient resources are made available to pursue necessary research, and to implement protective measures and conservation management of natural habitats of cats, often because of failure to recognize their ecological significance and through lack of political will.

HOW CATS CAN BE CONSERVED

Protected habitats of sufficient size and productivity to support viable populations of cats must be preserved, and linking corridors maintained wherever possible. The distribution of each species and

FIGURE 7.1   Young cougar (*Puma concolor*) in northern Mexico (Photograph Copyright Tim E. Fulbright).

the habitat available to it needs to be established in detail down to the level of discrete populations. Legislation to ensure long-term conservation of cat species and their prey, including controls on trade, national and international, must be passed and enforced.

Conservation of cats has to be reconciled with the needs of humans. Some conflict may be inevitable in areas where agriculture or livestock farming impinges on cat habitats, but it should be minimized by appropriate management measures. For many cats, and particularly large cats, parks and reserves may not be adequate. Land-use patterns in adjacent areas need to be designed so that they are compatible with use by both humans and cats. All these measures should be included in an overall conservation strategy for each species to ensure its survival (IUCN 1987).

Species need not be lost, provided action is taken to conserve them. Experience has shown that seemingly desperate situations can be reversed, if protection is given to species and their ecosystems. There is evidence that cat populations can rise, literally, from near extinction. Bobcats (*Felis rufus*) in the United States were reduced to dangerously low numbers in the 1970s because of high pelt prices. Emergency regulations were placed in effect; today bobcat populations are flourishing throughout their range. The common leopard (*Panthera pardus*) thrives in many areas of its vast range in the eastern hemisphere even though its numbers were extremely low in some regions 30 years ago. Cougars (*Puma concolor*), hunted to extinction in much of their range, have rebounded to perhaps historic highs in western North America because of the regulation of killing and high prey numbers — the result of good game management (Figure 7.1). Cheetahs (*Acinonyx jubatus*) flourish in some huge private holdings and in parks and reserve in parts of Africa, again as the result of proper management.

COUGAR CONSERVATION

THEORY AND MANAGEMENT

Linking ecological theory to management of all cat species would be difficult and cumbersome. For purposes of this discussion, it appears more productive to select a representative species, on which quantitative data are available, to illustrate the concept.

Therefore, this discussion will review cougar theory, management, and conservation. The cougar qualifies as a "large" carnivore and thus a key or indicator species, but it also is representative of many other species of cats. It is adaptable, it takes a wide range of prey animals, and it faces most of the problems facing cats worldwide. Further, cougars have been the objects of more research in recent years than have most other big cats, and this has provided for more management options. Finally, the cougar — among all the big cats — has made the most remarkable recovery in the past 30 years.

HISTORIC MANAGEMENT CONTROL

Cougar "management" in North America has ranged from efforts to completely exterminate the species to current programs aimed at conserving, even fostering, cougar populations. The theory, or belief, held by early settlers was that carnivores were direct competitors, killing desirable wildlife species and domestic livestock. Once one of the most widely distributed large carnivores in the world, cougars were eradicated in much of their historic range in North America, surviving only in remote areas of the West. Even here, they were under great pressure from predator control programs. Economic incentives in the form of bounties were initiated in the late 1800s and persisted until the 1960s in several western states (Young and Goldman 1946; Nowak 1976).

While bounties appeared to be effective in reducing cougar numbers, theory concerning all bounties was changing in the mid-1900s. Many wildlife managers had long assumed that the elimination of bounties would result in a widespread cougar population increase. However, Torres et al. (2004) tested this hypothesis and found that cougar removals did not decrease after government-subsidized bounties ceased. Cougar numbers, circa 1900, may have been low due to unregulated cougar removal, market hunting of ungulates, competition from domestic livestock, habitat changes during Euro-American settlement, and incidental killing of cougars from poisoning programs targeting wolves (*Canus lupus*) and coyotes (*C. latrans*) (Robinette et al. 1959). Therefore, any cougar population increases that may have occurred in the latter part of the twentieth century cannot be attributed solely to the abolition of bounties. Cougar increases, if they occurred, might more plausibly be attributed to increases in numbers and distributions of deer (*Odocoileus* sp.) and elk (*Cervus elaphus*) (Cougar Management Guidelines Working Group 2005).

Thus, controlling the numbers of cougars, classified everywhere as vermin, was the only management practiced until the 1960s. The environmental movement and changing public attitudes toward all carnivores were bringing about changes in management. Colorado in 1965 reclassified the cougar from vermin to game animal. All other western states followed Colorado's lead and, by 1972, the cougar was officially a game species throughout its range at that time. All states permitted hunting seasons but killing was restricted.

These protective measures have helped immensely in the increase in cougar numbers throughout their range. Prey numbers — deer and elk — have likewise increased and have contributed not only to cougar population increases, but also to the recolonization of many historic cougar ranges. Cougar populations are probably at historic highs in many part of their range in North America, and therefore more refined management is required.

CURRENT MANAGEMENT — PROBLEMS AND STRATEGIES

Management of cougars is difficult for several reasons: cougars are secretive, they exist at low population densities, they impact wild and domestic prey, they can threaten human safety, and public attitudes about them differ widely. Habitat and human attitudes are changing rapidly and bringing new management challenges. Ultimately, human values will determine management objectives and the means used to achieve them (Cougar Management Guidelines Working Group 2005).

Control of cougar numbers is still a major issue and is practiced in areas where livestock, public safety, or other wild species are threatened. All western states and provinces allow the killing of individual cougars that threaten people, livestock, or pets (Braun 1991; Cougar Management Guidelines Working Group 2005), and most states and provinces attempt to control cougar numbers through sport hunting.

Predation on game species, principally big game animals, has been a principal reason for controlling cougar numbers. The guiding theory was that unless cougars were controlled, they would increase in numbers and their effect on prey would be even greater. Research in the 1960s and 1970s showed that while food (prey) was the ultimate controlling factor, strong territorialism acted to keep cougar numbers in check (Hornocker 1969, 1970; Seidensticker 1973; Sweanor 1990). Subsequent research has supported this concept (Hopkins 1989; Lindzey et al. 1994; Logan et al. 1996; Murphy 1998; Murphy et al. 1999; Ruth 2001).

Pierce et al. (2000) and Logan and Sweanor (2001) discounted the importance of territorialism in cougar population regulation, concluding that prey availability alone limits cougar numbers. More research is needed, and this research should be long term, with both cougars and their prey studied concurrently. Experimental design, including manipulating both predator and prey populations, is required to more clearly define each situation.

The effects of cougar predation on prey populations may vary with ecological conditions (Logan and Sweanor 2000). In Idaho, California, Utah, and the northern Yellowstone ecosystem, cougars did not limit elk and deer populations (Hornocker 1970; Hopkins 1989; Lindzey et al. 1994; Murphy 1998; Murphy et al. 1999). In New Mexico, where mule deer were the major prey, cougar predation on deer increased significantly during a drought that reduced carrying capacity and fawn production. Although cougars contributed directly to the deer population decline, habitat quality was the ultimate cause for the decline because of increased vulnerability of deer to predation, malnutrition, and disease (Logan et al. 1996).

The effect of cougar predation on small populations of ungulates may vary. Cougars may have caused a small population of bighorn sheep to abandon winter range in the Sierra Nevada Mountains (Wehausen 1996). Individual cougars preyed heavily on bighorn sheep in Alberta (Ross et al. 1997). Cougar predation on female pronghorn in Arizona was high enough to stabilize or decrease the pronghorn population in rugged, brushy terrain, but cougar predation is probably insignificant in rolling grassland habitat where little cover exists (Ockenfels 1994). Cougar predation on foals stabilized a feral horse population on the California–Nevada border (Turner et al. 1992).

Cougar–ungulate management strategies must consider all factors operating in individual situations, whatever be the management objectives. Theory developed over time by rigorous research in different situations and different regions must be applied in management programs. And this management must be flexible so it can be changed in response to changed environmental conditions.

Livestock depredation has increased with the increase in cougar populations (Logan and Sweanor 2000). Northern states and provinces have relatively low rates of livestock depredation, but other areas have reported increasing trends in losses and complaints (Tully 1991; Padley 1997). These increased losses are a result of increased cougar numbers, whereas depredation on pets — especially dogs and cats — is related to increased human development in cougar habitat (Torres et al. 1996). Management strategies related to protection of private property should stress on selective control — targeting problem individuals. Education can play a key role — promoting better husbandry practices such as the use of guard dogs and herders, better fencing, vegetation removal and manipulation, and choosing appropriate livestock (Papouchis 2004).

Concurrently, public safety is an important concern (Torres 1997). Encounters and attacks have increased in the past 30 years (Aune 1991; Beier 1991; Green 1991; Torres et al. 1996). Such encounters are rare — 15 deaths and 59 nonfatal injuries in the United States and Canada between 1890 and 1996 (Beier 1991; Logan and Sweanor 2000). Young, inexperienced cougars are most often involved in attacks (Aune 1991; Beier 1991; Ruth 1991). Managers must respond not only to address the problem, but also to maintain credibility with the public. Targeting problem individuals and taking swift, decisive action strengthens management strategy. Communicating that strategy coupled with a vigorous education program has proved successful (Torres 1997). And it has proved useful in gaining public support for population management. Sport or recreational hunting can meet different objectives: (1) it can aid in control programs and possibly lessen effects of cougar depredations on livestock and pets, (2) it can possibly reduce human–cougar encounters, and (3) it can help stabilize local and regional cougar populations (Anderson and Lindzey 2005).

Most western states and provinces list the cougar as a game animal and have established regulations that have contributed to current viable populations. Quotas are used by most management agencies to regulate and direct hunting pressure. Females are given more protection under most quota systems. Further, quota systems may improve quality of the hunt by regulating pressure and by sustaining cougar numbers and harvest (Ross et al. 1996; Cougar Management Guidelines Working Group 2005).

Biological effects of hunting on cougar populations may be relatively benign if the kill is compensatory (Logan and Sweanor 2000). But excessive hunting can depress cougar populations if the kill is additive and especially if females and dependent cubs are taken (Lindzey et al. 1994; Logan et al. 1996; Murphy et al. 1999; Anderson and Lindzey 2005). Further, orphaned cubs have poor survival rates and may become nuisance animals (Barnhurst 1986; Maehr et al. 1989; Logan et al. 1996). Pursuit seasons where hunters are allowed to use dogs to capture cougars but not kill them were initiated to appease both the hunters and those opposed to killing. But accidental mauling of cubs and sometimes adults may be significant in high-use areas and contribute to mortality (Roberson and Lindzey 1984; Barnhurst 1986; Harlow et al. 1992).

Sport hunting is capable of depressing local cougar populations temporarily — those populations in a specific drainage or isolated mountain range. But most managers agree that sport hunting has not depressed cougar populations in a regional sense anywhere in western North America.

Most cougar researchers and managers agree that habitat loss and fragmentation are the greatest long-term threats to cougar conservation (Cougar Management Guidelines Working Group 2005). The West is the fastest growing region in the United States, resulting in increasing fragmentation and loss of cougar habitat. These changes are critical because large expanses of habitat and natural linkages are essential for the maintenance of individual cougar populations (Beier 1993; Maehr and Cox 1995; Logan et al. 1996). Evidence for the effects of habitat degradation includes increased cougar–human encounters, nuisance cougars, cougar deaths on highways, pet depredation, and disruption of natural dispersal patterns (Beier 1993; Logan et al. 1996; Torres et al. 1996). Ultimately populations could become isolated and depressed to the point of extinction (Beier 1993; Maehr and Caddick 1995; Hedrick 1995). The extreme isolation of the Florida panther appears to have resulted in reduced fitness that threatens to cause extinction of the subspecies (O'Brien et al. 1990; Barone et al. 1994; Hedrick 1995).

Self-sustaining cougar populations depend on large expanses of habitat and natural landscape linkages. Public lands presently provide most cougar habitat, and management of these lands should consider the needs of cougar populations — prey cover, security, and linkages (Logan and Sweanor 2000). Private lands are also important because in many regions of the West, low-elevation private lands provide winter range for ungulate prey. Economic incentives may be needed to induce private landowners to conserve cougar habitat in the future (Maehr and Cox 1995; Logan and Sweanor 2000).

Nuisance individuals are becoming more of a problem, particularly in urban areas where human development is encroaching on cougar habitat. Translocation has been used to deal with the problem, but rarely has the success of this option been monitored. Ruth et al. (1998), in the only intensive study on translocation of wild cougars with known life histories, found that cougars of dispersal age (less than 27 months) were the best candidates for success. Further, they found the size and character of the release areas were important in any attempt to reestablish populations.

The development of quick-response teams to deal with potentially dangerous nuisance individuals shows promise. Such a program has been very successful in Far Eastern Russia with Siberian tigers. Problem individuals are tranquilized, removed promptly, and either released elsewhere or placed in an appropriate facility. This approach is also very favorable from a public relations standpoint.

CONSERVATION AND THE FUTURE

Conservation of cougars in the future will require more refined management throughout much of their range. Research should be designed to address both theoretical and practical issues to form the basis for any management option. Logan and Sweanor (2000) have outlined conservation strategies clearly and concisely.

"Focal topics for cougar research include population dynamics, genetics, interactions with prey, habitat use, interactions with humans, population-monitoring techniques, and methods of live-stock husbandry. Studies of cougar population dynamics could help managers develop population objectives, assess impacts of harvest and control, and provide an understanding of how cougar meta-populations function. Genetic studies could illuminate cougar taxonomy, population structure, and potential human impacts (i.e., population isolation and reduction) on populations and subspecies. Long-term (greater than 10 years) experimental research will be needed to determine to what extent cougar predation limits or regulates prey populations (Sinclair 1989). To date, studies have not been long enough to include all of the phases of fluctuation of a population of cougars and its ungulate prey. Knowing how cougars use habitat should assist land managers to identify potential degradation and fragmentations, locations of dispersal and migration corridors, and areas of cougar–human conflict. Behavioral studies of cougar and people may shed light on how population or behavioral modification may minimize dangerous encounters. Because of the difficulty in censusing cougars, reliable, cost-effective methods are needed by managers to track population trends. Livestock growers in cougar habitat may benefit from alternate livestock types (e.g., cattle instead of sheep) or husbandry practices (e.g., guard dogs and herders) that are practical and cost effective. Particularly for population and cougar–prey interaction studies, *a priori* multifactorial hypotheses should be tested and empirical data should be collected for inclusion in models. Habitat models could be developed using Geographic Information System technology. In all research, key patterns should be identified to develop theory and to guide future research.

The key to the success of any cougar management approach involves education, and this should include both the interested public and wildlife managers. Information on how people should behave while living or recreating in cougar habitat has been effective in reducing the probability of dangerous encounters (Torres 1997). The public, informed about cougar biology, is likely to take a more active or understanding role in how cougars are managed. Likewise, educated agency personnel may function in a much more effective manner.

All of these strategies could be incorporated into an adaptive management plan similar to the one recently developed for New Mexico (Hornocker Wildlife Institute/New Mexico Department of Game and Fish 1997). The plan provides for the needs of people, while also considering the biological needs for cougar conservation. This is accomplished using a zone management approach that includes experimental cougar control in localized problem areas to protect private property, humans, and endangered species; sport hunting sustained by harvest quotas and protection of females and cubs; and a long-term conservation strategy that uses large protected areas or refuges. Monitoring the effects of management on representative zones is fundamental to this plan to determine if objectives are being met.

Zone management programs as outlined above should not only result in meeting the needs of humans and attaining true cougar conservation, but also result in maintaining the natural biodiversity of the ecosystems involved. Thus, cougar conservation can assist in achieving these other long-term conservation goals.

Eisenberg (1989) supports the key or umbrella species concept:

Full recognition must be made that top carnivores play an important role in structuring communities. The removal of a top carnivore from an ecosystem can have an impact on the relative abundance of herbivore species within a guild. In the absence of predation, usually one or two species come to dominate the community. The consequence of this is often a direct alteration of the herbaceous vegetation fed on by the herbivore guild or assemblage. Top carnivores have an important role to play in the structuring of communities and, ultimately, of ecosystems. Thus, the preservation of carnivores becomes an important consideration in the discipline of conservation biology.

A similar strong case is made by others (Ross et al. 1996; Terborgh et al. 1999, 2001; Mills 2005; Ray (2005) and Ross et al. (1996). Miquelle et al. (2005) state that although large carnivore conservation may not be synonymous with biodiversity conservation, the charisma, large area requirements (related to prey requirements), and plasticity in habitat requirements provide a mechanism for achieving other conservation objectives.

The roles played by large carnivores in structuring and maintaining biodiversity may vary. Different species of carnivores may exert different effects and influences on prey populations. Wolves have been shown to depress prey populations, sometimes severely (Murie 1944; Mech 1981; Bergerud et al. 1983; Gasaway et al. 1992; Ballard et al. 1997; Kunkel 1997; Berger and Smith 2005). The effect of cougar predation varies with ecological conditions. In Idaho, California, Utah, and in the Yellowstone ecosystem, cougars did not limit elk and deer populations (Hornocker 1970; Hopkins 1989; Lindzey et al. 1994; Murphy 1998). Cougar predation can, however, have a much greater effect on relatively small or isolated prey populations (Ockenfels 1994; Wehausen 1996; Ross et al. 1997). Difference in effects and influences may be attributed to relative densities of wolves and cougars, methods of hunting (coursing versus stalking), and the kind of individual prey animal taken (selectivity versus random).

Ruth (2001, 2005) has shown that wolves dominate cougars in Glacier and Yellowstone national parks. Wolves kill cougars, steal their kills, and have forced cougars to change their ranges. While cougars, before wolf reestablishment in Yellowstone, did not influence elk behavior to any extent or depress elk numbers (Murphy 1998), wolves have brought about change. Elk behavior and habitat use has changed, obviously in response to wolf predation, and evidence suggests that elk numbers have declined (Smith et al. 2003; Smith and Ferguson 2005).

Mech (1981) stated that few carnivores can compete successfully with the wolf. Miquelle et al. (2005), however, point out that the Siberian tiger (*Panthera tigris altaica*) is an exception. Evidence from the Russian Far East suggests that tigers depress wolf numbers either to the point of localized extinction or to such low numbers as to make them functionally insignificant in the ecosystem. Top-down influences of tigers and wolves differed significantly. Because of different hunting methods (cursorial versus ambush) and different prey selectivity, Miquelle et al. (2005) concluded that "limitation of prey populations by wolves is likely to be considerably higher than by tigers." They further summarize that "higher kill rates (a consequence of sociality) and the associated greater travel distances as well as greater energy expended in cursorial hunting (greater chase distances) and social interactions, all result in greater energy demands for wolves" (Miquelle et al. 2005).

Data from the Russian Far East strongly support the contention that wolves can limit prey to a much greater degree than can tigers (Miquelle et al. 2005). In Russia, where there is a strong hunting tradition based on maximum sustainable yield, large carnivores are viewed as competitors. Empirical evidence shows that tiger predation is unlikely to limit ungulate populations to the same extent as wolf predation (Miquelle et al. 1996). And since tigers exclude wolves, a case can be made for management in favor of tigers. Paraphrasing one argument for tiger conservation, local biologists have proposed to local hunters that "while tigers may not be desirable, they prevent wolves from becoming abundant ... and we all know that wolves are worse than tigers in depressing prey numbers, so it is to your advantage to tolerate the tiger" (Miquelle et al. 2005). This "backdoor" rationale for tiger conservation does have biological basis, but education and communication must precede the management plan if it is to be successfully implemented.

Wolves have become an important component of the ecosystem in the northern Rockies and their presence and influence must be considered in cougar management. Theory developed from research on both wolves and cougars is important in structuring management programs where both species occur. This is important regardless of the management objective — wolf or cougar conservation, prey population increase or decrease, or overall biodiversity conservation.

CONCLUSION

I cannot improve on Logan and Sweanor's (2000) summation concerning cougar conservation:

"the cougar's broad geographic distribution, solitary nature, and presence in some of the most rugged and remote habitats helped it escape the regional extinctions that befell the other large carnivores. The recovery of the cougar in the West has occurred only during the last three decades, the equivalent of three cougar lifetimes. Not only is the recovery an indication of our management successes, but also of the resiliency of the cougar. Today, cougar populations are "high" relative to our collective memories. What they were historically is conjecture. But cougars today also are facing a completely different world than they would have encountered even 100 years ago. Because of our growing human population, cougar habitats and landscape linkages are continually shrinking. Consequently, the recent increases in cougar populations may not be sustainable. The ecological role of cougars, including their ability to help dampen oscillations in prey populations, structure biological communities, and direct the evolution of their prey are all reasons why cougars should be conserved (Hornocker 1970; Logan et al. 1996). Moreover, they can be used as umbrella species to define minimum areas required to preserve ecologically intact ecosystems (Clark et al. 1996; Noss et al. 1996). In the long run, if humans are to successfully conserve cougars in self-sustaining populations, then people living in or impacting their wild environments will have to be educated and caring. Furthermore, wildlife managers will require a thorough understanding of the animal and potential methods for achieving success in dealing with short-term problems and long-term conservation goals."

While cougar management strategy cannot be applied across the board to other big cats worldwide, it can serve as a guide. All cats face a myriad of problems, which can be resolved only by developing theoretical and empirical evidence and applying that evidence through management programs. These programs must address, in addition to the biological and ecological concerns, the human elements if they are to be successful. All the economic, political, and cultural realities involved must be considered, and credibility must be established for the management strategy locally and regionally. Thus, an active information and education program is an essential part of the overall conservation program, whatever its objectives.

ACKNOWLEDGMENTS

I thank my former students and colleagues K. Logan, D. Miquelle, K. Murphy, H. Quigley, T. Ruth, and L. Sweanor for input, direct and indirect, in the development of this chapter. I have also drawn from interactions and discussions with fellow members of the Cougar Management Working Group. Special thanks to M. Tewes, another former student, for his encouragement and advice.

REFERENCES

Anderson, C. R., and F. G. Lindzey. 2005. Experimental evaluation of population trend and harvest composition in a Wyoming cougar population. *Wildl. Soc. Bull.* 33:179.

Aune, K. E. 1991. Increasing mountain lion populations and human–mountain lion interactions in Montana. In *Mountain Lion–Human Interaction: Symposium and Workshop*, C. E. Braun (ed.). Denver: Colorado Division of Wildlife, p. 86.

Ballard, W. B., et al. 1997. Ecology of wolves in relation to a migratory caribou herd in northwest Alaska. *Wildl. Monogr.* 135.

Barnhurst, D. 1986. Vulnerability of cougars to hunting. MS Thesis, Utah State University, Logan.

Barone, M. A., et al. 1994. Reproductive characteristics of male Florida panthers: Comparative studies from Florida, Texas, Colorado, Latin America, and North American zoos. *J. Mammal.* 75:150.

Beier, P. 1991. Cougar attacks on humans in the United States and Canada. *Wildl. Soc. Bull.* 19:403.

Beier, P. 1993. Determining minimum habitat areas for cougars. *Conserv. Biol.* 6:94.

Berger, J., and D. W. Smith. 2005. Restoring functionality in Yellowstone with recovering carnivores: Gains and uncertain. In *Large Carnivores and the Conservation of Biodiversity*, J. Ray, et al. (eds). Washington, DC: Island Press.

Bergerud, A. T., W. Wyett, and J. B. Snider. 1983. The role of wolf predation in limiting a moose population. *J. Wildl. Manage.* 47:977.

Braun, C. E. (ed.). 1991. *Mountain Lion–Human Interaction: Symposium, and Workshop.* Denver: Colorado Division of Wildlife Service.

Clark, T. W., P. C. Paquet, and A. P. Curlee. 1996. Introduction: Large carnivore conservation in the Rocky Mountains of the United States and Canada. *Conserv. Biol.* 10:940.

Cougar Management Guidelines Working Group. 2005. *Cougar Management Guidelines*, 1st edn. Bainbridge Island, Washington, DC: Wild Futures.

Eisenberg, J. F. 1989. *An Introduction to the Carnivora, Carnivore Behavior, Ecology, and Evolution*, J. L. Gittleman (ed.). Ithaca, NY: Cornell University Press.

Gasaway, W. C., et al. 1992. The role of predation in limiting moose at low densities in Alaska and Yukon and implications for conservation. *Wildl. Monogr.* 120.

Green, K. S. 1991. Summary: Mountain lion–human interaction questionnaires. In *Mountain Lion–Human Interaction: Symposium and Workshop*, C. E. Braun (ed.). Denver: Colorado Division of Wildlife.

Harlow, H. J., et al. 1992. Stress response of cougars to nonlethal pursuit by hunters. *Can. J. Zool.* 70:136.

Hedrick, P. W. 1995. Gene flow and genetic restoration: The Florida panther as a case study. *Conserv. Biol.* 9:996.

Hopkins, R. A. 1989. Ecology of the Puma in the Diablo Range, California. MS Thesis, University of California, Berkeley.

Hornocker, M. G. 1969. Winter territoriality in mountain lions. *J. Wildl. Manage.* 33:457.

Hornocker, M. G. 1970. An analysis of mountain lion predation upon mule deer and elk in the Idaho Primitive Area. *Wildl. Monogr.* 21.

Hornocker Wildlife Institute Report to New Mexico Department of Game and Fish. 1997. Long range plan for the management of cougar in New Mexico, 1997–2004. Federal Aid in Wildl. Rest., W-93-R-38, Proj. 1, Job 5.5, New Mexico Department of Game and Fish, Santa Fe.

IUCN Species Survival Commission. 1987. Cat Specialist Group, P. Jackson, Chairman, Manifesto on Cat Conservation. *Cat News* 6.

Kunkel, K. E. 1997. Predation by wolves and other large carnivores in northwestern Montana and southeastern British Columbia. PhD Thesis, University of Montana, Missoula.

Lindzey, F. G., et al. 1994. Cougar population dynamics in southern Utah. *J. Wildl. Manage.* 58:619.

Logan, K. A., and L. L. Sweanor. 2000. Puma. In *Ecology and Management of Large Mammals in North America*, S. Demarias, and P. Krausman (eds). Upper Saddle River, NJ: Prentice Hall.

Logan, K. A., and L. L. Sweanor. 2001. *Desert Puma: Evolutionary Ecology and Conservation of an Enduring Carnivore.* Washington, DC: Island Press.

Logan, K. A., et al. 1996. Cougars of the San Andres Mountains, New Mexico. Final report, Federal Aid in Wildl. Rest. Proj. W-128-R, New Mexico Department of Game and Fish, Santa Fe.

Maehr, D. S., and J. A. Cox. 1995. Landscape features and panthers in Florida. *Conserv. Biol.* 9:1008.

Mech, L. D. 1981. *The Wolf: The Ecology and Behavior of an Endangered Species.* Minneapolis: University of Minnesota Press.

Mehr, D., and G. B. Caddick. 1995. Demographics and genetic introgression in the Florida panther. *Conserv. Biol.* 9:1295.

Mills, M. G. L. 2005. Large carnivores and biodiversity in African savanna ecosystems. In *Large Carnivores and the Conservation of Biodiversity*, J. Ray, et al. (eds). Washington, DC: Island Press.

Miquelle, D. G., et al. 1996. Food habits of Amur tigers in Sikhote-Alin Zapovednik and the Russian Far East, and implications for conservation. *J. Wildl. Res.* 1:138.

Miquelle, D. G., et al. 2005. Tigers and wolves in the Russian Far East: Competitive exclusion, functional redundancy, and conservation implications. In *Large Carnivores and the Conservation of Biodiversity*, J. Ray, et al. (eds). Washington, DC: Island Press.

Murie, A. 1944. *The Wolves of Mt. McKinley*. Fauna Series No. 5. Fauna of the National Parks of the U.S.

Murphy, K. M. 1998. The ecology of the cougar (*Puma concolor*) in the northern Yellowstone ecosystem: Interactions with prey, bears, and humans. PhD Thesis, University of Idaho, Moscow.

Murphy, K. M., P. I. Ross, and M. G. Hornocker. 1999. The ecology of anthropogenic influences on cougars. In *Carnivores in Ecosystems: The Yellowstone Experience*, T. W. Clark, et al. (eds). New Haven, CT: Yale University Press.

Nowak, R. M. 1976. *The Cougar in the United States and Canada*. U.S. Department of the Interior, Fish and Wildlife Service, Washington, DC and New York Zoological Society, New York.

O'Brien, S. J., et al. 1990. Genetic introgression within the Florida panther, *Felis concolor coryi. Natl Geog. Res.* 6:485.

Ockenfels, R. A. 1994. *Factors Affecting Adult Pronghorn Mortality Rates in Central Arizona. Arizona Game and Fish Dept., Wildl. Digest* 16.

Padley, W. D. (ed.). 1997. *Proceedings of the Fifth Mountain Lion Workshop*, San Diego, CA.

Papouchis, C. M. 2004. Conserving mountain lions in a changing landscape. In *People and Predators*, N. Fascione, A. Delach, and M. E. Smith (ed.). Defenders of Wildlife, Washington, DC: Island Press.

Pierce, B. M., V. C. Bleich, and R. T. Bowyer. 2000. Social organization of mountain lions: Does a land-tenure system regulate population size? *Ecology* 81:1533.

Ray, J., (ed). 2005. *Large Carnivores and the Conservation of Biodiversity*. Washington, DC: Island Press.

Roberson, J., and F. Lindzey, (eds). 1984. *Proceedings of the Second Mountain Lion Workshop*. Logan: Utah Div. Wildl. Res. and Utah Coop. Wildl. Res. Unit.

Robinette, W. L., J. S. Gashwiler, and O. W. Morris. 1959. Food habits of the cougar in Utah and Nevada. *J. Wildl. Manage.* 23:261.

Ross, P., M. G. Jalkotsky, and M. Festa-Bianchet. 1997. Cougar predation on bighorn sheep in southwestern Alberta during winter. *Can. J. Zool.* 774:775.

Ross, P. I., M. G. Jalkotzy, and J. R. Gunson. 1996. The quota system of cougar harvest management in Alberta. *Wildl. Soc. Bull.* 24:490.

Ruth, T. K. 1991. Cougar use in an area of high recreation development in Big Bend National Park, Texas. MS Thesis, Texas A&M University, Collage Station, TX.

Ruth, T. K. 2001. *Cougar–Wolf Interactions in Yellowstone National Park: Competition, Demographics, and Spatial Relationships*. Ann. Tech. Rep., Wildl. Cons. Soc./Hornocker Wildl. Inst.

Ruth, T. K., et al. 1998. Evaluating cougar translocation in New Mexico. *J. Wildl. Manage.* 62:1264.

Seidensticker, J. C.,IV, M. G. Hornocker, M. V. Wiles, and J. P. Messick. 1973. Mountain lion social organization in the Idaho Primitive Area. *Wildl. Monogr.* 35.

Sinclair, A. R. E. 1989. Population regulation in animals. In *Ecological Concepts*, J. M. Cherett (ed.). Oxford: Blackwell Scientific Publications.

Smith, D. W., and G. Ferguson. 2005. *Decade of the Wolf: Returning the Wild to Yellowstone*. CT: Guilford, Globe-Pequot Press.

Smith, D. W., R. O. Peterson, and D. B. Houston. 2003. Yellowstone after wolves. *BioScience* 53:330.

Sweanor, L. L. 1990. Mountain lion social organization in a desert environment. MS Thesis, University of Idaho, Moscow.

Terborgh, J. J., et al. 1999. The role of top carnivores in regulating terrestrial ecosystems. In *Continental Conservation*, M. E. Soule, and J. Terborgh (eds). Washington, DC: Island Press.

Terborgh, J., et al. 2001. Ecological meltdown in predator-free forest fragments. *Science* 294:1923.

Torres, S. 1997. *Mountain Lion Alert*. Helena, Montana: Falcon Publishing Company.

Torres, S. G., H. Keough, and D. Dawn. 2004. *Puma Management in Western North America: A 100 Year Retrospective*. Seventh Mountain Lion Workshop, Jackson Hole, Wyoming.

Torres, S. G., et al. 1996. Mountain lion and human activity in California: Testing speculations. *Wildl. Soc. Bull.* 24:451.

Tully, R. J. 1991. Results, 1991 questionnaire on damage to livestock by mountain lion. In *Mountain Lion-Human Interaction: Symposium and Workshop*, C. E. Braun (ed.). Denver: Colorado Div. Wildl.

Turner, J. W., M. L. Wolfe, and J. F. Kirkpatrick. 1992. Seasonal mountain lion predation on a feral horse population. *Can. J. Zool.* 70:929.

Wehausen, J. D. 1996. Effects of mountain lion predation on bighorn sheep in the Sierra Nevada and Granite Mountains of California. *Wildl. Soc. Bull.* 24:471.

Young, S. P., and E. A. Goldman. 1946. *The Puma, Mysterious American Cat*. Washington, DC: The American, Wildlife Institute.

8 Effects of Drought on Bobcats and Ocelots

Michael E. Tewes and Maurice G. Hornocker

CONTENTS

A majority of the 36 species of wild cats are small felids that inhabit a wide variety of habitats and biomes (Sunquist and Sunquist 2002). With the exception of the bobcat (*Lynx rufus*) in North America (Lariviere and Walton 1997) and lynx (*Lynx* spp.) in the northern hemisphere (Anderson and Lovallo 2003), most small cats have received little ecological research. Typically, previous research on *Lynx* spp. has been short duration and constrained by many factors. These factors include nonrepresentative study animals or populations monitored over a confined geographic area under a limited array of environmental states. Conducting long-term field studies with large sample sizes of free-ranging wild cats that encompass the various temporal, spatial, and environmental elements typical of complex ecological systems is difficult and expensive. Experiments with replications are even more difficult to design and conduct with most secretive, nocturnal felids that occur at low densities in dense cover. Consequently, we often are forced to make critical management decisions based on uncertain or incomplete information.

Optimal foraging theory can help researchers understand foraging patterns of wild cats, particularly patterns associated with varying prey availability. Natural selection favors individuals with the highest fitness, thus maximizing efficiency in prey acquisition is important in contributing to reproductive success. By hunting efficiently, a wild cat has more time for other important activities related to fitness, including identifying and defending optimal habitat, mating, and successfully raising young. If a predator can improve its reproduction or survival by hunting more efficiently, then natural selection will favor an efficient predator (Krebs 1979).

A predator incurs a "cost" in terms of the energy and time required to search and acquire prey, and a "benefit" from the energy obtained. The net energy expended during prey capture divided by handling time is a measure of the profitability of a prey type. Predators can be viewed as time-minimizers by reducing the time needed to acquire a certain amount of energy (i.e., quickly finding abundant prey), or as energy-maximizers if they maximize the amount of energy acquired in a certain amount of time (i.e., consuming large prey) (Krebs 1979). Optimal foraging theory predicts that wild cats should choose profitable prey.

In this chapter, we use bobcats as a conceptual model for understanding wildcat response to prey declines, particularly as related to droughts in a semiarid temperate environment. Because of the broader information base available for bobcat and lynx, we are able to identify many of these cat–environment relationships. We apply concepts of optimal foraging theory to interpret bobcat use of a few profitable prey species during normal or wet conditions, and prey switching to less profitable species during drought conditions. This knowledge of bobcat–prey relationships in drought is used to conjecture ocelot (*Leopardus pardalis*) response to prey in a drought. Using this model and integrating our limited information of ocelot ecology, we describe the effects of drought on the endangered ocelot in the United States. Finally, predictions of prey and ocelot response to drought form the basis for management recommendations to benefit ocelot conservation.

The classic 8 to 11 year Canada lynx (*Lynx canadensis*) cycles related to patterns of snowshoe hare (*Lepus americanus*) abundance are well established (Elton and Nicholson 1942; Keith 1963, 1983; Brand et al. 1976; Brand and Keith 1979). During years with abundant hare, lynx populations often function according to optimal foraging theory. Lynx exhibit a numerical response by changes in their rates of recruitment, movement, and survival (Keith 1963; Koehler and Aubrey 1994), and exhibit a functional response by consuming more hares during hare abundance (Keith et al. 1977; O'Donoghue et al. 1998; Mowat 2000). Following hare declines, lynx populations decline through reduced recruitment and survival (Keith 1963, 1983; Mowat 2000). Lynx abundance generally reflects hare abundance with a short time lag, generally about a year.

Lagomorphs are important food items for bobcats (Bailey 1974, 1979; Berg 1979; Dibello et al. 1990) with rodents also playing a major role in bobcat diets through much of the bobcat range (McCord and Cardoza 1982; Rolley 1987; Thornton et al. 2004). In the southeastern United States, hispid cotton rat (*Sigmodon hispidus*) and eastern cottontail (*Sylvilagus floridanus*) are dominant components in bobcat diets (Fritts and Sealander 1978; Miller and Speake 1978; Story et al. 1982; Wassmer et al. 1988), with similar patterns in Texas (Beasom and Moore 1977; Blankenship 2000).

Hispid cotton rats (hereafter, cotton rats) and eastern cottontails (hereafter, rabbits) represent a large prey size, relative to the other potential prey of ground-dwelling birds and smaller rodents, with rabbits weighing 1 to 2 kg and cotton rats about 150 g. In addition, both prey can irrupt with high densities during normal or wet periods (Chapman and Litvaitis 2003; Bradley et al. 2006). These attributes of large prey size and periodic high abundance make these food items a profitable prey with high-energy content and reduced search and handling times for the bobcat.

Small mammals are the dominant ocelot prey in Central and South America (Bisbal 1986; Mondolfi 1986; Emmons 1987; Tewes and Schmidly 1987; Murray and Gardner 1997). In Venezuela, rodents occur >80% of the scats with *Sigmodon alstoni* and eastern cottontail included as prey (Ludlow and Sunquist 1987). Spiny rats (*Proechimys* sp.) are a common part of ocelot diet in Peru and Venezuela. Opossums (*Didelphis marsupialis* and *Philander opossum*) are the most common prey in Belize (Konecny 1989).

Considerable dietary similarity exists between sympatric ocelot and bobcat in Texas based on examination of scats with cotton rats and rabbits important to both cat species (M. Tewes and J. Young, unpublished data, 2006). Noteworthy to this discussion later, opossum (*Didelphis virginianus*) and feral hog (*Sus scrofa*) were found in 13 and 21% of the ocelot scats, respectively.

PREY DECLINE WITH DROUGHT

RODENTS

Understanding the relationship of wild cats with their prey is essential to predicting potential impacts of drought. Declines of small mammals during drought and recovery with precipitation have been observed many times in North America (Mutze et al. 1991; Madsen and Shine 1999; Ernest et al. 2000; Brown and Ernest 2002; Morrison et al. 2002). Drought effects are related to reproduction failure, embryo reabsorption, and smaller litter size in prey species (Bradley et al. 2006). Prey abundance is an important proximate factor affecting cat response to drought, whereas the ultimate factor seems to be reduced precipitation effects on primary production and related elements (e.g., cover) important to prey abundance.

Biologists studying periodic irruptions of rodent populations in semiarid regions of western South America have suspected the causal role of increased precipitation (Fulk 1975; Pearson 1975; Péfaur et al. 1979; Jiménez et al. 1992; Jaksic et al. 1996, 1997). Unusually high rainfall and a related increase in primary production in semiarid regions are correlated with rodent irruptions (Jiménez et al. 1992; Meserve et al. 1995; Jaksic et al. 1997). The El Niño Southern Oscillation (ENSO) disturbances cause variability in rainfall, which influences primary productivity (herbage and seeds) and related rodent abundance through food availability (Gutiérrez et al. 1993; Jaksic et al. 1997).

For example, the leaf-eared mouse (*Phyllotis darwini*) occupies a semiarid region of western South America that is subjected to ENSO-related precipitation. This 50-g cricetid undergoes periodic irruptions apparently triggered by unusually high precipitation and increased primary production (Lima and Jaksic 1998, 1999).

Cotton rats and rabbits are susceptible to drought effects. The boom–crash pattern and irruptive capability of cotton rat populations has been known for a long time (Strecker 1929; Stoddard 1931; Odum 1955), and is often related to dry periods that can cause nutritional stress and reduced reproduction (Layne 1974; Cameron and Spencer 1981).

A population irruption of cotton rats followed one of the most severe droughts during the twentieth century (1950–56) in Texas (Norwine and Bingham 1985; Davis and Schmidly 1994). Davis and Schmidly (1994, 189) stated:

> The size of the population is correlated with the amount of suitable habitat, and suitable habitat in turn is correlated with the amount of rainfall … . During the 7-year drought that began about 1950, cotton rat populations in central Texas were low because there were few places where they could live in numbers. Ground cover was sparse or even absent… . When the rains came in 1957 … ground cover increased, providing better cover and more nutritious green food, and the cotton rat population took off. By late May 1958, they were found in unbelievable numbers in especially favorable areas. Estimates were as high as several hundred rats per hectare (Davis and Schmidly 1994, 189).

Cotton rats declined at the onset of below-normal precipitation during a rodent study in southern Texas (Bradley et al. 2006). However, the drought-ending rainfall was followed by a rapid and dramatic irruption of cotton rats in only a few months. The rapidity of rodent response likely depends on amount of rainfall, soil texture, time of year, and other factors (Bradley et al. 2006).

RABBITS

Cottontails use a wide diversity of disturbed, early successional or shrubland habitats that provide abundant forage and dense understory cover (Beckwith 1954; Anderson and Pelton 1976; Chapman and Litvaitis 2003). Although eastern cottontails breed throughout the year in southern Texas, rainfall and temperature are the primary environmental factors that impact breeding activity in this region (Bothma and Teer 1977). Rainfall produces succulent vegetation and a beneficial nutritional environment for this lagomorph (Bothma and Teer 1977). Similarly, Ecke (1955) believed that availability of succulent vegetation was important to rabbit breeding. In addition to drought periods, digestible energy may be a limiting nutritional component for rabbits in winter and early spring (Rose 1973). The decline in essential amino acids often associated with mature or drought-stressed herbaceous communities may be another reason rabbits prefer recently disturbed habitats with new herbaceous growth (Lockmiller et al. 1995).

The eastern cottontail is the most fecund *Sylvilagus* spp. (Conaway et al. 1963, 1974; Trethewey and Verts 1971). Rainfall may increase productivity and survival of the first litter in early spring, which can impact the number of autumn litters contributed by these early-born young.

In summary, two important prey species (i.e., cotton rats and rabbits) used by bobcats and ocelots in Texas have documented declines during drought, and dramatic increases with subsequent rainfall. The two proximate factors that affect prey survival are nutritional deficiencies associated with poor-quality forage and loss of concealing or escape cover at the herbaceous stratum.

Prey declines during drought may result from diminished energy, protein and, in particular, certain amino acids absent from the dry herbaceous community that provides forage for cotton rats and rabbits. Many thorn shrub species defoliate during drought and cease production of berries and seeds during dry conditions. Nutritional deficiencies can cause stress and reduced reproduction in rodents and rabbits with eventual decline in population sizes.

Another factor associated with drought is loss of screening or escape cover for small mammals. The primary mortality factor for rabbits is predation, and it is the primary direct cause of cottontail population regulation (Chapman and Litvaitis 2003). Reduced herbaceous cover during dry periods can expose cotton rats and rabbits to the diverse predator community inhabiting southern Texas, including coyote (*Canis latrans*), raccoon (*Procyon lotor*), opossum, long-tailed weasel (*Mustela frenata*), red-tailed hawk (*Buteo jamaicensis*), Harris' hawk (*Parabuteo unicinctus*), great horned owl (*Bubo virginianus*), barn owl (*Tyto alba*), and several snake species (Chapman et al. 1982).

Cats Decline with Drought

Drought can reduce prey abundance, which in turn can affect bobcat prey switching, home range and habitat use, health, reproduction, and survival. The following is a discussion of the effects of low prey abundance on these five life history categories.

Optimal foraging theory predicts that prey switching will be observed when use of highly profitable prey decreases and the predator increases the number of less profitable prey in the diet. Prey switching has been observed on several occasions in lynx and bobcats (Brand and Keith 1979; Bailey 1981; Maehr and Brady 1986; Mowat et al. 2000).

Although the snowshoe hare represent the primary prey for lynx (Brand and Keith 1979; Mowat et al. 2000), red squirrel (*Tamiasciurus hudsonicus*) and ruffed grouse (*Bonasa umbellus*) serve as alternate prey for lynx in some areas following declines in snowshoe hares (Brand and Keith 1979; O'Donoghue et al. 1998; Mowat et al. 2000). During low snowshoe hare densities in one study, all tagged resident lynx either dispersed or died by the end of the first winter (Poole 1994).

Beasom and Moore (1977) examined 51 bobcat stomachs during a dry year (1971) in southern Texas, and compared it with 74 stomachs during a wet year (1972). During the dry year, bobcats consumed 21 different prey species. Although cotton rats and rabbits were not commonly observed in the wild in 1971, they still represented the dominant prey in the bobcat diet with 38 and 18% volume of stomachs, respectively. In 1972, following rainfall, at least 60 cotton rats were observed over a kilometer on multiple occasions. Because of the abundance of cotton rats and rabbits following rainfall, bobcat diet contained these two species almost exclusively. In 1972, 96% of the total volume of bobcat stomachs was cotton rats and rabbits. Stomachs containing solely these two species comprised 93% of the samples during the wet year (1972), but only 33% during the dry year (1971). This information suggests that when these two prey species are relatively scarce, bobcats rely on a more varied diet. The bobcat population apparently responded to change in prey availability by concentrating their food-gathering efforts on cotton rats and rabbits when these prey were abundant (Beasom and Moore 1977).

In a 7-year study of bobcat–prey relationships in Texas, Blankenship (2000) found bobcats were consistent with optimal foraging theory by switching to less preferred mammals and birds following a decline in their primary prey of cotton rats, rabbits, and southern plains wood rats (*Neotoma micropus*). Normally, birds represent a prey with low profitability for a bobcat and usually occur infrequently in bobcat diets throughout North America (Tewes et al. 2002), but were found in over 90% of bobcat scats following the decline of the primary prey in Texas (Blankenship 2000).

Wild cats occupying southern temperate ecosystems usually encounter much greater prey diversity and density than found in northern temperate or boreal environments. This diversity of prey may facilitate prey switching to ameliorate declines of one or two primary prey, thus buffering cat declines.

Dense understories occupied by snowshoe hare in Maine were preferred habitats for bobcats (Litvaitis et al. 1986). Similarly, there was a relationship between vegetation density, prey densities, and bobcat use in Montana (Knowles 1985).

In southern Texas, bobcats increased their home range size about 100% during low prey abundance (Blankenship 2000). In addition, bobcats increased use of grasslands and areas around water because alternate prey species (e.g., wading birds and waterfowl) were located in these areas (Blankenship 2000). Pocket gophers (*Geomys* sp.) represented an alternate bobcat prey that used the grasslands and were a low-profit prey because of the waiting time required for capture.

Following a prey decline in Idaho, some bobcats expanded their home range and made long extraterritorial forays, presumably in search of prey (Knick and Bailey 1986; Knick 1990). Bobcats increased dispersal during low prey abundance in southern Texas (Blankenship 2000).

Physiological changes recorded in bobcats following a decline in lagomorphs included higher hemoglobin levels, erythrocyte counts, and packed cell volume reflecting hemoconcentration and lower insulin and phosphorus levels. Decreased levels of serum insulin were likely associated with prey scarcity, and fewer and more widely spaced feeding events (Knick et al. 1993). Reduced levels of phosphorous suggested decreased food consumption even though bobcats were using alternative food sources. Three starving bobcats (which died 3–21 days following sampling) exhibited reduced cholesterol levels, elevated triglyceride levels, and elevated serum urea nitrogen, suggesting that these individuals were catabolizing body proteins and fat reserves (Knick et al. 1993). In addition, lower progesterone levels were observed in females during the period of prey scarcity, and it supported field observations that reproduction was reduced (Knick et al. 1993).

Reproduction can be reduced in small cats during low prey levels. Pregnancy rates for lynx ranged from 73 to 92% for adults and from 33 to −77% for yearlings during periods of abundant hares (*Lepus* spp.); whereas, pregnancy rates of only 33–64% in adults and 0–10% in yearlings were observed during low-hare periods (Koehler and Aubrey 1994). Also, some lynx may only have litters in alternate years during periods of low hare abundance (Saunders 1961).

Bobcat pregnancy rates and average litter size seem to be affected by low prey availability (Anderson and Lovallo 2003). Pregnancy rates of bobcats in Idaho dropped from 100 to 12%, and the bobcat population declined following a decline in their primary prey of black-tailed jackrabbits (*Lepus californicus*), an apparent response to the prey shortage (Knick 1990).

Drought-induced declines in prey availability reduced pregnancy rates of bobcats in Oklahoma (Rolley 1985). Blankenship (2000) found that the number of litters and kittens decreased as prey declined, and no female bobcats were observed raising kittens on the study site during 3 years of low prey abundance in southern Texas.

Reduced survival of lynx kittens during low prey abundance is reflected in the lynx harvest (Anderson and Lovallo 2003). The percentage of kittens and yearlings in the harvest decreased from 40% for both age groups during a period of high hare abundance to 0 and 8%, respectively, during low hare abundance (O'Conner 1984).

Prey abundance also strongly affects survival of bobcat kittens (Anderson and Lovallo 2003). During periods of scarce prey, adult female bobcats are believed to feed themselves before feeding their kittens (Bailey 1974). No bobcat kittens survived into the fall in Idaho when prey was scarce; whereas, kitten survival was high during previous years with high prey abundances (Bailey 1974; Knick 1990). Blankenship (2000) found 94% survival rate of bobcats during high prey abundance, declining to 44% during low prey abundance. Lower cat survival may vary with social class, with transient bobcats (Blankenship et al. 2006) and ocelots (Haines et al. 2005a) experiencing lower survival than resident conspecifics.

Bobcat starvation may be more likely during winter and early spring, because rodent and lagomorph populations are at their seasonal lowest. The three, tagged adult bobcats that starved to death during low prey had body mass at the final capture 64–69% of a previous capture mass (Knick 1990; Knick et al. 1993). In the Texas study, two bobcats starved during low prey periods, and another in poor condition was probably killed by coyotes (Blankenship 2000).

In addition to failure to acquire sufficient energy and protein during prey declines, another mechanism that may interact with this bobcat–prey relationship to cause declines in bobcat populations is insufficient Vitamin A. Domestic cat (*Felis catus*) and wild cats seem incapable of transforming β-carotene into retinol or Vitamin A (Scott 1968). Consequently, wild cats must be able to obtain sufficient Vitamin A from the kidneys, lungs, adrenal, and liver of their prey (Scott 1968). Insufficient Vitamin A can adversely affect egg implantation, and cause other physiological problems. Prey switching to a variety of smaller, less-profitable prey items may cause difficulty in bobcats obtaining minimum levels of Vitamin A. Acquisition of larger organs and sufficient Vitamin A is more easily achieved with large, abundant prey such as rabbits and cotton rats.

Although there is little doubt that cats and their prey are affected by moderate to severe droughts, the thresholds at which effects become significant are less understood. The onset of drought may be gradual and difficult to identify in the initial stages.

In addition to moisture, there are many interacting abiotic, biotic, and historical factors that impinge on the prey community. Subsequent response of wild cats to prey responses represents another layer of ecological complexity that must be considered by biologists prescribing management actions. Nonetheless, using bobcat and prey literature to understand potential ocelot responses to drought has merit.

Ocelot and Drought

How much of the bobcat–prey–drought model is applicable to ocelots? We will discuss the current status of the ocelot, and the ecological similarities and differences between bobcat and ocelot. Finally, management strategies to mitigate some drought effects on ocelot are described.

Ocelots are listed as an endangered species with the only documented resident population in the United States occurring in southern Texas (Tewes and Everett 1986). Less than 100 individuals occur in two separate populations (Tewes and Everett 1986). Because of the rarity and the endangered status of these cats, researchers have difficulty in conducting long-term studies with large sample sizes. The types of studies needed to detect the effects of drought and other stochastic or episodic events are difficult to implement.

Ocelot persistence in the United States is threatened for a variety of reasons (Haines et al. 2005b, 2006), including limited and diminishing habitat (Tewes and Everett 1986), high road mortality (Tewes and Hughes 2001), and other anthropogenic factors (Haines et al. 2005a, 2006). Recovery biologists do not have the luxury of time and money to examine in detail the ocelot–prey–drought relationship. Thus, we suggest the application of the bobcat–prey–drought model, where appropriate, for ocelot conservation.

Several anecdotal observations of ocelots are consistent with the pattern of bobcat response to drought, including reduced survival and reproduction, increased transient or dispersal behavior, and greater starvation. However, low sample sizes during the few intense drought periods over the past 25 years preclude strong inferences about the effects of drought on ocelot conservation.

During a drought from January 2000 to December 2002 in southern Texas, unusual patterns were observed with the ocelot population on the Laguna Atascosa National Wildlife Refuge. Culminating in April 2002, biologists noted several ocelot deaths, including three from unknown causes, two roadkills, and one starvation (Laack, pers. comm., 2002; U.S. Fish and Wildlife Service). The annual monitoring efforts recorded only two new captured ocelots from the previous winter and four ocelots with existing collars, both unusually low numbers.

Consequently, we examined ocelot telemetry data using the Palmer Modified Drought Index (PMDI) to assess the impacts of drought on ocelots. Survival data for 80 ocelots (72 resident and 20 transient) were evaluated from 1982 to 2001. During this period, two moderate droughts were identified from January 1989 to May 1991 (PMDI $= -2.42$) and January 2000 to December 2002 (PMDI $= -2.48$) (Haines et al. 2005a).

There was a decrease in survival of resident ocelots during drought ($S = 0.87$ in normal years and $S = 0.77$ in drought years), and a larger decrease occurred in transient ocelots ($S = 0.57$ in normal years and $S = 0.13$ in drought years). Although this study represents the largest sample size for a small cat other than a *Lynx* spp., with 80 ocelots monitored over 20 years, it still had too few transients monitored over too few drought years to be statistically robust. This difficulty reflects the challenge of building a large information base to develop confident management recommendations on wild cats.

Bobcats and ocelots in southern Texas show many ecological similarities (Figure 8.1). The home range size and polygynous social system are similar, and both display primarily nocturnal activity patterns with crepuscular peaks (Tewes 1986; Laack 1991). Prey use by co-occurring bobcat and ocelot in southern Texas also reveals many similarities. Both diets include frequent use of cotton rats, rabbits, and birds (Tewes and Young, unpublished data, 2006).

Both felids can breed through much of the year in southern Texas. However, bobcats tend to have distinctive parturition peaks in March and April, whereas ocelot parturition is more evenly distributed through the year with several litters observed in the fall (Blankenship and Swank 1979; Laack et al. 2005).

The ocelot and bobcat exhibit a few ecological differences, which are important in this discussion about drought impacts. In southern Texas, ocelots use a narrow range of cover types, generally characterized by extremely dense thorn shrub cover with >95% horizontal cover of the shrub layer. Although some use of 75–95% shrub cover occurs, there is almost no use of more open areas (Tewes 1986; Horne 1998; Shindle and Tewes 1998; Harveson et al. 2004). The dense spotting pattern of ocelots enables the ocelot to operate well in the sun-dappled environment found below the dense canopy.

In contrast, bobcats use a greater variety of cover types in Texas with considerable use of intermediate cover categories (Tewes 1986; Horne 1998; Harveson et al. 2004). Bobcat pelages also exhibit wide variation in color and spots. Some pelages are brown or tan with little spotting, others exhibit moderate spotting, and other phenotypes have a dense pattern of blotches and circles similar to ocelots. This broad diversity of coat phenotypes likely enables the bobcat to exploit various cover types and may be one reason bobcats operate successfully as a habitat generalist.

Consequently, a bobcat population typically occupies different habitats with varying prey availabilities. This variation enables bobcats to use a flexible foraging strategy for prey in different habitat types during a drought. In contrast, ocelots have few habitat options during a drought and are limited to the prey species occurring in the dense thorn shrub.

Drought Strategies for Ocelot Management

One of the primary reasons ocelots are endangered in the United States is the cover type preferred by ocelots is rare, covering <1% of southern Texas (Tewes and Everett 1986). This relict ocelot population is threatened with extinction over the next few decades by a variety of risk factors (Haines et al. 2005a). Consequently, management of ocelot habitat that is currently occupied, and expansion of new ocelot habitat near core populations, will contribute toward ocelot recovery in the United States.

Using optimal foraging theory and the bobcat–prey–drought model, reasonable predictions of ocelot response to periodic drought can be developed. Because the ocelot population in the United States is small and isolated, effects of a severe, prolonged drought have major implications for ocelot viability.

FIGURE 8.1 The bobcat (a) and ocelot (b) are two wildcats that share several physical and ecological traits in their sympatric range in southern Texas. Much of the previous research on bobcats can provide guidance for ocelot management, particularly during droughts. (Photographs Copyright Larry Ditto.)

The locations of many habitat tracts occupied by ocelots are known to recovery biologists, thus providing an opportunity for targeted management. From the previous discussion, we outlined how the decline of prey, particularly during droughts, can have a major impact on wildcat populations. Consequently, management strategies that increase prey populations during these stressful periods have merit for ocelot conservation. These strategies can be grouped into the following categories: management of prey habitat, competitor removal, prey supplementation, and ocelot habitat restoration.

Management of Prey Habitat

Thorn shrub patches used by ocelots often have distinct boundaries adjacent to herbaceous communities or mixed herbaceous–woody communities. These adjacent communities should be

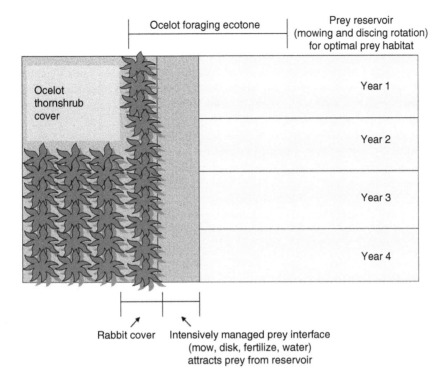

FIGURE 8.2   Active management of prey habitat adjacent to occupied ocelot thorn shrub may reduce the detrimental effects of drought on ocelots. Intensive management of the thorn shrub–herbaceous interface can create a foraging ecotone for ocelots that attracts prey from the surrounding prey reservoir.

intensively managed to increase prey populations, particularly in preparation for drought or stressful periods.

A managed prey population in optimal habitat should enter a drought-associated decline at higher levels and decline at a slower rate compared to a prey population occupying poor habitat with no management. Rabbits and cotton rats occupying mature herbaceous communities with little nutritious growth would succumb faster to the effects of drought on forbs and grasses. In contrast, a managed prey population should benefit ocelots during a drought, and delay or ameliorate drought impacts on ocelot nutrition, reproduction, and survival.

The interface of ocelot cover with prey habitat can be intensively managed using mowing, discing, fertilizer, and water supplementation to create a narrow zone of a nutrient-rich herbaceous environment for prey (Figure 8.2). Rodents and rabbits foraging in this interface use the adjacent thorn shrub for cover, and thereby become available to foraging ocelots. Ocelots have also been observed, particularly during the cover of darkness, to move into the adjacent herbaceous communities, presumably foraging on rabbits and rodents.

Dense ocelot thorn shrub is often associated with fertile soil series (Harveson et al. 2004), which is also beneficial for management of prey habitat. Fertile soils facilitate the nutritious herbaceous growth following manipulation. Consequently, management of the thorn shrub–herbaceous interface should provide profitable foraging zones for ocelots.

Another management tactic would be to maintain a "prey reservoir" around the habitat patches used by ocelots. Application of prescribed fire or a mowing-discing regime to these larger adjoining areas would help maintain a mosaic of early successional communities with succulent herbaceous regrowth. Although a more extensive area would be managed, management activities would occur on a less frequent schedule compared with the more intensively managed thorn shrub–herbaceous interface. A program of disturbing different large patches on a rotational schedule of 3–5 years would create a mosaic of different early seral stages, which could be occupied by prey from adjoining populations (Figure 8.2).

Managers should consider soil moisture, soil texture, and time of year to efficiently use these management applications to produce maximum benefits for the prey populations. For example, timing of prescribed burns, mowing, or discing should occur with good soil moisture, if possible, to maximize regrowth of desirable forbs.

These larger prey populations would be connected to the ocelot foraging ecotone, and sustain prey availability for ocelots further into a drought period (Figure 8.2). The intensively managed thorn shrub–herbaceous interface with succulent, nutritious vegetation would attract prey from the larger reservoir, particularly as conditions worsened during the drought.

If moderate to severe drought threatens survival of a small ocelot population, then extraordinary measures may be implemented for ocelot management until the drought abates. Options may include removal of potential competitors and possible supplementation of high-value prey in strategic locations.

Competitor Removal

Various predators that occur sympatric with ocelots consume rabbits and rodents in southern Texas. Two potential competitors that merit consideration for possible reduction include the bobcat and raccoon.

The bobcat is ecologically similar to ocelots and both felids share considerable dietary overlap. If prey is abundant, then predator coexistence may not be a problem. However, competition may occur when a resource is limited, and prey species become significantly reduced during drought.

Although raccoons are omnivorous and have less dietary overlap with ocelots, their abundance in ocelot habitat may be a cause of concern. For example, an increase of raccoons may affect competitive interactions within the tropic level of medium-sized carnivores and omnivores (e.g., opossums) (Kasparian et al. 2004). Following raccoon removal, opossums exhibited a niche shift, and a change in niche overlap between opossums and raccoons supported the possibility of competition (Ginger et al. 2003). Opossums are food items for ocelots that should be managed.

Reduction of a predator from the community can have unanticipated consequences (Henke and Bryant 1999). Consequently, research on the effects of removing competitors from the community inhabited by ocelots should be evaluated before implementing a broader removal program. If competitor removal is determined to benefit ocelot conservation, then removal trials may be initiated to evaluate effectiveness, particularly during crucial ecological conditions associated with drought.

Prey Supplementation

Another strategy to consider is the direct supplementation of prey into ocelot habitat during critical drought periods when prey populations are significantly reduced. Generally, prey supplementation is an unproven tactic that often is not ecologically effective or economically feasible. It has been attempted with feral hogs released within the large home ranges of Florida panthers (Maehr et al. 1989). In contrast, ocelots can occur at high-density within confined areas, increasing the opportunity for targeted delivery of supplemental prey. Consequently, prey supplementation should be studied to determine if it has value for ocelot conservation.

Small rodents may be difficult to obtain in substantial quantities. In contrast, relatively large-bodied prey such as rabbits (cottontails) can serve as a profitable prey for ocelots with reduced risk of unanticipated outcomes. Opossums are also large prey that can be easily obtained from natural areas or through opossum control efforts in urban areas. The health of released prey must be screened to resolve disease threats, but supplementation of these large prey, particularly to core areas used by ocelots should be evaluated as a potential source of energy, protein, and Vitamin A.

Feral hogs occur in ocelot scats from southern Texas, suggesting that adult ocelots consume immature hogs. Feral hogs are often considered a problem species and are removed liberally by hunters in many landscapes in Texas. However, excessive hunting or widespread removal of feral hogs should be avoided in areas occupied by ocelots to ensure the presence of this food item.

We must emphasize that, although the strategies of competitor removal or prey supplementation during critical drought periods should be explored, neither have been attempted or demonstrated as a feasible alternative. However, the urgency of ocelot recovery should warrant the evaluation of broader solutions, even if some methods may be considered unorthodox.

Ocelot Habitat Restoration

An important conservation strategy for ocelot recovery is to maintain existing habitat and expand or restore new habitat, particularly near the existing core ocelot populations in Texas. A larger ocelot population would have a greater opportunity to withstand the effects of drought or another ecological catastrophe.

Ultimately, additional thorn shrub habitat is required to increase the ocelot population in the United States and reduce its vulnerability. Restoration of thorn shrub communities with appropriate shrub density and composition is one possibility. Because funds are generally limited for expensive restoration associated with ocelot habitat, strategies are needed to maximize cost effectiveness for restoration efforts that are undertaken.

Restoring "islands" of thorn shrub habitat for ocelots in a "sea" of herbaceous habitat used by prey is one possible strategy (Figure 8.3). Patches of 40 to 100 ha of dense thorn shrub embedded in grassland or old agricultural fields would produce such a composition where ocelot habitat patches are surrounded by a large prey reservoir. A site-specific restoration map can be designed identifying ocelot habitat patches and connecting corridors to enable inter-patch movements within a herbaceous matrix.

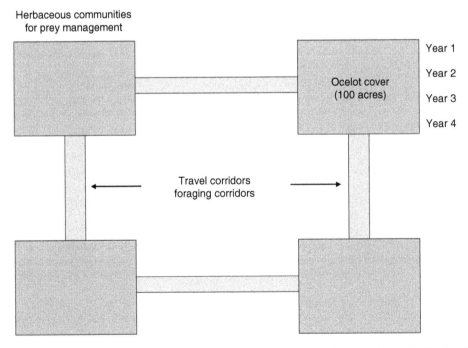

FIGURE 8.3   Restoration of ocelot "habitat islands" in a "sea of prey habitat" should help buffer the effects of drought on ocelot. The importance of nearby prey habitat should be considered prior to restoration of ocelot thorn shrub.

In addition, we have observed ocelots traveling along corridors connected to occupied habitat patches, presumably foraging during these corridor movements. Ocelots appear to use even "dead-end" corridors for foraging, likely benefiting from prey availability along the interface of the thorn shrub–herbaceous community. Use of forage corridors can provide ocelots with profitable foraging opportunities.

An additional strategy for ocelot recovery is establishment of another ocelot population in a location with appropriate habitat quality and quantity, preferably an area geographically separated from the existing ocelot population. Establishing a new ocelot population further north on the Texas coast would spread the threat of drought, which often has regional variation in intensity, and reduce susceptibility of the ocelot population to a catastrophic event.

The high risk of loss for the ocelot population requires urgent remedial actions based on the best information currently available. Recovery biologists need to develop a drought management plan for ocelots and have it ready before onset of the next drought. This plan should have considerable details that cover seasonal and annual management activities for normal and wet years, and selected activities that are implemented when certain drought thresholds have been reached.

ACKNOWLEDGMENTS

We thank C. A. Tewes for help with literature and development of figures. C. D. Reed helped maintain the cat library that we used extensively. We thank L. I. Grassman for help with figures and literature. Manuscript reviews were generously provided by T. L. Blankenship and L. I. Grassman. Substantive discussions about cat biology were provided by our cat research team, including L. I. Grassman, A. Caso, A. M. Haines, T. L. Blankenship, and J. H. Young.

REFERENCES

Anderson, B. F., and M. R. Pelton. 1976. Movements, home range, and cover use: Factors affecting the susceptibility of cottontails to hunting. *Proc. Southeast. Assoc. Game and Fish Commissioners* 30:525.

Anderson, E. M., and M. J. Lovallo. 2003. Bobcat and lynx. In *Wild Mammals of North America: Biology, Management, and Conservation*, G. A. Feldhamer, B. C. Thompson, and J. A. Chapman (eds). Baltimore: The Johns Hopkins University Press, p. 758.

Bailey, T. N. 1974. Social organization in a bobcat population. *J. Wildl. Manage.* 38:435.

Bailey, T. N. 1979. Den ecology, population parameters and diet of eastern Idaho bobcats. In *Proceedings of the 1979 Bobcat Res. Conf. (Science and Technology Series 6)*, P. C. Escherich, and L. Blum (eds). Washington, DC: National Wildlife Federation, p. 62.

Bailey, T. N. 1981. Factors of bobcat social organization and some management implications. In *Proceedings of the Worldwide Furbearer Conferences*, J. A. Chapman, and D. Pursley (eds). Frostburg, p. 984.

Beasom, S. L., and R. A. Moore. 1977. Bobcat food habit response to a change in prey abundance. *Southwest. Nat.* 21:451.

Beckwith, S. L. 1954. Ecological succession on abandoned farmlands and its relationship to wildlife management. *Ecol. Monogr.* 24:349.

Berg, W. E. 1979. Ecology of bobcats in northern Minnesota. In *Proceedings of the 1979 Bobcat Res. Conf. (Science and Technology Series 6)*, P. C. Escherich, and L. Blum (eds). Washington, DC: National Wildlife Federation, p. 55.

Bisbal, F. J. 1986. Food habits of some Neotropical carnivores in Venezuela (Mammalia, Carnivora). *Mammalia* 50:329.

Blankenship, T. L. 2000. Ecological response of bobcats to fluctuating prey populations on the Welder Wildlife Foundation Refuge. PhD dissertation, Texas A&M University — Kingsville, Kingsville, and Texas A&M University, College Station.

Blankenship, T. L., and W. G. Swank. 1979. Population dynamic aspects of the bobcat in Texas. In *Proceedings of the 1979 Bobcat Res. Conf. (Science and Technology Series 6)*, P. C. Escherich, and L. Blum (eds). Washington, DC: National Wildlife Federation, p. 166.

Blankenship, T. L., et al. 2006. Comparing survival and cause-specific mortality rates between resident and transient bobcats *Lynx rufus*. *Wildl. Biol.* 12:297.

Bothma, J. du P., and J. G. Teer. 1977. Reproduction and productivity in South Texas cottontail rabbits. *Mammalia* 41:253.

Bradley, R. D., et al. 2006. Rapid recovery of rodent populations following severe drought. *Southwest. Nat.* 51:87.

Brand, C. J., and L. B. Keith. 1979. Lynx demography during a snowshoe hare decline in Alberta. *J. Wildl. Manage.* 43:827.

Brand, C. J., L. B. Keith, and C. A. Fischer. 1976. Lynx responses to changing snowshoe hare densities in Alberta. *J. Wildl. Manage.* 40:416.

Brown, J. H., and S. K. M. Ernest. 2002. Rain and rodents: Complex dynamics of desert consumers. *Bioscience* 52:979.

Cameron, G. N., and S. R. Spencer. 1981. *Sigmodon hispidus*, *Mammal. Species* 158:1.

Chapman, J. A., J. G. Hockman, and W. R. Edwards. 1982. Cottontails (*Sylvilagus floridanus* and allies). In *Wild Mammals of North America*, J. A. Chapman, and G. A. Feldhamer (eds). Baltimore: Johns Hopkins University Press, p. 83.

Chapman, J. A., and J. A. Litvaitis. 2003. Eastern cottontail. In *Wild Mammals of North America: Biology, Management, and Conservation*, G. A. Feldhamer, B. C. Thompson, and J. A. Chapman (eds). Baltimore: The Johns Hopkins University Press, p. 101.

Conaway, C. H., H. M. Wight, and K. C. Sadler. 1963. Annual production by a cottontail population. *J. Wildl. Manage.* 27:171.

Conaway, C. H., K. C. Sadler, and D. H. Hazelwood. 1974. Geographic variation in litter size and onset of breeding in cottontails. *J. Wildl. Manage.* 38:473.

Davis, W. B., and D. J. Schmidly. 1994. *The Mammals of Texas*. Austin: Texas Parks and Wildlife Department.

Dibello, F. J., S. M. Arthur, and W. B. Krohn. 1990. Food habits of sympatric coyotes, *Canis*, red foxes, *Vulpes vulpes*, and bobcats, *Lynx rufus*, in Maine. *Canad. Field-Natural.* 104:403.

Ecke, D. H. 1955. The reproductive cycle of the Mearns cottontail in Illinois. *Am. Midl. Natural.* 53:294.

Elton, C., and M. Nicholson. 1942. The ten-year cycle in numbers of lynx in Canada. *J. Anim. Ecol.* 11:215.

Emmons, L. J. 1987. Comparative feeding ecology of felids in a Neotropical rainforest. *Behav. Ecol. Sociobiol.* 20:217.

Ernest, S. K. M., J. H. Brown, and R. R. Parmenter. 2000. Rodents, plants, and precipitation: Spatial and temporal dynamics of consumers and resources. *Oikos* 88:470.

Fritts, S. H., and J. A. Sealander. 1978. Diets of bobcats in Arkansas with special reference to age and sex differences. *J. Wildl. Manage.* 42:533.

Fulk, G. W. 1975. Population ecology of rodents in the semiarid shrublands of Chile. *Occasional Papers, The Museum, Texas Tech University,* 33:1.

Ginger, S. M., et al. 2003. Niche shift by Virginia opossum following reduction of a putative competitor, the raccoon. *J. Mammal.* 84:1279.

Gutiérrez, J. R., et al. 1993. Structure and dynamics of vegetation in a Chilean arid thornscrub community. *Acta Oecol.* 14:271.

Haines, A. M., M. E. Tewes, and L. L. Laack. 2005a. Survival and sources of mortality in ocelots. *J. Wildl. Manage.* 69:255.

Haines, A. M., et al. 2005b. Evaluating recovery strategies for an ocelot population in the United States. *Biol. Conserv.* 126:512.

Haines, A. M., et al. 2006. Habitat based population viability analysis of ocelots in southern Texas. *Bio. Conserv.* 132:424.

Harveson, P. M., et al. 2004. Habitat use by ocelots in south Texas: Implications for restoration. *Wildl. Soc. Bull.* 32:948.

Henke, S. E., and F. C. Bryant. 1999. Effects of coyote removal on the faunal community in western Texas. *J. Wildl. Manage.* 63:1066.

Horne, J. S. 1998. Habitat partitioning of sympatric ocelot and bobcat in southern Texas. Masters Thesis, Texas A&M University — Kingsville, Kingsville.

Jaksic, F. M., P. Feinsinger, and J. E. Jiménez. 1996. Ecological redundancy and long-term dynamics of vertebrate predators in semiarid Chile. *Conserv. Biol.* 10:252.

Jaksic, F. M., et al. 1997. A long-term study of vertebrate predator responses to an El Niño (ENSO) disturbance in western South America. *Oikos* 78:341.

Jiménez, J. E., P. Feinsinger, and F. M. Jaksic. 1992. Spatiotemporal patterns of an irruption and decline of small mammals in north central Chile. *J. Mammal.* 73:356.

Kasparian, M. A., E. C. Hellgren, and S. M. Ginger. 2004. Food habits of the Virginia opossum during raccoon removal in the Cross Timbers ecoregion, Oklahoma. *Proc. Okla. Acad. Sci.* 82:73.

Keith, L. B. 1963. *Wildlife's Ten-Year Cycle*. Madison: University of Wisconsin Press.

Keith, L. B. 1983. Role of food in hare population cycles. *Oikos* 40:385.

Keith, L. B., et al. 1977. An analysis of predation during a cyclic fluctuation of showshoe hares. *Proc. Int. Cong. Game Biol.* 13:151.

Knick, S. T. 1990. Ecology of bobcats relative to exploitation and a prey decline in southeastern Idaho. *Wildl. Monogr.* 108:1.

Knick, S. T., and T. N. Bailey. 1986. Long distance movements by two bobcats from southeastern Idaho. *Am. Midl. Nat.* 116:222.

Knick, S. T., E. C. Hellgren, and U. S. Seal. 1993. Hematologic, biochemical, and endocrine characteristics of bobcats during a prey decline in southeastern Idaho. *Can. J. Zool.* 71:1448.

Knowles, P. R. 1985. Home range size and habitat selection of bobcats, *Lynx rufus*, in north-central Montana. *Can. Field-Nat.* 99:6.

Koehler, G. M., and K. B. Aubrey. 1994. Lynx. In *The Scientific Basis for Conserving Forest Carnivores: American Marten, Fisher, Lynx, and Wolverine in the Western United States (General Technical Report RM-254)*, L. F. Ruggiero, et al. (eds). U.S. Forest Service, p. 74.

Konecny, J. J. 1989. Movement patterns and food habits of four sympatric carnivore species in Belize, Central America. In *Advances in Neotropical Mammalogy*, K. H. Redford, and J. F. Eisenberg (eds). Gainesville: The Sandhill Crane Press, P. 243.

Krebs, J. R. 1979. Optimal foraging: Decision rules for predators. In *Behavioural Ecology: An Evolutionary Approach*, J. R. Krebs, and N. B. Davies (eds). Oxford: Blackwell Scientific Publications, p. 23.

Laack, L. L. 1991. Ecology of the ocelot (*Felis pardalis*) in south Texas. Masters Thesis, Texas A&I University — Kingsville, Kingsville.

Laack, L. L., et al. 2005. Reproductive life history of ocelots *Leopardus pardalis* in southern Texas. *Acta Theriol.* 50:505.

Lariviere, S., and L. R. Walton. 1997. *Lynx rufus, Mammal. Species* 563:1.

Layne, J. N. 1974. Ecology of small mammals in a flatwoods habitat in northcentral Florida, with emphasis on the cotton rat (*Sigmodon hispidus*). *Am. Mus. Novitates* 2544:1.

Lima, M., and F. M. Jaksic. 1998. Delayed density-dependent and rainfall effects on reproductive parameters of an irruptive rodent in semiarid Chile. *Acta Theriol.* 43:225.

Lima, M., and F. M. Jaksic. 1999. Population dynamics of three Neotropical small mammals: Time series models and the role of delayed density-dependence in population irruptions. *Austral. J. Ecol.* 24:25.

Litvaitis, J. A., J. A. Sherburne, and J. A. Bissonette. 1986. Bobcat habitat use and home range size in relation to prey density. *J. Wildl. Manage.* 50:110.

Lockmiller, R. L., et al. 1995. Habitat-induced changes in essential amino-acid nutrition in populations of eastern cottontails. *J. Mammal.* 76:1164.

Ludlow, M. E., and M. E. Sunquist. 1987. Ecology and behavior of ocelots in Venezuela. *Natl Geograph. Res.* 3:447.

Madsen, T., and R. Shine. 1999. Rainfall and rats: Climatically-driven dynamics of a tropical rodent population. *Austral. J. Ecol.* 24:80.

Maehr, D. S., and J. R. Brady. 1986. Food habits of bobcats in Florida. *J. Mammal.* 67:133.

Maehr, D. S., et al. 1989. Fates of wild hogs released into occupied Florida panther home ranges. *Florida Field Nat.* 17:42.

McCord, C. M., and J. E. Cardoza. 1982. Bobcat and lynx: *Felis rufus* and *F. lynx*. In *Wild mammals of North America: Biology, management, and economics,* J. A. Chapman, and G. A. Feldhamer (eds). Baltimore: John Hopkins University Press, p. 728.

Meserve, P. L., et al. 1995. Heterogeneous responses of small mammals to an El Niño southern oscillation event in north central semiarid Chile and the importance of the ecological scale. *J. Mammal.* 76:580.

Miller, S. D., and D. W. Speake. 1978. Prey utilization by bobcats on quail plantations in southern Alabama. *Proc. of the Annual Conf. of the Southeast. Assoc. of Fish and Wildl. Agencies* 32:100.

Morrison, M. L., et al. 2002. Habitat use and abundance of rodents in southeastern Arizona. *Southwest. Nat.* 47:519.

Mondolfi, E. 1986. Notes on the biology and status of the small wild cats of Venezuela. In *Cats of the World: Biology, Conservation, and Management*, S. D. Miller, and D. D. Everett (eds). Washington, DC: National Wildlife Federation, p. 125.

Mowat, G., K. Poole, and M. O'Donoghue. 2000. Ecology of lynx in northern Canada and Alaska. In *Ecology and Conservation of the Lynx in the United States*, L. F. Ruggiero, et al. (eds). Boulder: University Press of Colorado, p. 265.

Murray, J. L., and G. L. Gardner. 1997. *Leopardus pardalis, Mammal. Species* 548.

Mutze, G. J., B. Green, and K. Newgrain. 1991. Water flux and energy use in wild house mice (*Mus domesticus*) and the impact of seasonal aridity on breeding and population-levels. *Oecologia* 88:529.

Norwine, J., and R. Bingham. 1985. Frequency and severity of droughts in south Texas: 1900–1983. In *Proceedings of a Workshop on Livestock and Wildl. Manage. During Drought*, R. D. Brown (ed.), Caesar Kleberg Wildlife Research Institute, Texas A&M University — Kingsville, Kingsville, p. 1.

O'Conner, R. M. 1984. Population trends, age structures, and reproductive characteristics of female lynx in Alaska, 1961 through 1973. Masters Thesis, University of Alaska, Fairbanks.

O'Donoghue, M., et al. 1998. Functional responses of coyotes and lynx to the snowshoe hare cycle. *Ecology* 79:1193.

Odum, E. P. 1955. An eleven year history of a *Sigmodon* population. *J. Mammal.* 36:368.

Pearson, O. P. 1975. An outbreak of mice in the coastal desert of Peru. *Mammalia* 39:375.

Péfaur, J. E., J. L. Yanez, and J. M. Jaksic. 1979. Biological and environmental aspects of a mouse outbreak in the semiarid region of Chile. *Mammalia* 43:313.

Poole, K. G. 1994. Spatial organization of a lynx population. *Can. J. Zool.* 73:632.

Rolley, R. E. 1985. Dynamics of a harvested bobcat population in Oklahoma. *J. Wildl. Manage.* 49:283.

Rolley, R. E. 1987. Bobcat. In *Wild Furbearer Management and Conservation in North America*, M. Novak, et al. (eds). Ontario: Ministry of Natural Resources, p. 670.

Rose, G. B. 1973. Energy metabolism of adult cottontail rabbits, *Sylvilagus floridanus*, in simulated field conditions. *Am. Midl. Nat.* 89:473.

Saunders, J. K. 1961. The biology of the Newfoundland lynx. Doctoral dissertation, Cornell University, Ithaca.

Scott, P. P. 1968. The special features of nutrition in cats, with observation on wild Felidae nutrition in the London zoo. *Symp. Lond. Zool. Soc.* 21:21.

Shindle, D. B., and M. E. Tewes. 1998. Woody species composition of habitats used by ocelots (*Leopardus pardalis*) in the Tamaulipan Biotic Province. *Southwest. Nat.* 43:273.

Stoddard, H. L. 1931. *The Bobwhite Quail: Its Habits, Preservation, and Increase*. New York: Charles Scribner's Sons.

Story, J. D., W. J. Galbraith, and J. T. Kitchings. 1982. Food habits of bobcats in eastern Tennessee. *J. Tenn. Acad. Sci.* 57:29.

Strecker, J. K. 1929. Notes on the Texas cotton and Attwater wood rats in Texas. *J. Mammal.* 10:216.

Sunquist, M., and F. Sunquist. 2002. *Wild Cats of the World*. Chicago and London: The University of Chicago Press.

Tewes, M. E. 1986. Ecological and behavioral correlates of ocelot spatial patterns. Doctoral dissertation, University of Idaho, Moscow.

Tewes, M. E., and D. D. Everett. 1986. Status and distribution of the endangered ocelot and jaguarundi in Texas. In *Cats of the World: Biology, Conservation, and Management*, S. D. Miller, and D. D. Everett (eds). Washington, DC: National Wildlife Federation, p. 147.

Tewes, M. E., and R. W. Hughes. 2001. Ocelot management and conservation along transportation corridors in southern Texas. In *Proceedings of the International Conference on Ecology and Transportation*, G. L. Evink (ed.). Keystone, p. 559.

Tewes, M. E., J. M. Mock, and J. H. Young. 2002. Bobcat predation on quail, birds, and mesomammals. In *Quail V: Proceedings of the Fifth National Quail Symposium*, S. J. DeMaso, et al. (eds). Austin: Texas Parks and Wildlife Department, p. 65.

Tewes, M. E., and D. J. Schmidly. 1987. The Neotropical felids: Jaguar, ocelot, margay, and jaguarundi. In *Wild Furbearer Management and Conservation in North America*, M. Novak, et al. (eds). Concord, ON: Ministry of Natural Resources, Concord, Ontario, p. 697.

Thornton, D. H., M. E. Sunquist, and M. B. Main. 2004. Ecological separation within newly sympatric populations of coyotes and bobcats in south-central Florida. *J. Mammal.* 85:973.

Trethewey, D. E. C., and B. J. Verts. 1971. Reproduction in eastern cottontails in western Oregon. *Am. Midl. Nat.* 86:463.

Wassmer, D. A., D. D. Guenther, and J. N. Layne. 1988. Ecology of the bobcat in south-central Florida. *Bull. Florida State Mus., Biol. Sci.* 33:159.

9 Seeing the World through the Nose of a Bear — Diversity of Foods Fosters Behavioral and Demographic Stability

David L. Garshelis and Karen V. Noyce

CONTENTS

Diversity of ecological systems has long been thought to increase the stability of those systems. Observations and intuitive reasoning by Odum (1953), MacArthur (1955), Elton (1958), and other early ecologists indicated that complex communities were more resistant or resilient to perturbations, meaning in essence that with more interacting species, the biomass, composition, and productivity of a community were less variable. This seemed logical in that a very simple (nondiverse) system relies on all parts functioning normally; a disturbance to one key element (species or group of species) is likely to upset the balance and function of many other components. Diverse systems are more failsafe because of internal redundancy, or many species serving similar roles, thus providing "insurance" against a widespread catastrophe, unless all species in a certain guild (or functional type; Hooper et al. 2005) react similarly to a perturbation. Moreover, unlike simple systems, diverse systems have many linkages among species, so change in abundance of one species would have less drastic effects on the others.

Early computer modeling of ecosystems initiated by May (1973), however, indicated that this intuitive relationship between diversity and stability was incorrect, or at least not as simple as originally thought. In fact, results of May's work and further modeling convinced many ecologists that the original paradigm was wrong, although empirical evidence still seemed to support it. Later this paradox was resolved when it was shown that modeling based on random processes did not accurately mimic actual diverse communities, which are not just a set of multiple interacting species, but species with complex and varying interactions among each other as well as varying responses to environmental fluctuations (reviewed by McCaan 2000). The complexity of natural systems, though, makes it difficult to evaluate the diversity–stability hypothesis empirically; some carefully controlled experiments provide qualified support for this relationship (reviewed by Schläpfer and Schmid 1999; Cottingham et al. 2001).

Reasons for increased stability stemming from diversity fall into several categories (Petchey 2000; Srivastava and Vellend 2005), two of which are most pertinent here. First is the "insurance effect" related to species redundancy within functional groups. This has been likened to redundancy incorporated into engineering projects to improve reliability (Naeem 1998). Second is what has been called statistical averaging, or the "portfolio effect" — just as in a diverse financial portfolio, the ups and downs of many individual components tend to be averaged out, resulting in greater overall stability (Doak et al. 1998; Lhomme and Winkel 2002). Notably, whereas the system tends to be stabilized and as a whole may even be more productive with increased diversity, the individual components of that system, each less abundant with more competing species, may be more variable (Lehman and Tilman 2000).

Here, we extend the diversity–stability principal to the case of a single species, the American black bear (*Ursus americanus*), arguing that greater stability arises when its food base is more diverse. Notably, MacArthur (1955, 535), in his seminal paper on community stability, commented that a "restricted diet lowers stability." We investigate stability in terms of four variables, two demographic (population size and reproductive rate), and two behavioral, with demographic consequences (attraction to human-related foods, which directly affects mortality rates, and seasonal movements). If year-to-year fluctuations in these variables are low, the "system" of bears is considered stable. A chief goal of management for many species is to maintain stability. Specifically in the case of bears, population stability is sought to balance recreational hunting opportunities (more hunter satisfaction with more bears) against potential adverse interactions with humans (less nuisance problems with fewer bears).

Because variability scales with the value of the mean, the coefficient of variation (CV = 100% $[SD/\bar{x}]$) is often used as a metric of stability (Lehman and Tilman 2000). Unfortunately, even this does not completely control for the association between the variance and the mean, but ecologists have not yet found a satisfactory way of dissociating the two (McArdle et al. 1990). This may confound some interpretations of diversity–stability relationships, including those presented here.

Mathematically, Doak et al. (1998) showed that variability should decline asymptotically with increased diversity, but the shape of the curve and the level of the asymptote depend on two factors: the equality of the various components comprising the diversity and correlations in their fluctuations (i.e., their covariance). To relate this to bears, we suggest that not only would more varied foods promote more stability, but that greater stability would arise with (1) increasing evenness in the abundance of these diverse foods (i.e., a system that was not heavily dominated by just a few key food species), and (2) decreasing similarity of responses of these various foods to environmental fluctuations (Figure 9.1). If all fruit-producing species failed under the same set of environmental circumstances (high covariance), then their functional diversity would be low (in terms of the portfolio effect, analogous to investing just in stocks of one category).

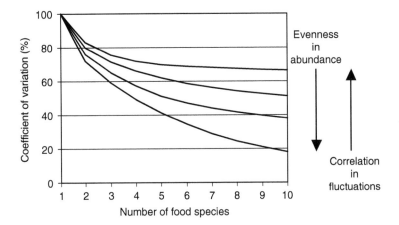

FIGURE 9.1 Ecological theory predicts increased community stability with increased species diversity. Stability is often measured in terms of the coefficient of variation (lower CV represents greater stability). Stability increases with more evenness in abundance among species, but declines with increasing correlation in the response of these species to environmental factors (redrawn from Doak, D. F., et al. 1998. *Am. Nat.* 151:264.). Here the relationship pertains to demographic and behavioral stability of bear populations with varying degrees of diversity, evenness, and correlation in productivity of fruit-bearing species.

RELEVANT LIFE HISTORY CHARACTERISTICS OF BLACK BEARS

American black bears are perhaps the quintessential generalist, opportunist, omnivores. They consume herbaceous vegetation, roots, buds, numerous kinds of fleshy fruits, nuts, insects in life stages from egg to adult, and vertebrates from fish to mammals, including their own kills as well as carrion. Moreover, they readily consume various sorts of human-related foods, from garbage and birdseed to a variety of agricultural products, including standing corn and oats just before harvest, apples, and honey and brood in commercial apiaries. The ability of black bears to vary their diet with the circumstances has enabled them to inhabit areas as different as dry Mexican scrub forests, Louisiana swamps, Alaskan rainforests, and Labrador tundra. Between these extremes, they occupy assorted forest types throughout the United States and Canada, each providing a different array of foods.

An important constraint, throughout most of their range (except the most southerly latitudes), is that these foods entirely disappear during winter. Black bears adapt to this by fasting for up to 7 months. They do so through a complex physiological process that (among bear biologists at least) is recognized as a form of hibernation (Hellgren 1998). During hibernation, bears sustain themselves by metabolizing body fat and possibly some smooth muscle and organ tissue, but they seem to be able to conserve most of the mass and strength in their skeletal muscles. This feat, matching the sustained fasting periods of all but a few extreme hibernators (Neuhaus 2000; Michener 2002), is reliant on the prehibernation accumulation of adequate fat reserves. In autumn, bears thus seek food sources that are not only high in fat, but also voluminous, so they can maximize their rate of food intake. In essence, they attempt to overeat (become hyperphagic).

Mature females have an additional constraint in producing and nursing cubs during hibernation. When emerging from their dens in the spring, cubs are nearly 3 months old and able to travel with their mother, but continue to nurse until autumn. Females retain cubs through the following winter (then called yearlings). Family break-up occurs just before mating in June. Females thus typically maintain a 2-year reproductive cycle.

When bears emerge from dens, they seek to replenish losses, and those that have not yet obtained their maximum size resume growth. For muscular and skeletal growth, they require protein, in addition to the fats and carbohydrates that provide metabolizable energy to fuel protein synthesis and support other body functions. Succulent green vegetation, usually the main component of the spring diet of black bears, provides a good source of protein (Noyce and Garshelis 1998), but may be low in energy. Bears continue to metabolize stored body fat until fruit, the staple of most summer bear diets, becomes available. Fruit, in contrast to the spring diet of vegetation, is rich in carbohydrate but depauperate in protein. In fact, bears on a strictly fruit diet may gain total mass but lose lean mass. As a strategy to overcome this, bears seem to seek mixed diets, incorporating insects, vegetation, meat if they can, or certain high-protein fruits (Rode and Robbins 2000; Felicetti et al. 2003).

Hence, in any given year, bears diversify their diet to match their seasonal life history stage. Additionally, among the various foods that black bears eat, fluctuations occur within and across years, prompting them to frequently substitute dietary constituents (Eiler et al. 1989; Schooley et al. 1994a; Iverson et al. 2001), or add new types of food that become available (Coop et al. 2005). A diverse array of potential foods thus provides a more reliable food base, ensuring more stability in the demographics and behavior of bears. We demonstrate this with a case study from Minnesota, the site of one of the longest-running research projects on black bears, and moreover a project closely tied to bear management.

BEAR MANAGEMENT IN MINNESOTA

Minnesota has long been a stronghold for black bears, although bear numbers likely fluctuated with changing public attitudes (Kinsey 1965). Magazine articles from the early 1900s, including some published by the predecessor of what is now the Minnesota Department of Natural Resources (MDNR; the agency responsible for bear management), warned that bears were becoming scarce and should be protected. Beginning in 1917, and for several decades thereafter, various laws prohibiting the taking of bears were established, rescinded, and reinstituted.

The 1940s were a low time for Minnesota black bears, as they could be shot, trapped, or poisoned at any time. In 1945, they were bountied, mainly to alleviate bear predation on domestic sheep. Legislation authorized that any person who killed a bear would be monetarily "rewarded," provided the person had not "spared the life of any other such bear he could have killed." Later, bears were protected in northeastern Minnesota because of their popularity among tourists. This dichotomy in how bears are viewed persists to the present among people from different walks of life, but it is clear that the overriding trend has been toward greater tolerance and appreciation of this species.

By the late 1950s and through the 1960s, black bears were still common in some parts of Minnesota. As such, they provided a source for bear restocking programs in Arkansas and Louisiana (Clark et al. 2002). The Minnesota bears were often captured in urban areas, although they were not habitual nuisance animals (Smith and Clark 1994).

In 1971, bears in Minnesota were classified as big game animals. Thereafter, they could be legally killed only by licensed hunters during a designated bear season (except nuisance animals, which could be killed at other times of year as well). However, by the early 1980s, annual increases in the number of bear hunters led to concern that bear numbers were continuing to decline. This prompted the establishment of quotas on sales of hunting licenses in 1982, a system that remains in effect today. This, the principal means of managing bears in Minnesota, enabled the population to increase both numerically and geographically. Bears now occupy approximately the northern two-fifths of the state, having expanded westward almost to the North Dakota border, and southward nearly to the suburbs of Minneapolis–St. Paul.

As their numbers increased, people encountered bears more often, passing through their backyards, rummaging through their garbage, or pulling down their birdfeeders. Black bears also became somewhat of a tourist attraction in northern Minnesota where they are considered by some to be the icon of a large wilderness carnivore — this despite the fact that they are neither mainly carnivorous, nor restricted to wilderness. In fact, their adaptability, both in terms of diet and habitat use, is the reason for their renewed range expansion. Ironically, farm crops, such as corn and oats, which are consumed and damaged by bears, often resulting in them being killed, also enhance their reproductive rates (Elowe and Dodge 1989; McLaughlin et al. 1994; McDonald and Fuller 2001); hence, human crops may sustain artificially high bear populations. This is clearly the case in Minnesota, where bear numbers at the agricultural fringe of the range have not been reduced even with targeted, heavy hunting (Kontio et al. 1998; Garshelis et al. 1999).

Hunting quotas in Minnesota are established separately for each of 11 bear management units (BMUs); a largely agricultural zone around the periphery of the primary range has no hunting quotas. Managers adjust quotas in response to indications of change in population size, hunter satisfaction, and level of bear nuisance activity.

DATA ON MINNESOTA BEARS

We have been studying black bears in Minnesota since 1981 (25 years as of this writing). Our study was prompted by concern that the bear population was being over-harvested. The chief goals of this research were to obtain data that could provide better guidance for bear management, namely estimates of population size, trend, sustainable harvest rate and sources of mortality, and ecological information that would enhance understanding of the variables affecting demography and nuisance behavior.

We captured and radio-collared over 500 individual bears on three study sites to monitor reproduction, survival, movements, habitat use, and diet. One study site was at the very northern edge of the Minnesota range (Voyagevrs National Park-VNP), one was near the southern periphery of the range (Camp Ripley), and one, our primary study area, was near the center of the range (Cippowa National Forest-CNF).

We monitored reproduction by tracking radio-collared females to their winter dens and examining their cub production. We also tagged over 400 cubs ($\sim$8 weeks old) in the dens of their mothers, and monitored their survival to the yearling age class by determining their presence or absence in the mother's den the next year. We monitored survival of older bears by periodically checking the status of their mortality-sensing radio collar from an airplane. Legal hunting caused most mortality, and hunters generally informed us when they killed a collared bear. We located radio-collared bears from airplanes to obtain information on seasonal movements and habitat use.

We evaluated diet by examining scats found mainly at trap sites. We also had the opportunity to closely follow and observe some wild, human-habituated bears foraging, which provided further information on diet (e.g., items like roots and tubers, which are unidentifiable in scats) and also insights into foraging strategies. We rigorously sampled all major forest types that bears used to quantify the production of fruits that bears ate (Noyce and Coy 1990). We did a separate evaluation of availability of ants by habitat type (Noyce et al. 1997).

We also monitored various aspects of the statewide bear population, including number of bears harvested, their sex and age, number of nuisance complaints, and availability of natural fruits and nuts. The tally of harvested bears was produced through mandatory registration, and hunters reported the sex of the bear they killed. Among collared bears killed by hunters, about 10% of females were reported as males, so we corrected for this reporting error in the total harvest. Hunters also were requested to submit a tooth, which was sectioned and stained for age analysis (Willey 1974); 70–80% of hunters complied.

Nuisance complaints were recorded by wildlife managers and conservation officers of the MDNR. Wild fruit production was assessed subjectively each year by 20–80 wildlife managers and field personnel from other agencies, in the course of their normal work. The list of monitored fruits ($n = 14$) was obtained from our analysis of bear diets. Participants in this survey rated both the prevalence and production of these fruiting species on a 0–4 scale (Noyce and Garshelis 1997).

A series of population estimates was produced for our study area near the center of the bear range during the 1980s by capturing and recapturing bears in baited barrel traps or foot snares. We marked bears with radio collars, which were used to estimate the proportion of time that they spent on the study area (trapping area); we then used that proportion to weigh each bear's contribution to the mark–recapture estimate (Garshelis 1992). We produced additional population estimates in the early 1990s using visually identifiable radio-collared bears as the marked sample and photographs at remote, baited stations for the recapture (Noyce et al. 2001).

Three statewide population estimates were obtained from 1991 to 2002 using tetracycline-laden baits to mark bears, and bones and teeth submitted by hunters as the recapture sample (Garshelis and Visser 1997; Garshelis and Noyce 2006). Tetracycline binds with calcium and produces a mark in bones and teeth that is visible on sectioned samples under ultraviolet light.

BEAR DIET, FORAGING STRATEGIES, AND HABITAT USE

Upon emerging from dens in spring (usually early April), few foods exist for Minnesota bears. They appear to be attracted to small wetlands with vernal ponds where they forage on succulent roots of aquatic grasses. Later, as the ground warms up and other herbaceous plants sprout, these form the bulk of the diet. Clover (*Trifolium* spp.) and dandelions (*Taraxacum officinale*), found in more open areas; wild calla (*Calla palustris*), in wet areas; and jewelweed (*Impatiens* sp.) and jack-in-the-pulpit (a tuber; *Arisaema triphyllum*), mainly in moist wooded areas, are commonly eaten. Over-wintered acorns (*Quercus* spp.) may be found in hardwood forests. As spring progresses, bears consume flowers and catkins from a variety of tree and shrub species, including willow (*Salix* spp.), red maple (*Acer rubrum*), and aspen (*Populus tremuloides, P. grandidentata*). Later, during leaf-out, bears consume emerging aspen leaves in large quantities. Thus, even in spring, when "green vegetation" comprises the bulk of their diet (Figure 9.2), bears seek foods in a diversity of habitats.

Previously, researchers concluded that spring diets of bears were but a holdover until fruits became available; accordingly, they called this the "negative foraging period" (i.e., a period of weight loss). In contrast, we found that most Minnesota bears, except lactating females and large adult males, gained weight, and young growing bears gained stature on this diet (Noyce and Garshelis 1998). Weight gains represented primarily increases in lean body mass; bears continued to lose body fat through the spring. Very large males may not be able to maintain weight on low-calorie spring foods, because they are constrained by their small bite size relative to body size (Rode et al. 2001). Also, various lines of evidence indicate that adult males eat relatively little during the June–July breeding season when they are actively seeking and courting estrus females. However, these large bears often have remaining fat stores from winter that can provide the energy they need in the spring and through the breeding season.

As spring progresses and these herbaceous foods become more fibrous and thus less digestible, bears in Minnesota turn largely to ants (Figure 9.2). At the same time (June and early July), ant eggs begin to develop into pupae, a rich meal for bears, lacking the acids, toxins, chitinous exoskeleton, mechanical defenses, and mobility of adult ants. Bears locate ant nests in decaying logs or stumps, or associated with rotting wood underground. They appear to use visual cues to locate potential sites, then olfaction to find nests in rotting wood.

Ants occur in a variety of habitat types in both uplands and lowlands, young forests and old. We did not detect consistent differences in overall ant nest density by habitat, probably due to extreme within-habitat variation (Noyce et al. 1997). In other studies, ants that were foraged upon by grizzly bears (*Ursus arctos*) were conspicuously more abundant in clear-cuts than in mature forests, due to increased sunlight and increased availability of downed wood (Nielsen et al. 2004b). In Minnesota, two of the ant species eaten by bears occurred most frequently in open habitats, but the species that comprised the bulk of the diet was similarly abundant across most of the habitats that we sampled.

For several weeks in late May and early June, Minnesota bears supplement their diet of greens and ants with newborn fawns of white-tailed deer (*Odocoileus virginianus*). As berries become widely available in mid-July, bears continue feeding on ants, but to a lesser extent. Ant feeding typically

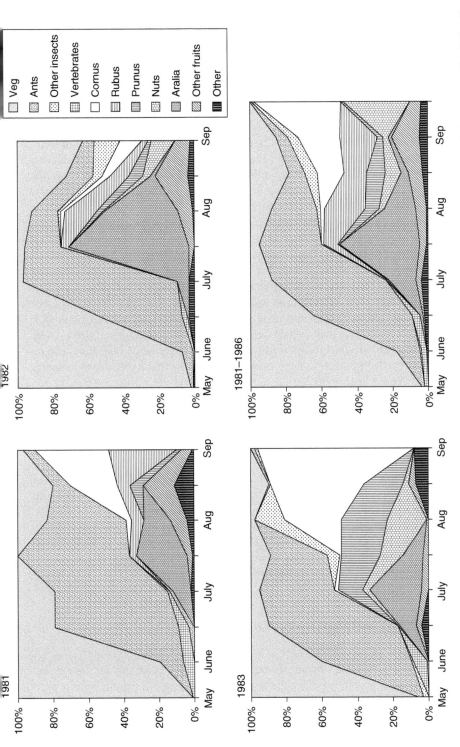

FIGURE 9.2 Bears in Minnesota varied their diet in response to changing food conditions, seasonally and yearly. Bottom-right panel shows generalized diet discerned from scats collected over a 6-year period, 1981–86. The first 3 years of this period are shown individually in the other panels. In all years, green vegetation comprised the bulk of the diet from May–early June, at which time ants became important. Ants remained the principal dietary item until mid-July, when ripe fruits became available. When fruiting was poor, as in 1981, ants continued as the principal food into August. Wild sarsaparilla (*Aralia*) was the first important fruit available; it was particularly abundant in 1982. Raspberries (*Rubus*) and cherries (*Prunus*) followed; these were particularly abundant in 1983. In late summer, bears subsisted on dogwood berries (*Cornus*; particularly abundant in 1983), and hazelnuts and acorns (*nuts*), which were more abundant in 1982 than the other years shown). These remained the most important dietary items through September (shown only in the 1981–86 panel), when bears began to den. If fruit and nut abundance in the fall were insufficient (as in 1982), bears resumed eating green vegetation.

subsides in late summer. Other insects, including bumblebees and yellow jackets, comprise a minor portion of the diet, but bears take advantage of periodic irruptions of defoliating caterpillars and june bugs.

Summer diet is dominated by fleshy fruits. We identified at least 19 shrub or herb-borne fruits that were commonly consumed by bears from summer to early autumn (Noyce and Coy 1990). Abundance and productivity of these different fruits varies by habitat; some favor upland forest types and some lowlands. In our study, the greatest fruit biomass was produced in conifer plantations, but this represented primarily three types of fruit, all of which thrive under sparse canopies, and one of which (raspberry, *Rubus idaeus* var. *strigosus*) seems not to be among the more highly sought-after foods by Minnesota bears. Conversely, certain key foods, including the earliest available summer fruit (wild sarsaparilla, *Aralia nudicaulis*; Figure 9.2), and two of only three staple autumn fruits (dogwood, *Cornus* spp., and hazelnut, *Corylus* spp.; Figure 9.2) are most abundant and productive under denser canopies. Abundance and productivity of different fruits also varies by age of the forest stand and silvicultural practices used in establishing or maintaining the stand, such as windrows, scarification, mechanical thinning, and use of herbicides.

Although our fruit sampling indicated that some habitat types generally produce relatively low overall biomass of fruit (Noyce and Coy 1990), our close-up observations of bears indicated that their foraging strategy differed significantly from our sampling strategy. We strived to situate sample plots randomly within forest stands, whereas bears use a "high-grading" strategy, quickly passing through areas with low-average fruit abundance to find rich patches of berries, which they seem to be able to smell from quite far away. We stratified sampling based on dominant tree type (cover type); whereas, bears choose feeding destinations mainly by the understory. Even cover types with consistently low overall fruit abundance may contain rich patches in forest openings and along roads, which we tended not to sample but which bears readily found. Dense patches where fruits can be consumed quickly are especially important to large bears that otherwise find it difficult to maintain weight on an all fruit diet (Welch et al. 1997).

Additionally, even where rich patches of fruit can provide more than enough biomass for sustained feeding, bears continue to use a variety of foods. We observed human-habituated bears purposefully switching, every few hours, between feeding on fruit to foraging for insects or vegetation. On these occasions, bears left sites where plenty of food remained to travel to a new feeding site in a different habitat (Figure 9.3).

Acorns are a notably important autumn food for black bears throughout much of their range in North America, with year-to-year variability affecting their reproduction, survival, and denning (Vaughan 2002). In Minnesota, oaks are scarce in the northern part of the range (along the Canadian border), patchy in the central range, and more prevalent in the southern part of the range. Bears in northern areas often move south in late summer for this reason (discussed later). Bears in the southern areas spend most of their time during late summer and autumn in hardwood stands, which, during other times of year, have low food availability (Noyce and Coy 1990).

TEMPORAL AND SPATIAL VARIATION IN FOOD PRODUCTION

Minnesota's forest landscape presents a variegated patchwork of lowland and upland cover types, characterized by different species and forest age-classes. This heterogeneous habitat holds a wide array of plants that produce fleshy fruits, plus several species of nut-bearing oak and hazel. By some measures, it appears to provide a very diverse habitat for bears, compared to other parts of North America, and a broad selection of foods. However, several features of these foods lessen the functional diversity of the system.

First, the overall prevalence of different fruit-bearing species on the landscape is highly uneven, thus diminishing the system's stability (Figure 9.1). Only 3 of 16 bear food species accounted for over

FIGURE 9.3 Photos of the same bear taken on the same day feeding on both ants (a) and blueberries (b). Although blueberries were sufficiently abundant to satiate the bear, it purposefully switched between dense patches of blueberries, where it could leisurely feed sometimes in a prone position, and far more sparsely distributed ants, which required extensive travel and probing into rotten logs, some of which had no ants.

two-thirds of the total ground cover of fruit-bearing plants in upland habitats (Noyce and Coy 1990). The relative biomass of fruits produced by the different species is similarly uneven. Also, distribution of individual species across the landscape is very patchy, with extreme site-to-site variation in fruit production, both within the same forest type and even within individual forest stands. For example, among 11 stands of similar-aged aspen (the most common forest type) observed in the same year, estimated abundance of wild sarsaparilla fruit was 0 kg/ha in five stands, 10 to 30 kg/ha in four stands, and >50 kg/ha in two stands. In the same stands, chokecherry production that year was 0 kg/ha in nine stands, but >100 kg/ha in two (not the same stands as those that produced abundant sarsaparilla).

Second, different plant species produce ripe fruits at different times through the year, and most species only produce for a few weeks. Thus, at any given time during the season, a relatively small number of species is in peak production. Moreover, after a bear feeds in a given patch, food availability there may replenish at a rate that varies with the type of fruit, type of habitat, and weather conditions at the time. The type and quantity of food available for bears thus literally changes from week to week [as shown by Davis et al. (2006) for another area]. For this reason, and the fact that different fruits vary in terms of the habitats in which they are most abundant, bears must track fluctuating foods by constantly probing different habitats and locations; however, they probably have a good sense of where to look, and may even be able to predict ripening of fruits based on weather conditions and other cues, as shown for other omnivores (Janmaat et al. 2006).

Third, from year to year, the relative abundance of different fruits varies markedly, as does total fruit production. During a 5-year span of our study, mean annual chokecherry production in pine plantations, the cover type with the most chokecherry, ranged from 6 to 222 kg/ha, and blackberry production ranged from 17 to 140 kg/ha. On our subjective rating scale, amalgamating 14 different fruits, the range-wide bear food production score averaged 61, but varied from the low 40s during 3 years to the 70s–80s in 4 years over a two-decade span.

Finally, year-to-year fluctuations in fruit production are not independent among species, thus diminishing their ability to buffer the system against environmental fluctuations, such as variation in temperature and precipitation. Range-wide food surveys across 20 years indicated that fruiting was strongly correlated ($r > .7$) among 5 of the 14 surveyed species; several other species showed weaker synchrony in fruiting.

For all of the above reasons, functional diversity of foods for bears — that is, the variety of food types available to them at any given time — is lower than it might at first appear. On the other hand, because bears are large and mobile, they can take advantage of landscape heterogeneity on a large scale. Across Minnesota's bear range, prevalent fruits differ. Some regions have generally better foods: More southerly areas have an average food production score of 63, whereas the northeastern and northwestern parts of the range average 56. In any given year, productivity of specific foods can vary regionally. For example, in 2005, blueberries, a highly preferred food, were of average abundance in western portions of the bear range, but well above average in the eastern part of the range. Two years before that, blueberries were average in the west but below average in the east. The ability of bears to find distant food sources potentially increases the seasonal array of foods that they can exploit beyond that available in their normal home range.

Central to our discussion here is whether diversity of the food base affects behavioral and demographic stability of Minnesota's black bears. This issue is difficult to address directly because we have little data on differences in food diversity, spatially or temporally, for which we can attempt to discern an effect. However, in the following sections, we are able to show strong behavioral and demographic effects of variations in food abundance. In other words, less variation in the food base would foster greater stability in the bear population. We rely on ecological theory, logic, and some direct evidence to link variations in food production with food diversity.

VARIATION IN MOVEMENTS

In many areas across North America, black bears make seasonal movements to exploit foods that are outside their normal home range. Such movements are particularly common in late summer and autumn. In areas where bears are protected, such as National Parks and other refuges, autumn movements outside the protected area put bears at risk of being hunted (Beeman and Pelton 1980; Pelton 1998; Samson and Huot 1998). This same situation occurs in Minnesota, where collared bears routinely traveled outside Voyageurs National Park (VNP) and Camp Ripley Military Reservation, two of our study sites. Hunting thus constitutes the main source of mortality for bears in both these areas (Table 9.1), even though the areas themselves are unhunted.

TABLE 9.1

Causes of Mortality of Radio-Collared Black Bears ≥1 Years Old from the Chippewa National Forest (CNF), Camp Ripley, and Voyageurs National Park (VNP), Minnesota, 1981–2005

Fate	CNF	Camp Ripley	VNP
Shot by hunter	211	9	10
Likely shot by hunter[a]	8	1	0
Shot as nuisance	22	2	1
Vehicle collision	12	5	1
Other human-caused death	9	0	0
Natural mortality	7	3	3
Died from unknown causes	3	1	0
Total deaths	272	21	15

[a] Lost track of during the hunting season.

Note: Bears did not necessarily die in the area where they usually lived, because they often moved extensively just prior to the opening of the bear hunting season in September. Hunting was not permitted within Camp Ripley or VNP, but hunters killed bears when they traveled outside these areas.

Bears from our northern and central study areas generally had few or no oaks in their summer home ranges and so were more prone to travel and to travel further during autumn than bears in the southern part of the geographic range, where oaks were prevalent. In our main study site near the center of the range, late summer movements (usually beginning in August) generally tended to be southward, toward better foods (mainly acorns and corn). Rogers (1987) reported that bears in northeastern Minnesota also tended to move southward, to areas with richer soils having more acorns and hazelnuts. He also noted that bears returned to areas that they had visited as cubs with their mother. We observed that young bears that had never made a seasonal excursion while with their mother nevertheless still left their summer range to find better autumn feeding areas. In some cases, unrelated radio-collared bears traveled large distances to virtually the same area, suggesting that they may have followed scent trails of other bears.

Others have reported a (logical) inverse relationship between the frequency of seasonal excursions of bears and prevailing food conditions (Beeman and Pelton 1980; Schooley et al. 1994a). "Conversely, we found a *positive* correlation" between acorn production and the likelihood of bears making late summer–autumn movements; when acorns were scarce, few bears moved. However, the magnitude of movements by adult males was greater in years when acorns were poor (median distance traveled = 64 km, n = 10) than in average or good food years (median distance = 28 km, n = 33). Rogers (1987), working on bears in northeastern Minnesota, found no association between food supply and frequency of late summer–autumn movements, but longer movements occurred in poorer food years. These regular seasonal travels by Minnesota bears were among the largest reported in any study of this species.

VARIATION IN NUISANCE ACTIVITY

During the 1980s, nuisance activity (measured in terms of number of complaints) by bears in Minnesota was among the highest in the United States. Surging numbers of complaints and the increasing time expended by DNR personnel to investigate and deal with those complaints were driving forces

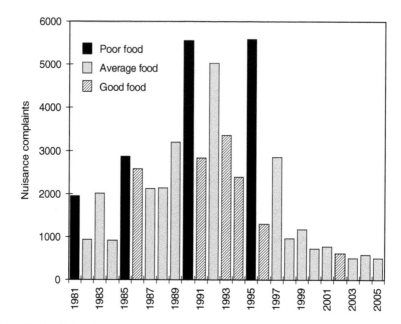

FIGURE 9.4   Number of nuisance bear complaints recorded by the Minnesota Department of Natural Resources, 1981–2005. Food availability was likely the driving factor in nuisance activity (i.e., bears' attraction to human food sources). However, no clear relationship was evident between natural food abundance (measured subjectively by field personnel) and number of complaints, except that more complaints were recorded in poor food years than adjacent years with better food. Complaints among the four worst food years showed an increasing trend, commensurate with increasing bear numbers from the early 1980s to the mid-1990s. Good food conditions kept nuisance activity relatively low during 3 of 4 years from 1991 to 1994. A change in management policy combined with average or above-average food conditions resulted in low numbers of complaints since 2000.

in bear management at that time (Garshelis 1989). During 1981–99, nearly 3000 nuisance bears were captured and translocated, and >3500 were known to have been killed (many more were likely killed and not reported).

High nuisance activity occurred in years of poor natural food production: 1981, 1985, 1990, and 1995. The level of nuisance activity among these 4 years showed an increasing trend (Figure 9.4), corresponding with an increasing bear population (both roughly tripling over this period). Recorded nuisance complaints in average or good food years showed no consistent trend through time, except that they were substantially lower than in adjacent poor food years.

A new policy for dealing with nuisance bears was adopted by the MDNR in 2000, mandating that bears would no longer be translocated, and agency personnel would deal with complaints mainly through telephone advice rather than on-site visits. As an apparent result, numbers of recorded complaints and numbers of nuisance bears killed have been lower during each year since 2000 than in any previous year. Additionally, average to good food conditions prevailed in all years since 2000.

VARIATION IN CUB PRODUCTION

The proportion of radio-collared bears that produced cubs varied year to year. Such variation was also evident in the age structure of the statewide harvest. Hunting cubs is not legal, but the proportion of yearlings in the harvest reflects cub production the previous year.

Since 1995, trends in proportion of yearlings in the statewide harvest sample matched trends in cub production by our radio-collared sample (Figure 9.5). Both 1993 and 1994 were particularly

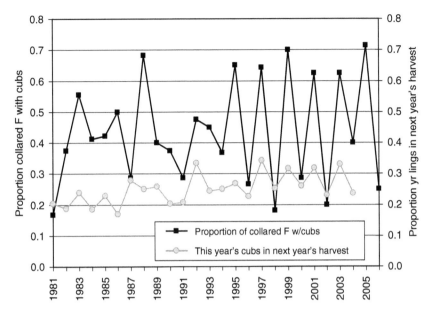

FIGURE 9.5 Year-to-year variability in cub production by Minnesota bears, measured as the proportion of adult (≥4 years old) radio-collared females with cubs (here showing data from the CNF study site, where on average 15 adult females were monitored each year) and the proportion of yearlings in statewide harvest samples (slid back 1 year to match the year they were born: e.g., yearlings in 2005 harvest corresponded with cubs that were born in 2004). Apparent synchrony in reproduction (high cub production in odd-numbered years) was initiated by back-to-back good food years in 1993 and 1994, enabling a large proportion of newly maturing bears to have cubs in 1995. Poor foods in 1995 dampened reproduction in 1996, followed by excellent foods in 1996, which reinforced high productivity in 1997. Most bears remained on a 2-year reproductive cycle after having their first cubs, enabling this synchrony to persist.

good food years (two of the four highest-rated years among the 22 years of monitoring), so this may have contributed to high cub production in January 1995. As bears cannot produce surviving cubs in two consecutive years, high reproduction in 1995 would be expected to result in lower reproduction in 1996; this effect could be obfuscated if 1995 was also a good food year, prompting many newly maturing bears to produce cubs. However, 1995 was among the three poorest food years on record, thus reinforcing 1996 as a poor reproductive year. Exceptionally good food production in 1996 and above-average production again in 1998, strengthened this 2-year synchronous rhythm, which has persisted to the present (Figure 9.5). This synchrony remained, because no food failures occurred during the past 7 years, enabling most mature females to produce a litter every other year ($\bar{x} = 2.06$ years, $n = 104$ litter intervals among collared bears). Reproductive synchrony among black bears instigated by food failures also has been observed in Tennessee, Florida, and Maine (Eiler et al. 1989; McLaughlin et al. 1994; Dobey et al. 2005); however, bears in Florida and Maine that had access to alternate, dependable foods did not exhibit such synchrony.

Reproductive synchrony should diminish as newly maturing bears of various ages produce cubs. The mean age of first reproduction in our central study area was 5.1 years (Garshelis et al. 1998), and 54% of females produced their first cubs at an odd-year age (3, 5, or 7 years). If these females were born during a reproductive peak year, they would produce cubs in a nonpeak year, thus dampening reproductive synchrony. However, if most females in other parts of the state first reproduce at an even-year age (4 or 6 years), this would help prolong the persistence of reproductive synchrony statewide.

Although reproductive synchrony starting in the mid-1990s occurred throughout the state, and was apparent among radio-collared bears on all of our study sites, reproductive output differed immensely by geographic region. Using the spacing of annuli in teeth from harvested bears to discern reproductive events (Coy and Garshelis 1992), Coy (1999) found that the percent of black bears that produced cubs by the time they were 4 years of age varied from <20% in the far north to >50% (up to 83%) in the southern part of the range, where foods are more plentiful. This same north–south trend in age of sexual maturity was evident among radio-collared bears. In our central CNF study area, 3% of bears produced their first cubs at 3 years old, 42% at four, and the rest at 5 to 10 years old (Garshelis et al. 1998). In contrast, in our northern (VNP) study area, no monitored bears produced cubs at either 3 or 4 years old. However, in our southern study area, five bears followed to maturity all had their first litters at age three (though one litter did not survive). Interval between litters (data from both radio-collared bears and annuli in teeth) and litter size (data just from collared bears) showed very little regional or year-to-year variation.

Age of first reproduction is affected by the growth rates and nutritional condition of bears (Noyce and Garshelis 1994, 2002). Larger mothers produce larger cubs, which, if living in a food rich area, grow faster and mature sooner. We found that nulliparous females must attain a mass of at least 41 kg in the March preceding the June breeding season to produce cubs the following January (although not all females that attain this weight produce cubs). Some minimum threshold summer and autumn mass might also be necessary for a successful first pregnancy.

A relationship between body mass and eventual age of first reproduction was apparent as early as 1 year old (Figure 9.6). With declining mass, age of first reproduction increased and became more variable. Regionally in Minnesota, yearlings were smaller in the northern study areas ($\bar{x} \cong 20$ kg) than at Camp Ripley, at the southern edge of the primary bear range ($\bar{x} \cong 30$ kg).

We observed the same trend at a local scale. A distinct ecotone divides our central study area, separating a hilly glacial moraine, characterized by varied topography and a diverse array of mostly upland forests, from a large glacial outwash plain, characterized by flat, poorly drained lowland forests of several types. Radio-collared bears living in the lowland forests had less plentiful and less diverse foods. They were lighter, at a given age, and began producing cubs more than 2 years later ($\bar{x} = 6.8$ years), on average, than bears in the immediately adjacent uplands ($\bar{x} = 4.5$ years). Yearling bears that were handled in winter dens in the lowland portion of this study area averaged

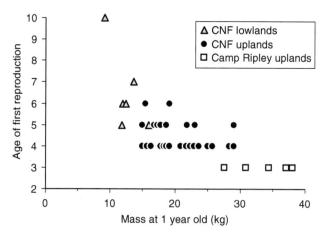

FIGURE 9.6　Relationship between mass of bears at 1 year of age and their eventual age of first reproduction: Heavier yearlings matured earlier and had less variability among their ages of first reproduction. Upland habitats in the central (CNF) and southern (Camp Ripley) parts of the Minnesota bear range produced heavier yearlings than lowland habitats, because foods in the uplands were more plentiful and more diverse; thus, productivity was higher and more consistent for bears living in the uplands, particularly the food-rich uplands at the southern part of the bear range. Many more yearling masses were measured, but only these 37 bears lived to reproductive age and were observed with cubs.

5 kg less than yearlings from the uplands (Figure 9.6); moreover, masses of lowland yearlings were more variable (CV = 36% lowland versus 25% for upland yearlings). Fifty percent of lowland yearlings, but only 10% of upland yearlings, weighed ≤15 kg; eventual ages of maturity for bears this light were highly variable (Figure 9.6). Of nine yearling bears that weighed <10 kg in the den, eight were offspring of females (four different mothers) that resided in the lowlands; nearly 90% of yearlings this small later died of nutrition-related problems.

VARIATION IN HARVEST AND HARVEST COMPOSITION

Most hunters in Minnesota (>80%) use bait to attract bears. Thus, they are in effect trying to lure bears away from their natural feeding places to an artificial food source. The better the natural food supply, the less apt hunters are to attract bears to their baits. That is, hunting success is inversely related to autumn food abundance (Noyce and Garshelis 1997). We believe there are three reasons for this. First, for the most part, bears seem to prefer natural foods to human-related foods. Second, bears that locate a good feeding area would likely not move much and thus may be unaware of hunters' baits. Third, bears may be wary of the scent of humans or other bears around bait sites. Food availability is such a driving force in hunting success that we can reasonably predict yearly harvests (±15% on average) with just two variables: number of hunters and availability of three autumn foods (Figure 9.7).

The sex structure of the harvest is also greatly affected by autumn foods. The harvest tends to be somewhat male-biased, probably because (1) males move more, and hence encounter more baits; (2) males may be aggressive toward females, and thus may discourage females from approaching concentrated food sources, such as baits; and (3) females appear to be more wary and thus less prone

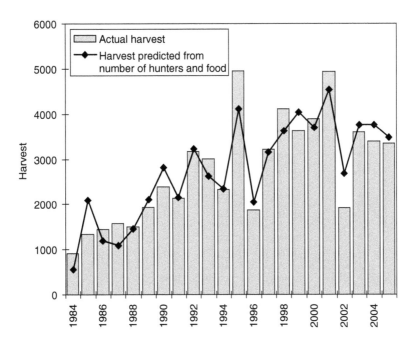

FIGURE 9.7 Harvests of black bears in Minnesota increased with increasing numbers of hunters, which were controlled by a quota on license sales. However, harvests were higher than normal when natural foods were poor and bears were more easily attracted to hunters' baits. Conversely, bears were less readily attracted to baits when natural food was abundant. Including both number of hunters and a subjective rating of abundance of three key fall foods in a regression model accounted for much of the variation in the harvest ($r^2 = .88$).

to approach a site with human odors. However, poor autumn food abundance results in harvest sex ratios that are either less skewed to males, or even female dominated (Noyce and Garshelis 1997). Apparently, hunger makes the behavior of the sexes less disparate, and hence more reflective of the living population.

Statewide, the harvest sex ratio has ranged from 41 to 60% male. Sex ratio varied even more within individual BMUs (36 to 68% male in one BMU in consecutive years). Northern BMUs showed more variability in harvest sex ratios than BMUs in the southern portion of the bear range (Figure 9.8), presumably related to differences in variability in the natural food base. Annual variation

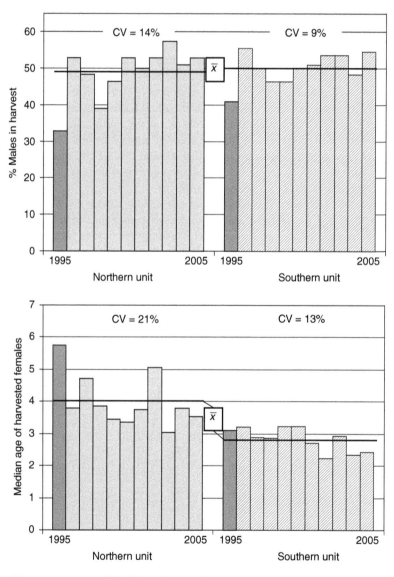

FIGURE 9.8 Sex–age structure of black bear harvests in the northern part of the Minnesota bear range were more variable year-to-year than in more southerly parts of the range, because natural foods in the north were more variable. Shown is a comparison of harvest data (sex ratio and median age of females) from a BMU along the Canadian border to a BMU in the southern part of the range. Coefficients of variability in sex and age structures were higher in the north with either similar (top) or differing (bottom) means. An exceedingly poor food year in 1995 (highlighted) had an especially large impact on the harvest sex ratio and age structure in the northern unit.

in productivity of both summer and autumn foods, as measured through our annual bear food surveys, was higher in the northern bear range (CV = 21% summer, 25% autumn) than in the south (CV = 18% summer, 14% autumn).

Harvest age structure is also affected by food availability. Juvenile bears (2–3-year olds) are normally more vulnerable to hunter harvest than are adults, but in years with low autumn food abundance, the harvest age structure was older, especially among females (Noyce and Garshelis 1997). In BMUs in northern Minnesota, where the median age of harvested females averages 3–4 years, the median age jumped to over 5 years (>7 years in some BMUs) during the 1995 food failure. Again, more southerly BMUs showed less variability, and less effect of the food failure (Figure 9.8).

VARIATION IN POPULATION SIZE AND COMPOSITION

Inasmuch as hunting is by far the largest source of mortality for Minnesota black bears, variations in harvest cause variations in population size. Moreover, the strong association between autumn food abundance and harvest (Figure 9.7) means that variations in food abundance directly affect population size. The killing of nuisance bears strengthens this link between population dynamics and foods: just as with the harvest, more bears are killed as nuisances when natural foods are poor. However, we cannot empirically demonstrate this relationship between population size and food, because variations in food also cause varying biases in our population estimates (see next section).

Foods also influence large-scale movements of bears, thereby altering local densities. Because bears from northern Minnesota tend to move southward, and move further when foods are poor, regional densities vary year-to-year with food availability. In northern parts of the state, we have evidence that bears from Canada, also moving southward, periodically enter Minnesota, thus temporarily augmenting our autumn bear population. This situation would be most notable in poor food years.

Variations in food abundance not only affect the size but also the composition of the living population through previously discussed effects on female productivity and the sex–age structure of the harvest. In particular, a series of poor food years would result in poor reproduction and high harvests of older females, thus reducing the population growth rate.

CAUSES OF BIAS IN POPULATION ESTIMATES

Capture heterogeneity, related mainly to differential attraction to baits, is a common problem associated with mark–recapture population estimates for bears. Specifically, problems are not caused by heterogeneity in capture vulnerability *per se*, but rather heterogeneity that persists unchanged from the initial marking to subsequent recapture samples. For example, if the marked sample is biased toward a certain sex–age group, then biases arise in the population estimate if that same sex–age group is also more likely to be recaptured. We commonly found such linkages between bears that were marked and those that were recaptured, due to the use of bait to obtain both samples. That is, certain sex–age groups or individuals that were initially attracted to bait and thus marked, tended to be more apt than other bears to approach another bait and thus become recaptured. This situation occurred whether the initial marking ensued from a trap capture or a tetracycline-laced bait, and whether the recapture was via a trap, a camera, or a hunter (Garshelis and Visser 1997; Noyce et al. 2001; Garshelis and Noyce 2006).

Because bears' attraction to bait varies with natural food conditions, the linkage between marking and recapturing events also relates to food availability. Hence, two population samples taken the same season (with the same food conditions) were strongly linked, leading to a severe underestimate of population size (Noyce et al. 2001). Marking in summer, followed by recapture that autumn (via the harvest) were also linked, but the extent of this linkage varied with food. Poor food conditions led to a sampling of bears with wide variability in their tendency to be attracted to bait, whereas good food conditions attracted a narrower sampling of only the most bait-attractable bears. Hence, a statewide summer marking (with tetracycline) and autumn hunting recapture during a year with abundant foods in both summer and autumn produced a highly biased population estimate (Garshelis and Noyce 2006).

Spatial heterogeneity was another confounding variable in population estimation. Food conditions varied across the bear range, so during our statewide estimates a higher proportion of bears were marked in geographical regions where food was less plentiful, either due to the particular situation that year, or to the general abundance of food-producing plants in that area. If the same conditions prevailed during the autumn hunter-kill sample, the statewide estimate would be strongly influenced by this spatially biased marking and recovery. Conceivably, this might be rectified by partitioning the range into regions, and obtaining separate estimates for each. This would be particularly beneficial if those regions were BMUs, so estimates could better guide management. However, bears moved among BMUs in response to food. In 1991, although bears took tetracycline baits during the summer in all portions of the bear range, with no obvious clustering, the autumn hunter-killed sample yielded no marked bears within 50 km of the Minnesota–Canada border. This indicated that a large number of bears that were marked in the northernmost part of the range during the summer, moved south in autumn (and were likely replaced by unmarked bears from Canada). Notably different spatial patterns of marking and recovery occurred in other years (Figure 9.9). The large scale of this spatial heterogeneity negates the possibility of obtaining reliable population estimates for individual BMUs.

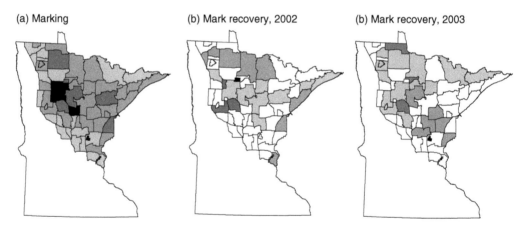

(a) Marking   (b) Mark recovery, 2002   (b) Mark recovery, 2003

FIGURE 9.9 Heterogeneous spatial distribution of marking and recovery of marks in Minnesota hampered population estimation for specific regions (e.g., BMUs) within the state. Tetracycline-laced baits (~3000), used to mark bears, were distributed fairly uniformly across the bear range in 2002, but regional differences in bear density and food conditions affected the number that were taken. Panel (a) shows the varying density of marking (consumed baits) among harvest blocks (regions smaller than BMUs): lowest marking density (<5 marks/1000 km^2) corresponds with lightest shade of gray; black shaded blocks had the highest marking (>15 marks/1000 km^2). Panels (b) and (c) show recovery of marked bone and tooth samples (obtained from hunters), which varied regionally with differing densities of marked bears (a function of local bear densities and the proportion that were marked) and the harvest rate by hunters. These panels also show different recovery patterns in 2002 and 2003 for bears that were tetracycline-marked in 2002 (marks persist >2 years). White blocks indicate no marked samples recovered. Lightest shade of gray corresponds with 0.1–0.9 marks/1000 km^2. Black shade indicates >2.7 (up to 12.7) marks/1000 km^2. Relatively high recovery of marks occurred in the northwestern and extreme eastern parts of the state in 2002, versus in the south-central part of the bear range in 2003. Differing patterns of mark recovery between years stems from fall movements that altered the distribution of bears and from regionally varying food conditions that altered their attraction to hunters' baits.

Within individual study areas, we attempted to deal with the issue of geographic closure (bears transgressing study area boundaries) by using radio collars to ascertain the proportion of time that each marked bear spent in the defined area (Garshelis 1992). Thus, any marked bears that left and were unavailable for recapture were omitted from the marked sample; only those that spent all of their time in the trapping area were considered full-time residents of that area, and others were weighted as partial residents. Hence, although we tended to capture more males than females, in part because males had larger home ranges that encompassed the trapping area, these males also spent more time outside the study area, so were weighted less. Nevertheless, in 1985, a perplexing result occurred, where our estimate of the number of males occupying the study was greater than the number of females (Figure 9.10) — a result inconsistent with a male-biased harvest. A conceivable explanation for this counter-intuitive result is that the particularly low food availability in 1985 caused many males living or traveling near the edges of our study area to be drawn in by our trapping baits; nearly one-third of males (but essentially no females) spent their third summer, at 2 years old, traveling widely, apparently looking for potential areas to settle. It thus appears that bait, a necessary instrument for studying bears, interacts with fluctuating food resources to affect measurements of population size.

MANAGEMENT IMPLICATIONS

Stability is desirable in wildlife populations because stable systems are more predictable and hence more manageable. Cottingham et al. (2001, 72) commented that "variability can ...cloud our perceptions of how nature operates and prevent us from making definitive conclusions about ecological systems." Commonly used indices of population change for bears may be affected more by year-to-year changes in food supply than by changes in bear numbers (Garshelis 1991; Clark et al. 2005).

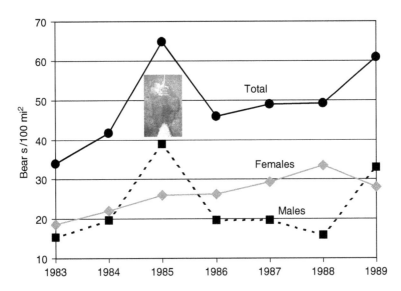

FIGURE 9.10 Mark-recapture estimates of black bear density on a study area in the central bear range of Minnesota. Geographic closure was accounted for by weighting bears by the proportion of time they spent on the trapping area (as per Garshelis 1992). Thus, the spike in the number of males in 1985 represents real bears residing in the area that year, not bears that just passed through. We posit that our trap baits spurred wide-ranging males, which population-wide were less abundant than females, to be attracted to and remain on the study area that year, because natural foods were scarce.

Severe food failures create particular difficulties for managers of bear populations because they foster the perception of instability, and even mismanagement. The sudden appearance of large numbers of nuisance bears and unusually high hunting success may suggest to managers and the public that the population had dramatically increased, or at least was higher than previously thought. Ironically, food failures may actually result in excessive bear mortality, especially among reproductive females, possibly engendering a population decline. In contrast, a stretch of years with above-average foods, and thus relatively few observations of bears by hunters and the public, may prompt the opinion that the population was in decline, or lower than previously thought. Few management agencies have rigorous statewide population estimates for black bears, and among those that do (such as Minnesota), confidence intervals are so wide as to allow for many differing interpretations of population size and trend (Garshelis and Hristienko 2006; Garshelis and Noyce 2006).

A reasonable response from managers is a quest to provide better habitat for bears, to enhance population "vigor" and reduce fluctuations in nuisance activity and harvest. Some wildlife managers in Minnesota have suggested that, through further research, we should ascertain the best habitats for bears and then lobby to create more of those types, so bears have better foods. Like farmers, ranchers, and foresters, wildlife managers have traditionally tended to follow prescribed patterns that favor uniformity over diversity because "creating more of that which produces the highest yield" not only makes intuitive sense, but it alleviates complexity and thus better enables fine-tuning of the prescription; moreover, in the short term, it probably does result in higher production (Hooper et al. 2005). In the case of bears, however, which use a variety of foods and habitats through the year, with different combinations of these resources across years (Figure 9.2), it would be very difficult to determine which specific habitats contributed most to their reproduction and survival (Garshelis 2000). For example, Davis et al. (2006) found that sites commonly used by black bears in British Columbia were not well characterized by traditional models based on cover type and canopy closure; hence, evaluating that the bears' selection criteria was much more complex, including phenological adjustments for berry production, phenological adjustments for succulent green vegetation, horizontal cover, proximity to high- and low-traffic roads (the former decreasing, the latter increasing the probability of bear use), and availability of salmon. All of these would likely be important factors to Minnesota bears as well, except that ants would substitute for salmon. Taking these variables one step further, from bear use to their value in terms of bear reproduction and survival, would be daunting.

Even if certain habitats were deemed especially beneficial, on average, we believe it would be unwise to manage for these at the expense of habitat diversity. The patchwork of different forest types and ages so prevalent in northern Minnesota provides bears with many foraging options. Demographic and behavioral fluctuations would surely be greater in a more homogeneous forest.

Certainly, the extensive pine forests that existed in much of northern Minnesota upon the arrival of early European settlers were not conducive to stable bear populations. Even though records of wildlife from that time are poor, evidence of seasonal movements of bears outside their normal range exists (Bailey 1929). Swanson (1940, 188) collected information from the mid-1800s indicating large bear "assemblages" in areas where food was locally abundant. In neighboring Wisconsin, Schorger (1949) described large periodic concentrations of bears and extraordinary numbers killed during specific years in the 1700s and 1800s, apparently due to widespread food failures. In the 1940s, when bears in Minnesota were killed for bounty, large surges in these kills and in the extent of nuisance activity were reported (Schorger 1946; Petraborg et al. 1949). Continued fluctuations occurred during the 1950s–1970s, but by this time the bear population was so low that encounters between people and bears were far less likely (Garshelis 1989).

Chances are that the current mix of habitats in Minnesota's northern forests is not ideal for bears, but determining the precise, ideal mix is likely not possible. Moreover, the natural heterogeneity of soils, topography, and microclimate across the Minnesota landscape, each combination of which favors different vegetation assemblages, limits the ability to create specific forest types at will, even if there existed a desire to do so. We can, however, employ research findings to make recommendations relative to timber harvest and silvicultural practices that would encourage within-stand productivity and diversity of fruit and nut-producing species. To accomplish meaningfully higher diversity at the local scale would require attention toward diversification at the landscape scale (Waide et al. 1999; Hooper et al. 2005). This is particularly relevant for Minnesota bears, which travel widely over the landscape.

Striving for forest diversity, rather than focusing on individual habitats, should be beneficial to bears and many other species to which Minnesota's forests are home. The ecological principal of diversity promoting stability applies on the level of the community and for individual species. During the 25-year period of our research, four disruptive food failures were witnessed, initially occurring at intervals of about 5 years, with correspondingly high levels of nuisance activity and spikes in the harvest. However, no food failures occurred during the past 10 years, and nuisance activity has remained low. Whether this relates to the escalating levels of forest cutting that have occurred in recent decades, providing greater stand age diversity and thus more variety of potential foods, is as yet unknown. Nevertheless, it seems logical that this sort of forest management may indeed benefit bears, even though it is controversial (Nielsen et al. 2004a,b; Davis et al. 2006).

An enigmatic result of food diversity may be increased bear abundance. Just as a diverse financial portfolio is more likely to gain in value — and show less year-to-year variability — than a narrower assortment of investments, so too would a bear population likely respond to a diverse array of foods. That is, greater diversity of foods should yield less demographic stochasticity for bears, and hence higher population growth (Boyce et al. 2006). In this sense, diversity results in greater short-term stability, but longer-term instability (growth) in population size. For bears, this also may create management challenges (increased nuisance activity and vehicle collisions), although these are probably more desirable than trying to manage a system with large year-to-year fluctuations. Notably, discernible within the year-to-year variability in the proportion of yearlings in the Minnesota bear harvest (Figure 9.5) is a slow but significant upward trend ($r^2 = .3$, $P = .005$).

A key conclusion from our research is that variation in the food supply has not only behavioral and demographic consequences for bears, but also consequences in terms of our ability to monitor and manage the bear population. Managers in Minnesota have requested reliable bear population estimates for individual BMUs: We cannot provide these, because population estimates for small areas are too greatly affected by capture heterogeneity and lack of closure (movements), both of which are related to the food supply. Managers would like to employ indicators of population trend, such as hunting success, harvest sex and age structure, and nuisance activity: these, though, are unlikely to be useful population trend indicators, except over the long-term, because they fluctuate too dramatically with varying food conditions. Managers want to fine-tune harvests through adjustments in quotas on license sales: However, food availability has as much influence on harvest as the number of hunters (Figure 9.7). Managers seek a male-skewed harvest, and indeed some states set goals and rules-of-thumb specifically related to harvest sex ratio and age structure (Garshelis 1991); in Minnesota, however, the sex and age structure of the harvest is mainly at the mercy of autumn food conditions.

Knowledge of food conditions has been instrumental in understanding and hence managing the Minnesota bear population. A simple food survey (Noyce and Garshelis 1997) has provided sufficient information to explain peaks and troughs in bear population-related statistics, thus alleviating "knee-jerk" management responses. Much of the variability in the food supply is out of our control and so uncertainty will always be present. However, the uneasiness related to unpredictable events that could severely impact the bear population (e.g., Pelton 1998) may be mitigated to an extent by striving for habitat diversification, thus buffering against widespread food failures.

POSTSCRIPT: DIVERSITY VERSUS KEY FOODS

Notwithstanding the greater stability conferred by diverse food resources, not all foods are equivalent in terms of their impacts on bear populations. Certain key foods have especially large influences on bear behavior and demography. For example, figs (*Ficus* spp.) are a key food for several populations of sun bears (*Helarctos malayanus*; Wong et al. 2005; Fredriksson et al. 2006), and salmon (*Oncorhynchus* spp.) are a driving force for many coastal brown bear populations (Hilderbrand et al. 1999; Mowat and Heard 2006). Pine nuts (*Pinus albicaulis*) and acorns strongly influenced the original geographic distribution and current population dynamics of interior brown (grizzly) bears (Mattson and Jonkel 1990; Pease and Mattson 1999; Mattson and Merrill 2002). Acorns [or in some places, beechnuts (*Fagus* sp.)] are of paramount importance to many populations of both American and Asiatic black bears (*Ursus thibetanus*), affecting movements, activity patterns, denning chronology, reproduction, survival and even social interactions of these species (Garshelis and Pelton 1981; Eiler et al. 1989; Pelton 1989; McLaughlin et al. 1994; McLean and Pelton 1994; Schooley et al. 1994b; Samson and Huot 1998; Huygens and Hayashi 2001; Costello et al. 2003; Doan-Crider 2003; Hwang 2003; Oka et al. 2004; Hellgren et al. 2005). We once observed two females fighting to feed on the acorns that had fallen under a single oak tree. Samson and Huot (2001) recorded similar aggressive interactions among bears foraging in proximity in a beech stand, and both these authors and Garshelis and Pelton (1981) observed evidence of mortality resulting from such interactions in preferred stands of hard mast.

The acorn crop has a particularly large effect on reproduction because it is the last food bears eat before hibernation, and because it may provide a sizeable proportion of the total annual caloric intake. In an average year in the Southern Appalachian Mountains, for example, Inman and Pelton (2002) estimated that hard mast (mainly acorns) accounted for nearly 75% of annual caloric production among bear foods; one species of oak produced two-thirds of bear food calories. However, they reported a 17-fold difference in calories contributed by oaks during years of optimal production versus years of mast failure. If female bears are able to find a rich source of this high-fat food, they are more likely to gain the weight sufficient to produce and sustain a litter. Therefore, in Minnesota (table 9.1) and elsewhere, bears may travel well outside their normal home range to seek productive stands of oaks, even though such travel may entail a heightened risk of mortality (Beeman and Pelton 1980; Beck 1991; Hellgren et al. 2005).

We were hampered in trying to assess the effects of food species diversity on Minnesota's bears because diversity, overall productivity, and availability of key foods all had significant, interrelated impacts. This situation is common in many ecological systems (Waide et al. 1999; Worm and Duffy 2003). The more southerly parts of the Minnesota bear range, where soils were richest, had the greatest diversity of fruit-producing species and the highest average fruit production, the lowest variability in nut production, and greatest abundance of oaks. This does not diminish the thesis that stability of bear populations is enhanced by a diversity of foods — but it complicates a direct investigation of cause and effect. Indeed, every ecological principal has its caveats.

ACKNOWLEDGMENTS

The Minnesota Department of Natural Resources provided funding for our long-term bear study. We thank P. L. Coy for her extensive contributions to the fieldwork.

REFERENCES

Bailey, B. 1929. Mammals of Sherburne County, Minnesota. *J. Mammal.* 10:153.
Beck, T. D. I. 1991. Black bears of west-central Colorado. Technical Publication No. 39, Colorado Division of Wildlife, Fort Collins, CO.
Beeman, L. E., and M. R. Pelton. 1980. Seasonal foods and feeding ecology of black bears in the Smoky Mountains. *Int. Conf. Bear Res. Manage.* 4:141.
Boyce, M. S., et al. 2006. Demography in an increasingly variable world. *Trends Ecol. Evol.* 21:141.
Clark, J. D., D. Huber, and C. Servheen. 2002. Bear reintroductions: Lessons and challenges. *Ursus* 13:335.

Clark, J. D., F. T. van Manen, and M. R. Pelton. 2005. Bait stations, hard mast, and black bear population growth in Great Smoky Mountains National Park. *J. Wildl. Manage.* 69:1633.

Coop, J. D. et al. 2005. Black bears forage on army cutworm moth aggregations in the Jemez Mountains, New Mexico. *Southwest. Nat.* 50:278.

Costello, C. M., et al. 2003. Relationship of variable mast production to American black bear reproductive parameters in New Mexico. *Ursus* 14:1.

Cottingham, K. L., B. L. Brown, and J. T. Lennon. 2001. Biodiversity may regulate the temporal variability of ecological systems. *Ecol. Lett.* 4:72.

Coy, P. L. 1999. Geographic variation in reproduction of Minnesota black bears. MS Thesis, University of Minnesota, Minneapolis, MN.

Coy, P. L., and D. L. Garshelis. 1992. Reconstructing reproductive histories of black bears from incremental layering in dental cementum. *Can. J. Zool.* 70:2150.

Davis, H., et al. 2006. Influence of phenology on site selection by female American black bears in coastal British Columbia. *Ursus* 17:41.

Doak, D. F., et al. 1998. The statistical inevitability of stability–diversity relationships in community ecology. *Am. Nat.* 151:264.

Doan-Crider, D. L. 2003. Movements and spatiotemporal variation in relation to food productivity and distribution, and population dynamics of the Mexican black bear in the Serranias Burro, Coahuila, Mexico. PhD dissertation, Texas A&M University–Kingsville, Kingsville, TX.

Dobey, S., et al. 2005. Ecology of Florida black bears in the Okefenokee–Osceola ecosystem. *Wildl. Monogr.* 158:1.

Eiler, J. H., G. W. Wathen, and M. R. Pelton. 1989. Reproduction of black bears in the Southern Appalachian Mountains. *J. Wildl. Manage.* 53:353.

Elowe, K. D., and W. E. Dodge. 1989. Factors affecting black bear reproductive success and cub survival. *J. Wildl. Manage.* 53:962.

Elton, C. S. 1958. *Ecology of Invasions by Animals and Plants.* London: Chapman & Hall.

Felicetti, L. A., C. T. Robbins, and L. A. Shipley. 2003. Dietary protein content alters energy expenditure and composition of the mass gain in grizzly bears (*Ursus arctos horribilis*). *Physiol. Biochem. Zool.* 76:256.

Fredriksson, G. M., S. A. Wich, and Trisno. 2006. Frugivory in sun bears (*Helarctos malayanus*) is linked to El Niño-related fluctuations in fruiting phenology, East Kalimantan, Indonesia. *Biol. J. Linn. Soc.* 89:489.

Garshelis, D. L. 1989. Nuisance bear activity and management in Minnesota. In *Bear–People Conflicts. Proceedings of a Symposium on Management Strategies*, M. Bromley (ed.). Yellowknife, Northwest Territories, Canada: Northwest Territories Department of Renewable Resources, p. 169.

Garshelis, D. L. 1991. Monitoring effects of harvest on black bear populations in North America: A review and evaluation of techniques. *East. Workshop Black Bear Res. Manage.* 10:120.

Garshelis, D. L. 1992. Mark-recapture density estimation for animals with large home ranges. In *Wildlife 2001: Populations*, D. R. McCullough, and R. H. Barrett (eds). London: Elsevier Applied Science, p. 1098.

Garshelis, D. L. 2000. Delusions in habitat evaluation: Measuring use, selection, and importance. In *Research techniques in animal ecology*, L. Boitani, and T. K. Fuller (eds). New York: Columbia University Press, p. 111.

Garshelis, D. L., and H. Hristienko. 2006. State and provincial estimates of American black bear numbers versus assessments of population trend. *Ursus* 17:1.

Garshelis, D. L., and K. V. Noyce. 2006. Discerning biases in a large scale mark-recapture population estimate for black bears. *J. Wildl. Manage.* 70:1634.

Garshelis, D. L., K. V. Noyce, and P. L. Coy. 1998. Calculating average age of first reproduction free of the biases prevalent in bear studies. *Ursus* 10:437.

Garshelis, D. L., and M. R. Pelton. 1981. Movements of black bears in the Great Smoky Mountains National Park. *J. Wildl. Manage.* 45:912.

Garshelis, D. L., et al. 1999. Landowners' perceptions of population trends, crop damage, and management practices related to black bears in east-central Minnesota. *Ursus* 11:219.

Garshelis, D. L., and L. G. Visser. 1997. Enumerating megapopulations of wild bears with an ingested biomarker. *J. Wildl. Manage.* 61:466.

Hellgren, E. C. 1998. Physiology of hibernation in bears. *Ursus* 10:467.

Hellgren, E. C., D. P. Onorato, and J. R. Skiles. 2005. Dynamics of a black bear population within a desert metapopulation. *Biol. Conserv.* 122:131.

Hilderbrand, G. V., et al. 1999. The importance of meat, particularly salmon, to body size, population productivity, and conservation of North American brown bears. *Can. J. Zool.* 77:132.

Hooper, D. U., et al. 2005. Effects of biodiversity on ecosystem functioning: A consensus of current knowledge. *Ecol. Monogr.* 75:3.

Huygens, O. C., and H. Hayashi. 2001. Use of stone pine seeds and oak acorns by Asiatic black bears in central Japan. *Ursus* 12:47.

Hwang, M.-H. 2003. Ecology of Asiatic black bears and people-bear interactions in Yushan National Park, Taiwan. PhD dissertation, University of Minnesota, Minneapolis, MN.

Inman, R. M., and M. R. Pelton. 2002. Energetic production by soft and hard mast foods of American black bears in the Smoky Mountains. *Ursus* 13:57.

Iverson, S. J., J. E. McDonald, Jr., and L. K. Smith. 2001. Changes in the diet of free-ranging black bears in years of contrasting food availability revealed through milk fatty acids. *Can. J. Zool.* 79:2268.

Janmaat, K. R. L., R. W. Byrne, and K. Zuberbühler. 2006. Primates take weather into account when searching for fruits. *Curr. Biol.* 16:1232.

Kinsey, C. 1965. The black bear in Minnesota. In *Big Game in Minnesota*, J. B. Moyle (ed.), Technical Bulletin Number 9, p. 179. St. Paul, MN: Minnesota Department of Conservation, Division of Game and Fish.

Kontio, B. D., et al. 1998. Resilience of a Minnesota black bear population to heavy hunting: Self-sustaining population or population sink? *Ursus* 10:139.

Lehman, C. L., and D. Tilman. 2000. Biodiversity, stability, and productivity in competitive communities. *Am. Nat.* 156:534.

Lhomme, J.-P., and T. Winkel. 2002. Diversity–stability relationships in community ecology: Re-examination of the portfolio effect. *Theor. Popul. Biol.* 62:271.

MacArthur, R. 1955. Fluctuations of animal populations, and a measure of community stability. *Ecology* 36:533.

Mattson, D. J., and C. Jonkel. 1990. Stone pines and bears. In *Symposium on Whitebark Pine Ecosystems: Ecology and Management of a High-Mountain Resource*, W. C. Schmidt, and K. J. McDonald (eds). U.S. For. Serv. Gen. Tech. Rep. INT-270, p. 223.

Mattson, D. J., and T. Merrill. 2002. Extirpations of grizzly bears in the contiguous United States, 1850–2000. *Conserv. Biol.* 16:1123.

May, R. M. 1973. *Stability and Complexity in Model Ecosystems*. Princeton, NJ: Princeton University Press.

McArdle, B. H., K. J. Gaston, and J. H. Lawton. 1990. Variation in the size of animal populations: Patterns, problems and artifacts. *J. Anim. Ecol.* 59:439.

McCaan, K. S. 2000. The diversity–stability debate. *Nature* 405:228.

McDonald, J. E., and T. K. Fuller. 2001. Prediction of litter size in American black bears. *Ursus* 12:93.

McLaughlin, C. R., G. J. Matula, Jr., and R. J. O'Connor. 1994. Synchronous reproduction by Maine black bears. *Int. Conf. Bear Res. Manage.* 9:471.

McLean, P. K., and M. R. Pelton. 1994. Estimates of population density and growth of black bears in the Smoky Mountains. *Int. Conf. Bear Res. Manage.* 9:253

Michener, G. R. 2002. Seasonal use of subterranean sleep and hibernation sites by adult female Richardson's ground squirrels. *J. Mammal.* 83:999.

Mowat, G., and D. C. Heard. 2006. Major components of grizzly bear diet across North America. *Can. J. Zool.* 84:473.

Naeem, S. 1998. Species redundancy and ecosystem reliability. *Conserv. Biol.* 12:39.

Neuhaus, P. 2000. Timing of hibernation and molt in female Columbian ground squirrels. *J. Mammal.* 81:571.

Nielsen, S. E., M. S. Boyce, and G. B. Stenhouse. 2004a. Grizzly bears and forestry. I. Selection of clearcuts by grizzly bears in west-central Alberta, Canada. *For. Ecol. Manage.* 199:51.

Nielsen, S. E., et al. 2004b. Grizzly bears and forestry. II. Distribution of grizzly bear foods in clearcuts of west-central Alberta, Canada. *For. Ecol. Manage.* 199:67.

Noyce, K. V., and P. L. Coy. 1990. Abundance and productivity of bear food species in different forest types of north central Minnesota. *Int. Conf. Bear Res. Manage.* 8:169.

Noyce, K. V., and D. L. Garshelis. 1994. Body size and blood characteristics as indicators of condition and reproductive performance in black bears. *Int. Conf. Bear Res. Manage.* 9:481.

Noyce, K. V., and D. L. Garshelis. 1997. Influence of natural food abundance on black bear harvests in Minnesota. *J. Wildl. Manage.* 61:1067.

Noyce, K. V., and D. L. Garshelis. 1998. Spring weight changes in black bears in northcentral Minnesota: The negative foraging period revisited. *Ursus* 10:521.

Noyce, K. V., and D. L. Garshelis. 2002. Bone prominence and skin-fold thickness as predictors of body fat and reproduction in American black bears. *Ursus* 13:275.

Noyce, K. V., D. L. Garshelis, and P. L. Coy. 2001. Differential vulnerability of black bears to trap and camera sampling and resulting biases in mark-recapture estimates. *Ursus* 12:211.

Noyce, K. V., P. B. Kannowski, and M. R. Riggs. 1997. Black bears as ant-eaters: Seasonal associations between bear myrmecophagy and ant ecology in north-central Minnesota. *Can. J. Zool.* 75:1671.

Odum, E. P. 1953. *Fundamentals of Ecology*. Philadelphia, PA: Saunders.

Oka, T., et al. 2004. Relationship between changes in beechnut production and Asiatic black bears in northern Japan. *J. Wildl. Manage.* 68:979.

Pease, C. M., and D. J. Mattson. 1999. Demography of the Yellowstone grizzly bears. *Ecology* 80:957.

Pelton, M. R. 1989. The impacts of oak mast on black bears in the Southern Appalachians. In *Proceedings of Workshop on Southern Appalachian Mast Management*, C. E. McGee (ed.). Knoxville, TN: University of Tennessee, p. 7.

Pelton, M. R. 1998. Tennessee bear harvest issue. *Int. Bear News* 7:26.

Petchey, O. L. 2000. Prey diversity, prey composition, and predator population dynamics in experimental microcosms. *J. Anim. Ecol.* 69:874.

Petraborg, W. H., M. H. Stenlund, and V. E. Gunvalson. 1949. General observations on numbers, distribution, size, weight, and predation by bears. *Minnesota Wildl. Res. Quart.* 9:142.

Rode, K. D., and C. T. Robbins. 2000. Why bears consume mixed diets during fruit abundance. *Can. J. Zool.* 78:1640.

Rode, K. D., C. T. Robbins, and L. A. Shipley. 2001. Constraints on herbivory by grizzly bears. *Oecologia* 128:62.

Rogers, L. L. 1987. Effects of food supply and kinship on social behavior, movements, and population growth of black bears in northeastern Minnesota. *Wildl. Monogr.* 97:1.

Samson, C., and J. Huot. 1998. Movements of female black bears in relation to landscape vegetation type in southern Quebec. *J. Wildl. Manage.* 62:718.

Samson, C., and J. Huot. 2001. Spatial and temporal interactions between female American black bears in mixed forests of eastern Canada. *Can. J. Zool.* 79:633.

Schooley, R. L., et al. 1994a. Spatiotemporal patterns of macrohabitat use by female black bears during fall. *Int. Conf. Bear Res. Manage.* 9:339.

Schooley. R. L., et al. 1994b. Denning chronology of female black bears: Effects of food, weather, and reproduction. *J. Mammal.* 75:466.

Schorger, A. W. 1946. Influx of bears into St. Louis County, Minnesota. *J. Mammal.* 27:177.

Schorger, A. W. 1949. The black bear in early Wisconsin. *Trans. Wisconsin Acad. Sci., Arts Lett.* 39:151.

Schläpfer, F., and B. Schmid. 1999. Ecosystem effects of biodiversity: A classification of hypotheses and exploration of empirical results. *Ecol. Appl.* 9:893.

Smith, K. G., and J. D. Clark. 1994. Black bears in Arkansas: Characteristics of a successful translocation. *J. Mammal.* 75:309.

Srivastava, D. S., and M. Vellend. 2005. Biodiversity-ecosystem function research: Is it relevant to conservation? *Annu. Rev. Ecol. Evol. Syst.* 36:267.

Swanson, E. B. 1940. The use and conservation of Minnesota game 1850–1900. PhD dissertation, University of Minnesota, Minneapolis, MN.

Vaughan, M. R. 2002. Oak trees, acorns, and bears. In *Oak forest ecosystems: Ecology and Management for Wildlife*, W. J. McShea, and W. M. Healy (eds). Baltimore: Johns Hopkins University Press, p. 224.

Waide, R. B., et al. 1999. The relationship between productivity and species richness. *Annu. Rev. Ecol. Syst.* 30:257.

Welch, C. A., et al. 1997. Constraints on frugivory by bears. *Ecology* 78:1105.

Willey, C. H. 1974. Aging black bears from first premolar tooth sections. *J. Wildl. Manage.* 38:97.

Wong, S. T., et al. 2005. Impacts of fruit production cycles on Malayan sun bears and bearded pigs in lowland tropical forest of Sabah, Malaysian Borneo. *J. Trop. Ecol.* 21:627.

Worm, B., and E. Duffy. 2003. Biodiversity, productivity and stability in real food webs. *Trends Ecol. Evol.* 18:628.

10 Metapopulations, Food, and People: Bear Management in Northern Mexico

David Glenn Hewitt and Diana Doan-Crider

CONTENTS

The end of the twentieth century ushered in new challenges in the management of large predator populations. Through much of the 1900s, management of wolves (*Canis lupus*), cougars (*Felis concolor*), and bears (*Ursus* spp.) consisted of reducing conflicts with people, often through lethal means, and unregulated sport hunting. This management was effective in eliminating large predators in North America from large expanses of their range in the early and mid-1900s. Changes in public attitudes, government predator policy, and environmental laws since the 1960s have enabled populations of many large predators to expand. Increasingly, management of these species is focusing on coexistence between predators and people.

Black bears (*Ursus americanus*) in northern Mexico and western Texas were an exception to this pattern only in that they were not extirpated entirely. Bear populations were severely reduced by shooting and poisoning, but remnant populations persisted in the remote mountains of northern

FIGURE 10.1   Black bear distribution in Mexico as reported by biologists participating in a black bear workshop in 2005. The Serranias del Burro are located southeast of the Big Bend Region of the Mexico–Texas border.

Mexico (Doan-Crider and Hellgren 1996). Changes in Mexican law and public attitudes during the 1960s and 1970s enabled these bear populations to increase (Medellin et al. 2005). For example, by the 1980s, bear populations in the Serranias del Burro of northern Mexico (Figure 10.1) had attained high densities (Doan-Crider and Hellgren 1996). This resurgence of bear populations has fueled cultural pride among inhabitants of northern Mexico, signified by private landowners actively supporting recovery of bear populations and the declaration of the black bear as a National Priority Species by the federal government (SEMARNAP 1999). During this period, breeding populations of black bears reestablished in the Big Bend region of Texas through natural recolonization from Mexico (Onorato and Hellgren 2001; Onorato et al. 2004). Breeding populations of black bears are now found in several mountain ranges of northern Mexico (Figure 10.1).

Management objectives for black bears in northern Mexico are broadly designed to allow bears to colonize suitable habitat and to maintain viable bear populations in these areas (SEMARNAP 1999). To meet these objectives, it is necessary to understand ecological and economic features of northern Mexico and how these features interact with the biology of black bears. Thus, our objective in this chapter is to describe the ecology and economy of northern Mexico as it relates to black bears, and illustrate how metapopulation theory should be used as a basis for bear management in northern Mexico.

ECOLOGICAL AND ECONOMIC FEATURES OF NORTHERN MEXICO

There are several key ecological and economic features that define this region as it relates to bears. First, black bear habitat in this area is the montane oak-pine forests found primarily above 1500 m

FIGURE 10.2 Bear habitat in northern Mexico is found primarily in the mountains that are surrounded by grassland and Chihuahuan desert.

elevation. Areas below this altitude are typically xeric Chihuahuan desert, grassland, or Tamaulipan thornscrub. Although bears may use some of these lower-elevation areas seasonally, they are typically poor bear habitat (Onorato et al. 2003). Because bear habitat in northern Mexico is found primarily in the mountains, and because desert and arid steppes often separate mountain ranges in this area from one another, bear habitat in this region is naturally patchy (Figure 10.2).

A second feature of the area is that there is wide variation in size of mountain ranges containing potential bear habitat. Large patches, such as the Sierra del Carmen, Serranias del Burro, and the Sierra Madre south of Monterrey, Mexico may encompass >1000 km^2, whereas smaller mountain ranges scattered throughout the area, such as the Chisos mountains in Texas and the Sierra Picachos in Nuevo Leon, may be <200 km^2.

A third feature of the area is that the climate varies from arid to subhumid, depending on the elevation and distance from the eastern coast (Ferrusquia-Villafranca et al. 2005). Temperature regimes vary from temperate, with warm summers and cool winters, to hot, with hot summers and mild winters (Ferrusquia-Villafranca et al. 2005). Precipitation can vary substantially among years. During wet years, the montane communities where bears live are very productive. During dry years, food production declines dramatically (Doan-Crider 2003). Drinking water also becomes scarce during drought, because in many areas permanent water is only provided by a small number of natural springs and livestock watering troughs. These water sources may not persist during dry periods.

Aridity limits large-scale agriculture to coastal areas of Tamaulipas, Sonora, and Sinaloa, and to areas where irrigation is possible, particularly along the Rio Grande River (Stoleson et al. 2005). Although timber harvest has not been widespread, it has been an important activity at high elevations in the Sierra Madre Occidental. The agricultural activity affecting the greatest land area in northern Mexico is range production of livestock (Stoleson et al. 2005). Cattle are the most common livestock species, but small producers often raise goats, sheep, and pigs. Land redistribution from the 1930s through the 1970s brought many people to the arid rangelands of northern Mexico where large tracts of land were subdivided into small communal sections. Most of these communal enterprises failed to thrive, and the remaining settlements are small and widely dispersed. Industry is concentrated along the northern border with the United States and in a small number of urban areas, such as Monterrey and Saltillo. These patterns of economic activity have resulted in the seemingly contradictory pattern of high rates of human population growth, but low human population density across much of the area (Stoleson et al. 2005). Since the discontinuation of expropriation of private lands due to a 1991 Constitutional amendment to the Agrarian Reform Act (Foley 1995), communal lands are now being returned into private ownership. Mountainous areas are increasingly being purchased by affluent Mexicans and are being managed not only for cattle production, but also for outdoor recreation. These changing motivations for landownership have benefited bear populations because of the return to large, contiguous tracts of land and an increase in people's tolerance of bears.

BLACK BEAR ECOLOGY

Knowledge of ecologic and socioeconomic characteristics of an area enables development of management strategies for wildlife. However, to interpret the implications of these factors for wildlife, they must be placed in the context of the species' ecology. Nutrition and demography are aspects of black bear ecology that are particularly useful in interpreting the effects of different ecologic and socioeconomic factors.

NUTRITIONAL ECOLOGY

Because foraging is a fundamental interaction between an animal and its environment, understanding the nutritional ecology of a species is a prerequisite to effective management. This is especially true of bears, because of constraints resulting from their evolutionary history and their unique reproductive strategy.

Black bears are in the family Carnivora. Similar to most other carnivores, the bear's digestive tract is relatively simple, with essentially no fermentation capability (Pritchard and Robbins 1990). This is important because it means bears do not have symbiotic microbes that can help digest plant structural carbohydrates. Thus, bears are limited to obtaining energy from easily digested components of their food. Other carnivores deal with this constraint by consuming animal tissue, which is composed of highly digestible protein and fat. Bears are not effective predators and only include large amounts of animal matter in their diet when they have access to carrion or particularly vulnerable prey, such as newborn ungulates, spawning fish, larva of colonial insects, and unhealthy adult prey (Pelton 2003). For this reason, black bear food habits are dominated by plant material (Pelton 2003). When consuming plants, bears must consume high-quality plants or plant parts, such as succulent grasses and forbs, fruits, and nuts.

Because black bears are temperate species, winter is a period when the high-quality food they require is not available. Their strategy to survive this period of low food abundance is to enter a den and hibernate. During hibernation, bears do not eat or drink (Hellgren 1998). They recycle most of the nutrients their body requires, such as protein (Lundberg et al. 1976). Energy, however, cannot be recycled. Instead, bears meet their energy needs by catabolizing fat reserves accumulated before entering hibernation. In preparation for hibernation, bears consume large amounts of food and specifically seek easily digested, high-energy foods that enable them to accumulate the necessary fat reserves.

In southern portions of the black bear's range, such as northern Mexico, there may be sufficient food for bears to remain active all winter. However, even in these areas, bears still increase food intake in autumn and accumulate large stores of body fat. They may not enter a den, but they often reduce their activity and cease feeding for extended periods of time (Doan-Crider and Hellgren

1996). This behavior is true of all bears except pregnant females, which enter a den and hibernate, even if food is available. They do this because their cubs are born in late January or February and are highly altricial, weighing only 200–300 g (Pelton 2000). The female remains with the cubs in the den for 2–3 months, during which time she is lactating and the cubs grow quickly. By the time they leave the den in April or May, the cubs weigh 3–5 kg and are capable of following their mother and climbing trees if confronted with danger. This reproductive strategy requires pregnant females to den over winter, and thus they must accumulate body fat and protein to support their own needs during hibernation and to produce sufficient milk for growing cubs.

The bear's simple digestive system, combined with their need to hibernate during winter, forces them to find large quantities of high-quality food during autumn. Typically, black bears meet this requirement with fruits and nuts. In years of poor mast production, bears expand (Bartoskewitz 2001; Doan-Crider 2003; Dobey et al. 2005; Moyer et al. 2006) or temporarily leave (Garshelis and Pelton 1981; Chapter 9, this volume) their summer home range in search of food needed to accumulate fat stores for winter.

DEMOGRAPHY

There are two key aspects of black bear demography that must be considered in managing black bear populations in variable environments. First, black bears are a large mammal with a relatively low reproductive rate. The age at first reproduction for females varies from 3 to 9 years of age, depending primarily on nutritional resources (Noyce and Garshelis 1994; Costello et al. 2003). Furthermore, females raise litters every other year if food resources are adequate, but will not reproduce if food resources are poor (Rogers 1987; Elowe and Dodge 1989; Costello et al. 2003). Because females do not become reproductively active until three or more years of age and because annual cub production per female is low, adult female survival is a particularly important determinant of bear population growth rates (Eberhardt 1990; Hebblewhite et al. 2003).

Bear dispersal patterns are the second demographic parameter of importance in managing an expanding bear population in a highly variable, patchy environment. Juvenile male bears readily disperse (Rogers 1987; Schwartz and Franzmann 1992; Lee and Vaughan 2003), traveling widely in search of suitable habitat in which to establish a home range. In contrast, juvenile female bears typically establish home ranges adjacent to their natal area (Schwartz and Franzmann 1992; Moyer et al. 2006). These dispersal patterns are a problem in areas where management goals seek to have bears colonize suitable, but unoccupied habitat (Swenson et al. 1998). Young male bears often disperse into such habitat, but females typically do not. Thus, for bears in northern Mexico to occupy and maintain populations in all suitable habitats, some mechanism is necessary to trigger female dispersal.

METAPOPULATION PROCESSES

Although behavior and productivity of individual animals are important, populations are the fundamental units of wildlife management. For this reason, a great deal of research has focused on structure and dynamics of populations. In the past 25 years, scientists have realized that populations of a species are not necessarily independent entities, but may interact with nearby populations of the same species through exchange of individuals. These aggregations of local populations connected by emigration and immigration, such that local dynamics are affected, are referred to as metapopulations (Hanski and Simberloff 1997). In the original formulation of the metapopulation idea, Levins (1969, 1970) proposed a model that included local population extinction rates and movements of individuals among local populations. Thus, in a broad sense, the balance of local population persistence and colonization from adjacent local populations determines the fate of metapopulations.

POPULATION PERSISTENCE

Many factors may influence persistence of local populations. For example, the size of the habitat patch in which the population occurs can influence persistence. Thus, ability of bear populations to persist may be lower in small mountain ranges than in larger mountain ranges. This effect probably results from larger population size, which reduces the effect of environmental and demographic stochasticity, and the more diverse habitat likely to be found in larger areas (Hanski 1997). Resource availability is also important because of its effects on demographic rates and on population density. Thus, low-density populations in poor habitat are less likely to persist than higher density populations in favorable areas. Finally, fluctuation in habitat patch size and resources influence persistence. Such fluctuations are most likely to result from stochastic events, such as storms, drought, and fire, but may also occur as a result of changes in populations of predators or competitors.

MOVEMENT AMONG POPULATIONS

The second key feature of metapopulation processes is movement of individuals among local populations. Such movement serves several critical functions (Stacey et al. 1997). Movement among populations can maintain genetic diversity within local populations. Populations with low persistence can be periodically rescued by immigration from adjacent populations. Populations with negative population growth rates, commonly referred to as sinks, can be maintained by immigration from adjacent source areas.

Because of the importance of animal movement in metapopulation function (Ims and Yoccoz 1997), a greater understanding of this process is needed. There are three distinct steps that must occur for successful movements between populations: emigration, transfer, and immigration (or colonization). In other words, an animal must have motivation to leave the area in which it lives, must be able to travel across the landscape to a new area, and must then choose to settle in the new area and be able to survive there.

Dispersal is a common motivation for emigration, however, and as discussed above, dispersal is not likely to result in emigration of female bears. Although movement of males provides crucial genetic exchange among populations, it does not provide the reproductive potential necessary to rescue populations that have gone extinct or to bolster sink populations. Thus, some mechanism must exist to periodically spur emigration of females.

For a metapopulation to function properly, movement of individuals among populations is critical. Successful transfer relies on habitat through which the animal is capable and comfortable traversing, making the matrix between populations an important component of the system (Weins 1997). The matrix must contain sufficient resources and low mortality potential for an emigrating animal to stay alive while moving between patches of suitable habitat. Water and cover are two resources that could influence the ability of bears to successfully traverse the matrix between suitable patches of habitat, particularly in arid environments. Low human density or high human acceptance of bears is also critical in successful movement of bears among populations, because human-caused mortality is the most common cause of death in many black bear populations (Hebblewhite et al. 2003; Koehler and Pierce 2005; Chapter 9, this volume).

Factors influencing an immigrating animal's decision to settle in a new area are poorly understood (Ims and Yoccoz 1997). Density of and interactions with conspecifics, habitat suitability, and human actions probably play key roles in the ability of bears to successfully settle in a new area.

FOOD AND BEAR METAPOPULATION PROCESSES IN NORTHERN MEXICO

The importance of food in the ecology and behavior of bears makes food a critical element in bear population processes. This fact is true in both heavily forested, productive environments

(Elowe and Dodge 1989; Chapter 9, this volume), and in arid, naturally fragmented habitats of northern Mexico and the southwestern United States (Costello et al. 2003). The role of food in bear population processes in northern Mexico is magnified by the substantial temporal and spatial variation in food resources in this area. Annual variation appears to be influenced by mast production cycles and variation in factors such as precipitation and insect infestation (Doan-Crider 2003; Hellgren et al. 2005). Spatial variation is influenced by variation in types of plant communities, which are a function of precipitation patterns, disturbance events such as fire and grazing, soils, and elevation. Quantifying this complex pattern of variation in bear foods is notoriously difficult, because bears eat a wide variety of foods and the substitution value of one food, for example, madrone (*Arbutus xalapensis*) berries, for another food, such as acorns (*Quercus*) and prickly pear (*Opuntia*) fruits, is not known. To quantify food resources of black bears in northern Mexico, Doan-Crider (2003) chose to focus, not on the biomass of food available, but on a critical nutrient that bears require. Because of the importance of energy to bears when preparing for hibernation during autumn, she chose digestible energy as a currency for quantifying autumn bear foods. Based on this approach, Doan-Crider (2003) documented a dramatic difference in the amount of digestible energy available in the Serranias del Burro in 1999 and 2000 (Figure 10.3). These digestible energy maps not only demonstrate the dramatic variation in food resources annually, but also show clearly that food resources vary spatially. Areas of high food production were plant communities dominated by oaks; whereas, areas of low food production, at least during autumn, were grassland and mixed brush plant communities. Because oak communities are often found at higher elevations in this region, low-elevation mountain ranges are not likely to produce as much digestible energy for bears in autumn. These low-elevation communities, however, may be important in other seasons.

The substantial temporal and spatial variation in bear food resources in northern Mexico has important ramifications for bear metapopulation dynamics. Black bear reproduction is influenced by food resources (Rogers 1987; Elowe and Dodge 1989; Costello et al. 2003; Dobey et al. 2005), implying black bear reproductive success varies spatially and temporally. Thus, birth rates of bear populations in small, low-elevation mountain ranges probably is lower than in bear populations occupying larger, high-elevation, diverse mountain ranges. If mortality rates in small mountain ranges are higher than birth rates, these mountain ranges would become population sinks, and bears would only be expected in these mountain ranges if there were a nearby source population and movement between the populations was possible (Doak 1995).

Average and annual variation in food production may influence the number of bears available to emigrate from source populations through its effect on bear reproduction. Cub production in the Serranias del Burro was high in 1991–1993 (Doan-Crider and Hellgren 1996) and 1998–1999 (Doan-Crider 2003). These were periods of average or above-average precipitation (85 cm in 1992; 70 cm in 1998–99; unpublished data, E. S. Sellers Rancho La Escondida). In 2000–2001, when precipitation was much lower (38.9 cm; unpublished data, E. S. Sellers Rancho La Escondida), cub production dropped precipitously due to a lower proportion of females reproducing and a dramatic decline in cub survival (Doan-Crider and Hellgren 1996; Doan-Crider 2003). These changes in reproduction rates would be expected to reduce the capacity for this large bear population to serve as a source for smaller patches of bear habitat nearby.

Another influence of food resources on metapopulation processes of black bears is through its effect on long distance bear movements. Such movements occur during autumn and usually involve bears moving to feeding areas during late summer or early autumn and returning to their summer home ranges to den (Garshelis and Pelton 1981; Hellgren and Vaughan 1990). Movement distances average 22.1 km in Colorado (Beck 1991) and 18.5 km in Arizona (LeCount et al. 1984) and occurred irrespective of autumn mast production. Bears in Minnesota show increased movements in years of poor food production (Chapter 9, this volume). Female bears in Florida expanded their autumn home ranges fivefold in a year of poor mast production compared with a year of good mast production (Dobey et al. 2005).

An unusual event was described by Hellgren et al. (2005) in which radio-collared bears in the Big Bend region of Texas moved extraordinary distances, presumably because of a drought and food failure in autumn 2000. Of bears that moved away from the study area and did not return, females moved an average of 76 km and males an average of 92 km. Three additional bears returned after traveling from 154 to 214 km. No movements outside the study area were observed in the previous two years. Although production of bear foods was not measured in the Big Bend study site, food production in autumn 2000 was low 90 km to the southeast in the Serranias del Burro (Figure 10.3).

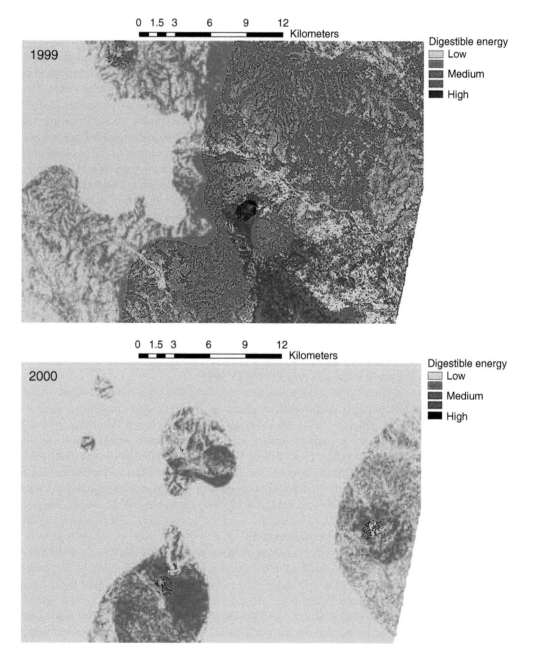

FIGURE 10.3 Digestible energy density of black bear foods during autumn in the Serranias del Burro, Mexico in 1999 (a), which was a year of average precipitation, and in 2000 (b), a drought year.

No radio-collared bears in the Serranias del Burro were noted leaving the study area, but drought conditions and a wildfire on the study area caused female bears to shift home ranges an average of 10.5 km, which doubled their annual home range size relative to previous years in which precipitation was average or above average (Bartoskewitz 2001). Because both male and female bears move in response to low food availability, we are faced with the intriguing possibility that periodic food shortages may be a cause of population decline, especially in small habitat patches where there may be less diversity to support bears in drought, while simultaneously providing the impetus for female bears to leave their natal area and travel to patches of distant bear habitat.

A final critical effect of fluctuating food resources is on bear mortality rates. Although it is not clear how often bears succumb to starvation, especially in southern portions of their range, where winters are mild, it is clear that when natural food resources fail, bear interactions with people increase (Mattson et al. 1992; Peine 2001; Chapter 9, this volume), and this can lead to greater mortality from legal harvest, wildlife damage control measures, or accidents, such as being struck by vehicles (Hebblewhite et al. 2003; Oka et al. 2004; Dobey et al. 2005; Koehler and Pierce 2005; Goldstein et al. 2006). Mortality of black bears in the Big Bend region of Texas increased dramatically in a drought year of low food production compared to two previous years (Hellgren et al. 2005). In many instances, increased mortality occurs, because bears are unable to find sufficient natural food and are attracted to foods around human habitations, including garbage, agricultural crops, and food intended for livestock, pets, or other wildlife. In some instances, bears may prey on livestock when their normal foods become scarce. This occurred in the Serranias del Burro during spring 2000 when predation on cow calves increased from an average of <10/year to >60 bear kills (Doan-Crider and Hewitt 2007). In rare instances, bears may even prey on people (Herrero 1985). During 2000 in the Serranias del Burro, a laborer sleeping outside a small hut was attacked by a black bear (Doan-Crider and Hewitt 2000). This bear was captured the next day and euthanized.

Average quality and quantity of food available and annual fluctuations in food resources can influence bear reproductive, mortality, and emigration rates in an area. Thus, food resources have a substantial effect on metapopulation dynamics of bears in northern Mexico by potentially influencing movements of bears among local populations and the persistence of local populations.

HUMANS AND BEAR METAPOPULATION PROCESSES IN NORTHERN MEXICO

Despite the low human density in much of northern Mexico, the impact of people is still evident throughout the area in livestock and other agricultural operations, development of water sources, habitation, and roads. Furthermore, northern Mexico has rapidly growing metropolitan areas adjacent to bear habitat, such as Monterrey and Ciudad Victoria. As in other places where bears and people interact, there are both positive and negative impacts of people on bears (Figure 10.4). These impacts have important implications for metapopulation dynamics of black bear populations.

HUMAN SOURCES OF FOOD AND WATER

In the arid environments of northern Mexico, food resources may vary dramatically (Figure 10.3) and water sources may disappear as natural springs quit flowing and water basins dry up. Such processes can be highly disruptive; particularly for bear populations inhabiting small, isolated habitat patches (Hellgren et al. 2005). Although very little is known of bear water requirements, it is likely bears require at least periodic access to drinking water. As a result of livestock production and human habitation, permanent water sources have been built and are available to bears in areas that may have been inhospitable otherwise. Black bears were routinely observed using cattle watering troughs in the Serranias del Burro, and on one occasion during a prolonged drought, 14 bears were observed using a water source in a single afternoon (D. L. Doan-Crider, pers. observ., 2000). Black bears in

FIGURE 10.4 Human activities can have both positive and negative effects on bear demographics. The food in this deer feeder will probably increase the nutritional status of these bears, potentially improving their reproduction and survival, but their use of this food source could put them in danger of being trapped or killed if the landowner is not tolerant of their activities.

the Chisos Mountains frequent the water treatment pond of a campground in Big Bend National Park when natural water sources become dry (J. R. Skiles, pers. comm., 2006).

Bears with access to human foods are often larger and more productive than bears consuming only natural food. For example, the only radio-collared bear of nine potentially reproductive females to successfully conceive cubs during a drought year in the Serranias del Burro was a female whose home range included a human habitation and its open garbage pit (Doan-Crider 2003). Corn provided as bait by deer hunters resulted in smaller home range sizes and larger body weights for bears in Florida (Dobey et al. 2005). In Massachusetts, bears with access to agricultural crops had higher reproductive rates than bears in nonagricultural areas, especially in years when mast production was low (Elowe and Dodge 1989). Bears living in the urban–wildlands interface of Lake Tahoe, California had higher reproductive rates and body mass than bears living in the surrounding forests largely because of a sustained high nutritional plane resulting from human derived foods (Beckmann and Berger 2003).

Persistence of bear populations in stochastic environments, like northern Mexico, may be increased by reliable sources of food and water provided by people. Greater persistence of bear populations promotes more stable metapopulations and may increase frequency and length of occupancy of smaller, less viable habitat patches. Greater population persistence and higher occupancy rates could be especially important if ability of bears to move among patches of habitat was in some way restricted by humans, as described below.

HUMAN SOURCES OF MORTALITY

Reliable food and water sources are not a net benefit to bears if these resources regularly bring bears into contact with people and result in greater bear mortality. Because bears are opportunistic and can sometimes pose a serious risk, greater mortality may result from people protecting their property and lives against real and perceived threats by bears. Even if bears do not come into direct conflict with people, bear mortality may increase as a result of collisions with vehicles (Hebblewhite et al. 2003) and from recreational hunting, whether that hunting is legal or not.

Human-induced mortality of bears is influenced by many ecological and sociological factors. As already discussed, variation in food resources influences bear susceptibility to human-caused mortality. Location of natural food sources relative to human habitation and activity and amount and distribution of hiding cover are other ecological factors that may bring bears into close contact with humans. Tolerance for wildlife damage is influenced by many sociological factors, including livelihood, income, and past experiences (Conover 2002). A change in attitude by the public and most agricultural producers, along with a shift in land ownership patterns from communal and subsistence agricultural production to larger operations and outdoor recreation contribute to increased tolerance for bears and their activities. Such changes are occurring in northern Mexico and Texas, and have contributed to increasing bear populations.

Governmental regulations and social norms influence how people address damage by bears and help determine if lethal means are used to alleviate the damage (Conover 2002). Bear populations declined during much of the twentieth century as a result of unregulated sport hunting and killing bears because of actual or potential damage (Leopold 1959). Changes in governmental regulations and people's attitudes appear to have reduced bear mortality rates in many areas of Mexico. Currently it is illegal to kill a bear in Mexico, even a bear causing damage, without prior authorization and proper permits. Such permits are currently available in very limited numbers and only in a small number of areas.

HUMAN EFFECTS ON BEAR MOVEMENTS

Movement of bears among patches of suitable habitat is essential for bear populations to thrive in northern Mexico. A variety of human impacts may reduce the ability of bears to traverse the low-elevation desert and steppe habitat in between mountain ranges. The most dramatic impact limiting bear movement is urban development. Although viable bear populations can exist on the outskirts of urban areas, cities in themselves are a barrier to bear movement. While there are few large urban areas in northern Mexico, the places where they occur could limit bear movement, potentially affecting persistence of bear populations in adjacent patches of otherwise suitable habitat.

Another important human impact that is becoming more prominent as the economic vitality of northern Mexico increases is transportation corridors. Bears are capable of crossing large highways and rail lines (Gibeau and Herrero 1998), but are susceptible to mortality from vehicle collisions (Comly 1993; Hebblewhite et al. 2003). Furthermore, bears avoid areas around roads (Kasworm and Manley 1990), which is likely to limit their movement across a landscape fragmented by roads. It is important to note, however, that while Mexico has many heavily trafficked transportation corridors, no highway mortalities have been reported to date despite the fact that some of these corridors traverse bear populations.

Finally, even in areas of low human densities, bears may be limited from traversing areas between mountain ranges in northern Mexico because of the increased likelihood of bears being killed by poachers. Of 13 bears that left the mountains of the Big Bend region in Texas, at least three were suspected to have been poached in the sparsely inhabited deserts of northern Mexico (Hellgren et al. 2005). Human habitations have many food sources that could attract a bear traveling nearby. Once a bear is attracted to a human settlement, the chances of survival drop dramatically, especially in areas where subsistence agriculture is the main economic activity. Even bears attempting to avoid human settlements may be shot if poaching is not controlled or there is a perceived threat of bears causing damage to property or livestock.

IMPLICATIONS FOR MANAGEMENT

Metapopulation theory suggests that persistence of a species living in fragmented habitat depends on maintaining populations in multiple patches of habitat and ensuring that individuals are able to move among those patches. Recognition of the existence of metapopulations has expanded the scale of wildlife management and has become increasingly important as habitats are fragmented by human activities. In this chapter, we have presented evidence that black bears in northern Mexico have a metapopulation structure and that this structure has important management implications.

Scale of Management

The first management implication from metapopulation theory is that bear management should occur on a scale much greater than the area occupied by a single bear population. Therefore, decisions concerning management of one bear population ultimately influence bear populations in the entire region. This distinction is critical to recognize, because political entities, which are the level at which people make and implement decisions, are small compared to the area used by bears. For northern Mexico, these political entities include private landowners, communal lands, towns and cities, states, and countries. Not only are geographic political boundaries transcended by a landscape view of bear management, but agency boundaries are also transcended. In the United States, for example, bear management decisions could potentially be made by local (city or county), state (Texas Parks and Wildlife Department), or federal (National Park Service or USDA-Wildlife Services) personnel. Management of black bears in northern Mexico will require coordination among similarly diverse political entities.

This broadening of the scale of bear management has some consequences that may not be readily apparent. One is that bear managers need to be concerned with areas of potential bear habitat that are not currently occupied in addition to areas where bears are currently found. Such areas may become inhabited in the future, if properly managed. Even patches of bear habitat too small to support a population over the long term and distant from occupied habitat may be important in enabling bears to move from one large habitat patch to another.

Another consequence of broadening the scale of bear management is that managers must work on much greater temporal scales. Bear management needs to be concerned with the ebb and flow of bear populations over decades. Without a long-term approach to management, small, incremental changes in habitat may seem insignificant. Over the long term, however, such changes can have dramatic effects on the persistence of bear populations and the ability of bears to move among populations. Furthermore, managers planning over large areas and long periods will not be overly concerned with annual fluctuations due to natural causes in any one bear population, because they know that as long as regional populations remain healthy, bears will colonize areas with declining or extirpated populations. In fact, such natural declines in one population may be the impetus for female dispersal, which is necessary to establish populations in previously unoccupied habitat. A final consequence of managing bear populations on large spatial and temporal scales is that managers may pay special attention to a population that is seemingly stable but is isolated in such a way that emigration from other populations is unlikely.

Barriers to Movement

A second implication of a metapopulation structure for bears in northern Mexico is that the ability of bears to move among populations is essential. For this reason, bear management plans must be concerned with the matrix between patches of habitat, not just the bear habitat itself. Land use patterns in the matrix are critical (Aberg et al. 1995; Kupfer et al. 2006). One reason bear populations in northern Mexico have been able to successfully recover from dramatic declines in the mid-1900s is that much of the matrix is dedicated to range production of livestock. Ranching operations maintain wildlife habitat better than many other forms of economic activity. Another encouraging trend is toward the use of rangelands in this region for outdoor recreation. These and other economic enterprises that maintain bears' ability to move across the landscape should be promoted.

Roads are a human impact that may influence bears' ability to move among habitat patches. Whereas part of the impact of a road may be bears avoiding the road itself, the greatest effect from roads is likely the increased exposure to humans and illegal hunting. Because roads are generally built in low-lying areas between mountain ranges, these areas may become mortality sinks. In northern Mexico, for example, there is limited access to private ranches in mountainous areas, but roads traversing between mountain ranges to and from small townships serve as important routes for mining and timber operations, cargo transport, and in some cases, illegal immigrant and drug smuggling. Ranchers often complain of poaching problems along these roads. Recent attempts have been made to increase wildlife law enforcement, but government funds are limited and poaching continues to be problematic in many areas.

Although it is unrealistic to expect placement of highways to be decided primarily in the interest of bear management, it is realistic in some instances for modifications to be made for wildlife, such as inclusion of wildlife underpasses or modifying location of a road to minimize disturbance of critical habitat. Furthermore, interagency collaboration between transportation and wildlife departments should be emphasized and the needs of bears should be added to the variety of factors involved in transportation decisions.

COEXISTENCE OF PEOPLE AND BEARS

Metapopulation theory predicts that a species will cease to exist in an area if extinction rates of populations are higher than colonization rates. The greatest threat to black bear population persistence and movement of individual bears among populations in Mexico is probably not habitat loss or barriers to movement, but conflict with people. Increased conflict as both bear and human populations expand will not only increase bear mortality but can undermine conservation efforts through diminished public support. For this reason, one of the greatest challenges to bear management in Mexico is reducing conflicts between people and bears and promoting tolerance of bears by people living in bear habitat and in the matrix, which bears must traverse between patches of habitat.

Because of bears' strong desire to consume high-quality foods, conflict between bears and humans is generally initiated by attractants, which are potential sources of food to which bears are drawn. Agricultural products produced for consumption by people are often a good source of food for bears, and thus are a source of conflict. Bears commonly cause problems with fruit orchards, grain fields, and apiaries. Much of the damage occurs from consumption of the crop, but bears also cause damage by breaking tree limbs, flattening grain plants, and destroying bee boxes. Conflicts with livestock are generally associated with drought, but in some cases, bears can become conditioned to feeding on livestock and continue to cause problems even when natural foods are available. In the Serranias del Burro, bears began using watering areas that were located in the center of calving herds and were feeding on supplemental feed distributed for cattle (Doan-Crider and Hewitt 2007). Because bears more frequently killed calves in areas with heavy vegetation compared to open areas, and also killed more frequently at night (Doan-Crider and Hewitt 2007), ranchers altered their cattle management by moving calving areas into open grasslands and posting cowboys for nighttime watches. These changes in cattle management greatly reduced bear predation on calves. Thus, management strategies are available for use in areas where bears come into conflict with livestock production.

It is essential to provide legal means for all agricultural producers to reduce bear damage if bears arc to exist in areas of agricultural production. In the United States, governmental agencies work with producers to develop protective or exclusion devices, provide technical assistance, or deal directly with problem animals. In Mexico, wildlife or animal damage control staff is minimal, and training is insufficient to deal with conflicts. Building the wildlife management and extension capacity of agencies in Mexico is an important step in enabling people and bears to coexist.

Inappropriate garbage disposal in both urban and rural areas has led to increased conflict by attracting and conditioning bears to human food sources. In addition to garbage, bears may be attracted to human dwellings to eat domestic animal food, domestic animals (dogs, pigs, goats, and chickens), lard, birdseed, and hummingbird food. Numerous reports have been received of bears entering houses in the presence of humans (D. L. Doan-Crider, pers. observ., 2006), sometimes leading to serious human injury (Doan-Crider and Hewitt 2000). An active campaign to promote bear-proof garbage containers in bear-prone areas and to inform rural and urban dwellers on the importance of waste and attractant disposal could be an effective management action that would ultimately benefit bear conservation.

In localized areas, intentional feeding of bears has become problematic, also leading to human conditioned bears, property damage, and human injury. In some instances, fed bears have become more aggressive in approaching human dwellings, leading to break-ins and other property damage (D. L. Doan-Crider, pers. observ., 2000). Currently, there are no restrictions on feeding bears in Mexico. In the United States, bear feeding is strictly prohibited in national parks, and in most states. Prohibiting feeding of wild bears in Mexico would significantly reduce bear–human conflicts, particularly as interest in photography and eco-tourism increase.

Education programs directed at both rural and urban areas in Mexico are essentially nonexistent, and implementation of such programs should be emphasized by wildlife management agencies. Other areas of high bear/human densities in North America have been successful at minimizing conflict by implementing bear management plans that incorporate education and prevention (Davis et al. 2002). As conflicts increase, public perception about bears can become negative, thus increasing the potential for poaching, and reducing support for bear conservation in public planning efforts. While relocation of problem bears is a management option, it is costly and often results in the bear dying (Clark et al. 2002). Furthermore, relocation does not resolve the root cause of the problem, and often moves the problem along with the bear.

CONCLUSION

Although many management challenges remain, the immediate future of bears in northern Mexico is bright. The bear population appears to be expanding and large areas of bear habitat remain. Public opinion concerning bears is positive and through education and management, designed to prevent conflicts, these positive attitudes can be bolstered. Long-term security for black bears in northern Mexico will involve implementation of large-scale, cooperative management programs that allow metapopulation processes, as described in this chapter, to function.

ACKNOWLEDGMENTS

Many people have contributed to the Mexico Black Bear Project during the past 15 years. Financial and logistical support were provided by David Garza Laguera, Guillermo Osuna Saenz, Felipe and Sandra Holschneider, Charlie and Elizabeth Sellers, and the ranchers of CONECO, A. C. Additional funding was provided by the Lyndhurst Foundation, Mrs Daphne Vaughan, DuPont de Mexico, SA, Unidos Para la Conservación, A.C., the National Fish & Wildlife Foundation, U.S. Fish & Wildlife Service, National Rifle Association, Boone & Crockett Club, George Baker Trust, Mr Juan Brittingham, Robin Brittingham, the Caesar Kleberg Wildlife Research Institute, the Owsley Foundation, the Wray Trust, the Turner Foundation, and the Berryman Institute. Field assistance was provided by Charity Kraft-Lawson, Jonas Delgadillo Villalobos, Oscar Castillo, Sandra Aguilar de Castillo, Barbara Alejandro, Sergio Ledesma Pineda, Esteban Pinedo, Francisco Reyes, and Cody

Crider. Eric Redeker provided assistance in the development of the GIS maps and digestible energy analyses. Eric Hellgren served as advisor for the study during 1991–95, and reviewed a draft of the manuscript. The Secretaria de Medio Ambiente, Recursos Naturales y Pesca granted permission and support for this work to be conducted in Mexico. Rodrigo Medellin, Tom Lacher, Ben Wu, Tim Fulbright, Dave Garshelis, and Steve Herrero provided advice on study design and technical issues.

REFERENCES

Aberg, J., G. Jansson, J. E. Swenson, and P. Angelstam. 1995. The effect of matrix on the occurrence of hazel grouse (*Bonasa bonasia*) in isolated habitat fragments. *Oecologia* 103:265.

Bartoskewitz, C. A. 2001. Spatial relationships related to reproductive status of female black bears in the Serranias del Burro, Coahuila, Mexico. Thesis, Texas A&M University — Kingsville, Kingsville, TX.

Beck, T. D. I. 1991. Black bears of west-central Colorado. Technical publication No. 39, Colorado Division of Wildlife, Denver.

Beckmann, J. P., and J. Berger. 2003. Using black bears to test ideal-free distribution models experimentally. *J. Mammal.* 84:594.

Clark, J. D., D. Huber, and C. Servheen. 2002. Bear reintroductions: Lessons and challenges. *Ursus* 13:335.

Comly, L. M. 1993. *Survival, Reproduction, and Movements of Translocated Nuisance Black Bears in Virginia.* Thesis, Virginia Polytechnic Institute and State University, Blacksburg, VI.

Conover, M. R. 2002. *Resolving Human–Wildlife Conflicts: The Science of Wildlife Damage Management.* Boca Raton, FL: CRC Press.

Costello, C. M., D. E. Jones, R. M. Inman, K. H. Inman, B. C. Thompson, and H. B. Quigley. 2003. Relationship of variable mast production to American black bear reproductive parameters in New Mexico. *Ursus* 14:1.

Davis, H., D. Wellwood, and L. Ciarniello. 2002. *"Bear Smart" Community Program: Background Report.* Victoria: British Columbia Ministry of Water, Land and Air Protection.

Doak, D. F. 1995. Source–sink models and the problem of habitat degradation: General models and applications to the Yellowstone grizzly bear. *Conserv. Biol.* 9:1370.

Doan-Crider, D. L. 2003. Movements and spatiotemporal variation in relation to food productivity and distribution, and population dynamics of the Mexican black bear population in the Serranias del Burro, Coahuila, Mexico. Dissertation, Texas A&M Univeristy — Kingsville, Kingsville, TX.

Doan-Crider, D. L., and E. C. Hellgren. 1996. Population characteristics and winter ecology of black bears in Coahuila, Mexico. *J. Wildl. Manage.* 60:398.

Doan-Crider, D. L., and D. G. Hewitt. 2000. Northern Coahuila predatory black bear attack. *Int. Bear News* 9 (4):22.

Doan-Crider, D. L., and D. G. Hewitt. 2007. Black bear predation upon cattle in the Serranias del Burro, Coahuila, Mexico. *Ursus* In review.

Dobey, S., D. V. Masters, B. K. Scheick, J. D. Clark, M. R. Pelton, and M. E. Sunquist. 2005. Ecology of Florida black bears in the Okefenokee–Osceola ecosystem. *Wildl. Monogr.* 158:1.

Eberhardt, L. L. 1990. Survival rates required to sustain bear populations. *J. Wildl. Manage.* 54:587.

Elowe, K. D., and W. E. Dodge. 1989. Factors affecting black bear reproductive success and cub survival. *J. Wildl. Manage.* 53:962.

Ferrusquia-Villafranca, I., L. I. G. Guzman, and J. E. Cartron. 2005. Northern Mexico's landscape, Part I: The physical setting and constraints on modeling biotic evolution. In *Biodiversity, Ecosystems, and Conservation in Northern Mexico*, J. E. Cartron, G. Ceballos, and R. S. Felger (eds). New York: Oxford University Press, p. 11.

Foley, M. W. 1995. Privatizing the countryside: The Mexican peasant movement and neoliberal reform. *Latin American Perspectives, Labor and the Free Market in the Americas* 22:59.

Garshelis, D. L., and M. R. Pelton. 1981. Movements of black bears in the Great Smokey Mountains National Park. *J. Wildl. Manage.* 45:912.

Gibeau, M. L., and S. Herrero. 1998. Roads, rails and grizzly bears in the Bow River Valley, Alberta. In *Proceedings of the International Conference on Wildlife Ecology and Transportation*, G. L. Evink, P. Garrett, D. Zeigler, and J. Berry (eds). Tallahassee: Florida Department of Transportation, p. 104.

Goldstein, I., S. Paisley, R. Wallace, J. P. Jorgenson, F. Cuesta, and A. Castellanos. 2006. Andean bear-livestock conflicts: A review. *Ursus* 17:8.

Hanski, I. A. 1997. Metapopulation dynamics: From concepts and observations to predictive models. In *Metapopulation Biology: Ecology, Genetics, and Evolution*, I. A. Hanski, and M. E. Gilpin (eds). San Diego: Academic Press, p. 69.

Hanski, I. A., and D. Simberloff. 1997. The metapopulation approach, its history, conceptual domain, and application to conservation. In *Metapopulation Biology: Ecology, Genetics, and Evolution*, I. A. Hanski, and M. E. Gilpin (eds). San Diego: Academic Press, p. 5.

Hebblewhite, M., M. Percy, and R. Serrouya. 2003. Black bear (*Ursus americanus*) survival and demography in the Bow Valley of Banff National Park, Alberta. *Biol. Conserv.* 112:415.

Hellgren, E. C. 1998. Physiology of hibernation in bears. *Ursus* 10:467.

Hellgren, E. C., D. P. Onorato, and J. R. Skiles. 2005. Dynamics of a black bear population within a desert metapopulation. *Biol. Conserv.* 122:131.

Hellgren, E. C., and M. R. Vaughan. 1990. Range dynamics of black bears in Great Dismal Swamp, Virginia-North Carolina. *Proc. Ann. Conf. Southeast Assoc. Fish Wildl. Agen.* 44:268.

Herrero, S. 1985. *Bear Attacks*. New York: Nick Lyons Books.

Ims, R. A., and N. G. Yoccoz. 1997. Studying transfer processes in metapopulations: Emigration, migration, and colonization. In *Metapopulation Biology: Ecology, Genetics, and Evolution*, I. A. Hanski, and M. E. Gilpin (eds). San Diego, CA: Academic Press, p. 247.

Kasworm, W. F., and T. L. Manley. 1990. Road and trail influences on grizzly bears and black bears in northwest Montana. *Int. Conf. Bear Res. Manage.* 8:79.

Koehler, G. M., and D. J. Pierce. 2005. Survival, cause-specific mortality, sex, and ages of American black bears in Washington State, USA. *Ursus* 16:157.

Kupfer, J. A., G. P. Malanson, and S. B. Franklin. 2006. Not seeing the ocean for the islands: The mediating influence of matrix-based processes on forest fragmentation effects. *Global Ecol. Biogeogr.* 15:8.

LeCount, A. L., R. H. Smith, and J. R. Wegge. 1984. Black bear habitat requirements in central Arizona. Special Report 14, Arizona Game and Fish Department, Phoenix.

Lee, D. J., and M. R. Vaughan. 2003. Dispersal movements by subadult American black bears in Virginia. *Ursus* 14:162.

Leopold, A. S. 1959. *Wildlife of Mexico: The Game Birds and Mammals*. Berkeley: University of California Press.

Levins, R. 1969. Some demographic and genetic consequences of environmental heterogeneity for biological control. *Bull. Entomol. Soc. Am.* 15:237.

Levins, R. 1970. Extinction. In *Some Mathematical Problems in Biology*, M. Gerstenhaber (eds). Providence: American Mathematical Society, p. 75.

Lundberg, D. A., R. A. Nelson, H. W. Wahner, and J. D. Jones. 1976. Protein metabolism in the black bear before and during hibernation. *Mayo Clinic Proc.* 51:716.

Mattson, D. J., B. M. Blanchard, and R. R. Knight. 1992. Yellowstone grizzly bear mortality, human habituation, and white barked pine seed crops. *J. Wildl. Manage.* 56:432.

Medellin, R. A., C. Manterola, M. Valdez, D. G. Hewitt, D. Doan-Crider, and T. E. Fulbright. 2005. History, ecology, and conservation of the pronghorn antelope, bighorn sheep, and black bear in Mexico. In *Biodiversity, Ecosystems, and Conservation in Northern Mexico*, J. E. Cartron, G. Ceballos, and R. S. Felger (eds). New York: Oxford University Press, p. 387.

Moyer, M. A., J. W. McCown, T. H. Eason, and M. K. Oli. 2006. Does genetic relatedness influence space use pattern? A test on Florida black bears. *J. Mammal.* 87:255.

Noyce, K. V., and D. L. Garshelis. 1994. Body size and blood characteristics as indicators of condition and reproductive performance in black bears. *Int. Conf. Bear Res. Manage.* 9:481.

Oka, T., S. Miura, T. Masaki, K. Suzuki, K. Osumi, and S. Saitoh. 2004. Relationships between changes in beechnut production and Asiatic black bears in northern Japan. *J. Wildl. Manage.* 68:979.

Onorato, D. P., and E. C. Hellgren. 2001. Black bear at the border: Natural recolonization of the Trans-Pecos. In *Large Mammal Restoration: Ecological and Sociological Challenges in the 21st Century*, D. S. Maehr, R. F. Noss, and J. L. Larkin (eds). Washington, DC: Island Press, p. 245.

Onorato, D. P., E. C. Hellgren, F. S. Mitchell, and J. R. Skiles. 2003. Home range and habitat use of American black bears on a desert montane island in Texas. *Ursus* 14:120.

Onorato, D. P., E. C. Hellgren, R. A. Van Den Bussche, and D. L. Doan–Crider. 2004. Phylogeographic patterns within a metapopulation of black bears (*Ursus americanus*) in the American southwest. *J. Mammal.* 85:140.

Peine, J. D. 2001. Nuisance bears in communities: Strategies to reduce conflict. *Human. Dimensions Wildl.* 6:223.

Pelton, M. R. 2000. Black bear. In *Ecology and Management of Large Mammals in North America*, S. Demarais, and P. R. Krausman (eds). Upper Saddle River, NJ: Prentice Hall, p. 389.

Pelton, M. R. 2003. Black bear. In *Wild Mammals of North American: Biology, Management, and Conservation*, 2nd edn, G. A. Feldhamer, B. C. Thompson, and J. A. Chapman (eds). Baltimore, MD: John Hopkins University Press, p. 547.

Pritchard, G. T., and C. T. Robbins. 1990. Digestive and metabolic efficiencies of grizzly and black bears. *Can. J. Zool.* 68:1645.

Rogers, L. L. 1987. Effects of food supply and kinship on social behavior, movements, and population growth of black bears in northeastern Minnesota. *Wildl. Monogr.* 97:1.

Schwartz, C. C., and A. W. Franzmann. 1992. Dispersal and survival of subadult black bears from the Kenai Peninsula, Alaska. *J. Wildl. Manage.* 56:426.

SEMARNAP (Secretaria de Medio Ambiente, Recursos Naturales, y Pesca). 1999. *Programa de Conservacion de la Vida Silvestre y Diversificacion Productiva en el Sector Rural.* Impresora Grafica Publicitaria.

Stacey, P. B., V. A. Johnson, and M. L. Taper. 1997. Migration within metapopulations: The impact upon local population dynamics. In *Metapopulation Biology: Ecology, Genetics, and Evolution*, I. A. Hanski, and M. E. Gilpin (eds). San Diego, CA: Academic Press, p. 267.

Stoleson, S. H., R. S. Felger, G. Ceballos, C. Raish, M. F. Wilson, and A. Burquez. 2005. Recent history of natural resource use and population growth in northern Mexico. In *Biodiversity, Ecosystems, and Conservation in Northern Mexico*, J. E. Cartron, G. Ceballos, and R. S. Felger (eds). New York: Oxford University Press, p. 52.

Swenson, J. E., F. Sandegren, and A. Soderberg. 1998. Geographic expansion of an increasing brown bear population: Evidence for presaturation dispersal. *J. Anim. Ecol.* 67:819.

Weins, J. A. 1997. Metapopulation dynamics and landscape ecology. In *Metapopulation Biology: Ecology, Genetics, and Evolution*, I. A. Hanski, and M. E. Gilpin (eds). Diego, CA: Academic Press, p. 43.

11 Ecology, Evolution, Economics, and Ungulate Management

Marco Festa-Bianchet

CONTENTS

Modern ungulate management must have three major components: research to increase its knowledge base, a choice of management objectives dictated in part by societal choices, and use of scientific knowledge to achieve those objectives. In addition to being valuable for management, research on ungulates has made major contributions to the development of ecological theory. The study of ungulates is particularly important, because their longevity, strong iteroparity, and overlapping generations produce unique patterns of population dynamics and life-history evolution (Gaillard et al. 2000, 2001). Wildlife management is motivated by human activities: wildlife would not need managers if it was not because of society's wishes to either exploit it, to minimize human impacts on ecosystems, or to avoid wildlife impacts on humans. As human populations expand, use more resources, and increasingly affect ecosystem functions, the need for scientific information to guide wildlife management increases. The diversity of wildlife management issues also increases. A few decades ago, ungulate management mostly involved setting hunting seasons and quotas to avoid overexploitation. In many cases, managers were reintroducing ungulates in areas where they had been extirpated (Komers and Curman 2000). Usually, the main preoccupation was that there were too few ungulates, not too many. The "client" of the fledgling wildlife management profession was the sport hunter, and some countries had no tradition of professional wildlife management based on ecological research.

Over the past 20 years, ungulate management has evolved. Conservation remains a guiding principle, and sport hunters remain a major user of ungulate populations, but ungulate overabundance is now an ecological and economic preoccupation in many parts of the world (Côté et al. 2004; Gordon et al. 2004). Large predators are recolonizing areas from where they had been absent for decades or centuries, some exotic ungulates are now widespread, and some ungulate populations have become a concern for human safety (through disease transmission or vehicle accidents), and for local economies (through damage to crops and forests, or disease transmission to livestock) (Gordon et al. 2004). On the other hand, some species that were abundant a few decades ago, such as woodland caribou (*Rangifer tarandus*) in North America, *Hippocamelus* deer in South America (Saucedo and Gill 2004),

183

and several ungulates in Central Asia are now being threatened with extinction, because human-induced habitat changes (including forest harvests and exotic species) have modified both forage availability and predator–prey relationships (Wittmer et al. 2005a). Other species, such as chiru (*Pantholops hodgsonii*) (Li et al. 2000) and musk deer (*Moschus* spp.) (Yang et al. 2003), are over-exploited to obtain commercial products. Although the number of sport hunters is rapidly decreasing in many countries, "high-end" tourist hunting for trophy males is expanding and generating new ecological, social, and economic challenges and opportunities (Hofer 2002). "Alternative" hunting products such as penned hunts and hunts for exotic ungulates on game ranches are proliferating. These activities provide economic diversification and substantial gain for a few individuals, but have negative impacts on biodiversity and provide choice fodder for antihunting groups.

Wildlife scientists are increasingly preoccupied with habitat fragmentation, climate change (Thomas et al. 2004), and the negative impacts of high ungulate densities on biodiversity. Mounting public interest in conservation means that the "clients" of wildlife managers are now a very diverse group, often with conflicting values or objectives. Wildlife managers also face new legal obligations, such as endangered species legislations and requirements to consult with Aboriginal Peoples and various stakeholder groups. These changes require new ideas and initiatives from wildlife and social scientists. Ungulate managers are evolving from providers of hoofed targets to stewards of ecosystems.

Here, I briefly examine a few examples of how ungulate management has changed over the past few decades, then examine how advances in our understanding of ungulate population dynamics, population genetics, and evolutionary ecology could improve management. I will also discuss new challenges for wildlife managers over the next few decades. Most of those challenges hinge more on improving communication than on increasing our knowledge base. I suspect that managers often know what needs to be done to both manage ungulates and conserve biodiversity, but cannot do it because of social, political, or economic constraints.

PREDATORS, EQUILIBRIA, AND LACK THEREOF

Ungulate management has always had a difficult relationship with large predators. "Predator control" used to be an acceptable part of management when predators were seen as competitors for sport hunters. The pendulum then swung to the opposite side (mostly pushed by people living in urban areas or in regions without large predators), and predator control became a controversial issue. Some populations of large predators now enjoy high levels of protection and are increasing in both abundance and geographical range. As society's awareness of the value of biodiversity increased, a combination of interest by nonhunters in conservation and a near-religious belief in the "Balance of Nature" contributed to elevate the wolf (*Canis lupus*), in some countries (including those whose last wolf was shot long ago), to a level of social reverence shared only by whales and baby seals. Increasing ungulate populations, changes in societal attitudes, and in some cases abandonment of rural areas have recently allowed large predators to reoccupy areas from where they had disappeared. Wolves were reintroduced in Yellowstone National Park (United States) (Vucetich et al. 2005), and have increased substantially in both numbers and range in the north-central United States (Harper et al. 2005) and in both southern and northern Europe (Valière et al. 2003; Vilà et al. 2003). Brown bears (*Ursus arctos*) have increased their range in Europe [partly through reintroductions (Apollonio et al. 2003)], North America, and in Hokkaido in Japan. Cougars (*Puma concolor*) have increased in numbers in much of western North America and may be spreading eastward, possibly supplemented by illegal releases of captive animals (Scott 1998). European lynx (*Lynx lynx*) [which, unlike Canadian lynx (*Lynx canadensis*), are effective predators of small- and medium-sized ungulates] have expanded their range in Scandinavia and have been reintroduced in the Alps (Molinari-Jobin et al. 2002).

Despite these welcome cases of recolonization or reintroductions, however, on a global scale the conservation status of large carnivores is deteriorating. Many populations face an uncertain future,

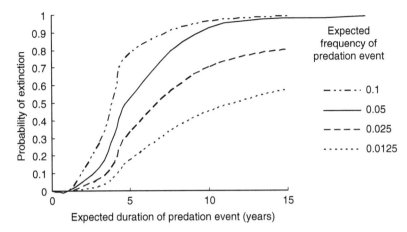

FIGURE 11.1 An example of apparently unsustainable predation: Cougars and bighorn sheep at Ram Mountain, Alberta. The probability of population extinction increased with the frequency of occurrence of individual cougars that specialize on predation on bighorn sheep and with the duration of the predation "event." (From Festa-Bianchet et al. 2006. *Proc. R. Soc. B* 273:1537. With permission.)

and several species are threatened with extinction or extirpation over wide areas (Brashares 2003; Proctor et al. 2005). Even in countries where large carnivores have recently made some gains, the land surface from which they have been eliminated over the past two centuries is typically much greater than the area that they have reoccupied (Leonard et al. 2004). Nevertheless, some expanding carnivore populations are having an impact on ungulates and represent a challenge for wildlife managers, sometimes because of a public perception that large carnivores are under threat, which is true in much of the world (Cardillo et al. 2004), but not everywhere.

There are numerous recent examples of unsustainable predation on ungulates, leading to drastic declines or local extirpations (Figure 11.1). Woodland caribou are disappearing in the face of wolf and sometimes cougar predation (Wittmer et al. 2005b), bighorn sheep (*Ovis canadensis*) populations can be decimated by cougar predation (Festa-Bianchet et al. 2006), and several species of ungulates in the Kruger National Park in South Africa are declining rapidly, apparently because of lion (*Panthera leo*) predation (Owen-Smith et al. 2005).

Restoration of large carnivores is highly desirable for conservation of biodiversity (Berger et al. 2003). Sport hunting should be curtailed when predation increases natural mortality, and sometimes may have to cease. For example, sport hunting of nonmigratory woodland caribou is no longer permitted in most of Canada. In some cases, the impact of returning predators has been moderate. Clearly, more research is needed on how best to adjust ungulate harvests in the presence of predators (Nilsen et al. 2005), and the importance of other factors cannot be discounted: ungulate declines are not necessarily due to the return of large predators (Vucetich et al. 2005). But what should be done when ungulate populations are driven to extinction by predation, and more importantly, why do ungulate and predator populations sometimes not reach equilibrium? With the exception of some island populations, all extant species of ungulates were exposed to predation during their evolutionary history. It is normal for ungulates and large predators to coexist. Why then do we see cases where predation is a threat to the persistence of ungulate populations? I suggest that the answer lies in considering predator–prey equilibria over appropriate spatial and temporal scales. Unfortunately, the temporal scale at which equilibrium is likely to occur may not be acceptable to society, and the spatial scale required may no longer exist because of anthropogenic habitat modifications.

Managers could react to the threat of extinction caused by predation by removing some predators, but that measure is seldom taken (Ernest et al. 2002; Courchamp et al. 2003) because of social opposition. The problem in these situations is not the lack of tools to protect disappearing ungulate populations from large predators, but rather an inability to convince the public that predator–prey equilibrium is not always possible. For example, public opinion in Canada generally opposes removal of wolves and cougars to protect endangered Vancouver Island marmots (*Marmota vancouverensis*), even though both predators are plentiful (on Vancouver Island and in many other areas) and the marmot is so rare (less than 40 in the wild) that it could go extinct within a few years (Bryant 1997). Most people typically oppose predator control because, intuitively, predator–prey equilibria seem inevitable, otherwise either the prey or the predator would have gone extinct. Equilibria over hundreds of years and thousands of square kilometers, however, do not imply short-term and small-scale equilibria, especially where habitats or community dynamics have been artificially modified (Darimont et al. 2005; Whitehead and Reeves 2005).

Over the long term, predators and prey typically reach an equilibrium, and many ungulate populations coexist with large predators, especially in remote areas (Messier 1994; Sinclair et al. 2003). There is no theoretical justification, however, to expect a fixed balance between prey and predators, particularly over small areas or short time spans (Sinclair and Pech 1996). Local extinctions due to predation, followed by recolonization, sometimes over a vast spatial scale, can be part of a long-term equilibrium even in pristine conditions (Kraus and Rödel 2004). Today, some ungulate species are rare because of habitat modifications through human activities, as is likely the case with caribou (that are easily killed by wolves and need old-growth forests) living in areas where populations of moose (*Alces alces*) (that wolves find difficult to kill and benefit from forestry operations) have increased partly because of forestry practices (Messier 1995; Stuarth-Smith et al. 1997; Schaefer et al. 1999) but see Hayes et al. (2000). In other places, the small amount of remaining habitat, or the lack of connectivity between habitat patches, make it unlikely that predators and prey will reach an equilibrium: that may be the current situation in the Kruger National Park in South Africa and that until recently was mostly fenced (Owen-Smith et al. 2005). In these cases, predator control may be an acceptable stopgap measure, but it is not a solution to the problem. If a predator–prey disequilibrium results from changes in the ecosystem, then the ecosystem requires management, not just the predators. Restoration of woodland caribou in Canada will require the restoration of mature forests. Over the long term, caribou conservation will involve cutting fewer trees, reclaiming roads, and limiting snowmobile access. It will not be accomplished just by killing wolves. Over the short term, however, some populations of caribou will disappear under the current level of predation (Wittmer et al. 2005b).

Recolonizing populations of large carnivores are a welcome development for biodiversity, but they present ungulate managers with several challenges. The widely held belief that predators and prey will reach equilibrium is fundamentally correct, but it requires an understanding of the temporal and spatial scales of predator–prey dynamics. For large mammals, those scales may involve decades or centuries, and thousands of square kilometers. Much of the public, however, expects that predator–prey equilibria will always occur, regardless of how small the area considered, how short the time frame, or how much the ecosystem has been compromised by human activities. Wildlife managers need more information on the effects of large predators on small populations of ungulates and on the interactions between predators, habitat changes, fragmentation, and the availability of alternative prey (including introduced exotic herbivores). Predator control on a large scale neither is nor should be socially acceptable, but an unjustified belief in an unfailing "Balance of Nature" may hamper conservation measures required to preserve populations threatened by predation (Courchamp et al. 2003).

EXPONENTIAL POPULATION GROWTH AND UNGULATE IMPACTS ON BIODIVERSITY

Many ungulate species in Europe and North America are probably as numerous today as they have ever been. They live in areas where land-use practices have created good habitat, and are typically

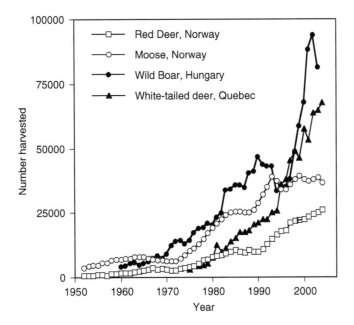

FIGURE 11.2   Examples of increasing sport harvests of selected ungulates in Europe and North America.

harvested conservatively. Where large predators have been eliminated, ungulates suffer little if any predation, particularly on adults. The recent increase in ungulate density is typically viewed as a success story by sport hunters, and many hunting statistics suggest a continuing increase in the number of ungulates harvested over the past few decades (Figure 11.2). The number of red deer (*Cervus elaphus*) harvested in different European countries has increased by 400–700% over the past 30 years, yet in most countries that increase in harvest has not prevented an increase in both numbers and geographical range of red deer (Milner et al. 2006). The range of most species has expanded, and many are now sustaining sport harvests in areas where they were absent 15 or 20 years ago. For example, roe deer (*Capreolus capreolus*) are now found almost everywhere in Europe, including large open agricultural areas and mountains over 2000 m (Andersen et al. 1998). Unfortunately, the same is true for several exotic species, such as red deer in South America, New Zealand, and Australia, or sika deer (*Cervus nippon*) in Europe (Coomes et al. 2003; Pitra et al. 2005).

In this section, I will frequently refer to ungulate impacts on biodiversity as a bad thing, and I should explain why. Just like wildlife management "problems" would not exist if people did not create them, most "impacts on biodiversity" should not be seen as a bad thing unless they were caused by human actions. Over time, ecosystems change, species go extinct, distribution patterns change, and new species evolve. For example, over the past few hundreds of thousands of years, changes in ocean level allowed repeated waves of Old World species to enter North America. Those invasions certainly led to major ecological impacts and species extinctions, but there is no reason to see those events as negative: it is just what happens. If, however, deer browsing modifies habitats and causes local extinctions because people have removed the predators, built barriers to natural dispersal or migration, or introduced deer outside their natural range, then we have a problem and must fix it. There is accumulating evidence that higher biodiversity leads to greater ecosystem productivity and stability (Tilman et al. 1996; Sankaran and McNaughton 1999; Hughes et al. 2005) and, therefore, managers should attempt to limit the human-caused impacts of ungulates on other species. Because ecosystems naturally vary over time, it is always difficult to determine what a "normal" range of impacts is, particularly over short time scales. A useful rule of thumb is that impacts that can be attributed to human actions should generally be seen as negative.

When resources are not limiting, animal populations typically increase exponentially and not linearly. Consequently, protected or lightly hunted predator-free ungulate populations can increase rapidly and reach very high densities (Caughley 1970). Harvest quotas are often not increased sufficiently to contain expanding populations (Milner et al. 2006), because managers have to work with outdated population estimates (Fryxell et al. 1991; Solberg et al. 1999), sport hunters typically like more ungulates rather than fewer, and in some cases hunters simply cannot harvest enough deer (Giles and Findlay 2004). Ungulates can have substantial impacts on vegetation (Gordon et al. 2004). Those impacts are perhaps most obvious following artificial introductions, such as the effects of both wild and domestic ungulates in Australia, New Zealand, and many other islands (Choquenot 1993; Fraser et al. 2000). Although introduced bovids can be the focus of concern, such as mountain goats (*Oreamnos americanus*) in Olympics National Park, United States (Hutchings 1995), most of the current preoccupation with impacts of ungulates on biodiversity focuses on effects of browsing by cervids on the regeneration of woody species or on the persistence of rare plants (Côté et al. 2004). Not all populations of ungulates without either predation or harvests expand uncontrollably: Red deer on the unhunted part of the Isle of Rum or alpine ibex (*Capra ibex*) in the Gran Paradiso National Park in Italy seem to have reached an equilibrium with their environment (at least over a scale of decades) (Coulson et al. 2004; Jacobson et al. 2004). In general, negative impacts of ungulates on biodiversity are greater for browsers than for grazers (Mysterud 2006). Here I am concerned with cases where a herbivore-vegetation equilibrium does not appear possible without human intervention.

Wildlife managers have long been aware of the potential impacts of ungulates on vegetation, but that impact has greatly increased over the past two decades, prompting much concern and research, as recently reviewed by Côté et al. (2004). Extremely high deer densities are reached on islands and in protected areas, but densities sufficiently high to affect succession and prevent forest regeneration are now common over very wide areas. Unfortunately, deer can prevent regeneration of forests at densities well below their short-term carrying capacity (Coomes et al. 2003; Côté et al. 2004). The impacts of high deer density on biodiversity have been reviewed elsewhere in considerable detail (Augustine and McNaughton 1998; Berger et al. 2003; Côté et al. 2004) and I will only briefly summarize them here. Most of the attention has been devoted to impacts on vegetation structure and composition, and on bird diversity, because browsing removes nesting habitat and food resources. In addition, deer likely affect nutrient cycling and soil moisture characteristics, and modify the abundance and distribution of other animal groups such as invertebrates. Wild boar damage to many ecosystems and cultivations is increasing in many areas where boars or feral pigs are introduced exotics, and includes predation on nesting birds, small vertebrates, and invertebrates (Geisser and Reyer 2004). Additional undesirable consequences of high ungulate density are increased vehicle collisions and transmission of diseases to wild and domestic ungulates and to humans, including Lyme disease, brucellosis, tuberculosis, giant liver fluke, and chronic wasting disease (Brownstein et al. 2005). There is no question that ungulates can have a negative impact on biodiversity and on economic activities in areas where they are abundant. The question is what to do about it.

Two aspects of high-density populations of ungulates are particularly relevant to their management: Negative impacts on biodiversity are often evident at densities where ungulate populations still grow rapidly (Coomes et al. 2003), and the density required to limit environmental or economic damage is typically much lower than the density favored by sport hunters (Côté et al. 2004). Consequently, ungulate populations may not stabilize at levels where forest regeneration is possible. Management decisions to maintain low ungulate density in favor of biodiversity are likely to result in substantially lower harvest and therefore unsatisfied sport hunters, an important group of users of wildlife. Alternatives such as exclosures to allow forest regeneration are being experimented with. In parts of Europe, there is a long tradition of artificial feeding of deer in winter, and in some central European countries landowners have a legal obligation to provide supplementary winter food (Putman and Staines 2004). In many countries, hunters also pay compensation for forest and agricultural damage caused by wild ungulates. In France, in recent years, about US $22 million a year have been paid to landowners and farmers in compensation for damage caused by wild ungulates

(Office National de la Chasse 2003). In extreme situations, deer are baited into large enclosures and artificially fed for much of the winter. These practices may allow both maintaining a high deer density and limiting impacts on biodiversity, but they tend to be costly, produce an artificial system that requires continued human intervention, and are typically only feasible over a small scale.

Although several studies have documented severe impacts of high deer density on vegetation [reviewed in Côté et al. (2004)], few have directly addressed its long-term effects on both deer and biodiversity. Over a few decades, even if forest regeneration is prevented by browsing, high densities of deer may be sustained by litter fall and blowdown of mature trees (Tremblay et al. 2005). That system is unsustainable, however, because the mature trees will eventually die, presumably leaving a high-density deer population with no winter forage. Ungulate populations using resources that are not renewing themselves will eventually decline. An unresolved important concern is whether corrective measures (such as a drastic and sustained decrease in deer density) will result in a return to the original vegetation community, or whether some of the changes caused by deer are irreversible (Côté et al. 2004). There is an urgent need for research on both long-term impacts of high ungulate densities on ecosystems, and effectiveness of possible remedial measures (Coomes et al. 2003).

Because negative ecological consequences of overbrowsing often occur at much lower ungulate densities than those that lead to density-dependent reduction in population growth, and because high ungulate density can be sustained by "ecological subsidies" such as litter fall from mature trees (or artificial feeding programs), a superficial assessment of the situation may lead to a false sense of security, particularly by people strongly influenced by ideas about harmony in nature. Wildlife managers may not have an easy task in arguing for ungulate culls when the public thinks that there is no problem, such as in the case of protected areas with no large predators, where hunting is forbidden. That situation is akin to the person falling from a 25-floor building: asked how he was doing as he flew past the tenth floor, he answered "so far, so good." Unfortunately, much of the public still equates "protection of biodiversity" with "no hunting," because long-term ecosystem deterioration is difficult to portray in a 10-sec TV news item.

WHITHER THE BALANCE OF NATURE?

I trust that by now readers will wonder how I can first warn about unsustainable predation on ungulates by large carnivores and immediately after lament the negative impacts of predator-free, overabundant ungulates on biodiversity. I argued that ungulates do not need wildlife managers in the absence of human-caused problems. What I have illustrated since is a series of human-caused problems: predator extirpation followed by ungulate overabundance, complicated by habitat alterations and occasionally a return of predators to landscapes modified by habitat fragmentation, changes in land-use practices, and sometimes the introduction of exotic species. Under those circumstances, a "hands-off" approach is inexcusable, particularly when other sectors of society, such as agriculture, resource extraction industries, land developers, and various recreational industries, are not keeping their hands off. Wildlife managers face two challenges: to defeat the simplistic expectation of a short-term and ubiquitous balance between predators and prey, herbivores and forage, which is so ingrained in much of society, and to base management decisions on scientific knowledge. Management actions that involve killing either predators or ungulates typically face public opposition. They must be based on a combination of solid scientific evidence and professional integrity. There is a fine line between killing a wolf to save endangered caribou and a smokescreen to hide inaction on protecting caribou habitat from logging, snowmobiles, hydrocarbon exploration, and expanding road networks. When controversial policies are justified by the conservation of biodiversity, however, wildlife managers cannot simply shirk away from them just because they may be unpopular. Unfortunately, most politicians will typically select the path of least resistance and opt for policies guided by public opinion rather than by science. Consequently, it may fall upon wildlife scientists in academic institutions, rather than those working for government agencies, to provide an independent assessment of the scientific basis of unpopular management decisions.

SELECTIVE HUNTING SELECTS!

Over the past few years, realization that humans can affect evolution of harvested species has become established in the fisheries literature (Rochet 1998). This realization also led to changes in fishery management practices in the small subset of cases where fishery management is driven partly by scientific knowledge and not just by short-term political objectives (Hutchings et al. 1997; Olsen et al. 2004). Many sport-fishing regulations, for example, now emphasize the importance of protecting large individuals, and harvest of some species is regulated through maximum size limits rather than minimum size limits. Evolutionary effects of overfishing preoccupy fisheries managers, because if fish are artificially selected to reproduce at an earlier age and at a smaller body size, then both their fertility and the fish biomass available to be exploited will decrease (Hutchings 2004). Natural mortality may also increase if earlier reproduction lowers life expectancy and decreases reproductive success (Walsh et al. 2006). If fishing mortality is extremely high, however, reduced life expectancy will be irrelevant, as most fish die young, scooped up in a net.

In many populations of ungulates, most adult mortality is due to hunting (Langvatn and Loison 1999; Ballard et al. 2000; Biederbeck et al. 2001; Nixon et al. 2001; Bender et al. 2004). Hunters are typically selective of the sex–age or morphological characteristics of what they harvest, either because of hunting regulations or of social preferences (Hartl et al. 1995; Maher and Mitchell 2000; Solberg et al. 2000; Strickland et al. 2001; Martinez et al. 2005). Therefore, it is important that wildlife scientists examine the effects of sport harvest on evolution of exploited populations (Law 2001; Harris et al. 2002; Festa-Bianchet 2003).

Two characteristics of sport hunting of ungulates are particularly likely to lead to artificial selection: the preference for hunters to shoot males with large horns or antlers ("trophy" males), and the reduction in age-specific survival imposed by hunting. I have considered elsewhere the potential selective effects of sport hunting on life history strategies (Festa-Bianchet 2003) and I will only briefly summarize them here. I underline, however, that very few studies have addressed this issue. Therefore, although sport hunting may be a selective pressure, there are very few data available to assist scientists in assessing the extent (if any) of artificial selection in ungulates.

In an isolated population of bighorn sheep at Ram Mountain, Alberta, Canada, three decades of trophy hunting selected for rams with genetically smaller horns, mostly by creating a negative correlation between horn size and male reproductive success (Figure 11.3). This result was greeted with indignation by some hunter groups, skepticism by some managers, much interest by other managers, and a yawn by many evolutionary ecologists, who thought that the outcome was rather obvious. In bighorn sheep, horn length affects mating success for mature rams that can defend estrous ewes, but not for rams younger than about 7 years (Coltman et al. 2002). Young rams use alternative mating tactics and father some lambs (Hogg and Forbes 1997), but neither horn size nor social dominance appear to affect their mating success (Hogg and Forbes 1997; Coltman et al. 2002). Presumably, a subordinate ram's mating success is determined by his speed, agility, and willingness to risk being hit by other rams. Because bighorn rams complete much of their horn growth by 5 years of age (Jorgenson et al. 1998), however, Alberta's hunting regulations that require a minimum horn size of 4/5 curl (Figure 11.4) allow fast-growing rams to be shot at age 4. Therefore, large horns will increase a male's mating success from age 7, but will put him at risk of being shot from age four. With a harvest rate of about 30% for "legal" rams, about 10% natural mortality (Jorgenson et al. 1997), and a prerut hunt, a male "legal" at age 4 has only about a 15% chance of surviving to rut as a 7 year old. Males with small horns that never reach legal size see most of their potential competitors eliminated by hunters, and consequently father many lambs (Coltman et al. 2003).

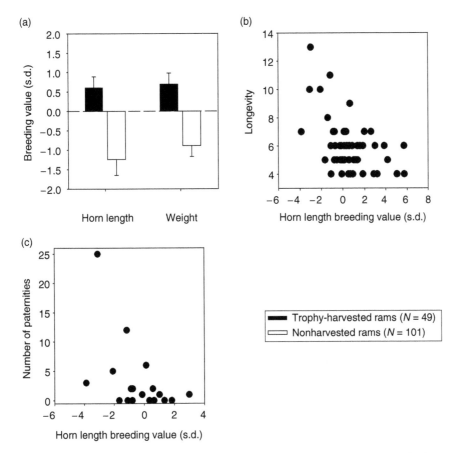

FIGURE 11.3 At Ram Mountain, Alberta, Canada, selective removal of large-horned rams by hunters led to rams with high breeding values for both horn length and body weight having a greater probability of being shot (a), a negative relationship between horn length breeding value and longevity (b), and a negative correlation between horn length breeding value and lifetime reproductive success measured by the number of lambs fathered as determined by DNA analysis (c). Breeding value is a representation of the genetic component of a given trait. (From Coltman et al. 2003. *Nature* 426:655. With permission.)

Artificial selection through trophy hunting is possibly more likely at Ram Mountain than elsewhere, because the population is isolated. There is no immigration from protected refugia such as national parks, where rams with rapidly growing horns should have high mating success, because they will become dominant at a younger age (Pelletier and Festa-Bianchet 2006). During the rut, high- but not top-ranking rams from protected populations may move to areas where many of their potential competitors are removed through trophy hunting, undertaking "breeding commutes" over linear distances of up to 50 km (Hogg 2000). Consequently, a network of protected areas may retard or possibly even negate the selective effects of trophy hunting, if those areas can serve as a source of immigrants. In addition, lower levels of harvest will presumably result in a lower (and possibly negligible) evolutionary impact. There is little reliable information on the harvest rate of "trophy" males in any ungulate population, but the 30% rate measured at Ram Mountain is probably typical for mountain sheep in areas with limited access (Festa-Bianchet 1989). In populations that are easily accessible and where there are no limits to the number of permits issued, it is likely that most rams are shot the year they become "legal."

Ungulate managers are mostly concerned with population dynamics and habitat characteristics that may affect productivity of ungulate populations and, therefore, the sustainable level of hunting.

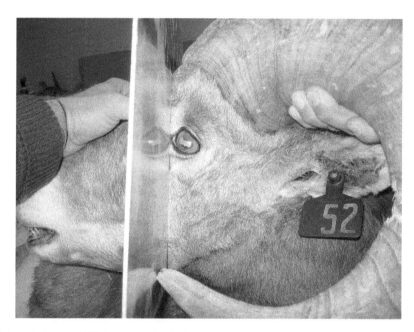

FIGURE 11.4   A 4-year-old bighorn ram shot in 2005 in southwestern Alberta. A line (represented here by the Plexiglas sheet) drawn from the base of the horn to the tip of the eye intercepts the tip of the horn, making this ram legal for harvest. Had its horn been one cm shorter, it would have been illegal to shoot it. Only rams with exceptionally large horns become "legal" at four years of age.

In many populations, sport hunting is the main source of adult mortality, and much of the harvest is selective, either through regulations or through hunter preferences. Therefore, managers should be concerned about potential evolutionary impacts of sport hunting (Festa-Bianchet 2003). Selective hunting could have several undesirable consequences on morphology, life history traits, and eventually population performance of ungulates. The evidence of artificial selection for small horns in trophy-hunted bighorn sheep, and the genetic correlation between traits that favor large horns and fitness-related traits in both sexes (Coltman et al. 2005) is both a conservation and an economic concern. Similar conclusions have recently been reached for fish under very high levels of experimental selective harvest (Walsh et al. 2006). These studies suggest that a harvest regime that targets the largest individuals can quickly lead to negative demographic consequences, as low-quality individuals are left to do most of the breeding.

When the artificial selective pressure is strong, evolution can happen surprisingly quickly. In the study by Walsh et al. (2006), major differences in reproductive performance were induced by just five generations of selection. It is, therefore, urgent to obtain empirical data on intensity of artificial selective pressures caused by sport hunting. An excellent research opportunity is provided by the ongoing drive towards "quality deer management" and by the patchwork of different hunting regulations over different geographical areas (Bishop et al. 2005). An assessment of the selective effects of hunting regulations is likely to be rewarding from both an applied and a fundamental viewpoint, and offers great opportunities for collaborations between wildlife managers and academic scientists.

Vast sums of money are generated through tourist hunting of trophy ungulates, yet a high level of selective removal of "trophy" males may have negative consequences. Hunters are willing to pay large amounts of money for the opportunity to shoot a large-horned male; therefore, a management regime that selects small-horned males appears rather counterproductive. Ecologically sensible harvest schemes, however, are unlikely to generate the same revenues given the current social preferences of trophy hunters. The person who pays $40,000 to shoot an argali (*Ovis ammon*) ram is unlikely to pay that much to shoot a lamb, even if harvesting juveniles would mimic natural mortality and allow a much greater harvest rate. In those cases where some of the revenue generated through trophy hunting is used for conservation (Harris and Pletscher 2002), the potential loss of that revenue through ineffective management would be a serious conservation (as well as economic) concern. Unfortunately, however, while many trophy-hunting programs in developing countries claim to contribute to conservation, most of them only contribute to trophy hunting. Currently, very little, if any, of the money generated through most trophy-hunting of mountain ungulates in Asia benefits either conservation or the local economy (Hofer 2002).

Socially, trophy hunting is less acceptable than other forms of sport hunting and "trophy hunters" are a favorite target of antihunting groups. Yet, if properly managed, trophy hunting can be sustainable and used to finance protection of biodiversity, particularly in developing countries (Leader-Williams et al. 2001). Over the short and medium term, the challenge for managers and researchers is to identify trophy-hunting management practices that do not affect evolution. Possible solutions include reductions in harvest of mature males, greater selectivity for those that have had opportunities to breed (rather than killing high-quality males before they can pass on their genes) and a network of protected areas to provide unselected immigrants. Over the long term, however, the greater challenge is to do away with the competitive aspect of trophy hunting. An end to "scoring" mentality (the "mine-is-bigger-than-yours" approach to hunting) and to bizarre traditions such as "slams" (the "stamp collecting" approach to hunting) would be a good start.

UNGULATE POPULATION DYNAMICS WITH AND WITHOUT HUNTING

Population dynamics and life history evolution are inevitably connected: reproductive strategies, mating systems, maternal investment strategies, and mate choice are all influenced by age-specific survival and reproduction probabilities, and by sex ratio (Stearns 1992). In turn, reproductive strategies can affect age- and sex-specific mortality. It has long been assumed that higher mortality of males than of females among adult ungulates is due to a greater reproductive effort of males during the rut, although the evidence supporting that contention is not very convincing (Toïgo and Gaillard 2003). In particular, it is far from clear that highly successful males suffer higher natural mortality than less successful ones (McElligott et al. 2002; Pelletier et al. 2006). Recent research has revealed that sex–age structure is a major determinant of population dynamics in ungulates (Coulson et al. 2001; Festa-Bianchet et al. 2003), which is not surprising, given how strongly the reproduction and survival probabilities of ungulates vary with age (Gaillard et al. 2000). If mortality induced by sport hunting differs from that due to natural causes, it will almost inevitably affect both evolution (as argued in the previous section) and population dynamics of ungulates.

Three characteristics of sport hunting mortality usually differ drastically from natural mortality: age, sex, and timing. These differences can increase population productivity but can also increase impacts of changes in both density and weather upon population growth rate.

There are now several unhunted or lightly hunted populations of ungulates in North America and Europe where the sex- and age-specific survival of marked individuals has been monitored for many cohorts. The results for females are remarkably similar (Figure 11.5): Juvenile survival is generally low (averaging about 50%) and very variable from year to year, yearling survival is typically 5–10% lower than the survival of adults, then there is a "prime-age" phase (typically from 2 to about 7–9 years of age) when female survival is very high (92–95% or higher) and stable from year to year, followed by a senescent phase where yearly survival gradually declines to about 50% at 17–19 years, an age reached by extremely few individuals (Gaillard et al. 2000; Loison et al. 1999a). Male age-specific survival is usually lower than female survival but follows a similar pattern, although with much more interspecific variability (Figure 11.5). A review of recent studies by Gaillard et al. (2000) revealed that these sex- and age-specific patterns of mortality apply to

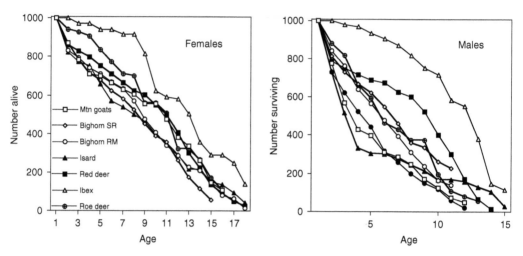

FIGURE 11.5   Natural local survival of cohorts of 1000 yearling females and males in unhunted populations of ungulates, based upon the age-specific survival measured through long-term monitoring of marked individuals: bighorn sheep at Sheep River and Ram Mountain, Alberta (Canada) (Loison et al. 1999a), mountain goats at Caw Ridge, Alberta (M. Festa-Bianchet and S. D. Côté, unpublished data), isard (*Pyrenean chamois*) at Orlu (France) (Loison et al. 1999a), fallow deer (males only, filled circles) at Phoenix Park, Ireland (McElligott et al. 2002), alpine ibex at Belledonne, France (C. Toïgo, unpublished data), red deer on Rum, Scotland (Catchpole et al. 2004) and roe deer at Chizé, France (J.-M. Gaillard, unpublished data). For mountain goat and isard males, some disappearances of animals aged 1–3 years are due to emigration rather than mortality.

most ungulate species: in unhunted populations most mortality affects young of the year, yearlings, and individuals aged 10 years or more (although mortality of males aged 6–10 years can also be substantial in several species; Figure 11.5). Most populations examined in that review had no large predators, but predation can lead to substantial adult mortality (Owen-Smith and Mason 2005). Often, however, predation on ungulates is most severe for very young and very old individuals (Kunkel et al. 1999). Gaillard et al. (1998) suggested that juveniles, yearlings, and possibly senescent individuals are most susceptible to yearly variation in mortality due to changes in weather, population density, and predation. On the other hand, very few studies that accounted for age effects on survival found significant yearly variability in the survival of prime-aged adults, particularly females. Evidence of weather or density effects on the survival of prime-aged adult female ungulates is particularly scarce. It appears that most ungulates evolved under conditions of low and variable juvenile survival and high and stable survival of presenescent adults (Gaillard and Yoccoz 2003).

In contrast, in most hunted populations of ungulates, mortality of prime-aged adults is high, so that few (if any) individuals reach senescence, or even the age of asymptotic mass or horn/antler size. In many hunted populations, mortality of very young males is extreme, leading to the somewhat bizarre situation where hunters think of a 4-year-old as an "old" male, or of a ratio of 5 bulls per 100 cows in elk as normal (Bender et al. 2002). Although there are few precise data on age-specific survival of ungulates in hunted populations, a harvest rate of 20% of adult females, assuming an additional 2% natural mortality, would mean that only about 38% of yearlings would reach the age of five years, half as many as the average of 75% (range 66–94%) in seven unhunted populations of ungulates monitored over the long term (Figure 11.5). For males, a harvest rate of 35% and natural mortality of 3% (both very conservative assumptions) would let only 10% of yearlings survive to age 5, much less than the average of 57% (range 31–93%) for the eight unhunted populations [including mountain goats and isard (*Rupicapra pyrenaica*), where some disappearances of young males were due to emigration] shown in Figure 11.5. Sport hunting of ungulates leads to a truncated age distribution and a strong female bias among adults (Solberg et al. 2002). I have discussed elsewhere the potential evolutionary consequences of shortened life expectancy and biased adult sex ratio (Festa-Bianchet 2003). Here, I will examine some potential consequences for population dynamics.

First, compared with unhunted populations of ungulates, many hunted populations contain few or no senescent individuals, almost no mature males (aged six years and older in most species), and a high proportion of young females that typically enjoy extremely high natural survival (Gaillard et al. 2000). Hunted populations should therefore experience low natural mortality of adults (Festa-Bianchet et al. 2003) and high productivity. Particularly if hunters avoid harvesting juveniles, however, a very high proportion of the post-hunt population will be made up of young of the year and yearlings, the age classes that are most sensitive to the effects of weather and population density (Gaillard et al. 1998). Therefore, heavily harvested populations may show greater changes in numbers according to winter weather than unhunted populations, particularly in interaction with high population density (Portier et al. 1998). A heavily juvenile-biased age structure should also result in stronger density-dependence of overall survival.

Second, male mortality is very high in hunted populations, usually through a combination of hunter preference and management regulations. Consequently, males have a very short life expectancy. In Norway, only 5% of male moose shot by hunters were aged 5 years or older, and 60% were yearlings (Mysterud et al. 2005). In hunted elk populations in Oregon, 7% or less of males survived to 4 years of age (Biederbeck et al. 2001). A female-biased adult sex ratio will increase overall "adult" survival (because males have lower natural survival than females) and increase recruitment compared with naturally regulated populations (Solberg et al. 1999). In extreme cases, however, a scarcity of reproductive males may lower recruitment or increase the proportion of females that fail to conceive in their first estrus (Milner-Gulland et al. 2003; Sæther et al. 2003). The resulting late-born juveniles are likely to experience high mortality (Festa-Bianchet 1988). Although it is sometimes assumed that males are "superfluous" in ungulate populations, theory and data both suggest that males play an important role in population dynamics (Gaillard et al. 2003; Mysterud et al. 2002). The age distribution of males may even affect juvenile sex ratio (Saether et al. 2004). Therefore, the effects of changes in adult sex ratio or male age structure in hunted populations are worth investigating.

Third, the timing of mortality is very different in hunted and unhunted populations. In unhunted populations of northern ungulates, most mortality occurs in late winter, when body condition is at its yearly minimum (Loison et al. 1999b). In hunted populations, much of the mortality is during the autumn hunting season, when animals are typically in peak body condition. In unhunted high-density populations, all animals compete with many conspecifics until late winter, when finally those in worse condition (typically juveniles and senescent individuals) die (Clutton-Brock et al. 1987). In contrast, in heavily hunted populations the survivors of the hunting season face winter with a much-reduced number of competitors. Presumably, the lower level of competition will improve overwinter survival. Consequently, the dynamics of a hunted and an unhunted population with the same summer density are likely to be very different. In the hunted population, survivors of both the hunting season and the winter should be in better condition in the spring and may have higher reproductive success than survivors from the unhunted population that faced a higher level of intraspecific competition during winter (Boyce et al. 1999).

Because of opposing (and possibly interacting) effects of biases in sex and age structure and of difference in timing of natural and hunting mortality, it is difficult to predict how the dynamics of hunted and unhunted populations may differ in the face of changes in weather or density. Demographic and life history theories developed for unhunted populations may see their basic assumptions (sex–age distribution, timing of mortality, strength of the effects of weather, and density-dependence) violated in hunted populations. If harvest levels of adults of both sexes are very high, and higher for males than for females, the surviving population would include mostly juveniles, yearling males, and females aged 1–3 years. Such a population would be highly productive but also very vulnerable to harsh winter weather. On the other hand, a heavy harvest of males and of juveniles of both sexes could maintain a highly productive population with high survival, as intraspecific competition during winter will be lowered by autumn harvests. Finally, the sudden cessation of harvests (as may occur following changes in land tenure) could lead to unpredictable changes in density and sex–age structure (Coulson et al. 2004).

CONCLUSIONS: WHAT TO DO?

Wildlife management should minimize the impacts of humans (including hunters) on biodiversity. Sport hunting must be sustainable, but "sustainable" should not simply mean that enough animals are left to hunt again the following year. It should also mean a management regime that will not drastically alter either the selective pressures acting on wild ungulates, or their impact on the ecosystem. The potential consequences of alternative management strategies must be considered over the long term, because both the ungulates and some of the species they interact with have long generation times. Management leading to artificial selection, or to negative impacts of ungulates on biodiversity, is not sustainable even if it fulfills the short-term goal of providing recreational opportunities.

Sport hunting is and should be a component of conservation, because it generates both interest and education in biodiversity, and income that can be used for conservation. Providing sustainable sport hunting opportunities is an acceptable goal of a democratic society. Inevitably, hunted populations will differ from those that are not hunted. Wildlife managers must know what those possible differences may be, and select a strategy that provides recreational opportunities while minimizing any consequences that are undesirable from an ecological or societal viewpoint.

Sometimes ungulate hunting is necessary, for example, to limit their impacts on ecosystems where large predators have been removed, or to control exotic species. In many other cases, hunting is tolerable, because it does not severely affect biodiversity. Ungulate removals are at times necessary in protected areas where ungulates alter ecosystem functions. Sport hunting of exotic ungulates is encouraged in national parks in New Zealand and may be required in some parks elsewhere if natural predators have been eliminated and cannot be reintroduced.

Sustainable sport harvest should not be selective for morphological attributes and should attempt to mimic natural mortality. Ideally, it would mostly remove young of the year and older individuals. In many species, however, hunters cannot distinguish juveniles from adults, and in most species, they cannot recognize senescent individuals (especially females). There is often a cultural resistance to killing juveniles. When asked about the best sport-hunting strategy for mountain goats, I reply that hunters should only shoot kids. That is because mountain goats have a late age of primiparity, low recruitment, and are highly susceptible to the harvest of adults, especially females (Côté and Festa-Bianchet 2003). Given that mountain goat kids are as cute as baby seals, however, my suggestion usually elicits a negative reaction. Hunters should always be encouraged to harvest juveniles rather than adult females or prime-aged males. Directing most of the harvest to young of the year has been used very successfully for Scandinavian moose (Nilsen et al. 2005) and other cervids (Milner et al. 2006). Although I consider trophy hunting to be ecologically undesirable, it is not realistic to advocate its immediate end, and some harvest of adult males is sustainable. The tradition of seeking to harvest males with large horns or antlers is deeply engrained in the social fabric, and generates revenues that could be (and occasionally are) directed to conservation (Leader-Williams et al. 2001; Harris and Pletscher 2002). More research is required to establish what trophy hunting programs are sustainable and do not affect the evolution of harvested populations. Males with large horns or antlers should only be harvested after they have had a chance to benefit from those large weapons by obtaining a high mating success. For most species, it means harvesting males at least 9–10 years old (Clutton-Brock et al. 1988; Coltman et al. 2002). Clearly, this will require a reduction in the numbers that can be shot. For the eight unhunted populations illustrated in Figure 11.5, the average male survival from yearling to 5 years is 57%, but survival to 10 years averages only 27%.

Wildlife managers and scientists should speak out against the emphasis on trophy size over everything else. That emphasis is motivated by economic gain and does not serve either sport hunting or conservation. It is reflected in attempts to artificially feed ungulates to increase trophy size, the popularity of penned "hunts," and the increasing efforts of outfitters and providers of private "pay-per-hunt" facilities to boost trophy size (Geist 1994). Market-driven hunting industries predictably seek to maximize revenue, and they can best do so by encouraging a hunting ethic that identifies horn or antler size with hunt satisfaction or with personal prestige. Those interested in promoting hunting as a conservation tool rather than as a source of income, however, must advocate ecologically responsible hunting practices. This includes both a limitation of ungulate density to prevent negative effects on biodiversity, and harvest plans that attempt to mimic natural mortality. It also requires policies that maintain sport hunting within the reach of most sectors of society, not restrict it to a small elite group (Geist 1992, 1994).

Although they share many broad similarities in population ecology and life history strategies, not all ungulates have identical ecological attributes, and basic knowledge of the biology of each species (or of the same species in different ecological situations) is required to manage them sustainably. Although age-specific survival of adult females in different species appears broadly similar, for males it can be very different (Figure 11.5). One should not manage the harvest of mountain goats or ibex based on the assumption that they have sex- and age-specific schedules of survival and reproduction similar to those of most cervids. Recent examples of research on marked individuals that underlined interspecific differences among ungulates include the extremely high susceptibility of mountain goats to sport harvest (Hamel et al. 2006), possibly due to a very late age of first reproduction (Côté and Festa-Bianchet 2001), and the unusually high survival of adult male ibex (Toïgo et al. 1997). Already, management of mountain goats based on the assumption that they were similar to other ungulates led to overharvesting: goats may be the only North American ungulate for which sport hunting led to local extirpations or drastic declines (Côté and Festa-Bianchet 2003). For alpine ibex, the high survival rate of males means that harvest of young and middle-aged males would produce an artificial age structure with many fewer older males than in unhunted populations. The same harvest practices would have a much lower impact on the age structure of moose, deer, or mountain sheep.

I conclude with a plea for research on marked individuals in hunted populations. Long-term monitoring of marked individuals is not always possible and presents many challenges. It requires access to a study area that does not undergo drastic changes in accessibility, land tenure, or administration for decades. For long-lived species such as ungulates, however, it is the best way to document basic biological attributes that are relevant to management, such as age-specific survival and reproduction, or how density-dependent and density-independent factors affect reproduction, growth, and survival of different sex–age classes. Results from long-term studies of marked individuals have been instrumental in affecting ungulate management policies (Gordon et al. 2004), yet basic biological information is still lacking for many sport-hunted species. For example, I am unaware of data on sex- and age-specific mortality of white-tailed deer or caribou that could be comparable to those available for roe deer, red deer, or bighorn sheep (Gaillard et al. 1998). There is almost no comparable information for any ungulate species from Asia, Africa, or South America (or for macropod marsupials). Both wildlife management and ecological theory stand to benefit from long-term studies of marked individuals in hunted populations, because hunting affects both population and evolutionary ecology. Most long-term research on marked ungulates investigated populations that were either lightly hunted or not hunted at all. Because most ungulate populations are subject to sport hunting, however, questions remain about the applicability of results from long-term, individual-based research to hunted populations (Festa-Bianchet 2003). In particular, heavy sport harvest may select for different reproductive strategies in both sexes, which could have important (but currently unknown) effects on both population ecology and evolution.

ACKNOWLEDGMENTS

Many current and previous collaborators and students helped me develop ideas about ungulate evolutionary ecology and management, but any remaining bits of incongruence are entirely of my own making. In particular, I thank Steeve Côté, Tim Coulson, Jean-Michel Gaillard, Jack Hogg, Jon Jorgenson, Fanie Pelletier, Kathreen Ruckstuhl, and Bill Wishart. I am grateful to Tim Fullbright, David Hewitt, Jos Milner, and Atle Mysterud for comments on earlier drafts of this manuscript. My long-term research on the ecology of mountain ungulates over the past 16 years was supported by the Natural Sciences and Engineering Research Council of Canada.

REFERENCES

Andersen, R., P. Duncan, and J. D. C. Linnell (eds). 1998. *The European Roe Deer: The Biology of Success.* Oslo: Scandinavian University Press.

Apollonio, M., B. Bassano, and A. Mustoni. 2003. Behavioral aspects of conservation and management of European mammals. In *Animal Behavior and Wildlife Conservation*, M. Festa-Bianchet, and M. Apollonio (eds). Washington, DC: Island Press, p. 157.

Augustine, D. J., and S. J. McNaughton. 1998. Ungulate effects on the functional species composition of plant communities: Herbivore selectivity and plant tolerance. *J. Wildl. Manage.* 62:1165.

Ballard, W. B., et al. 2000. Survival of female elk in northern Arizona. *J. Wildl. Manage.* 64:500.

Bender, L., et al. 2002. Effects of open-entry spike-bull, limited-entry branched-bull harvesting on elk composition in Washington. *Wildl. Soc. Bull.* 30:1078.

Bender, L. C., et al. 2004. Survival, cause-specific mortality, and harvesting of male black-tailed deer in Washington. *J. Wildl. Manage.* 68:870.

Berger, J., et al. 2003. Through the eyes of prey: How the extinction and conservation of North America's large carnivores alter prey systems and biodiversity. In *Animal Behavior and Wildlife Conservation*, M. Festa-Bianchet, and M. Apollonio (eds). Washington, DC: Island Press, p. 133.

Biederbeck, H. H., M. C. Boulay, and D. H. Jackson. 2001. Effects of hunting regulations on bull elk survival and age structure. *Wildl. Soc. Bull.* 29:1271.

Bishop, C. J., et al. 2005. Effect of limited antlered harvest on mule deer sex and age ratios. *Wildl. Soc. Bull.* 33:662.

Boyce, M. S., A. R. E. Sinclair, and G. C. White. 1999. Seasonal compensation of predation and harvesting. *Oikos* 87:419.

Brashares, J. S. 2003. Ecological, behavioral, and life history correlates of mammal extinctions in West Africa. *Conserv. Biol.* 17:733.

Brownstein, J. S., et al. 2005. Forest fragmentation predicts local scale heterogeneity of Lyme disease risk. *Oecologia* 146:469.

Bryant, A. A. 1997. *Update Cosewic Status Report on the Vancouver Island Marmot Marmota vancouverensis in Canada.* Committee on the status of endangered wildlife in Canada.

Cardillo, M., et al. 2004. Human population density and extinction risk in the world's carnivores. *PLoS Biol.* 2:909.

Catchpole, E. A., et al. 2004. Sexual dimorphism, survival and dispersal in red deer. *J Agri. Biol. Ecol. Stat.* 9:1–26.

Caughley, G. 1970. Eruption of ungulate populations, with emphasis on Himalayan Thar in New Zealand. *Ecology* 51:53.

Choquenot, D. 1993. Growth, body condition and demography of wild banteng (*Bos javanicus*) on Cobourg peninsula, Northern Australia. *J. Zool.* 231:533.

Clutton-Brock, T. H., S. D. Albon, and F. E. Guinness. 1988. Reproductive success in male and female red deer. In *Reproductive Success*, T. H. Clutton-Brock (ed.). Chicago: University of Chicago Press, p. 325.

Clutton-Brock, T. H., M. Major, S. D. Albon, and F. E. Guinness. 1987. Early development and population dynamics in red deer. I. Density-dependent effects on juvenile survival. *J. Anim. Ecol.* 56:53.

Coltman, D. W., P. O'Donoghue, J. T. Hogg, and M. Festa-Bianchet. 2005. Selection and genetic (co)variance in bighorn sheep. *Evolution* 59:1372.

Coltman, D. W., P. O'Donoghue, J. T. Jorgenson, J. T. Hogg, C. Strobeck, and M. Festa-Bianchet. 2003. Undesirable evolutionary consequences of trophy hunting. *Nature* 426:655.

Coltman, D. W., M. Festa-Bianchet, J. T. Jorgenson, and C. Strobeck. 2002. Age-dependent sexual selection in bighorn rams. *Proc. R. Soc. Lond. B* 269:165.

Coomes, D. A., et al. 2003. Factors preventing the recovery of New Zealand forests following control of invasive deer. *Conserv. Biol.* 17:450.

Côté, S. D., and M. Festa-Bianchet. 2001. Reproductive success in female mountain goats: The influence of maternal age and social rank. *Anim. Behav.* 62:173.

Côté, S. D., and M. Festa-Bianchet. 2003. Mountain goat, *Oreamnos americanus*. In *Wild Mammals of North America: Biology, Management, Conservation*, G. A. Feldhamer, B. Thompson, and J. Chapman (eds). Baltimore: John Hopkins University Press, p. 1061.

Côté, S. D., et al. 2004. Ecological impacts of deer overabundance. *Ann. Rev. Ecol. Syst.* 35:113.

Coulson, T., et al. 2001. Age, sex, density, winter weather, and population crashes in soay sheep. *Science* 292:1528.

Coulson, T., et al. 2004. The demographic consequences of releasing a population of red deer from culling. *Ecology* 85:411.

Courchamp, F., R. Woodroffe, and G. Roemer. 2003. Removing protected populations to save endangered species. *Science* 302:1532.

Darimont, C. T., et al. 2005. Range expansion by moose into coastal temperate rainforests of British Columbia, Canada. *Divers. Distrib.* 11:235.

Ernest, H. B., E. S. Rubin, and W. M. Boyce. 2002. Fecal DNA analysis and risk assessment of mountain lion predation of bighorn sheep. *J. Wildl. Manage.* 66:75.

Festa-Bianchet, M. 1988. Birth date and survival in bighorn lambs (*Ovis canadensis*). *J. Zool.* 214:653.

Festa-Bianchet, M. 1989. Survival of male bighorn sheep in southwestern Alberta. *J. Wildl. Manage.* 53:259.

Festa-Bianchet, M. 2003. Exploitative wildlife management as a selective pressure for the life history evolution of large mammals. In *Animal Behavior and Wildlife Conservation*, M. Festa-Bianchet, and M. Apollonio (eds). Washington, DC: Island Press, p. 191.

Festa-Bianchet, M., et al. 2006. Stochastic predation events and population persistence in bighorn sheep. *Proc. R. Soc. Lond. B* 273:1537.

Festa-Bianchet, M., J.-M. Gaillard, and S. D. Côté. 2003. Variable age structure and apparent density-dependence in survival of adult ungulates. *J. Anim. Ecol.* 72:640.

Fraser, K. W., J. M. Cone, and E. J. Whitford. 2000. A revision of the established ranges and new populations of 11 introduced ungulate species in New Zealand. *J. R. Soc. New Zealand* 30:419.

Fryxell, J. M., et al. 1991. Time lags and population fluctuations in white-tailed deer. *J. Wildl. Manage.* 55:377.

Gaillard, J. M., M. Festa-Bianchet, and N. G. Yoccoz. 1998. Population dynamics of large herbivores: Variable recruitment with constant adult survival. *Trends Ecol. Evol.* 13:58.

Gaillard, J.-M., M. Festa-Bianchet, and N. G. Yoccoz. 2001. Not all sheep are equal. *Science* 292:1499.

Gaillard, J.-M., M. Festa-Bianchet, N. G. Yoccoz, A. Loison, and C. Toigo. 2000. Temporal variation in fitness components and population dynamics of large herbivores. *Ann. Rev. Ecol. Syst.* 31:367.

Gaillard, J.-M., A. Loison, and C. Toïgo. 2003. Variation in life history traits and realistic population models for wildlife management: The case of ungulates. In *Animal Behavior and Wildlife Conservation*, M. Festa-Bianchet, and M. Apollonio (eds). Washington, DC: Island Press, p. 115.

Gaillard, J.-M., and N. G. Yoccoz. 2003. Temporal variation in survival of mammals: A case of environmental canalization? *Ecology* 84:3294.

Geisser, H., and H. U. Reyer. 2004. Efficacy of hunting, feeding, and fencing to reduce crop damage by wild boars. *J. Wildl. Manage.* 68:939.

Geist, V. 1992. Deer ranching for products and paid hunting: Threat to conservation and biodiversity by luxury markets. In *The Biology of Deer*, R. D. Brown (ed.). New York: Springer, p. 554.

Geist, V. 1994. Wildlife conservation as wealth. *Nature* 368:491.

Giles, B. G., and C. S. Findlay. 2004. Effectiveness of a selective harvest system in regulating deer populations in Ontario. *J. Wildl. Manage.* 68:266.

Gordon, I. J., A. J. Hester, and M. Festa-Bianchet. 2004. The management of wild large herbivores to meet economic, conservation and environmental objectives. *J. Appl. Ecol.* 41:1021.

Hamel, S., et al. 2006. Population dynamics and harvest potential of mountain goat herds in Alberta. *J. Wildl. Manage.* 69:1044–1053.

Harper, E. K., W. J. Paul, and L. D. Mech. 2005. Causes of wolf depredation increase in Minnesota from 1979–1998. *Wildl. Soc. Bull.* 33:888.

Harris, R. B., and D. H. Pletscher. 2002. Incentives toward conservation of argaili *Ovis ammon*: A case study of trophy hunting in western China. *Oryx* 36:373.

Harris, R. B., W. A. Wall, and F. W. Allendorf. 2002. Genetic consequences of hunting: What do we know and what should we do? *Wildl. Soc. Bull.* 30:634.

Hartl, G. B., et al. 1995. Allozymes and the genetics of antler development in red deer (*Cervus elaphus*). *J. Zool.* 237:83.

Hayes, R. D., et al. 2000. Kill rate by wolves on moose in the Yukon. *Can. J. Zool.* 78:49.

Hofer, D. 2002. The lion's share of the hunt–Trophy hunting and conservation: A review of the legal Eurasian tourist hunting market and trophy trade under CITES. *Traffic Europe.*

Hogg, J. T. 2000. Mating systems and conservation at large spatial scales. In *Vertebrate Mating Systems*, M. Apollonio, M. Festa-Bianchet, and D. Mainardi (eds). Singapore: World Scientific, p. 214.

Hogg, J. T., and S. H. Forbes. 1997. Mating in bighorn sheep: Frequent male reproduction via a high-risk "unconventional" tactic. *Behav. Ecol. Sociobiol.* 41:33.

Hughes, T. P., et al. 2005. New paradigms for supporting the resilience of marine ecosystems. *Trends Ecol. Evol.* 20:380.

Hutchings, J. A. 2004. Evolutionary biology — The cod that got away. *Nature* 428:899.

Hutchings, J. A., C. Walters, and R. L. Haedrich. 1997. Is scientific inquiry incompatible with government information control? *Can. J. Fish. Aquat. Sci.* 54:1198.

Hutchings, M. 1995. Olympic Mountain goat controversy continues. *Conserv. Biol.* 9:1324.

Jacobson, A. R., et al. 2004. Climate forcing and density-dependence in a mountain ungulate population. *Ecology* 85:1598.

Jorgenson, J. T., M. Festa-Bianchet, J.-M. Gaillard, and W. D. Wishart. 1997. Effects of age, sex, disease, and density on survival of bighorn sheep. *Ecology* 78:1019.

Jorgenson, J. T., M. Festa-Bianchet, and W. D. Wishart. 1998. Effects of population density on horn development in bighorn rams. *J. Wildl. Manage.* 62:1011.

Komers, P. E., and G. P. Curman. 2000. The effect of demographic characteristics on the success of ungulate re-introductions. *Biol. Conserv.* 93:187.

Kraus, C., and H. G. Rödel. 2004. Where have all the cavies gone? Causes and consequences of predation by the minor grison on a wild cavy population. *Oikos* 105:489.

Kunkel, K. E., et al. 1999. Winter prey selection by wolves and cougars in and near glacier national park, Montana. *J. Wildl. Manage.* 63:901.

Langvatn, R., and A. Loison. 1999. Consequences of harvesting on age structure, sex ratio, and population dynamics of red deer *Cervus elaphus* in Central Norway. *Wildl. Biol.* 5:213.

Law, R. 2001. Phenotypic and genetic changes due to selective exploitation. In *Conservation of Exploited Species*, J. D. Reynolds, et al. (eds). Cambridge: Cambridge University Press, p. 323.

Leader-Williams, N., R. J. Smith, and M. J. Walpole. 2001. Elephant hunting and conservation. *Science* 293:2203.

Leonard, J. A., C. Vilà, and R. K. Wayne. 2004. Legacy lost: Genetic variability and population size of extirpated us grey wolves (*Canis lupus*). *Mol. Ecol.* 14:9.

Li, Y. M., Z. X. Gao, and X. H. Li. 2000. Illegal wildlife trade in the Himalayan region of China. *Biodiv. Conserv.* 9:901.

Loison, A., et al. 1999a. Age-specific survival in five populations of ungulates: Evidence of senescence. *Ecology* 80:2539.

Loison, A., R. Langvatn, and E. J. Solberg. 1999b. Body mass and winter mortality in red deer calves: Disentangling sex and climate effects. *Ecography* 22:20.

Maher, C. R., and C. D. Mitchell. 2000. Effects of selective hunting on group composition and behavior patterns of pronghorn, *Antilocapra americana*, males in Montana. *Can. Field-Nat.* 114:264.

Martinez, M., et al. 2005. Different hunting strategies select for different weights in red deer. *Biol. Lett.* 1:353.

McElligott, A. G., R. Altwegg, and T. J. Hayden. 2002. Age-specific survival and reproductive probabilities: Evidence for senescence in male fallow deer (*Dama dama*). *Proc. R. Soc. Lond. B* 269:1129.

Messier, F. 1994. Ungulate population models with predation: A case study with the North American moose. *Ecology* 75:478.

Messier, F. 1995. Trophic interactions in two northern wolf-ungulate systems. *Wildl. Res.* 22:131.

Milner, J. M., et al. 2006. Temporal and spatial development of red deer harvesting in Europe: Biological and cultural factors. *J. Appl. Ecol.* 43:721–24.

Milner-Gulland, E. J., et al. 2003. Reproductive collapse in saiga antelope harems. *Nature* 422:135.

Molinari-Jobin, A., et al. 2002. Significance of lynx *Lynx lynx* predation for roe deer *Capreolus capreolus* and chamois *Rupicapra rupicapra* mortality in the Swiss Jura mountains. *Wildl. Biol.* 8:109.

Mysterud, A. 2006. The concept of overgrazing and its role in the management of large herbivores. *Wildl. Biol.* 12:129–41.

Mysterud, A., T. Coulson, and N. C. Stenseth. 2002. The role of males in the dynamics of ungulate populations. *J. Anim. Ecol.* 71:907.

Mysterud, A., E. J. Solberg, and N. G. Yoccoz. 2005. Ageing and reproductive effort in male moose under variable levels of intrasexual competition. *J. Anim. Ecol.* 74:742.

Nilsen, E. B., et al. 2005. Moose harvesting strategies in the presence of wolves. *J. Appl. Ecol.* 42:389.

Nixon, C. M., et al. 2001. Survival of white-tailed deer in intensively farmed areas of Illinois. *Can. J. Zool.* 79:581.

Office National de la Chasse. 2003. http://www.oncfs.gouv.fr/degats/index.php.

Olsen, E. M., et al. 2004. Maturation trends indicative of rapid evolution preceded the collapse of northern cod. *Nature* 428:932.

Owen-Smith, N., and D. R. Mason. 2005. Comparative changes in adult vs. juvenile survival affecting population trends of African ungulates. *J. Anim. Ecol.* 74:762.

Owen-Smith, N., D. R. Mason, and J. O. Ogutu. 2005. Correlates of survival rates for 10 African ungulate populations: Density, rainfall and predation. *J. Anim. Ecol.* 74:774.

Pelletier, F., and M. Festa-Bianchet. 2006. Sexual selection and social rank in bighorn rams. *Anim. Behav.* 71:649.

Pelletier, F., J. T. Hogg, and M. Festa-Bianchet. 2006. Male mating effort in a polygynous ungulate. *Behav. Ecol. Sociobiol.* 60:645–54.

Pitra, C., S. Rehbein, and W. Lutz. 2005. Tracing the genetic roots of the sika deer *Cervus nippon* naturalized in Germany and Austria. *Eur. J. Wildl. Res.* 51:237.

Portier, C., et al. 1998. Effects of density and weather on survival of bighorn sheep lambs (*Ovis canadensis*). *J. Zool.* 245:271.

Proctor, M. F., et al. 2005. Genetic analysis reveals demographic fragmentation of grizzly bears yielding vulnerably small populations. *Proc. R. Soc. B* 272:2409.

Putman, R. J., and B. W. Staines. 2004. Supplementary winter feeding of wild red deer in Europe and North America: Justifications, feeding practice and effectiveness. *Mamm. Rev.* 34:285.

Rochet, M. J. 1998. Short-term effects of fishing on life history traits of fishes. *Ices J. Mar. Sci.* 55:371.

Sæther, B. E., E. J. Solberg, and M. Heim. 2003. Effects of altering sex ratio structure on the demography of an isolated moose population. *J. Wildl. Manage.* 67:455.

Sæther, B. E., et al. 2004. Offspring sex ratio in moose *Alces alces* in relation to paternal age: An experiment. *Wildl. Biol.* 10:51.

Sankaran, M., and S. J. McNaughton. 1999. Determinants of biodiversity regulate compositional stability of communities. *Nature* 401:691.

Saucedo, C., and R. Gill. 2004. The endangered huemul or south Andean deer *Hippocamelus bisulcus*. *Oryx* 38:132.

Schaefer, J. A., et al. 1999. Demography of decline of the red wine mountains Caribou herd. *J. Wildl. Manage.* 63:580.

Scott, F. 1998. *Update Cosewic Status Report on the Eastern Cougar in Canada*. Committee on the status of endangered wildlife in Canada.

Sinclair, A. R. E., S. Mduma, and J. S. Brashares. 2003. Patterns of predation in a diverse predator–prey system. *Nature* 425:288.

Sinclair, A. R. E., and R. P. Pech. 1996. Density dependence, stochasticity, compensation and predator regulation. *Oikos* 75:164.

Solberg, E. J., A. Loison, T. H. Ringsby, B. E. Sæther, and M. Heim. 2002. Biased adult sex ratio can affect fecundity in primiparous moose *Alces alces*. *Wildl. Biol.* 8:117.

Solberg, E. J., A. Loison, B. E. Sæther, and O. Strand. 2000. Age-specific harvest mortality in a Norwegian moose *Alces alces* population. *Wildl. Biol.* 6:41.

Solberg, E. J., B.-E. Sæther, O. Strand, and A. Loison. 1999. Dynamics of a harvested moose population in a variable environment. *J. Anim. Ecol.* 68:186.

Stearns, S. C. 1992. *The Evolution of Life Histories*. Oxford: Oxford University Press.

Strickland, B. K., et al. 2001. Effects of selective-harvest strategies on white-tailed deer antler size. *Wildl. Soc. Bull.* 29:509.

Stuarth-Smith, A. K., et al. 1997. Woodland Caribou relative to landscape patterns in northeastern Alberta. *J. Wildl. Manage.* 61:622.

Thomas, C. D., et al. 2004. Extinction risk from climate change. *Nature* 427:145.

Tilman, D., D. Wedin, and J. Knops. 1996. Productivity and sustainability influenced by biodiversity in grassland ecosystems. *Nature* 379:718.

Toïgo, C., and J.-M. Gaillard. 2003. Causes of sex-biased adult survival in ungulates: Sexual size dimorphism, mating tactic or environment harshness? *Oikos* 101:376.

Toïgo, C., J.-M. Gaillard, and J. Michallet. 1997. Adult survival of the sexually dimorphic alpine ibex (*Capra ibex ibex*). *Can. J. Zool.* 75:75.

Tremblay, J. P., et al. 2005. Long-term decline in white-tailed deer browse supply: Can lichens and litterfall act as alternative food sources that preclude density-dependent feedbacks. *Can. J. Zool.* 83:1087.

Valière, N., et al. 2003. Long-distance wolf recolonization of France and Switzerland inferred from noninvasive genetic sampling over a period of 10 years. *Anim. Conserv.* 6:83.

Vilà, C., et al. 2003. Rescue of a severely bottlenecked wolf (*Canis lupus*) population by a single immigrant. *Proc. Roy. Soc. B* 270:91.

Vucetich, J. A., D. W. Smith, and D. R. Stahler. 2005. Influence of harvest, climate, and wolf predation on Yellowstone elk, 1961–2004. *Oikos* 111:259.

Walsh, M. R., et al. 2006. Maladaptive changes in multiple traits caused by fishing: Impediments to population recovery. *Ecol. Lett.* 9:142.

Whitehead, H., and R. Reeves. 2005. Killer whales and whaling: The scavenging hypothesis. *Biol. Lett.* 1.

Wittmer, H. U., et al. 2005a. Population dynamics of the endangered mountain ecotype of Woodland Caribou (*Rangifer tarandus caribou*) in British Columbia, Canada. *Can. J. Zool.* 83:407.

Wittmer, H. U., A. R. E. Sinclair, and B. N. Mclellan. 2005b. The role of predation in the decline and extirpation of woodland caribou. *Oecologia* 114:257.

Yang, Q. S., et al. 2003. Conservation status and causes of decline of musk deer (*Moschus* spp.) in China. *Biol. Conserv.* 109:333.

12 Density Dependence in Deer Populations: Relevance for Management in Variable Environments

Charles A. DeYoung, D. Lynn Drawe,
Timothy Edward Fulbright, David Glenn Hewitt,
Stuart W. Stedman, David R. Synatzske, and James G. Teer

CONTENTS

New-world deer of the genus *Odocoileus* are commonly assumed to respond to food shortage due to intraspecific competition by reduced recruitment, body mass, and other manifestations. McCullough's (1979) book on the George Reserve deer herd is commonly cited as the definitive work on density dependence in white-tailed deer (*O. virginianus*), and by extension, mule deer (*O. hemionus*). The George Reserve is a 464-ha, high-fenced property in Michigan, United States. Two males and four females were introduced in 1928, and by 1933, the population was estimated to be 160. The population fluctuated until 1952, when a series of experiments with deer density began. The population was reduced during this time in a series of steps and data collected to form a population model (McCullough 1979). Strong density dependence was evident in this model.

Reductions in the George Reserve population continued into 1975, when an estimated 10 deer remained (McCullough 1982). No deer were subsequently harvested for 5 years to allow the population to increase. McCullough (1982, 1983) concluded that the population response after 1975 was similar to the original increase from 1928 to 1933.

Downing and Guynn (1985) presented a generalized sustained yield table using McCullough (1979) as a starting point. Downing and Guynn (1985) used their experience and literature values to present a table scaled to percent *K* carrying capacity. McCullough (1979, 1982, 1983) and Downing and Guynn (1985) showed density-dependent responses across the population growth range from low density to *K*. We define *K* as the maximum sustainable population level where the deer are in approximate equilibrium with their food supply (Macnab 1985).

Downing and Guynn (1985) recognized that their generalized model might not apply to all deer populations. They wondered if their model would be applicable to low-density populations and areas with poor habitat, which precluded high rates of recruitment. They suggested ways their generalized model could be modified for populations that did not fit the mold of those where recruitment was consistently high and relatively stable across time.

McCullough (1984) also recognized that if environmental variation is great, density-dependent effects, while present in the mix of factors impinging on a population, may be masked. He suggested that these situations were rare and occurred at extreme fringes of whitetail range.

Mackie et al. (1990) questioned whether density-dependent models have utility for management. They presented data from three mule deer and two white-tailed deer populations in Montana, United States, and concluded that there was evidence of density-dependent behavior in one mule deer and one white-tailed deer population. They stated that western North America has a high degree of environmental variation resulting in fluctuating carrying capacity. They suggested that some Montana populations had declined, because expected density-dependent responses to harvest did not happen. Finally, they suggested that in variable environments, managers should employ techniques providing regular tracking of population size and performance and not depend on predictions from density-dependent models.

McCullough (1990) issued a strong caution to the conclusions of Mackie et al. (1990). He stressed that density-dependent behavior may be missed because it is obscured by environmental factors and sampling error. Also, experimental and statistical design frequently places the burden of proof on density dependence, that is, the null model is a lack of density dependence. Finally, time lags, study area scale, environmental homogeneity, life history, behavior, and predation may make density dependence difficult to detect when in fact it is present. Importantly, McCullough (1990) hypothesized that several of these factors, singularly or in combination, can result in a population of deer expressing no density-dependent response until very near K.

Fryxell et al. (1991) reported on a white-tailed deer population in southeastern Ontario, Canada, that fluctuated widely over 34 years. They concluded that variation in hunting effort strongly affected the fluctuation and that the population showed time-lagged density dependence.

McCullough (1992) again emphasized that environmental variation can obscure density-dependent responses in deer populations. He stated that this may require study of a population over a large range of densities to detect a density-dependent response. McCullough (1999) also reviewed and extended his concept of some populations of ungulates having a "plateau" of constant growth and then a "ramp" of declining growth in the graph of r on N (Figure 12.1B, b, C, and c). No density-dependent response would be observed in the plateau phase. He hypothesized that this model could fit more K-selected species with low reproductive rates. However, he speculated that a plateau and ramp model may fit *Odocoileus* deer in desert environments.

Bartmann et al. (1992) could not detect differences in fawn survival in response to experimental removal of 22 and 16% in consecutive years in a migratory Colorado, United States, mule deer population. They subsequently simulated density-dependent fawn mortality in enclosures with a wide range of density. Natural mortality of adult does was low in the free-ranging population but fawn mortality was relatively high and varied with winter severity. Relation of this population to K was unknown, but the authors assumed it to be near or at K.

Keyser et al. (2005) studied long-term data sets for nine white-tailed deer populations in the southeastern United States. They concluded that eight out of nine populations showed density-dependent responses, but that these responses frequently lagged 1 or 2 years. They stated that the population that did not show density-dependent responses occurred on exceptionally poor habitat.

Shea et al. (1992) collected data from a white-tailed deer population in Florida, United States, that declined 75% in density during a 10-year period. They found little difference in deer physiological indices during this period and concluded that the habitat, which was characterized by low-fertility soils, produced low amounts of high-quality forage and an abundance of poor-quality forage. Lack of nutritious forage, coupled with abundant poor-quality forage precluded a density-dependent response, because there was little opportunity for intraspecific competition, even when densities were high. Shea and Osborne (1995) discussed poor-quality habitat across North America. They surveyed

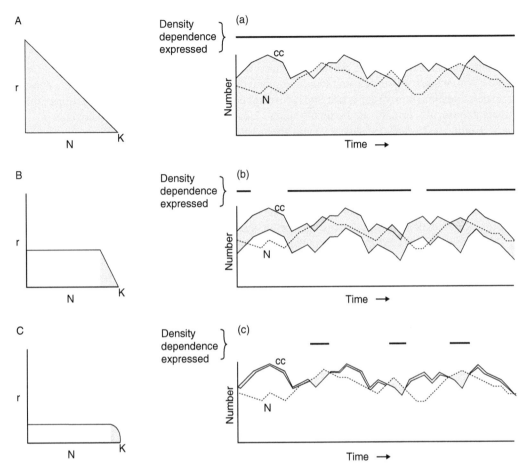

FIGURE 12.1 Plateau and ramp graphs (A, B, C) showing a range of deer population density-dependent responses and corresponding graphs (a, b, c) of carrying capacity variation in comparison to population level variation. (Adapted from D. R. McCullough. *J. Mammal.* 80:1132 and 1133:1999. With permission.)

state game departments and produced a map within whitetail range where density-dependent response would be lacking or masked.

Dumont et al. (2000) worked on white-tailed deer on the northern limit of their range in southeastern Quebec, Canada. They stated that severe winters were among the major factors limiting deer populations, but found density-dependent responses during mild winters.

Gilbert and Raedeke (2004) worked on Columbian black-tailed deer (*Odocoileus hemionus columbianus*) and found that minimum temperatures in May and the amount of precipitation in June affected fawn recruitment. However, they also reported that plant production was correlated with deer density in the same year. Also, their best models of fawn production included time-lagged density or forage terms. They concluded that the population was expressing density-dependent behavior during the study period.

McCullough (1999) cited the intrinsic rate of increase of the population, scale of area occupied by the population, heterogeneity of environment, and general quality of the habitat as factors that might explain why ungulates respond differently to a range of densities. Scale of the area occupied by the population refers to confined areas such as enclosures or islands where limitations on dispersal may change population growth rate. A heterogeneous environment allows ungulates more types of high-quality food, leading to competition among different classes of individuals as density increases.

Finally, high-quality habitats, including moderate temperatures and precipitation, lead to high plant production. On the contrary, habitats with strong limitations on plant growth may result in most forage being of low quality, except perhaps in occasional good years.

White-tailed deer have the potential to have a high rate of increase in rich habitats with stable environments, where female fawns commonly breed. Such populations may have a density-dependent function (r on N) that is all ramp, with no plateau (Figure 12.1A). Such populations express density dependence virtually all the time (Figure 12.1a), even though N may be well below K.

McCullough (1999) listed mule deer among the species that may exhibit a plateau-ramp density-dependent function (Figure 12.1B and b). He stated that these populations may not reach K very often, because they tend to live in variable environments, and have lower rates of increase as compared to white-tailed deer. Populations that fit the plateau-ramp hypothesis (Figure 12.1B and b) have a plateau of the density-dependent function where no density-dependent responses would occur. Only when these populations approach K are density-dependent effects observed. When these populations are significantly below K, density-dependent effects are not observed (Figure 12.1B and b).

McCullough (1999) felt that desert mule deer (*O. H. crooki*) may exhibit the density-dependent function shown in Figure 12.1C, based on the work of Short (1979). This hypothesis fits populations with low intrinsic rates of increase, homogenous habitats with mostly low-quality forage, and variable environments. Figure 12.1c shows that these populations only exhibit density-dependent responses during occasional favorable periods.

This review of literature shows that *Odocoileus* population dynamics are complex and frequently site specific. The complex nature of population dynamics in this genus makes formulating a general population model challenging. Almost without exception, researchers cite McCullough's (1979) George Reserve work as the conventional model for density dependence. However, as this review has shown, there are many situations where density-dependent behavior in *Odocoileus* populations cannot be detected. How widespread are habitats where the assumption of density-dependent behavior is not useful to management? Our objectives in this chapter are to (1) analyze three long-term sets of white-tailed deer counts in South Texas for density-dependent behavior, (2) suggest some unifying concepts for considering density-dependent and density-independent behavior of *Odocoileus* populations, and (3) suggest regions of deer range where population behavior cannot regularly be predicted with density-dependent models.

TESTING SOUTH TEXAS DEER COUNTS FOR DENSITY DEPENDENCE

Early European explorers in South Texas, United States, found a landscape that was mostly grassland, commonly interspersed with shrub communities (Inglis 1964; Fulbright 2001). White-tailed deer were present in wooded stream bottoms, shrub communities on upland areas, and on the open prairie. Little is known about deer populations from this period, except they were commonly mentioned in traveler's journals (Doughty 1983, 29; Fulbright 2001).

Cattle and horses were at least locally numerous by the mid-1700s (Lehmann 1969). Shrubs probably began to increase at this time, and this trend continued into the twentieth century (Jones 1975). Although famous for cattle ranching, the region harbored millions of domestic sheep in the latter part of the nineteenth century (Lehman 1969). Climatic change may have been a background condition influencing changes in plant ecology in the region, with grazing by domestic livestock being the driving force (Van Auken 2000). A cool, wet period lasted from about 1350 to 1850 (Foster 1998, 9). After 1850, the climate became warmer and dryer, which may also have influenced the increase in shrub density and distribution. Removal of fuel by livestock grazing and suppression by humans reduced or eliminated natural fires that inhibited the increase in woody plants during pre-Columbian times (Van Auken 2000). In the twentieth and early twenty-first centuries, the region has been covered by a canopy of shrubs, frequently in complex taxonomic mixes (Inglis 1964;

Jones 1975). Exclusive of the coastal sand plain and coastal prairie, over 90% of the region has been subjected to ≥ 1 attempts to reduce shrub density to increase cattle-carrying capacity (Davis and Spicer 1965).

Increased shrub density during at least the past two centuries may have facilitated increased deer populations. Deer did not become locally extinct in South Texas after European settlement as they did in much of North America. This was the result of low human density in the region, and large land ownerships. Roads and highways were scarce until the 1920s when oil exploration began in earnest. South Texas, particularly the King Ranch and the Aransas National Wildlife Refuge, provided deer for reestablishing populations elsewhere in the state.

Historically, medium- and large-sized predators of deer have included jaguar (*Panthera onca*), mountain lion (*Puma concolor*), bobcat (*Lynx rufus*), black bear (*Ursus americanus*), gray wolf (*Canis lupus*), red wolf (*Canis niger*), and coyote (*Canis latrans*). Of these, wolves and jaguar are extirpated. Black bears are limited to occasional dispersers from northern Mexico. Mountain lions are present in generally low, but apparently increasing density with high densities in localized areas (Harveson 1997). Coyotes and bobcats are present throughout the region, often at high population densities.

Mountain lions prey on deer in South Texas, but do not appear to exert a region-wide influence on populations. Bobcats kill deer but do not appear to be an important factor to deer populations (Blankenship 2000; Ballard et al. 2001). Studies in the 1960s and 1970s in eastern South Texas showed significant coyote predation on deer fawns (Cook et al. 1971; Beasom 1974; Carroll and Brown 1977; Kie and White 1985). Meyer et al. (1984) suggested that in addition to coyote predation, poor summer nutrition may be a strong factor in low South Texas fawn survival.

Even before there was any formal management of deer populations, South Texas was well known for producing large-antlered bucks (Brothers and Ray 1975; Helmer 2002). Large antlers are consistent with populations well below K carrying capacity (McCullough 1979). Examples of irruptive behavior in deer populations in the region are lacking, although an irruption was experimentally induced by Kie and White (1985). Fawn survival from birth to fall is erratic (Ginnett and Young 2000) and low compared to white-tailed deer populations in general (Downing and Guynn 1985). Unhunted and otherwise unmanaged deer populations persist in a generally healthy state on some large, remote ranches.

The region is virtually all private land, much of which is leased for hunting. Intense interest in deer management has developed among landowners and hunters during the past three decades. Wildlife biologists in the region commonly prescribe management practices for deer populations based on the assumption of density-dependent population behavior (Brothers and Ray 1975: 62). This is particularly true of prescriptions to harvest does, as there is a common belief that without significant doe harvest, deer populations will increase to undesirable levels.

In variable environments, K carrying capacity (Macnab 1985) varies from year to year. The same number of animals may be above K in dry years and below K in wet years (McCullough 1979: 156). The negative feedback of animals on food plants is less important as a population influence compared to the annual swings in K. For South Texas, the CV for annual rainfall varies from 29 to 41% (Norwine and Bingham 1985). Rainfall occurs throughout the year with statistical peaks in May and September. The average growing season is about 300 days; however, plant growth can occur any month when moisture and temperature permit (Box 1960; Ansotegui and Lesperance 1973).

We analyzed for density dependence in long-term time-series of deer counts on three study areas. The Faith Ranch ($28°15'$ N, $100°00'$ W) was 16,115 ha in the western portion of South Texas, the Chaparral Wildlife Management Area ($28°20'$ N, $99°25'$ W) consisted of 5,930 ha approximately in the center, and the Rob and Bessie Welder Wildlife Foundation Refuge ($28°6'$ N, $97°75'$ W) was 3,158 ha in the eastern portion of the region (Figure 12.2).

For the Faith Ranch, a single helicopter survey of deer was conducted annually during 1975–1977 and 1981–1997. This consisted of flying adjacent belt transects about 200 m wide, at a height of about 20 m and speed of about 55 km/h (DeYoung 1985; Beasom et al. 1986). Surveys encompassed various

FIGURE 12.2   Location of Faith Ranch, Chaparral Wildlife Management Area, and Rob and Bessie Welder Wildlife Refuge, South Texas, USA.

portions of the Faith Ranch over time (Table 12.1). The time series consisted of raw, unadjusted numbers of deer counted each year and classified as does, bucks, and fawns. Because of the variation in area flown and counted over the time series, deer numbers were transformed into deer density (deer/405 ha).

On the Chaparral Area, population estimates from 1969 to 1975 were made by spotlight counts following methods described by Fafarman and DeYoung (1986). A density estimate was calculated for the spotlight route on the Chaparral Area. The density estimate was subsequently projected to a population estimate for the entire area each year. During 1975–1997, a single, complete-coverage helicopter survey (DeYoung 1985; Beasom et al. 1986) was conducted on the entire area. The time series consisted of estimates of deer population size derived from spotlight surveys, or the raw, unadjusted number of deer counted by helicopter each year (Table 12.2). The change in census methods undoubtedly introduced additional variability into the time series. Fafarman and DeYoung (1986), working on the Welder Wildlife Refuge, reported that population estimates from spotlight counts were about 10% higher than raw winter helicopter surveys. Deer were classified as does, bucks, and fawns during counts by spotlight and helicopter. A few unidentified deer were also tallied, but excluded from analysis. Because the same area was counted throughout the time series, number of deer estimated or counted was used to form variables for analysis.

Census data were available on the Welder Wildlife Refuge from 1963 to 1997, except for 1964 and 1969. During 1963–1976, population estimates were made by spotlight counts (Fafarman and

TABLE 12.1

March–May Rainfall and Number of Deer by Class Counted during Fall Surveys by Helicopter on the Faith Ranch, South Texas, USA, 1974–97

Year	March–May rain (cm)[a]	Hectares surveyed	Does	Bucks	Fawns
1974	—	5,805	253	101	44
1975	—	5,805	261	118	141
1976	—	5,805	334	139	150
1977	—	5,805	298	147	98
1978	—	—	—	—	—
1979	—	—	—	—	—
1980	—	—	—	—	—
1981	—	6,049	184	110	104
1982	10.41	7,942	169	137	76
1983	3.53	9,670	196	136	27
1984	5.06	10,630	293	161	3
1985	16.51	10,630	277	113	113
1986	17.86	11,060	286	131	90
1987	24.71	11,320	323	217	150
1988	5.79	11,320	440	237	62
1989	4.88	11,320	356	276	9
1990	17.60	11,320	331	286	110
1991	16.59	11,320	265	217	78
1992	18.16	11,320	536	339	175
1993	11.63	11,320	453	260	81
1994	14.66	11,320	531	308	80
1995	17.63	11,320	444	378	184
1996	2.11	11,320	372	281	70
1997	22.10	11,320	436	300	296

[a] Mean March–May rain = 13.08; CV = 53.7.

DeYoung 1986). During 1977–1998, estimates were made by a single helicopter survey (DeYoung 1985; Beasom et al. 1986) conducted in January each year. The change in census method undoubtedly introduced additional variability into the time series, as noted for the Chaparral Area. Helicopter surveys were made using procedures similar to those described for the Faith Ranch and Chaparral Area, except belt transects were spaced to result in about 50% coverage of the Refuge. The unadjusted number of deer counted was used for all years (Table 12.3). Breakdowns by class of deer were not available for all years. However, because an estimate of recruitment was needed for some of the time series analysis, mean number of embryos per mature doe collected each year for scientific purposes were substituted for fawns counted during census (Table 12.3).

The first method used to test for density dependence was that suggested by White and Bartmann (1997, 128). This involved regressing the variable tested for density dependence (V) against the estimate of number of deer (N_t) and N_t^2 without an intercept as follows:

$$V_t = B_1 N_t + B_2 N_t^2$$

and then testing the null hypothesis of $B_2 = 0$. If the test rejects the null hypotheses and B_2 is less than 0, then V has been shown to be density dependent.

TABLE 12.2
March–May Rainfall and Number of Deer by Class, Counted during Fall Spotlight Counts (1969–75) or Survey by Helicopter (1976–97) on the Chaparral Wildlife Management Area, South Texas, USA

Year	March–May rain (cm)[a]	Does	Bucks	Fawns
1969	—	366	108	40
1970	17.55	302	155	158
1971	6.30	257	257	28
1972	22.33	323	245	81
1973	5.94	252	93	76
1974	19.53	250	157	43
1975	18.90	242	106	162
1976	19.99	470	151	84
1977	14.63	476	208	254
1978	15.42	357	198	140
1979	28.17	693	256	146
1980	20.20	262	197	10
1981	35.79	296	115	160
1982	16.92	627	369	113
1983	3.76	337	205	42
1984	5.98	237	177	20
1985	17.76	114	78	59
1986	13.98	195	100	131
1987	16.79	155	128	115
1988	6.22	220	153	81
1989	12.92	176	170	22
1990	20.58	139	118	83
1991	10.50	139	121	83
1992	18.38	191	127	67
1993	16.57	138	131	36
1994	26.38	153	103	35
1995	30.88	137	84	92
1996	5.36	208	116	62
1997	28.65	183	117	160

[a] Mean March–May rain = 17.01; CV = 48.4.

For the Faith Ranch time series, all years (21) available were used and V_t = fawns/405 ha, whereas N_t = does/405 ha. The Chaparral Area data consisted of 29 years of time series with V_t = number of fawns and N_t = number of does. For the Welder Refuge, we used 33 years of data (no counts were available for 1964 and 1969) and V_t = mean embryos/adult doe, whereas N_t = number of deer.

The second method used to test for density dependence was described by Dennis and Otten (2000). Because Ginnett and Young (2000) showed rainfall influencing fawn: doe ratios in South Texas, a rainfall term was included in the model. The model used was written as

$$N_t = N_{t-1} \exp(a + bN_{t-1} + cW_{t-1} + \sigma Z_t),$$

TABLE 12.3
March–May Rainfall, Deer Density Determined by Spotlight Counts (1963–76) or January Survey by Helicopter (1977–98), and Mean Number of Embryos in Adult Does Collected for Scientific Purposes on the Welder Wildlife Refuge, South Texas, USA

Year	March–May rain (cm)[a]	Deer/km^2	Embryos/adult does (n doe)
1963	1.25	36.27	1.42 (33)
1964	1.96	—	1.54 (13)
1965	18.24	41.37	1.94 (17)
1966	25.40	51.04	1.62 (13)
1967	16.03	34.63	1.52 (25)
1968	25.25	43.37	1.74 (23)
1969	28.58	—	1.89 (18)
1970	20.14	59.62	1.55 (20)
1971	20.65	47.87	1.72 (18)
1972	17.88	32.25	1.50 (14)
1973	38.25	30.25	1.20 (27)
1974	6.53	37.57	1.74 (19)
1975	15.01	39.38	1.60 (20)
1976	25.88	44.10	1.43 (30)
1977	32.00	41.37	1.22 (27)
1978	19.51	32.85	1.38 (13)
1979	24.99	34.69	1.44 (25)
1980	28.32	25.15	1.67 (24)
1981	21.44	28.26	1.47 (19)
1982	28.32	25.15	1.67 (3)
1983	25.17	33.07	1.37 (27)
1984	4.47	25.72	1.62 (29)
1985	29.36	21.29	1.58 (26)
1986	18.52	27.50	1.68 (19)
1987	13.08	33.07	1.64 (33)
1988	22.23	25.60	1.68 (19)
1989	16.33	16.35	1.70 (20)
1990	15.70	18.56	1.53 (19)
1991	14.78	21.35	1.75 (20)
1992	39.70	21.54	1.73 (11)
1993	48.79	31.11	1.72 (18)
1994	43.74	20.59	1.54 (13)
1995	15.60	21.04	1.78 (18)
1996	10.41	28.20	1.79 (14)
1997	27.38	30.29	1.75 (20)
1998	17.83	—	—

[a] Mean March–May rain = 21.63; CV = 49.7.

where N_t is deer abundance (density in the case of the Faith Ranch) at time t (year: $t = 0, 1, 2, ...,$ number of years in time series), W_t is spring (March, April, and May) rainfall total (cm) for time t, and Z_t is standard noise (with $Z_1, Z_2, ...$ uncorrelated). Unknown parameters to be estimated from the data were a, b, c, and σ. The random variables Z_t represent unpredictable fluctuations in growth rate (logarithmic) over and above fluctuations accounted for by density dependence and precipitation. Under this model, the population abundances N_t ($t = 1, 2, ...$) are random variables correlated through time, and N_o is fixed. See Dennis and Otten (2000) for details on methodology.

Four cases of the model were fitted to the data for each study area as separate statistical hypotheses: H_0: $b = 0$ and $c = 0$ (no density dependence, no rainfall effect); H_1: $b \neq 0$ and $c = 0$ (density dependence, no rainfall effect); H_2: $b = 0$ and $c \neq 0$ (no density dependence, rainfall effect); and H_3: $b \neq 0$ and $c \neq 0$ (density dependence, rainfall effect).

We calculated maximum-likelihood estimates of unknown parameters in the model for all four hypotheses using time series data from each study area in conjunction with rainfall data (Dennis and Otten 2000). Because of missing years in the time series, for the Faith Ranch we used data from 1982 to 1997 (16 years) and for the Welder Refuge data from 1970 to 1997 (28 years) were used. We tested for density dependence and rainfall effects on density (deer/405 ha) of total deer, adult deer, does, bucks, and fawns using the four hypotheses on the Faith Ranch. We tested for density dependence and rainfall effects on total number of deer, adult deer, does, bucks, and fawns for the Chaparral Wildlife Management Area. For the Welder Refuge, we tested for density dependence and rainfall effects on total number of deer and mean number of embryos per adult doe (although fall rather than spring rainfall may have more influence on embryos/doe).

Statistical hypotheses were tested using parametric bootstrapping (Dennis and Taper 1994). Four statistical hypothesis tests were conducted for the density dependence–rainfall model, as follows: H_0 versus H_1, H_0 versus H_2, H_1 versus H_3, and H_2 versus H_3. For these tests, the null model is contained within the alternative model as a special case, and is obtained by setting one parameter equal to 0. Details of this approach are in Dennis and Taper (1994) and Dennis and Otten (2000).

Analysis of the time series by the method suggested by White and Bartmann (1997) showed no density dependence on the Faith Ranch, but did indicate that density dependence was operating at Chaparral and Welder (Table 12.4). The model for Chaparral was heavily influenced by very high census counts in 1979 and 1982 (Table 12.2). If these data points are omitted, density dependence is not indicated ($P = .247$).

Covariance analysis by the method of Dennis and Otten (2000) showed a similar trend to the White and Bartmann (1997) model (Tables 12.5–12.8). For the Faith Ranch, hypothesis tests provided no support for density dependence either with a rainfall covariate (H_2 versus H_3, $P \geq .26$) or without the rainfall covariate (H_0 versus H_1, $P \geq .15$) for any response variable. For the Chaparral Area,

---

TABLE 12.4

Tests for Density Dependence in Time Series of Deer Abundance and Reproduction Using the Method Suggested by White and Bartmann (1987) for the Faith Ranch, Chaparral Wildlife Management Area, and Welder Wildlife Refuge, South Texas, USA

Study area	Number of years in series	Variables	N	N^2
Faith Ranch	21	$V_1 =$ fawn density	0.2063	0.0071
		$N_t =$ doe density		$t = 0.76$
				$P > 0.05$
Chaparral Area	29	$V_t =$ fawns	0.4438	-0.0004
		$N_t =$ does		$t = -1.99$
				$P = 0.028$
Welder Refuge	33	$V_t =$ embryos/ad. doe	0.0925	-0.0012
		$N_t =$ deer		$t = -9.85$
				$P < 0.0001$

TABLE 12.5

Maximum-Likelihood Estimates (a, b, c, o^2) of Parameters in a Density Dependence–Spring Rainfall Model, Generalized R^2, and Schwartz Information Criterion (SIC) for Four Model Hypotheses (H) Fitted to White-Tailed Deer Density Data Obtained by Helicopter Survey for 1982–97, Faith Ranch, South Texas, USA

Variable	H[a]	$\hat{a}$	$\hat{b}$	$\hat{c}$	$\hat{o}^2$	R^2	SIC
Total deer	H_0	.02039			.0701	.000	8.4
	H_1	.49285	−.01905		.0554	.206	7.4
	H_2	−.28192		.05198	.0454	.368	4.3
	H_3	.12259	−.01491	.05198	.0367	.485	3.6
Adult deer	H_0	.01819			.0500	.000	3.0
	H_1	.37703	−.01682		.0415	.352	2.8
	H_2	−.22287		.02546	.0453	.285	4.2
	H_3	.23395	−.01576	.022354	.0380	.418	4.2
Does	H_0	.01476			.0668	.000	7.6
	H_1	.48876	−.03679		.0520	.187	6.4
	H_2	−.0723		.01691	.0647	−.011	9.9
	H_3	.40280	−.03657	.016140	.0502	.219	8.6
Bucks	H_0	.02355			.0597	.000	5.9
	H_1	.33479	−.03680		.0510	.461	6.1
	H_2	−.17941		.039428	.0485	.466	5.3
	H_3	.10217	−.03008	.034140	.0430	.560	6.1
Fawns	H_0	.02621			2.2772	.000	64.1
	H_1	1.34218	−.38023		1.6934	−.153	62.2
	H_2	−1.98586		.339087	1.1841	−2.028	56.4
	H_3	−.88778	−.22992	.332133	.9954	−.641	56.4

[a] H_0 = No density dependence, no rainfall effect; H_1 = density dependence, no rainfall effect; H_2 = no density dependence, rainfall effect; and H_3 = density dependence, rainfall effect.

Note: See Dennis and Otten (2000) for details on methodology. Table 12.1 contains rainfall and census data.

hypothesis tests provided support for density dependence both with the rainfall covariate (H_2 versus H_3, $P \leq .06$) and without the rainfall covariate (H_0 versus H_1, $P \leq .06$) for all response variables (Table 12.8). Hypothesis tests also supported density dependence with and without the rainfall covariate on the Welder Refuge for total deer ($P = .05$) and for embryos/adult doe ($P \leq .002$).

Both methods of time series analysis suggested the same trend for data from the three study areas: no density dependence detected for the Faith Ranch, modest indications of density dependence on the Chaparral Area, and a stronger density-dependence indication for Welder Refuge. Shenk et al. (1998) criticized the methods of Dennis and Otten (2000) for detecting density dependence in time series data. They concluded that sampling error would result in a high probability of Type II error. There is without doubt much sampling error in the deer census data we collected. We did (unpublished) simulations of the regression approach of White and Bartmann (1997), which also showed a high propensity for Type II error. This is why we used two methods to analyze the time series and the fact that they yielded similar results was encouraging.

These results are correlated with a rainfall gradient with lower rainfall on the Faith Ranch (54.6 cm average annual) and higher rainfall as the coast is neared on the east (Welder Refuge = 88.9 cm annually). Ginnett and Young (2000) demonstrated correlations between spring–summer rainfall and

TABLE 12.6

Maximum-Likelihood Estimates (a, b, c, o^2) of Parameters in a Density Dependence–Spring Rainfall Model, Generalized R^2, and Schwartz Information Criterion (SIC) for Four Model Hypotheses (H) Fitted to White-Tailed Deer Abundance Data Obtained by Spotlight Counts (1969–75) or Surveys by Helicopter (1976–97), Chaparral Wildlife Management Area, South Texas, USA

Variable	H[a]	$\hat{a}$	$\hat{b}$	$\hat{c}$	$\hat{o}^2$	R^2	SIC
Total deer	H_0	.00396			.1168	.000	26.0
	H_1	.41524	−.0081		.857	.296	20.7
	H_2	−.22044		.03199	.1061	.070	26.6
	H_3	.24189	−.00074	.02019	.0817	.348	22.7
Adult deer	H_0	−.01634			.1403	.000	31.1
	H_1	.38517	−.00093		.1069	.262	26.9
	H_2	−.07342		.00843	.1395	−.053	34.3
	H_3	.40137	−.00094	.00200	.1069	.258	30.2
Does	H_0	−.02476			.1672	.000	36.1
	H_1	.36788	−.00143		.1263	.286	31.5
	H_2	−.15460		.01919	.1634	−.008	38.7
	H_3	.29833	−.00139	.00889	.1255	.304	34.7
Bucks	H_0	.00286			.1897	.000	39.6
	H_1	.57366	−.00368		.1306	.076	32.5
	H_2	.05956		−.00838	.1889	−.445	42.8
	H_3	.75981	−.00388	−.02285	.1254	.085	34.6
Fawns	H_0	.04951			1.2285	.000	90.9
	H_1	1.15667	−.01279		.7302	−.260	80.7
	H_2	−1.14187		.17609	.9051	−1.126	86.7
	H_3	.15371	−.01064	.12072	.5923	−.107	78.1

[a] H_0 = No density dependence, no rainfall effect; H_1 = density dependence, no rainfall effect; H_2 = no density dependence, rainfall effect; and H_3 = density dependence, rainfall effect.

Note: See Dennis and Otten (2000) for details on methodology. Table 12.2 contains rainfall and census data.

fawn:doe ratios followed a west–east gradient across Texas, and this may be a driving factor in the gradient of density dependence found in the time series analysis. They presented graphs showing the rainfall-fawn production correlation was strong in the west and declined in strength to become essentially nonexistent in east Texas. Thus, if it does not rain in the spring–summer in western South Texas (Faith Ranch), fawn survival declines. Perhaps this happens frequently enough to prevent populations with a plateau phase from building close to K and exhibiting density-dependent behavior. Following Ginnett and Young (2000), the rainfall–fawn survival relationship becomes weaker in eastern South Texas (Welder Refuge), but remains evident.

Mean March–May rainfall over a period of years on the Faith Ranch, Chaparral Area, and Welder Refuge was 13.08, 17.01, and 21.63 cm, respectively (Tables 12.1–12.3). We chose to work with March–May for a finer breakdown of the annual cycle versus the March–July period used by Ginnett and Young (2000). Interestingly, although the mean amount of rain declined from east to west, the variation in spring rainfall was nearly identical across the three areas. The CV for March–May rainfall was 53.7, 48.4, and 49.7% for Faith Ranch, Chaparral Area, and Welder Refuge, respectively (Tables 12.1–12.3). McCullough (1992) reviewed the impact of environmental

TABLE 12.7
Maximum-Likelihood Estimates (a, b, c, o^2) of Parameters in a Density Dependence–Spring Rainfall Model, Generalized R^2, and Schwartz Information Criterion (SIC) for Four Model Hypotheses (H) Fitted to White-Tailed Deer Density Data Obtained by Spotlight Counts (1963–1976) or January Surveys by Helicopter (1977–1998) or Embryos/Adult Doe, Welder Wildlife Refuge, South Texas, USA

Variable	H[a]	$\hat{a}$	$\hat{b}$	$\hat{c}$	$\hat{o}^2$	R^2	SIC
Total Deer	H_0	−.03851			.0611	.000	7.9
	H_1	.34037	−.01247		.0468	.419	3.8
	H_2	−.06762		.003319	.0609	.241	11.1
	H_3	.31865	−.01244	.002356	.4680	.442	7.0
Embryos/adult doe	H_0	−.00490			.0186	.000	−33.1
	H_1	.81994	−.51258		.0112	.032	−47.2
	H_2	.07818		−.00942	.0170	−.556	−32.0
	H_3	.81265	−.48286	−.00482	.0108	.053	−44.9

[a] H_0 = No density dependence, no rainfall effect; H_1 = density dependence, no rainfall effect; H_2 = no density dependence, rainfall effect; and H_3 = density dependence, rainfall effect.
Note: See Dennis and Otten (2000) for details on methodology. Table 12.3 contains rainfall, density, and embryo data.

stochasticity on density dependence in ungulate populations. Additionally, McCullough (2001) found density-dependent behavior in black-tailed deer occupying a variable environment but with a high average annual rainfall (95 cm).

As a hypothesis for future research, we attempted to fit our three South Texas study areas to McCullough's (1999) models (Figure 12.1). Fetal rates (Table 12.3) for the Welder Wildlife Refuge provided a means to scale this population relative to K over a long period. We used the generalized sustained yield table in Downing and Guynn (1985), which scales fawns/adult doe to various percentages of K. Plugging the fetal rates (Table 12.3) into Downing and Guynn's (1985) table indicated that the Welder Refuge deer population has exceeded 50% K over most of the time series, except during the late 1960s and most of the 1990s (Figure 12.3). We used 50% of K as an arbitrary cut-off, above which the population would express density-dependent responses and below which it would not. In other words, in the context of McCullough's (1999) models (Figure 12.1), we assumed a plateau below 50% K and a ramp above this level. Given these assumptions, and strong indications of density dependence in the time series analyses, we hypothesize that the Welder Refuge deer population approximates Figure 12.1B and b.

Although our time series analysis failed to detect density dependence at the Faith Ranch, the westernmost population we analyzed, we hypothesize that this population occasionally builds up enough to exhibit density-dependent behavior. Thus, we posit that the Faith Ranch population approximates the models in Figure 12.1C and c. This would leave the Chaparral Area population somewhere in the middle as far as density-dependent behavior is concerned.

FOUNDATION THEORY AND MANAGEMENT RELEVANCE

Researchers commonly debate whether deer populations are density dependent or density independent as if they are competing population models. McCullough (1992) observed that all populations

TABLE 12.8

Results of Statistical Hypothesis Tests of Influence of Density Dependence and March–May Rainfall on Time Series of Abundance in South Texas White-Tailed Deer Populations

Study area[a]	Test[b,c]	Stat.	Total deer	Adults	Does	Bucks	Fawns	Embryos/doe
F	1	t	−1.93	.412	.287	.481	.154	
		P	.310	−1.69	−1.99	−1.54	−2.20	
	2	t	2.76	1.20	.67	1.79	3.60	
		P	.015	.251	.517	.095	.003	
	3	t	2.57	1.10	.695	1.56	3.02	
		P	.047	.448	.599	.293	.012	
	4	t	−1.76	−1.59	−1.94	−1.30	−1.57	
		P	.286	.406	.265	.515	.261	
C	1	t	−3.07	−2.85	−2.90	−3.43	−4.24	
		P	.034	.057	.049	.016	.002	
	2	t	1.62	.37	.78	−.32	3.05	
		P	.118	.713	.441	.754	.005	
	3	t	1.11	−.10	.40	-1.02	2.41	
		P	.368	.934	.747	.376	.024	
	4	T	−2.73	2.76	−2.75	−3.56	−3.63	
		P	.064	.064	.061	.011	.006	
W	1	t	−2.81					−4.66
		P	.050					.001
	2	t	.29					−1.77
		P	.776					.086
	3	t	.23					−1.09
		P	.854					.307
	4	t	−2.75					−4.27
		P	.052					.002

[a] F = Faith Ranch; C = Chaparral Wildlife Management Area; and W = Welder Refuge.

[b] 1 = H_0 versus H_1; 2 = H_0 versus H_2; 3 = H_1 versus H_3; and H_2 versus H_3.

[c] H_0 = No density dependence, no rainfall effect; H_1 = density dependence, no rainfall effect; H_2 = no density dependence, rainfall effect; and H_3 = density dependence, rainfall effect.

Note: Tables 12.1–12.3 contain time series and rainfall data. Tables 12.4–12.6 contain maximum-likelihood estimates of parameters, along with generalized R^2 and Schwartz information criterion for a density dependence–rainfall model. See Dennis and Otten (2000) for details.

experience periods of density dependence and density independence. This is better than considering the two as competing models, but still leads to some confusion as to the underlying theoretical infrastructure. Because food-limited intraspecific competition has commonly been demonstrated, we posit that a food-limited, density-dependent model is the best theoretical underpinning for all *Odocoileus* populations. However, just because a population's behavior is being understood through a food-limited, density-dependent model does not mean that the population is always expressing density-dependent responses. In other words, a population does not have to be expressing density dependence to be understood within a density-dependent model context. This is a subtle but important point.

McCullough (1999) proposed hypotheses for populations that have a plateau and a ramp in the density-dependent function. He proposed that mule deer and desert mule deer are among the ungulates that may have populations with this type of curve. He stated that such populations may not reach *K* very often because of low intrinsic rate of increase, high environmental variability,

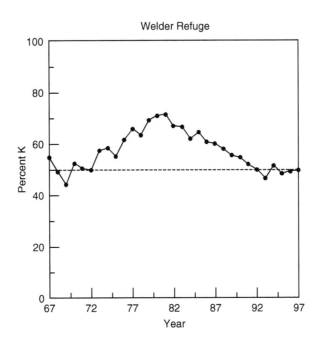

FIGURE 12.3   Five-year running average of percent K carrying capacity estimated for a white-tailed deer population on the Rob and Bessie Welder Wildlife Refuge, South Texas, USA, 1967–97. Percent K was estimated from fawns *in utero* of adult does collected annually on the refuge, using the percent K scaling table of Downing and Guynn (1985).

large home ranges in homogenous habitat, or habitats with low-quality forage. Such populations would not show density-dependent response at densities where they are on the plateau phase of the density-dependent function. We have proposed, based on analysis of the South Texas time series, that white-tailed deer populations in some environments will also exhibit a plateau and ramp function. However, during favorable times, such as a string of wet years, plateau and ramp populations can build close enough to K where they function as density dependent (Figure 12.1b and c).

We argue that it is better conceptually to consider plateau and ramp populations within a food-based, density-dependent model, recognizing that, for many reasons, deer may spend much of their time in the plateau phase. The frequency with which such populations occupy the ramp phase determines whether predictions based on a density-dependent model will be useful to managers. We believe that populations such as the white-tailed deer on the Welder Refuge could be managed with the expectation of density-dependent behavior. However, for populations such as the Chaparral Area, and certainly the Faith Ranch, a manager would seldom expect a density-dependent model to be predictive.

A smoothed plot of the three South Texas populations over the time series of population counts is shown in Figure 12.4. Deer density is always considerably higher on the more productive habitat of the Welder Refuge. Densities of all populations peaked in the late 1970s, which was the wettest string of years in the twentieth century for the South Texas region (DeYoung 2001). During this time of relatively high population density, all three populations may have been in the ramp phase. However, this rainy period was followed by more typical rainfall patterns in the 1980s and 1990s, when densities of all populations declined. Likely, the Faith Ranch and Chaparral Area were in the plateau phase during this time, and the Welder Refuge may also have been at times (Figure 12.4).

A frequently repeated statement in the literature is that density-dependent behavior, while present, is difficult for researchers to detect (McCullough 1999). Researchers can allocate more resources to population work than managers in most cases can. If researchers cannot detect density dependence,

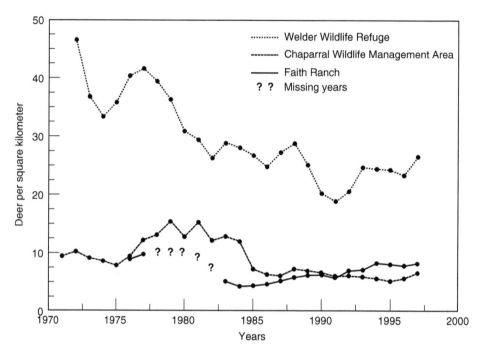

FIGURE 12.4   Three-year running average of counts of deer/km^2 on three South Texas, USA, study areas.

even though it is present, then a density-dependent model would be inadequate for managers to rely upon for such populations. Therefore, in addition to no density-dependent responses from populations in a plateau phase, there are presumably frequent circumstances where density dependence is present but masked. A density-dependent model is not useful to a manager in either situation except as an underpinning theory or a component of a more complex model.

This led us to wonder what a map of *Odocoileus* range would look like with regions for "density-dependent model likely predictive" and "density-dependent model likely not predictive." Shea and Osborne (1995) surveyed state and provincial wildlife departments within white-tailed deer range in the United States and Canada and asked them to identify "sub-optimal habitats." With these data, they produced a map of such habitats, where density-dependent population responses may not occur. We extended this map by also including habitats with high precipitation variability and habitats where occasional severe winters limit deer populations.

To construct the map, we first obtained ranges for white-tailed deer and mule deer from Hef-felfinger (2006) (Figure 12.5). For simplicity, only the ranges in the lower 48 states of the United States were used. We superimposed on the range map the suboptimal habitats from Shea and Osborne (1995). To approximate areas with variable environments, we mapped two variables. First, we constructed a grid with 100 squares across the United States. Then we selected the U.S. Weather Station nearest the center of each grid and obtained 30 years of annual precipitation records. We calculated the coefficient of variation (CV) from these data and superimposed on the deer range map areas with a CV $\geq$ 30% (Figure 12.5). This level of variation in annual precipitation was selected based on the approximate CV for the Faith Ranch, where we detected no density dependence in the time series.

The second variable we selected was a measure of winter severity. For simplicity, we used mean minimum air temperature for January, averaged over 30 years. We used temperature data from the U.S. Weather Stations of "cold grid squares" identified previously for precipitation variation. We arbitrarily selected mean minimum January air temperature of $\leq -12°C$ as the cut-off for the "cold winter" variable and superimposed these areas on the deer range map (Figure 12.5).

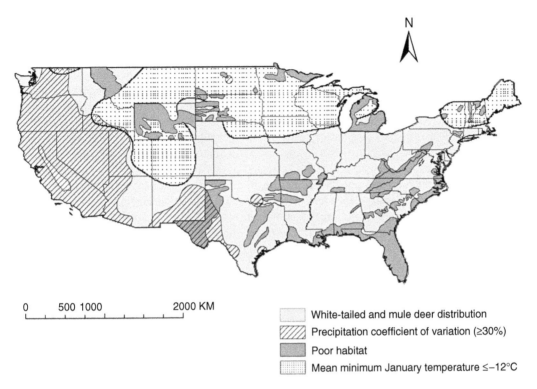

FIGURE 12.5   Map of lower 48 states of the United States showing areas where simple density-dependent population models may not be useful to managers. (*White-tailed Deer and Mule Deer Distribution*: From Heffelfinger, J. 2006. *Deer of the Southwest*. Texas A&M University Press. With permission; *Poor Habitat*: From Shea, S. M., and J. S. Osborne. 1995. Poor-quality habitats. In *Quality Whitetails: The Why and How of Quality Deer Management*, K. V. Miller, and R. L. Marchinton (eds). Mechanicsburg: Stackpole Books. With permission.)

The completed map (Figure 12.5) indicates that approximately 59% of *Odocoileus* deer range in 48 states of the United States may not be suitable for density-dependent management models. Obviously, this map is a crude first approximation. It is a hypothesis that needs refining with empirical data. There are certainly deer populations in the variable environments that exhibit density-dependent responses on a regular basis. However, the map also shows that there are substantial areas where density-dependent management models will not be predictive.

So, what is a deer manager in a variable environment to do? McCullough (1984) proposed an *ad hoc* strategy for such situations. He stated that management decisions would need to be more on a yearly basis in response to the immediate environmental conditions when lacking a predictive longer-term model to follow. Mackie et al. (1990) similarly advocated annually surveying population size and performance to guide management decisions. Hopefully, our technology for predicting the behavior of deer populations in variable environments can improve beyond an *ad hoc* strategy. A principle value of predictive density-dependent models is their simplicity. Future models should be developed by researchers for managers that apply to variable environments. A density-dependent variable that switches on in the model when a fluctuating population builds close to K should be basic. However, future models will almost assuredly need to be more complex, and they will likely need to include at least one other term such as a weather variable or a plant variable. There are empirical studies that have shown strong correlation between broad vegetation variables such as forb biomass and deer population performance or carrying capacity (Strickland 1998). Patterson and Power (2002) developed a model that explained 80% of population variation for white-tailed deer in Nova Scotia, Canada. Their model included a density-dependent term, harvest term, and a term for winter weather.

In summary, density-dependence effects in *Odocoileus* deer are complex. Some populations show effects most of the time, others very seldom. The pattern for a particular population may change over long time horizons. Arguing whether deer populations are density dependent or density independent is overly simplistic. Food-limited, density-dependent models are the simplest and most useful theoretical construct until modeling technology advances. However, because populations may be held in the plateau phase of the density-dependent function by density-independent factors, there may be no negative feedback from food competition. Such models, while theoretically useful, are not always predictive and thus useful to managers. Also, density dependence may be acting in the mix of factors impinging on a population, but be obscured. The result for managers is the same. Regions where simple density-dependent models are not useful to managers because of environmental variation may be more extensive than most have realized. More research is needed to support or refute this hypothesis. In the meantime, researchers should develop for managers more complex predictive models with a density-dependent factor and a least one factor that integrates as much of the environmental variation as feasible.

ACKNOWLEDGMENTS

Large-scale ecological research, such as that presented in this paper, cannot be done without the support of many individuals, including landowners, colleagues, and students. The authors thank all individuals who made this research possible and D. Guynn, who reviewed a draft of the paper.

REFERENCES

Ansotegui, R. P., and A. L. Lesperance. 1973. Effect of precipitation patterns on forage quality. *Proc. W. Sec. Am. Soc. Ani. Sci.* 24:229.

Ballard, W. B., et al. 2001. Deer–predator relationships: A review of recent North American studies with emphasis on mule and black-tailed deer. *Wildl. Soc. Bull.* 29:99.

Bartmann, R. M., G. C. White, and L. H. Carpenter. 1992. Compensatory mortality in a Colorado mule deer population. *Wildl. Monogr.* 121.

Beasom, S. L. 1974. Relationships between predator removal and white-tailed deer net productivity. *J. Wildl. Manage.* 38:854.

Beasom, S. L., F. G. Leon, III, and D. R. Synatzske. 1986. Accuracy and precision of counting white-tailed deer with helicopters at different sampling intensities. *Wildl. Soc. Bull.* 14:364.

Blankenship, T. L. 2000. Ecological response of bobcats to fluctuating prey populations on the Welder Wildlife Foundation Refuge. PhD Thesis, Texas A&M University, College Station and Texas A&M University - Kingsville, Kingsville, TX.

Box, T. W. 1960. Herbage production in forage range plant communities in South Texas. *J. Range Manage.* 13:72.

Brothers, A., and M. E. Ray, Jr. 1975. *Producing Quality Whitetails*, 1st edn. Laredo: Wildlife Services.

Carroll, B. K., and D. L. Brown. 1977. Factors affecting neonatal fawn survival in southern-central Texas. *J. Wildl. Manage.* 41:63.

Cook, R. S., et al. 1971. Mortality of young white-tailed deer fawns in South Texas. *J. Wildl. Manage.* 35:47.

Davis, R. B., and R. L. Spicer. 1965. Status of the practice of brush control in the Rio Grande Plain. Final Rep. Fed. Aid Proj. W-84-R, Texas Parks & Wildlife Department, Austin.

Dennis, B., and M. R. Otten. 2000. Joint effects of density dependence and rainfall on abundance of San Joaquin kit fox. *J. Wildl. Manage.* 64:388.

Dennis, B., and M. L. Taper. 1994. Density dependence in time series observations of natural populations: estimation and testing. *Ecol. Monogr.* 64:205.

DeYoung, C. A. 1985. Accuracy of helicopter surveys of deer in South Texas. *Wildl. Soc. Bull.* 13:146.

DeYoung, C. A. 2001. Predator control in deer management: South Texas. In *Proceedings of the Symposium on the Role of Predator Control as a Tool in Game Management*. College Station, TX: Tex. Ag. Exten. Serv.

Doughty, R. W. 1983. *Wildlife and Man in Texas*, 1st edn. College Station, TX: Texas A&M University Press.

Downing, R. L., and D. C. Guynn, Jr. 1985. A generalized sustained yield table for white-tailed deer. In *Game harvest management*. S. L. Beasom, and S. F. Robrson (eds). Kingsville: Caesar Kleberg Wildlife Research Institute, Texas A&M University - Kingsville, p. 95.

Dumont, A., et al. 2000. Population dynamics of northern white-tailed deer during mild winters: evidence of regulation by food competition. *Can. J. Zool.* 78:764.

Fafarman, K. R., and C. A. DeYoung. 1986. Evaluation of spotlight counts of deer in South Texas. *Wildl. Soc. Bull.* 14:180.

Foster, W. C. 1998. *The La Salle Expedition to Texas: The Journal of Henri Joutel 1684–1687*, 1st edn. Austin: Tex. State Hist. Assoc., Center for Studies in Tex. Hist., University of Texas at Austin.

Fryxell, J. M., et al. 1991. Time lags and population fluctuations in white-tailed deer. *J. Wildl. Manage.* 55:377.

Fulbright, T. E. 2001. Human induced vegetation changes in the Tamaulipan semiarid scrub. In *Changing Plant Life in La Frontera*, G. L. Webster, and C. J. Bahre (eds). Albuquerque: University of New Mexico Press, p. 166.

Gilbert, B. A., and K. J. Raedeke. 2004. Recruitment dynamics of black-tailed deer in the western Cascades. *J. Wildl. Manage.* 68:120.

Ginnett, T. F., and E. L. B. Young. 2000. Stochastic recruitment in white-tailed deer along an environmental gradient. *J. Wildl. Manage.* 64:713.

Harveson, L. A. 1997. Ecology of a mountain lion population in southern Texas. PhD Thesis, Texas A&M University - Kingsville, College Station.

Heffelfinger, J. 2006. *Deer of the Southwest*. College Station: Texas A&M University Press.

Helmer, J. A. 2002. Boone and Crockett whitetails: a geographic analysis. *Fair Chase.* 17:34.

Inglis, J. M. 1964. A history of vegetation on the Rio Grande Plain. Bulletin 45, Texas Parks and Wildlife Department, Austin.

Jones, F. B. 1975. *Flora of the Texas Coastal Bend*, 1st edn. Sinton: Rob and Bessie Welder Wildlife Foundation.

Keyser, P. D., D. C. Guynn, Jr., and H. S. Hill, Jr. 2005. Density-dependent recruitment patterns in white-tailed deer. *Wildl. Soc. Bull.* 33:222.

Kie, J. G., and M. White. 1985. Population dynamics of white-tailed deer (*Odocoileus virginianus*) on the Welder Wildlife Refuge, Texas. *Southwest. Nat.* 30:105.

Lehmann, V. W. 1969. *Forgotten Legions: Sheep in the Rio Grande Plains of Texas*. El Paso: Western Press.

Mackie, R. J., et al. 1990. Compensation in free-ranging deer populations. *Trans. N. Am. Wildl. Nat. Res. Conf.* 55:518.

Macnab, J. 1985. Carrying capacity and related slippery shibboleths. *Wildl. Soc. Bull.* 13:403.

McCullough, D. L. 1979. *The George Reserve Deer Herd*. Ann Arbor, MI: The University Michigan Press.

McCullough, D. L. 1982. Population growth rate of the George Reserve deer herd. *J. Wildl. Manage.* 46:1079.

McCullough, D. L. 1983. Rate of increase of white-tailed deer on the George Reserve: a response. *J. Wildl. Manage.* 47:1248.

McCullough, D. L. 1984. Lessons from the George Reserve, Michigan. In *White-tailed deer: Ecology and Management*, L. K. Halls (ed.). Mechanicsburg: Stackpole Books, p. 211.

McCullough, D. L. 1990. Detecting density dependence: filtering the baby from the bathwater. *Trans. N. Am. Wildl. Nat. Res. Conf.* 55:534.

McCullough, D. L. 1992. Concepts of large herbivore population dynamics. In *Wildlife 2001: Populations*, D. L. McCullough, and R. H. Barrett (eds). New York: Elsevier Applied Science, p. 967.

McCullough, D. L. 1999. Density dependence and life-history strategies of ungulates. *J. Mammal.* 80:1130.

McCullough, D. L. 2001. Male harvest in relation to female removals in a black-tailed deer population. *J. Wildl. Manage.* 65:46.

Meyer, M. W., R. D. Brown, and M. W. Graham. 1984. Protein and energy content of white-tailed deer diets in the Texas Coastal Bend. *J. Wildl. Manage.* 48:527.

Norwine, J., and R. Bingham. 1985. Frequency and severity of droughts in South Texas: 1900–1983. In *Proceedings of a Workshop on Livestock and Wildlife Management during Drought*, R. D. Brown (ed.). Kingsville: Caesar Kleberg Wildlife Research Institute, Texas A&M University - Kingsville, p. 1.

Patterson, B. R. and V. A. Power. 2002. Contributions of forage competition, harvest, and climate fluctuation to changes in population growth of northern white-tailed deer. *Oecologia* 130:62.

Shea, S. M., and J. S. Osborne. 1995. Poor-quality habitats. In *Quality Whitetails: The Why and How of Quality Deer Management*, K. V. Miller, and R. L. Marchinton (eds). Mechanicsburg: Stackpole Books, p. 193.

Shea, S. M., T. A. Breault, and M. L. Richardson. 1992. Herd density and physical condition of white-tailed deer in Florida flatwoods. *J. Wildl. Manage.* 56:262.

Shenk, T. M., G. C. White, and K. P. Burnham. 1998. Sampling-variance effects on detecting density dependence from temporal trends in natural populations. *Ecol. Monogr.* 68:445.

Short, H. L. 1979. Deer in Arizona and New Mexico: Their ecology and a theory explaining recent population decrease. Gen. Tech. Rep. RM-70, Department of Agriculture, Rocky Mountain Forest and Range Experiment Station, Washington, DC.

Strickland, B. K. 1998. Using tame white-tailed deer to index carrying capacity in South Texas. MS Thesis, Texas A&M University - Kingsville, Kingsville.

Van Auken, O. W. 2000. Shrub invasions of North American semiarid grasslands. *Annu. Rev. Ecol. Syst.* 31:197.

White, G. C., and R. M. Bartmann. 1997. Density dependence in deer populations. In *The Science of Overabundance*, W. J. McShea, H. B. Underwood, and J. H. Rappole (eds). Washington, DC: Smithsonian Institution Press, p. 120.

Part III

Habitat

13 From the Management of Single Species to Ecosystem Management

Jack Ward Thomas

CONTENTS

Wildlife management has been traditionally separated, though not too cleanly, into the management of populations and the management of habitats of targeted species (Leopold 1933). Most species of management attention were game species (species that were hunted), species trapped because of the value of pelts, or (much more recently) species determined to be threatened or endangered — or likely to be accorded such status. This separation between species management and habitat management can never be complete, because many species have the capability to dramatically influence their own habitats and, simultaneously, the habitats of myriad other species.

Early on it was recognized that wildlife management is a "decision science," wherein ecological understanding, public opinion, economic trade-offs, and compliance with laws and regulations are melded into management decisions and activities, as detailed by Giles (1978). These various aspects of wildlife conservation and management expanded over the years, and this growing recognition was well illustrated in The Wildlife Society's publication *Wildlife Conservation — Principles and Practices* in 1979.

During the period 1890–2006, in the United States, knowledge within the overall realm of ecology increased at an ever increasing rate — along with building legal pressures to apply this new knowledge in what was to become known as *multiple-use management*. Within some 50–60 years after 1890, this had led to dramatic alterations in the management of wildlife populations and, to some lesser extent, their habitats. The expanding knowledge base emanating from experience, research, and their integration helped produce a growing public and scientific concern relative to the welfare of wildlife and wildlife habitats. The end result was the passage of a plethora of laws focused on environmental concerns, with some focus on wildlife and ecological integrity in the late 1960s and 1970s (e.g., USDA Forest Service 1993).

Around the turn of the nineteenth century, roughly during 1890–1910, a rapidly increasing number of citizens of the United States were expressing concern and supporting action to protect remaining wildlife populations and begin restoration of wildlife to areas of extirpation. These efforts included increased involvement by scientists — often through or by members of the Boone and Crockett Club (Trefethen 1961), which was established in 1887. Accumulating knowledge and rising concerns combined, not always smoothly, to produce a half century (1900–1950) of intermittent but dramatic changes in the management of both plant and wildlife communities. It is interesting to note that such leaders as Aldo Leopold were seminal thinkers who, quite easily and naturally, crossed disciplinary lines. Leopold was trained as a forester but was both a founder and President of the Ecological Society and The Wildlife Society (Meine 1988).

As MacFayden (1963) put it

The ecologist is something of a chartered libertine. He roams at will over the legitimate pre-
serves of the plant and animal biologist, the taxonomist, the physiologist, the behaviorist, the
meteorologist, the geologist, the physicist, the chemist, and even the sociologist; he poaches
from all these and other disciplines. It is indeed a major problem for the ecologist, in his own
interest, to set bound to his divagations.

Smith (1995) and Golley (1993) did a masterful job of succinctly tracing the development of
ecological sciences. These innovations provided the building blocks for the establishment of the profession of wildlife management, which is, simply, application of ecological understanding coupled
with frequent mid-course corrections taken in response to legal requirements and in the spirit of
adaptive management. The following discussion of the early development of the science of ecology
as an underpinning for wildlife management borrows heavily from Smith's (1995) work.

In North America, Cowles (1899) and Clements (1916) were leaders in the development of
information and theories in the dynamics of plant communities including the concepts of plant
community succession. The trophic–dynamic concepts of ecology (Lindeman 1942) evolved from
the studies of E. A. Birge and C. Juday of the Wisconsin Natural History Survey in aquatic systems
and marked the beginnings of modern ecology. From these efforts evolved work by Hutchison (1957,
1969) and Odum (1969, 1970) on energy flow and energy budgets. The pioneering work on nutrient
cycling was done by Ovington (1961) in England and by Rodin and Bazilevic (1964) in Russia.

Early studies in animal ecology in Europe were mostly influenced by Elton's (1927) *Animal Ecology*. Tansley (1935), a British plant ecologist, advanced the concept of the ecosystem — and things
were never the same again. Shelford's (1913) *Animal Communities in Temperate North America* was
an early pioneer of ecological studies in North America. He stressed the relationships of plant and
animals, with emphasis on the concept of ecology as a science of communities. Allee et al. (1949)
came out with the encyclopedic *Principles of Animal Ecology*, which emphasized trophic structures
and energy budgets, population dynamics, natural selection, and evolution. This was accompanied
with the increasing understanding across ecological disciplines that plant and animal communities
are not and were never truly separate entities.

Lotka (1925) and Volterra (1926) developed theoretical approaches to ecological studies that
opened new experimental approaches. Gause (1934) investigated interactions of predators and prey
and competition between species. Andrewartha and Birch (1954) and Lack (1954) provided a foundation for studying regulation of populations. Niche theory evolved from these studies. Behavioral
ecology evolved from the seminal work on territoriality by Nice in the 1930s and 1940s (e.g., Nice
1941). The sub-field of behavioral ecology emerged from these beginnings. Seminal publications in
this arena included Wynne-Edwards' (1962) *Animal Dispersion in Relation to Social Behavior* and
Wilson's (1975) *Sociobiology*.

Population ecology, due to the requirements for qualitative approaches, led inexorably to the
development of theoretical mathematical ecology in the arenas of competition, predation, community
and population stability, cycles, community structure, community association, and species diversity
(Smith 1995). Mathematical modeling as an approach to ecological understanding and management
has continued to the present.

Pioneers in the arena of the application of ecological principles to the management of natural
resources — forests, rangelands, fish, and wildlife — could have been called applied ecologists.
But, instead, they are referred to as foresters, range managers, and fish and wildlife biologists. As
examples of early application of ecological principles and understanding, we have such pioneering
work as that of Stoddard (1931) who in, *The Bobwhite Quail*, put forward the concept of using fire to
hold back plant succession to favor the production of bobwhite quail (*Colinus virginianus*). Federal
land managers concerned with fire management only full grasped such concepts of fire in ecosystem
management with the past two decades.

Leopold (1933) put forward the application of ecological principles to the management of wildlife in the classic *Game Management*. Lutz and Chandler (1954), in *Forest Soils*, discussed nutrient cycles and the role they played in the forest ecosystem and their consideration in management. Leopold's (1949) *A Sand County Almanac* called for an ecological land ethic but did not receive wide attention from the general public until the mid-1970s.

Carson's (1962) *Silent Spring* brought the public's attention to environmental problems resulting from the careless use of chlorinated hydrocarbons and the effects on animal life. Ehrlich and Ehrlich (1981) brought attention to causes and consequences of extinctions. And, essentially, a revolution in public concern and attention to environmental problems resulted.

The sum effect of this accumulating knowledge and concerns over the environment led to the plethora of environmental laws that were enacted in the United States (and elsewhere in the world) in the 1960s and 1970s. The 1970s were to become known as the "environmental decade." These laws included Multiple-Use Sustained-Yield Act of 1960, Wilderness Act (1964), Wild and Scenic Rivers Act (1968), National Environmental Policy Act (1970), Endangered Species Act of 1973, Federal Land Policy and Management Act of 1976, and the National Forest Management Act of 1976 (USDA Forest Service 1993).

After 1960, a plethora of volumes began to appear related to the application of ecological understanding in wildlife management. A few of these were *Wildlife Biology* (Dasman 1964), *Wildlife Ecology: An Analytical Approach* (Moen 1973), *Wildlife Management* (Giles 1978), *Principles of Wildlife Management* (Bailey 1984), *Wildlife Ecology and Management* (Robinson and Bolen 1989), *Managing Our Wildlife Resources* (Anderson 1985), *A Review of Wildlife Management* (Peek 1986), and *The Philosophy and Practice of Wildlife Management* (Gilbert and Dodds 1987). Each was predicated on the inclusion of knowledge from the sister or codiscipline of ecology in forms that could be applied to wildlife management.

In the United States, among other nations, a "feed back loop" developed, wherein new concepts emanating from science, especially the ecologically based disciplines, produced demands from the scientific community and conservation-minded publics for incorporation of those new concepts into the laws and regulations outlining public policy relative to wildlife and land management. Beginning in the 1970s, politically, legally, and technically inadequate attempts by government entities (particularly public land management agencies) to comply with new laws and regulations demonstrated that there was inadequate information and applicable and defensible techniques to comply with legal requirements (Thomas 1979).

This proved especially onerous as the situation was constantly revised and redefined by legal opinions from the courts relative to challenges to agency actions by proliferating environmental organizations. These challenges were encouraged by the Equal Access to Justice Act (USDA Forest Service 1993), which not only allowed suits against federal agencies but provided for compensation to plaintiffs if they were victorious and excused them from liability if they lost.

In addition to incorporating new information and understanding from ongoing research and consultations with technical experts, land and wildlife management agencies were investing heavily in their own research and synthesis to enable them to comply with applicable laws. A number of examples, singly and in accumulating interactions, of this "feed back" phenomenon exist. I believe, perhaps because I was deeply involved as a scientist, planner, and head of the U.S. Forest Service with the dramatic changes in land and wildlife management on the National Forests, that this rapid evolution of management approaches, presents the best illustration of the transition to ecosystem management (Thomas 2004).

These changes in attitude, approach, and application of available scientific understanding came dramatically to bear in land and wildlife management on the National Forests over a period of less than a decade. This shift was facilitated by rapidly accumulating knowledge from ongoing research. New knowledge was accumulating faster than it was, or perhaps could be, incorporated into planning and management — that is, federal land and wildlife management agencies were increasingly vulnerable to challenge in court.

This new knowledge almost immediately interacted with a spate of new environmental laws and regulations — that is, it had to be incorporated more expeditiously into management to avoid legal challenge. The "petri dish" was to be the federal court system. That, in turn, focused public attention, increasingly expressed through legal challenges, on management of the National Forests. This attention was exacerbated by the insistence of the Reagan and G. H. W. Bush administrations, and reinforced by consistent budget direction from Congress, to "get out the cut," as the employees put it, and, to steal a phrase from Admiral Dewey, with the implication of "Damn the torpedoes — full speed ahead" (Hirt 1994).

From 1897 to 1960, the National Forests were managed under the sole direction of the Organic Administration Act of 1897. That Act stated that no Forest Reserves (the precursor to the National Forests, which were established in 1905 along with the simultaneous establishment of the U.S. Forest Service) could be set aside "except to improve and protect the forest ... or for the purpose of securing favorable conditions of water flows, and to furnish a continuous supply of timber ..." (USDA Forest Service 1993). It was not until the passage of the Multiple-Use Sustained-Yield Act of 1960 that the Forest Service was mandated to be concerned with sustained protection and production of wildlife along with timber, wood, water, recreation, and grazing. This Act legitimized the course of management that the Forest Service had already put into action — at least in theory. This Act essentially stopped the ongoing "raids" on National Forest System lands by the National Park Service (for National Parks) and the U.S. Fish and Wildlife Service (for wildlife refuges).

Improved understanding of ecological processes applicable in management and planning by wildlife and land management professionals was increasing rapidly in federal land management agencies. Newly hired fish and wildlife biologists, geologists, hydrologists, economists, social scientists, ecologists, and other technical specialists produced increased internal pressure for change in management direction and processes that were manifested in the "environmental laws" passed largely in the 1970s. The most prominent laws affecting how wildlife and their habitats were to be considered in planning and management were the National Environmental Policy Act of 1970, the Endangered Species Act of 1973 (which would become known as the "900-pound gorilla" of environmental laws), and the National Forest Management Act of 1976.

Though unclear at the time of their enactment, these three laws would ultimately interact to ensure that rapidly accumulating knowledge in the overall arena of ecology would more profoundly influence management of the National Forest System. These influences on the management of National Forests were, in less than a decade, rapidly spread — in varying degrees and aspects — to other federal lands and to state and local jurisdictions and ownerships.

The National Environmental Policy Act of 1970 required that all land management activities carried out with federal funds be subject to detailed analysis to determine the associated environmental effects of proposed alternative courses of action (including a "no action" alternative). Such analyses were required to take a "hard look" at environmental consequences of proposed actions. Just how "hard" that "look" had to be to meet the intent of the law continuously ratcheted up as decisions continued to emerge from the federal courts (i.e., as legal precedents were established). One of the initial sponsors of the National Environment Policy Act, Senator Henry "Scoop" Jackson of Washington, said that he expected the longest Environmental Impact Statement to be about 10 pages. Today, some such documents run to well over 1,000 pages.

The Endangered Species Act of 1973 also contained a "time bomb," which was unrecognized or unappreciated at the time of passage. This "time bomb" was hidden from plain sight and was clearly stated for all those who actually studied the wording of the proposed Act. The bomb did not explode upon the wildlife and land management scene until the early 1990s. But, when the explosion happened, land management was profoundly changed. When queried as to the stated purpose of the Act, natural resource management professionals, and ordinary citizens, will almost universally reply, even today, that the purpose is — pure and simple — the prevention of the extinction of individual species. That is true enough, but the stated purpose of the Act is broader than that — much broader than attention to individual species. The focus was on ecosystems. "The purposes of this Act are to provide a means whereby the ecosystems upon which endangered species and threatened species may be conserved, (and) to provide a program for the conservation of such endangered species and threatened species ..." (USDA Forest Service 1993).

Whether it was clear or not to those who drafted the Endangered Species Act, or to those who enacted it into law, this was a huge change in law, one of the most dramatic in history related to the management of natural resources. Ecosystems consist of interacting organisms (plant and animal) and the abiotic environment with which organisms interact (Smith 1995). Therefore, conserving a threatened species or an endangered species will, by law, involve consideration of the conservation of the ecosystem of which it is part. The general public and members of Congress — and even Presidents — exhibited no real understanding, or appreciation, of just what that meant and required in terms of changes in land management (Yaffee 1994). Conversely, the federal judiciary did and does understand. That understanding is consistently reflected in a consistent series of rulings from the federal bench. After all, the law means what judges say it means.

The focus of protective management actions in response to those court decisions is, increasingly, focused on the immediate and long-term sustenance of ecosystems upon which threatened species or endangered species depend. And, "species" has grown to include all species of animals (vertebrates and invertebrates) and plants (including even fungi and bryophytes). Many of the legislators that voted for the passage of this Act now maintain that they either misunderstood what was quite clearly stated or they had no appreciation of what the consequences of adherence to the letter of the law would produce in terms of social and economic effects. But, there have been no serious attempts to change the Endangered Species Act since its passage in 1973 — 34 years at this point.

The National Forest Management Act of 1976 provided, also quite inadvertently, another bold requirement for National Forest planning and management that has since spread to other federal lands. The Forest Service had, upon request, drafted the legislation that was sponsored by Senator Hubert Humphrey of Minnesota. The partial intent of that legislation was to restore prerogatives of agency professionals that had been lost in court decisions relative to clear cutting on the National Forests. The Forest Service had been practicing clear cutting even though the Organic Act called for the marking of individual trees for cutting. The new law (National Forest Management Act) also required that long-term (10–15 years) plans be prepared for each National Forest.

Details as to just how the law would be carried out were left to the Forest Service to describe in regulations. Another time bomb, probably inadvertently, would be tucked away in these new regulations. This obscure provision, coupled with the Endangered Species Act, would force a dramatic change in management direction for the public lands of the United States — first in the Pacific Northwest. That obscure provision in regulations required that "viable populations of all native and desirable non-native vertebrate species will be maintained well-distributed in the planning area." Given the spatial constraints involved (which did not exist in the Endangered Species Act), this was an even tougher requirement.

The federal lands of the United States and, most dramatically, the National Forest System have been profoundly impacted, over the period 1990–2005, by required changes in management approach. This resulted from dynamic interaction of new knowledge, the increased engagement of ecologically based professions, a more active and environmentally concerned citizenry, extant environmental laws, and increasing litigation (Hirt 1994). Application of the new knowledge resulting from scientific inquiry to federal land management has been most dramatic with lesser impacts on state and private lands. Why? Political decisions for the management of federal lands have been made, to the extent possible, to bear the economic consequences of adherence to the environmental laws on public lands (Forest Ecosystem Management Assessment Team 1993).

The Forest Service was, in spite of the directions in the Organic Act of 1897 to provide a continuing supply of timber to the American people, increasingly constrained by political and economic factors (1897–1946) from producing any appreciable amounts of timber. Powerful forces in the private sector did not, for obvious reasons, want the competition of timber from federal lands. Pressures to hold down production of timber from public lands were reinforced during the Great Depression (1929–40) when demand for wood products dropped dramatically. Then, the "cost plus 10%" contracts relative to the provision of materials needed for the war effort, during the run up to and through the end of World War II, provided all the timber required — even for war effort. This was simply too sweet a deal for private industry, and their representatives in Congress, to allow any significant competition from timber originating from federal lands. But, by the end of the war, timber supplies from private lands were running out (Hirt 1994).

American military personnel, most of who came to maturity during the Great Depression, started their return to civilian life in late 1945. A grateful nation welcomed them with the so-called GI Bill, which, among other features, facilitated home ownership. And, suddenly, the housing boom was on. There had been few houses constructed from the onset of the Great Depression in 1929 to end of World War II in 1945 — a 16-year hiatus. Demand for wood products suddenly exploded. New contracting rules were adopted that facilitated sales of timber from public lands. As a result of these factors, timbers sales from the National Forests grew steadily until, by the start of the 1990s, the timber cut from the National Forest System was near 13 billion board feet per year and the administrations in power were pushing, through the newly required forest planning process, for 25 billion board feet per year. The Forest Service was, creatively, dragging its collective feet from complying with these pressures (Hirt 1994).

To come into compliance with environmental laws, it was essential for the Forest Service and other federal land management agencies to step up the hiring of an array of specialists whose professions were based in applied ecology. Such specialists had, up to that time, been a relative rarity in a professional workforce dominated by foresters, range management specialists, and engineers. Included were wildlife biologists, fisheries biologists, ecologists of various types, soils scientists, hydrologists, landscape architects, economists, social scientists, and others (Hirt 1994; Yaffee 1994). Those who resented their increasing presence and influence referred to them, collectively, often with a sneer or worse, as "those *ologists*."

By this time, the late 1980s, hundreds of thousands of miles of logging roads snaked across the National Forest System and adjacent private lands. Clear cut timber harvest units fragmented the landscape into smaller and smaller patches of mature trees (Hirt 1994). At first, there was little concern as the guidelines instituted to maximize the positive effects and minimize the negative effects of these actions on "charismatic mega fauna" such as mule deer (*Odocoileus hemionus*), white-tailed deer (*Odocoileus virginanus*), and elk (*Cervus elaphus*) paid off. The new roads opened up many hundreds of thousands of acres to easy motorized human access. Species adapted to an interspersion of mature forest and openings created by timber harvest activities prospered.

Until the second half of the twentieth century, wildlife management was most commonly directed at species that were hunted or fished. That management took several basic forms. First and foremost was protection from overexploitation by hunters. Second was the control of predators (which might also prey on domestic livestock). Third was the protection of habitat usually in the form of "refuges" of one sort on another (Trefethen 1961).

Early texts, such as Leopold's (1933) *Game Management*, emphasized those factors and added insights such as that without habitat there would be no wildlife and that provision of "food, cover, and water" within the "home ranges" of the animals of interest was the essence of wildlife management. The economic depression that swept over the western world from 1929 to 1939 and the years of World War II from 1939 to 1945 put a damper on what had been the growing concerns with wildlife welfare. But the seminal thinkers in the field were, if anything, increasingly active and innovative.

The rapidly accumulating knowledge from the various fields of ecology was being ever more rapidly translated and transferred into the realm of wildlife management and recognized by the emerging philosophers of the role of ecological understanding in human affairs. And, some of those thoughts were to be proven both prescient and precursors to change that would reach far beyond wildlife management. A rising public understanding and concern over ecosystems was looming just over the horizon.

In 1942, Aldo Leopold simply stated that "Communities are like clocks, they tick best when possessed of all their cogs and wheels" (Leopold 1942). This now obvious but, then prescient statement, was the maxim that was to serve as the ecological underpinning of the Endangered Species Act in 1973.

Paul Ehrlich echoed that profound observation by saying "Everything is connected to everything else" and, therefore, "There is no free lunch." In other words, managers cannot make any change in the natural world that will not have ripple effects throughout plant and animal communities. Intensive management of any ecosystem for products desired by humans would, inevitably, have ecological consequences. That likewise prescient statement would find its way into law in the form of the National Environmental Policy Act of 1973.

Another early ecologist, Frank Egler (1954), observed that "Ecosystems are not only more complex than we think; they are more complex than we can think." Those simple maxims from early ecologists make it clear that "ecosystem management" — to the extent that anyone was thinking in those terms — was never anticipated to be easy.

The old debate of the early twentieth century relative to the management of the public domain was personified by John Muir (who favored "preservation") and Gifford Pinchot (who favored wise use guided by trained professionals) never abated and continues to rage today — though the animosity between Pinchot and Muir seems to have been much exaggerated (at least until the fight over Hetch Hetchy dam). Pinchot modified his views over the years and in his later years clearly recognized what we would come to call "ecosystem considerations and values" — so much so that a biographer (Miller 2001) would identify his significant role in the evolution of "modern environmentalism." Times and the understanding and opinions of recognized experts change — as should be expected.

But, the debate was enhanced and the focus changed with introduction of the concept of the retention of biodiversity as the key to preservation or retention of ecosystem function (Wilson 1986; Huston 1994). Some National Parks, and later wilderness areas, were originally set aside to protect them from alteration by human activity — an intuitive move to preserve what would first emerge as concerns over dampening of human impacts and evolve over the next half century into concerns over the retention of biodiversity. But, even these "set aside" areas were gradually changing, some quite dramatically, as fire exclusion, increasing human use, and — in some cases — overpopulations of ungulates dramatically altered plant communities that were evolving rapidly toward what was then thought of as "climax" conditions (in the Clementsian sense of the word: Clements 1916).

Forest and rangeland managers, from the early twentieth century until quite recently, relied on the theory of "ecological succession" (Clements 1916; Gleason 1917) in formulating management over time. Under that concept, a plant community reduced to essentially bare ground by some event (fire, logging, grazing, etc.) would, through an orderly and predictable process of clearly identifiable stages, ultimately reach a "climax state," which would be maintained until altered by a "catastrophic" (i.e., stand replacing) event (Oliver and Larson 1996).

Early wildlife managers quickly recognized that various species of wildlife were associated with various conditions related to successional stages of plant communities. Those managers also discerned that some wildlife species of particular interest to hunters were particularly plentiful where markedly different successional stages were juxtaposed — the so-called "edge effect" — providing simultaneous access to food and cover (Leopold 1933). Because game species such as deer, elk, ruffed grouse (*Bonasa umbellus*), turkeys (*Meleagris gallopavo*), and others seemed to favor such circumstances, the early assumption was that maximization of edge was a universally desirable habitat feature. Later research revealed that for many, perhaps most, species, various habitat conditions represented "sources" (survival and reproduction were high and in excess of replacement) and "sinks" (where survival and reproductive success were less than replacement). Interior forest bird species were found to be particularly susceptible to nest predation and nest parasitism in edge habitats [see discussion of edges in Hunter (1999)].

Other research (Gleason 1926; Oliver and Larson 1996) showed that the stages of succession, particularly as related to plant species composition, were not as mechanistic and predictable as originally thought. This adjustment in theory was not immediately accepted. However, even after it was accepted that succession was more dynamic than originally thought, structural relationships were at least somewhat predictable and the concept was still useful. When working in managed forest conditions, wildlife ecologists spend considerable effort in timing, sizing, and spacing of cutting (reforestation) units and subsequent stand treatments to maintain a mosaic of conditions most favorable to selected species, or a broad spectrum of species, depending on management objectives (Thomas 1979).

It also became evident, over time, that various species exhibited a positive correlation with increasing patch sizes (Leopold 1933; Thomas 1979), which conversely reduced the amount of edge. Then, it was wise to consider not only the patch size but the distance between patches and the condition of vegetation between patches, that is, "connectivity" between patches was of concern. These needs were found to be widely different among species depending on their mobility and plasticity to react to changing habitat conditions. Many of the effects were first noted in studying wildlife and the exchange of individuals between oceanic islands of various sizes and degrees of separation (Simberloff and Abele 1976a,b, 1982). This "island biogeographical theory" (Diamond 1975, 1976) was a key in forest planning in the Pacific Northwest to deal with the long-term survival of species adapted to "old growth" forest conditions — with emphasis on the northern spotted owl (*Strix occidentalis*) (Interagency Scientific Committee 1990).

In 1975, the fir (*Abies* sp.) forests in the Blue Mountains of Oregon and Washington had been severely damaged by an outbreak of Douglas-fir tussock moth (*Orgyia pseudotsugata*). The Forest Service was gearing up for a massive salvage effort. But this time, due to the combined consequences of the National Forest Management Act and the National Environmental Policy Act, National Forest System personnel were required to prepare massive Environmental Impact Statements before instituting any management actions related to insect control and timber salvage operations. Those completed environmental impact statements would have to meet the letter of the new regulations issued pursuant to the National Forest Management Act that required "viable populations of all vertebrate species" be maintained "well distributed in the planning area." Forest supervisors were at a loss for how to comply quickly with that requirement so as to salvage the dead timber while it was still economically valuable enough to offset costs. A team of Forest Service research biologists teamed with biologists from the National Forest system and the Oregon and Washington state wildlife agencies to produce a document (Thomas 1979) to support the needed assessments.

On the basis of the adaptation of guild theory pioneered by Haapanen (1965, 1966) to associate bird groups with forest type and successional stages, the document divided the 379 vertebrate species known to occur in the "planning area" into groups that exhibited affinity to various successional or structural stages of the various forest plant communities. Patch size, edge, and juxtaposition of structural stages were ecological theories upon which this approach was based (Leopold 1933). Chapters were written for species of special interest (deer, elk, and cavity nesting birds and mammals) and habitats of particular importance, such as riparian zones. Patch sizes, edges, and juxtaposition of stands in various successional states and conditions were considered (Thomas 1979). Over the years, numerous other planning documents structured on this first effort were compiled for other ecosystems elsewhere in North America and abroad (Morrison et al. 1998).

This effort begged a question that had been festering in the circles of ecologists and wildlife/land managers for some time — at least since Leopold (1949), in his essay *Green Fire*, pondered his experience with eradicating predators during his time with the Forest Service in the Southwest in the 1920s. How could "natural communities" and "intact ecosystems" be said to exist unless all of the species that evolved in that ecosystems were present and in ecologically effective numbers? These thoughts were later encapsulated in the developing concerns with the preservation of biodiversity in managed landscapes (Thomas 1979; Hunter 1999).

By the 1970s, concerns of ecologists and an increasing number of ecologically aware citizens began to be heard in the political arena. These concerns grew even faster than the populations of major predators — grizzly bears (*Ursus arctos horribilis*), black bears (*Ursus americanus*), wolves

(*Canis lupus*), coyotes (*Canis latrans*), mountain lions (*Puma concolor*), golden (*Aquila chrysaetos*) and bald eagles (*Haliaetus leucocephalus*), and others were being extirpated, or reduced to levels that precluded ecological effectiveness, over vast areas of North America.

These concerns were, by now, backed up by legal muscle and the impending, inevitable application of the Endangered Species Act (Snape and Houck 1996; Peterson 2002; Salzman and Thompson 2003) and the regulations issued pursuant to the National Forest Management Act. Concerns over retention of biodiversity in land management operations were coming to the fore (Huston 1994; Hunter 1999; Natural Resources Council 1999). It was clear that a day of reckoning was at hand for federal land managers. Yet, responsible federal agencies were slow to act. Why? Both Congress and the executive branch, regardless of the political party in power, were not eager to cross significant majorities of the agricultural and sport hunting constituencies, inflict social and economic costs in adjusting ongoing management, and then suffer the potential political consequences. So, pressure, in the form of instructions from the administrations and the Congress, expressed primarily through the budget, continued the push for a high level of timber production from National Forests.

The first big breaks fostering the "return of the natives" were precipitated by the inevitable listing of the grizzly bear, the bald eagle, and the wolf as threatened or endangered species under the auspices of the Endangered Species Act. The focus of the legally required recovery plans for all three species in the West was on public lands — for grizzlies and wolves this would center on the Greater Yellowstone Area (Yellowstone National Park and the surrounding National Forests) and the Glacier Area (Glacier National Park and the surrounding National Forests in northern Montana). Wolves were both reintroduced via transplants from Canada, and both wolves and grizzlies spread from existing pockets of occupancy. Grizzlies responded to enhanced protection and understanding and modification of limiting factors. All of those efforts have been successful to the point that serious consideration is now being given to the "delisting" of all three species as threatened species or endangered species.

Beginning in the mid-1980s, it was increasingly evident that the "old-growth" (late-successional) forests of the Pacific Northwest were being rapidly logged and fragmented as a habitat type. Almost all such forests on state and private lands had already been cut, and those stands remaining on National Forests and lands managed by the Bureau of Land Management were being steadily cut away with concurrent increasing fragmentation of old-growth habitats. Researchers from the Forest Service, Fish and Wildlife Service, Oregon and California state wildlife agencies, and universities were focusing on a cryptic sub-species of owl — the northern spotted owl — that was primarily associated with old-growth forest conditions in western Washington, Oregon, and northern California west of the crest of the Cascade Mountains. It was becoming increasingly obvious that the northern spotted owl would be, sooner or later, listed as a threatened species. That listing, in turn, could be anticipated to have dramatic negative consequences to the extremely politically and economically powerful timber industry — and its thousands of workers (Yaffee 1994).

The key to the northern spotted owl's survival was ever more clearly the "preservation of the ecosystem upon which..." it depended — and that was dominated by the most valuable timber in North America. The signs were clear — but elected and appointed officials could not bring themselves to face facts — at least not until after the next election, or the one after that, or the one after that (Yaffee 1994).

In 1992, President George H. W. Bush attended an international gathering in Brazil of world leaders concerned with the environment — the so-called Rio Summit. The pressure was on the United States to take a leadership role, and the Bush administration seemed equally determined not to take on any more constraints on U.S.-based businesses — for example, the wood products industry — than were absolutely politically necessary. Heat built quickly relative to the perceived intransigent behavior of the United States in this regard.

President Bush called back to Washington and asked for a bold statement or initiative that would play well in the world's press and not have any really serious impact on business as usual. After conference with technical staff, Forest Service Chief F. Dale Robertson suggested to the White House

Chief of Staff, John Sununu, that the president might announce that, henceforth, the public lands of the United States would be managed under the concept of "ecosystem management" (Robertson, pers. comm., 1993) or, as the concept was called in some quarters, "applied landscape ecology" (Liu and Taylor 2002).

From today's vantage point, it seems that all concerned with that decision had no real idea of what that commitment meant, or what it might come to mean, with the passage of time and the rapid expansion of knowledge relative to ecosystems. The Endangered Species Act's attention to "ecosystems," the Forest Service regulations issued pursuant to the National Forest Management Act to "maintain viable populations of vertebrates well-distributed in the planning area" were now backed up by a presidential commitment to install the practice of "ecosystem management" on the public lands of the United States.

A magical moment had arrived that would, ultimately, have dramatic consequences on the management of public lands in the United States. The still-evolving scientific concepts of ecosystems, the requirements of laws and regulations, the insistence of the courts on compliance, and political commitment by the president of the United States were in alignment. The nation had come to a fork in the road — a really big fork — relative to management of public lands. The president of the United States, whether with full understanding or not, had made the decision — and a commitment to the world's governments — of how federal land managers were to proceed. But, there was still struggle ahead. After all, just how do you actually execute "ecosystem management?" The first response was to institute what became known as "bioregional assessments" to lay the necessary foundations upon which to begin ecosystem management (Johnson et al. 1999).

Making the decision to proceed with ecosystem management was easier said than it was to make an operative reality. The focus, and test, of the commitment quickly settled on the Pacific Northwest, the northern spotted owl, and the fate of old-growth ecosystems that comprised its primary habitat.

The directors of the Fish and Wildlife Service, Bureau of Land Management, National Parks Service, and the chief of the Forest Service had sought to come to grips with the situation but failed, several times, when they simply could not bring themselves to take the political heat that would result from even a minor reduction in timber cut annually from federal lands in the Pacific Northwest. Out of frustration — with, perhaps, a dash of desperation thrown in — the four agency heads named a team (a.k.a. the Interagency Scientific Committee or ISC) of federal research scientists, assisted by observers from the timber industry, environmental community, and academia, to develop a plan for the long-term survival of the northern spotted owl.

There was no mention in the marching orders of "ecosystem management" or of producing an array of options for management. The team was given six months to do the job. The Committee's report (Interagency Scientific Committee 1990) was structured around controlling the developing mosaics of forest stand structures (e.g., Forman 1995) across the federal estate in western Oregon, Washington, and California. The suggested management plan landed like a grenade in the politics and life styles of the Pacific Northwest.

Attacks on the plan and the scientists that produced the plan came from every direction as politicians and political appointees tried to distance themselves from the report. But, the peer reviews upheld the validity of the approach and the suggested solution. The timber industry lobbyists raged, and the environmentalists licked their chops in anticipation of legal actions. To add to the drama, this explosion occurred in the midst of a hard-fought campaign for the presidency of the United States.

The administration, in a seeming act of desperation, essentially put the Interagency Scientific Committee "on trial" in an evidentiary hearing in front of the Endangered Species Committee — a group authorized in the Endangered Species Act and empowered to allow a species to drift into extirpation or extinction if the economic/social consequences of saving the species are judged too great to bear. The Interagency Scientific Committee and their report were, in essence, upheld both by the federal courts and an intense peer review. The Endangered Species Committee decision was along those same lines and was perceived as a humiliating defeat for the administration. This decision was even more stunning as it came from a committee whose members were all cabinet-and sub-cabinet-level political appointees of the George H. W. Bush administration.

The Federal Courts had put all timber sales on federal lands within the range of the northern spotted owl on hold pending adoption of a satisfactory plan by the George H. W. Bush administration. All further timber management actions involving old growth timber management (i.e., cutting of old growth) were held in abeyance until after the elections in November 1991 — that is, pending a potential "political fix." President G. H. W. Bush and H. Ross Perot ran their campaigns in the Pacific Northwest on promises to "adjust" the Endangered Species Act once they were elected. Candidate Governor William Clinton of Arkansas merely acknowledged that he understood the problem and empathized with the quandary of the people of the Pacific Northwest and promised to address the issue with a "forest summit" of stakeholders immediately after his election. Governor Clinton carried both Oregon and Washington with a minority of the votes cast — and was elected president.

As promised, President Clinton held the Forest Summit within a few months of his election. At the end of the meeting, he commissioned another team (the Forest Ecosystem Management Assessment Team) to develop options for his consideration to end the stalemate in the management of federal forests in the Pacific Northwest. Among his instructions (Forest Ecosystem Management Assessment Team 1993) to the team was that all developed options will be based on an "ecosystem management approach" — thereby keeping President Bush's commitment at the Rio Summit. The team ended up assessing the consequences of each of ten management options on over 1000 species, including vertebrates (including fishes), invertebrates (aquatic and terrestrial), and plants.

The option chosen by the president became known as the Northwest Forest Plan (Forest Ecosystem Management Assessment Team 1993). The option was significantly modified after its selection, by a second team, to assure more attention to questions of biodiversity retention and is still in effect at this writing. Ten years later, scientists convened in Vancouver, Washington, concluded that the plan had worked well in protecting and recruiting additional old growth; spotted owls had continued to decline at or near anticipated rates; and the anticipated timber targets had consistently remained unmet due to consistent resistance by the hardcore environmental organizations, that is, very few on either the environmental or timber industry sides of the ongoing issue seemed happy with the result.

Clearly, "ecosystem management" is a sound — or at least necessary — concept given current law in the United States. But, it remains somewhat nebulous in consistent application in management. Improved application of such theory requires improved definition within the bounds of time, space, actions, and social and economic consequences. Equally clearly, science can only serve to guide management because — in the end — applicable law, the courts, and ultimately, the people speaking through the Congress will decide what are acceptable approaches to management of natural resources, particularly on the public lands.

For example, once the Northwest Forest Plan was complete and in place, President William Clinton instructed the federal land management agencies to extend "ecosystem management" to the portions of Oregon and Washington east of the Cascades that contained streams that fed into the Columbia River and harbored threatened species of anadromous fish. Once it was explained to the president that the migrating fish did not stop at the state line between Oregon/Washington and Idaho but, rather, included the entire watershed of the Columbia River, the orders changed. An effort was launched to derive alternatives for the management of the federal lands of the Columbia River basin lying east of the Cascades.

Many of the residents of those areas, having witnessed the outcome of "ecosystem management" on federal lands west of the Cascades, were highly suspicious. Congressmen (mostly Republicans) from the affected regions were not supportive of the ongoing assessment and figured out that there could be no use of the assessments being developed if the project was not allowed to be completed. That was accomplished by holding completion of the effort in abeyance until the White House changed hands. So, that effort ended with the assessment phase, and the use of the data was limited to preparation of individual National Forest and Bureau of Land Management District Plans.

The Northwest Forest Plan had been at such a large scale (three states were involved) that traditional boundaries and traditional power brokers had been overwhelmed, that is, the scale of the plan exceeded extant human scale and ignored long-standing arrangements of governance worked out over 100 years or more. Some that disparaged the effort compared the Northwest Forest Plan to the old central planning of the Soviet Union.

At that point, it was clear that human beings were an essential part of the ecosystem management equation and that any ecosystem management plan that dictated or constrained human action without the consent of those most affected was in for trouble — or at least significant resistance. Application of any concept of stewardship across political and ownership boundaries would be fraught with problems that clearly necessitated the involvement of social scientists, economists, legal scholars, and newly developed political processes (Knight and Landres 1998).

That left open the question as to whether the ultimate result of the application of evolving ecological concepts (even when backed by statute and court decisions) can, should, or will maintain their viability in the face of increased economic costs and inherent social consequences. Clearly, as ecosystem management continues to be applied it will become obvious that, to ensure success, linkage of socio-economic and psychological theory into natural resources management will be required. After all, *Homo sapiens* is the dominant vertebrate species in most ecosystems and social, economic, and psychological mechanisms heavily influence our species' behavior. Future management paradigms will need to be partially guided by insights from the realms of social and political science (Grumbine 1994).

What makes ecosystem management different from previous approaches to land and wildlife management? Relative to traditional approaches, it seems that the scale of the effort is larger, more variables are considered (including human needs and desires), and the acceptance of new paradigms is involved. For example, the policy of the government of the United States to suppress immediately all wild fires is rapidly evolving to accepting that fire has a critical role to play in management — range, forestry, wildlife, and ecosystem management (e.g., Agee 1993; Carle 2002; Pyne 2004). Ecosystem management means nothing in application until the scale of a particular effort is specifically described, the variables are listed, and the time frame defined (Boyce and Haney 1997). But how is that to be done so as to be politically acceptable?

The boundaries of ecosystems can never, in reality, be established so as to be free of influences from outside that boundary. No matter how the variables of consideration are arrayed, the list will be incomplete and the interactions open to question. Time frames will be meaningful only in the sense of human understanding (Meffe et al. 2002).

Evolving understanding of ecosystems makes the application of ecosystem management much more difficult than originally anticipated. For example, Mann's (2005) treatise on the social and ecological ramifications of the arrival of Europeans in the Americas in 1492 added new caveats to the long-term application of ecosystem management. The world's ecosystems are becoming more and more rapidly homogenized in terms of plants and animals — which, in turn, has caused and is causing, ongoing alterations in those systems. Evidence is presented that *H. sapiens*, already resident in the Americas for thousands of years and in much larger numbers than previously estimated, had dramatically altered ecosystems before their populations declined precipitously between 1492 and mid-seventeenth century. "Ecosystems" in North America encountered by European explorers and settlers were, to a large extent, the result of many centuries of anthropogenic influence ranging from intensive agriculture to manipulations from intentional repeated burning of the landscape coupled with being the top predator of large mammals and some birds.

At this point, the observation of Frank Egler that "Ecosystems are not only more complex than we think, they are more complex than we can think" comes fresh to mind — and he was speaking only of the ecological concept of an ecosystem not including the ramifications of supporting human societies in various forms. Our understanding of the concept of ecosystems as a framework for land and wildlife management is rapidly improving, but it can never be complete or, perhaps, even adequate.

Where lands, such as those in much of mid- and southern Texas, have been subject to centuries of division into increasingly smaller ownerships and subjected to various management actions (including exclusion of stand-replacement wildfire and overgrazing by both domestic livestock and wild ungulates, also comprising species of cervids, antelopes, sheep, and goats imported from other hemispheres) it has led to sustained conversion from grasslands and grassland savannahs to forest/brush lands. In addition, each ownership has a different record — some quite dramatic — of manipulation of vegetation ranging from acceptance of the developing *status quo* to heroic efforts to control woody plants and imported noxious weeds to re-establish grasslands (albeit, sometimes with exotic grasses). Root plowing, chaining, application of herbicides, fire, and seeding of exotic grasses were included in this mix. Such circumstances were and are conducive to invasion of exotic vegetation. The resulting ecological conversions are variable but are both dramatic and ongoing. Chances of an ecological reversion to anything resembling "original ecological conditions" (conditions existing at the time of occupancy by Europeans) are remote.

Ecosystem management concepts can be best and most efficiently applied where it is practical to consider relatively extensive ownerships. Such conditions are most likely to occur on public lands (federal and state) and areas where private owners or cooperatives of multiple owners control tens of thousands of acres. As the scale of application decreases, difficulties in application will increase due to the highly variable conditions involved, reducing the operative factors that can be rationally considered in management. Individual ownerships have highly variable habitat conditions, individual desires and objectives of the owners, individually applicable economic assessments of cost and benefit, and productive capability of the land, coupled with — ordinarily — short tenure. Where property sizes are small, land managers are forced to be more site specific, dealing with realistic time frames, affordable practices, and the ability to react quickly to short-term situations (e.g., weather patterns). Management, therefore, occurs over shorter time frames with considerable flexibility in applying management approaches to meet management objectives. In such cases, the "old-fashioned" approaches to wildlife management discussed are more likely to be applied — and appropriately so.

In summary, there are applicable lessons that can be learned from recent attention to principles of ecosystem management. But, those approaches must be tailored to fit various landscapes and variables of purpose, resources, and inherent ecological conditions.

REFERENCES

Agee, J. K. 1993. *Fire Ecology of Pacific Northwest Forests.* Washington, DC: Island Press.

Allee, W. C., A. E. Emerson, O. Park, T. Park, and K. P. Schmidt. 1949. *Principles of Animal Ecology.* Philadelphia: Saunders.

Anderson, S. H. 1985. *Managing Our Wildlife Resources.* Columbus, Ohio: Charles E. Merrill Publishing Company.

Andrewartha, H. G., and L. C. Birth. 1954. *The Distribution and Abundance of Animals,* Chicago: University of Chicago Press.

Bailey, J. A. 1984. *Principles of Wildlife Management.* New York: John Wiley & Sons.

Boyce, M. S., and A. Haney. 1997. *Ecosystem Management: Applications for Sustainable Forest and Wildlife Resources,* New Haven: Yale University Press.

Carle, D. 2002. *Burning Questions: America's Fight With Nature's Fire,* Connecticut: Praeger Publishers.

Carson, R. 1962. *Silent Spring.* Boston: Houghton Mifflin.

Clements, F. C. 1916. *Plant Succession.* Washington, DC: Carnegie Institute, Publication No. 242.

Cowles, H. C.. 1899. The ecological relations of vegetation on the sand dunes of Lake Michigan, *Bot. Gazette,* 27:95.

Dasmann, R. F. 1964. *Wildlife Biology.* New York: John Wiley & Sons.

Diamond, J. M. 1975. The island dilemma: lesson of modern biogeographic studies for the design of natural reserves, *Biol. Conserv.,* 7:129.

Diamond, J. M. 1976. Island biogeography and conservation: strategy and limitations. *Science,* 193:1027.

Egler, F. E. 1954. Vegetation science concept. I. Initial floristic composition: a factor in old field vegetation development, *Vegetatio*, 4:412.

Ehrlich, P. R., and A. H. Ehrlich. 1981. *Extinction: The Causes and Consequences of the Extinction of Species.* New York: Random House.

Elton, C. S. 1927. *Animal Ecology*, London: Sidgwick and Jackson.

Forest Ecosystem Management Assessment Team. 1993. *Forest Ecosystem Management: An Ecological, Economic, and Social Assessment.* Washington, DC: U.S. Department of Agriculture, Forest Service.

Forman, R. T. T. 1995. *Land Mosaics: The Ecology of Landscapes and Regions.* Cambridge: Cambridge University Press.

Gause, G. F. 1934. *The Struggle for Existence.* Baltimore: Williams and Wilkins.

Gilbert, F. F., and D. G. Dodds. 1987. *The Philosophy and Practice of Wildlife Management.* Malabar: Krieger Publishing Company.

Giles, R. H, Jr. 1978. *Wildlife Management.* San Francisco: W. H. Freeman and Company.

Gleason, H. A. 1917. The structure and development of the plant association, *Bull. Torrey Bot. Club*, 44:463.

Gleason, H. A. 1926. The individualistic concept of the plant association, *Bull. Torrey Bot. Club*, 44:463.

Golley, F. B. 1993. *A History of the Ecosystem Management in Ecology: More than the Sum of the Parts.* New Haven: Yale University Press.

Grumbine, R. E. (ed.). 1994. *Environmental Policy and Biodiversity.* Washington, DC: Island Press .

Haapanen, A. 1965. Bird fauna of the Finnish forests in relation to forest succession I, *Ann. Zool. Fenn.*, 2:153.

Haapanen, A. 1966. Bird fauna in relation to forest succession. II., *Ann. Zool. Fenn.*, 36:176.

Hirt, P. W., *A Conspiracy of Optimism: Management of the National Forests Since World War II.* Lincoln: University of Nebraska Press.

Hunter, M. C., Jr. 1999. *Maintaining Biodiversity in Forest Ecosystems.* New York: Cambridge University Press.

Huston, M. A. 1994. *Biological Diversity: Existence of Species on Changing Landscapes.* New York: Cambridge University Press.

Hutchison, G. E. 1957. *A Treatise on Limnology, Vol. 1. Geography, Physics, Chemistry.* New York: Wiley.

Hutchison, G. E. 1969. Eutrophication, past and present, in *National Academy of Science, Eutrophication, Causes, Consequences, Corrections.* Washington, D.C., p. 12.

Interagency Scientific Committee. 1990. *A Conservation Strategy for the Northern Spotted Owl.* Portland: U.S. Department of Agriculture.

Johnson, K. J., F. Swanson, M. Herring, and S. Greesse (eds). 1999. *Bioregional Assessments: Science at the Crossroads of Management and Policy.* Washington, DC: Island Press.

Knight, R. L., and P. B. Landres (eds). 1998. *Stewardship Across Boundaries.* Washington, D.C.: Island Press.

Lack, D. L. 1954. *The Natural Regulation of Animal Numbers.* Oxford: Clarendon Press.

Lindeman, R. 1942. Trophic-dynamic aspects of ecology, *Ecology*, 23:399.

Liu, J., and W. W. Taylor. 2002. *Integrating Landscape Ecology Into Natural Resource Management.* Cambridge: Cambridge University Press.

Leopold, A. 1933. *Game Management.* New York: Scribner.

Leopold, A. 1942. The last stand, *Outdoor America*, 7:7.

Leopold, A. 1949. *A Sand County Almanac and Sketches Here and There.* New York: Oxford University Press.

Lotka, A. J. 1925. *Elements of Physical Biology.* Baltimore: Williams & Wilkins.

Lutz, H., and R. F. Chandler. 1954. *Forest Soils.* New York: Wiley.

Macfayden, A. 1963. *Animal Ecology: Aims and Methods*, 2nd edn. London: Pittman.

Mann, C. C. 2005. *1491: New Revelations of the Americas before Columbus.* New York: Alfred Knopf.

Meffe, G. K., L. A. Nielsen, R. L. Knight, and D. A. Schenborn. 2002. *Ecosystem Management: Adaptive, Community-Based Conservation.* Washington, D.C.: Island Press.

Meine, C. 1988. *Aldo Leopold: His Life and Work.* Madison: University of Wisconsin Press.

Miller, C. 2001. *Gifford Pinchot and the Making of Modern Environmentalism.* Washington, DC: Island Press.

Moen, A. N. 1973. *Wildlife Ecology: An Analytical Approach.* San Francisco: W. H. Freeman and Company.

Morrison, M. L., B. G. Marcot, and R. W. Mannan. 1998. *Wildlife–Habitat Relationships: Concepts and Applications.* Madison: University of Wisconsin Press.

National Resources Council. 1999. *Perspective on Biodiversity: Valuing Its Role in an Ever-Changing World.* Washington, DC: National Academy Press.

Nice, M. M. 1941. The role of territory in bird life, *Am. Midl. Nat.*, 26:441.

Odum, E. P. 1969. The strategy of ecosystem development, *Science*, 164:262.

Odum, H. T. 1970. Summary: an emerging view of the ecological system at El Verde, in *A Tropical Rain Forest*, H. T. Odum, and R. F. Pigeon (eds). Washington, DC: U.S. Atomic Energy Commission, p. 1.

Oliver, C. D., and B. C. Larson. 1996. *Forest Stand Dynamics*. New York: John Wiley & Sons.

Ovington, J. D. 1961. Some aspects of energy flow in plantation of *Pinus sylvestris, L. Ann. Bot., London*, 25:12.

Peek, J. M. 1986. *A Review of Wildlife Management*. Englewood Cliffs: Prentice Hall.

Petersen, S. 2002. *Acting for Endangered Species: The Statutory Ark*. Lawrence: University Press of Kansas.

Pyne, S. J. 2004. *Tending Fire: Coping with America's Wildland Fires*. Washington, D.C.: Island Press.

Robinson, W. L., and E. G. Bolen. 1989. *Wildlife Ecology and Management*. New York: Macmillan Publishing Company.

Rodin, L. E., and N. I. Bazilevic. 1964. The biological productivity of the main vegetation types in the Northern Hemisphere of the Old Word, *Forest. Abst.*, 27:369.

Salzman, J., and B. H. Thompson, Jr. 2003. *Environmental Law and Policy*. New York: Foundation Press.

Shelford, V. E. 1913. Animal communities in temperate North America as illustrated in the Chicago Region, No. 5, *Bulletin of the Geographical Society of Chicago*. Chicago.

Simberloff, D., and L. G. Abele. 1976a. Island biogeography theory and conservation practices. *Science*, 191:285.

Simberloff, D., and L. G. Abele. 1976b. Island biogeography and conservation: strategy and limitations, *Science*, 193:1032.

Simberloff, D., and L. G. Abele. 1982. Refuge design and island biogeography theory: effects of fragmentation, *Am. Nat.*, 120:141.

Smith, R. L. 1995. *Ecology and Field Biology*, 5th edn. New York: Harper and Row.

Snape, W. J. III, and O. L. Houck. 1996. *Biodiversity and the Law*. Washington, D.C.: Island Press.

Stoddard, H. L. 1931. *The Bobwhite Quail: Its Habits, Preservation, and Increase*. New York: Scribner.

Tansley, A. G. 1935. The use and abuse of vegetational concepts and terms, *Ecology*, 16:284.

Thomas, J. W. (ed.) 1979. *Wildlife Habitats in Managed Forests — The Blue Mountains of Oregon and Washington*. Washington, DC: U.S. Dept. of Agriculture, Agriculture Handbook No. 553.

Thomas, J. W. 2004. *Jack Ward Thomas: The Journals of a Forest Service Chief*. Seattle: University of Washington Press.

Trefethen, J. B. 1961. *Crusade for Wildlife*. Harrisburg: The Telegraph Press.

USDA Forest Service. 1993. *The Principal Laws Relating to Forest Service Activities*, Washington, DC: USDA Forest Service.

Volterra, V. 1926. Varigzione e fluttqzoni de numero d'individiu in specie animali conviventi, translated in Chapman, R. N., 1931, *Animal Ecology*. New York: McGraw-Hill.

Wilson, E. O. 1975. *Sociobiology*. Cambridge: Harvard University Press.

Wilson, E. O. (ed.) 1986. *Biodiversity*. Washington, D.C.: National Academy Press.

Wynne-Edwards, V. C. 1962. *Animal Dispersal in Relation to Social Behavior*. New York: Hofner.

Yaffee, S. L. 1994. *The Wisdom of the Spotted Owl: Policy Lessons for a New Century*. Washington, DC: Island Press.

14 Applying Ecological Theory to Habitat Management: The Altering Effects of Climate

Timothy Edward Fulbright, J. Alfonso Ortega-S.,
Allen Rasmussen, and Eric J. Redeker

CONTENTS

Ecological succession is one of the most important theories influencing wildlife management practices (Bolen and Robinson 2003: 43). Ecological theories providing the basis for predicting the outcome of habitat manipulations to benefit wildlife are often applied as generalizations across broad geographic regions. Many of the paradigms used in wildlife habitat management were developed in relatively mesic environments. These paradigms may not consistently predict outcomes of management efforts where low and erratic precipitation and periodic drought change, at least temporarily, the nature of vegetation dynamics. In arid and semiarid regions, variation in precipitation may modify or override anticipated effects of habitat manipulations (Fulbright 1999; Fuhlendorf et al. 2001; Rogers et al. 2004).

In arid and semiarid regions, results of management practices based on theory developed in more productive regions where precipitation is nonlimiting for plant growth may differ considerably from predicted outcomes. Arid zones are those in which the ratio of annual precipitation to potential evapotranspiration ranges from 0.05 to <0.2; the ratio is 0.2 to <0.45 in semiarid zones, 0.45 to <0.65 in dry subhumid zones, and >0.65 in humid zones (Fulbright and Ortega-S. 2006). Our objective in this chapter is to review selected paradigms derived from ecological theory that are commonly used as the basis for manipulating vegetation to benefit wildlife and suggest refinements based in part on research conducted in southern Texas.

Scale is an important variable in planning habitat manipulations for wildlife. Habitat manipulations have often been applied at the "pasture" or "management unit" scale without taking into account potential impacts on the landscape as a whole. Researchers and practitioners became more aware of the need to account for broader scales in management planning with the emergence of landscape ecology and patch theory (Pickett and White 1985; Turner 2005). We will refer in this chapter to a management unit such as an individual pasture, farm, or small (<500 ha) ranch as the "patch" scale relative to the broader landscape scale that consists of a mosaic of patches that vary in soil texture and plant communities. "Patch dynamics," the science of the interrelationships and interactions among patches, may provide conceptual models for habitat manipulations at the landscape scale (Pickett and White 1985). We question, however, the applicability of traditional patch dynamic theory across climatic regions, because endogenous factors, such as water and nutrient redistribution, may be more important than exogenous factors, such as natural disturbances, in the dynamics of some arid and semiarid ecosystems.

ECOLOGICAL THEORIES UNDERLYING HABITAT MANAGEMENT

Ecological succession and accompanying theory, including the concepts of retrogression, climax, and convergence, often serve as the theoretical underpinning for manipulations to improve wildlife habitat (Connel and Slatyer 1977; Van der Maarel 1988; Fulbright and Guthery 1996b). Manipulations to improve habitat for specific wildlife species are often done to create the stage of succession to which the species is perceived by wildlife managers to be best adapted. Northern bobwhites (*Colinus virginianus*), for example, are often classified as a "early succession stage" species, white-tailed deer (*Odocoileus virginianus*) as a "mid-succession species," and grizzly bears (*Ursus arctos*) as a "climax" species (Bolen and Robinson 2003: 43–44). Wildlife managers have traditionally applied various forms of disturbance to manipulate succession based on recognition that wildlife species vary in seral stages they are adapted to. A disturbance is any event, either natural or caused by humans, resulting in temporary or permanent alteration of the appearance, composition, function, or structure of organisms or their physical environment in a discrete area. Grazing by livestock, mechanical treatments, and prescribed fire are disturbances commonly used by wildlife managers as tools to manipulate ecological succession. In southern Texas, these habitat manipulations are commonly directed at game species such as northern bobwhites and white-tailed deer.

MANAGEMENT PRACTICES BASED ON DIRECTIONAL VEGETATION DYNAMICS

Maintaining the landscape in an early stage of ecological succession by cattle grazing is traditionally considered a means of habitat improvement for northern bobwhites (Wildlife Habitat Council 1999). Ecological theory underlying this recommendation for northern bobwhites is that grazing will cause retrogression of the plant community from a later stage of succession stage to an earlier stage of succession favored by bobwhites. Retrogression is vegetation change in the direction opposite of climax (Westoby 1980).

A test of this theory revealed that in a dry subhumid study site in the eastern portion of southern Texas, bobwhite densities were greater in areas where heavy grazing by cattle maintained an early stage of succession compared to areas in mid- and late-successional stages (Spears et al. 1993). In a semiarid study site in the western portion of southern Texas, in contrast, the stage of succession resulting from cattle grazing ranging from early to late did not influence bobwhite population densities. On the basis of these results, bobwhites appear to be a mid- to late-successional species in semiarid environments. This relationship, however, may be more a result of structural attributes of the vegetation than stage of succession. Structure of vegetation in late-succession semiarid environments where vegetation productivity is relatively low is similar to structure of vegetation in early-succession areas in high rainfall environments where productivity is relatively high. Structural attributes of the vegetation appear to be more important to bobwhites than changes in plant species composition resulting from ecological succession or retrogression.

EQUILIBRIUM AND NONEQUILIBRIUM VEGETATION DYNAMICS

Directional models of vegetation dynamics assume vegetation change following disturbance is toward the original climax plant community that existed before treatment. These models are commonly used as the basis for habitat manipulations such as grazing by livestock, prescribed burning, and mechanical manipulations such as discing and root plowing. Vegetation change in response to disturbance in arid and semiarid environments, however, may be nondirectional and nonreversible (Wiens 1984; Illius and O'Connor 1999). Lack of directional vegetation change results primarily from the extreme variation in precipitation characteristic of arid and semiarid environments (Sullivan and Rohde 2002).

Rangeland scientists have replaced traditional directional models of vegetation dynamics used in rangeland evaluation with state-and-transition models of vegetation dynamics (Briske et al. 2005). State-and-transition models were originally designed for arid and semiarid rangelands wherein vegetation change is nondirectional and nonreversible, but they were not intended to suggest that directional change does not occur. State-and-transition models assume that one or more compositionally different plant communities that are self-perpetuating, are relatively stable, and do not exhibit successional trends to the original community are possible for a site (Friedel 1991; Briske et al. 2005). These communities are called stable states. Thresholds are boundaries in space and time that separate stable states. Disturbance to the original plant community on a site may push the vegetation across a threshold to a new stable state if the disturbance is sufficiently intense. Plant species composition and other characteristics of the new stable state that develops in response to disturbance may depend on the nature of the disturbance, soil fertility, rainfall, and other environmental factors.

The response of vegetation to root plowing in southern Texas provides an example of the development of a new stable state. Root plowing is commonly used on semiarid rangelands in Texas to reduce abundance of woody plants and increase herbaceous plants. A root plow is a large, V-shaped blade pulled by a crawler tractor to sever the roots of woody plants. Woody plants become reestablished within 10–20 years following root plowing. The woody plant community that develops following root plowing of diverse, mesquite (*Prosopis glandulosa*)-mixed brush communities differs considerably from the original plant community.

On upland soils, root plowing pushes the diverse mesquite-mixed brush community across a threshold to a community with reduced woody plant species diversity (Mutz et al. 1978; Fulbright and Beasom 1987; Ruthven et al. 1993; Stewart et al. 1997) (Figure 14.1). In the eastern, dry subhumid portion of southern Texas, the community that reestablishes following root plowing is commonly dominated by a near monoculture of huisache (Mutz et al. 1978; Ruthven et al. 1993).

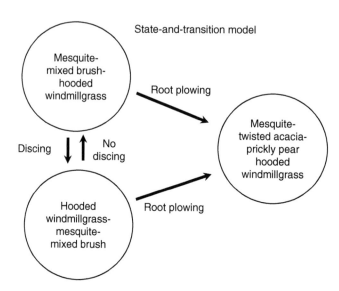

FIGURE 14.1   Hypothetical model of the effects of root plowing on vegetation in southern Texas. On upland soils, root plowing pushes the diverse mesquite-mixed brush community across a threshold to a community with reduced woody plant species diversity.

In the western, semiarid portion of south Texas mesquite, twisted acacia, and prickly pear are commonly the dominants of the new community (Fulbright and Beasom 1987; Stewart et al. 1997).

The huisache-dominated community and the mesquite-twisted acacia-prickly pear communities appear to be new stable states that do not show successional trends back to the original mesquite-mixed brush community, at least during timescales applicable to management (Stewart et al. 1997). Computer simulation models estimated that subordinate shrub species richness in the mesquite-twisted acacia-prickly pear communities on sites root plowed 25 years earlier would not return to pretreatment levels for 150–200 years (Fulbright and Guthery 1996a).

In semiarid environments, variation in precipitation may override effects of human-imposed perturbations, including those severe enough to push vegetation across thresholds to new stable states, on wildlife species abundance and richness. In the semiarid habitats where studies on long-term effects of root plowing on vegetation change were conducted, for example, white-tailed deer, small mammal, and herpetofauna populations appear to be more strongly influenced by variation in precipitation than by the transition from one stable state to another caused by root plowing. Although browse species important to white-tailed deer were either absent from the huisache communities that replaced mesquite-mixed brush communities or were present in greatly reduced numbers compared to the original mesquite-mixed brush community, nutritional condition and population status of white-tailed deer were similar in untreated sites and sites root plowed 17 years earlier in the eastern Rio Grande Plains (Ruthven et al. 1994). Changes in body condition, reproduction, and diet were associated with variation in precipitation rather than with plant community differences. Likewise, estimated carrying capacity of sites in the western South Texas Plains that had been root plowed was similar to that of untreated sites (Draeger 1996). Estimated carrying capacity based on digestible energy of deer diets appeared to be closely related to previous rainfall on root-plowed sites.

Small mammal and herpetofauna species richness and abundance were similar in untreated mesquite-mixed brush sites and on sites root plowed 36–39 years previously and dominated by a mesquite-twisted acacia-prickly pear community (A. Lozano-Cavazos et al. unpublished data, 2006). Variation in abundance of many of the lizard, amphibian, and small mammal species was related to annual variation in rainfall rather than to differences in plant communities.

Site productivity may influence whether vegetation change following disturbance on arid and semiarid environments is directional or nonequilibrium. Root plowing does not appear to push mesquite-mixed brush communities across a threshold to a different stable state in ephemeral drainages in southern Texas (Nolte et al. 1994). Woody and herbaceous species richness and diversity were similar when sites chained in 1950 then root plowed in 1960 were compared to untreated sites in 1993. Ephemeral drainages tend to be more productive than the upland sites where root plowing resulted in new stable states.

Rather than maintaining a perpetual nonequilibrium state, extreme seasonal and annual variation in rainfall may induce temporal switching between temporary equilibrium and nonequilibrium states (Fuhlendorf et al. 2001). Directional vegetation change may occur intermittently over relatively long periods rather than continually as predicted by traditional succession models. Disturbances such as grazing may establish the direction of change, but this change may be interrupted by drought, or low precipitation that may, at least, slow the rate and alter the trajectory of change. Episodes of directional change may thus be interrupted by stable states that persist for indefinite periods depending on length of the drought or dry cycle.

Discing to create sites for secondary succession is another example of the application of directional vegetation dynamics theory to habitat management. The predicted outcome of discing is an increase in early successional forbs, seeds, and insects eaten by wildlife (Wildlife Habitat Council 1999). In the mesic environment of Georgia and Florida, discing produces food plants for northern bobwhites, and northern bobwhite densities are no different from those produced by agricultural plantings to increase food availability (Brennan et al. 2000). In contrast, response of wildlife foods to discing in semiarid regions is variable, possibly reflecting a lack of directional vegetation change or interruption of directional change by erratic precipitation. For example, discing on sandy soils in northwestern Oklahoma did not increase forb biomass or seed production (Peoples et al. 1994). In southern Texas, effects of discing on forbs preferred by white-tailed deer were inconsistent from year to year because of variation in rainfall. Low rainfall in some years possibly inhibited disced plots from producing sufficient canopy cover of herbaceous plants to detect differences between disced and untreated plots (Fulbright 1999).

State-and-transition models and the idea that the extreme seasonal and annual variation in rainfall characteristic of semiarid environments may induce temporal switching between equilibrium and nonequilibrium states may provide a better conceptual framework for making wildlife habitat management decisions in semiarid environments than traditional models of succession. The National Resources Conservation Service in the United States has developed state and transition models for most of the ecological sites in Texas (Natural Resources Conservation Service 2003). However, the thresholds have not been empirically identified.

FROM THE PATCH TO THE LANDSCAPE

Scientists realized during the late twentieth century that management at the patch level rather than at the landscape level was often ineffective to maintain optimum population levels of many wildlife species. Management generalizations made at the patch scale may be inadequate, because they do not take into account habitat requirements of a species such as usable space, and effects of habitat fragmentation (Williams et al. 2004). For example, heavy grazing in areas of high rainfall may improve vegetation structure for movement and feeding by bobwhites, but the birds also need mid-grasses for nesting and woody vegetation for thermal cover (Hellickson and Radomski 1999). Neither of these may be available in a habitat patch maintained in an early seral stage of plant succession by grazing. The most favorable habitat for bobwhites is a landscape including patches of mid-grasses for nesting, relatively open areas for roosting and feeding, and patches of woody vegetation for thermal and hiding cover.

Fragmentation of habitat is an increasingly severe problem in wildlife conservation (Saunders et al. 1991). No matter how much a "patch," for example, a ranch or farm, is "improved" for a wildlife species, population densities of that species may be limited, or even prone to extinction, as a result of fragmentation in the surrounding landscape. Decreasing size of habitat patches in the surrounding landscape or increasing distance to other habitat patches may limit immigration of individuals into a patch, thereby increasing vulnerability of the population to extinction (Pulliam 1988). For example, natural mortality and hunting may reduce bobwhite coveys to relatively small size during winter (Williams et al. 2004). Individuals from the surrounding landscape may move to a patch, for example, a farm, to form adequate-sized coveys. The probability of extinction of bobwhites in the patch is increased if habitat fragmentation inhibits bobwhite movement from adjacent habitat patches.

Management based on the landscape level is more appropriate for many wildlife species than management at the patch level based solely on a single seral stage. For example, portions of the landscape in high rainfall areas may be heavily grazed and portions lightly grazed, or ungrazed, to create a mixture of feeding and roosting cover and nesting cover for bobwhites. Patches of woody vegetation may be interspersed within this mosaic to provide thermal and loafing cover. Applying patch dynamics as the theoretical underpinning of wildlife habitat management entails managing patches of habitat within a landscape mosaic to provide a variety of successional stages or plant communities that meet the varied habitat requirements of each species. New technological tools have provided significant advancement in our ability to integrate patch dynamic theory into applied management, and to improve our understanding of the habitat requirements of wildlife species.

LANDSCAPE ECOLOGY AND GEOSPATIAL TECHNOLOGIES

The primary technologies that have contributed to the advancement of landscape ecology are geographic information systems (GISs) and remote sensing (Egbert et al. 2002; Pearson 2002; Weiers et al. 2004; Turner 2005). These technologies provide a means to quickly and accurately analyze environmental variables at larger spatial extents that what was possible using traditional field sampling techniques (Turner 2005). The ever-expanding quantities of spatial data offer a means to test for the effects of scale by adjusting both the grain and extent of input data and modeling procedures (Pearson 2002). Although these technologies did not drive development of landscape ecology, they did provide the tools that were needed to conduct landscape ecological analysis (Turner 2005).

Modern geospatial technology has enhanced our ability to measure, quantify, and compare landscape level processes both spatially and temporally. Additionally, many software applications have been developed that work directly within a geospatial environment to analyze landscape metrics for characterizing landscape structure and function in addition to comparing similar landscapes or performing temporal change detection on an individual landscape (Egbert et al. 2002, Keane et al. 2002, Pearson 2002). The majority of these landscape ecological tools are based on the patch theory, which breaks a landscape into three categories of landscape elements: patches, corridors, and matrix (Forman and Godron 1981). These distinct landscape elements are easily identified in mesic and anthropogenic landscapes, but are often difficult to identify in native arid to semiarid rangeland (Pearson 2002). Traditional remote sensing approaches (supervised and unsupervised classification) classify native rangeland as a homogeneous landscape element. As a result, the use of traditional patch theory and remote sensing will often produce misleading results that underestimate heterogeneity in semiarid ecosystems. Incorporation of spatial statistical methods, like autocorrelation, into landscape analysis has enabled researchers to capture the true heterogeneity of these seemingly homogenous environments (Pearson 2002). The ability to identify heterogeneity within ecosystems that demonstrate smooth gradients rather than clear divisions between landscape elements is essential when evaluating landscape structure for grassland species that operate at a finer scale than is detectable by traditional remote sensing approaches (Pearson 2002).

Before the advent of geospatial technology it was not possible to extrapolate knowledge gained at the patch scale to the landscape scale. Numerous research projects have demonstrated that using a combination of patch scale field data collection and geospatial technology, it is possible to accurately extrapolate patch knowledge to the landscape scale. For example, landscape-scale digestible energy maps were developed by Doan-Crider (2003) for black bears in northern Mexico to mitigate human–bear conflict by estimating bear locations during different seasons and climatic conditions. This map was produced by a model that combined a thematic vegetation map generated from remotely sensed Landsat^{TM} satellite imagery, spatially explicit mast crop estimates by vegetation class for each season based on transect data, and laboratory analysis of digestible energy for each of the mast species.

In a second example of extrapolating patch knowledge to the landscape level, water budget data collected on the patch scale was applied to the landscape scale to estimate the impact of temporal changes in woody cover on regional water yield by Wu et al. (2001). Traditional remote sensing techniques were used to classify vegetation species composition at the landscape scale, using two sets of aerial photography that spanned 35 years. The classified vegetation data were then used as an input into the Simulation of Production and Utilization of Rangelands (SPUR-91) hydrologic model to estimate the landscape scale water budget. Analysis was performed for each individual range site to account for variability in soils and vegetation. Although this model was based on patch-scale empirical data, it was validated on numerous landscapes dominated by native rangeland and provided remarkably accurate results that were suitable for temporally assessing the impact of landscape scale vegetation change on water yield.

Arid and semiarid ecosystems that do not follow directional models of vegetation dynamics present unique challenges to simulation modeling. Simulation models are often used to predict the long-range outcome of management practices at the landscape scale. Within these models, succession is treated as a deterministic process and disturbance is treated as a stochastic process (Keane et al. 2002). However, when succession is nondirectional, developing accurate long-range simulation models becomes a much more difficult task. The development of accurate simulation models specifically designed for arid ecosystems that do not follow directional succession are essential if we are to make informed management decision within these environments.

NEW HABITAT MANAGEMENT PARADIGMS: INCREASING LANDSCAPE HETEROGENEITY

Greater emphasis on multi-species management and biodiversity has helped increase emphasis on landscape level management. New paradigms to promote landscape heterogeneity to increase biodiversity and improve wildlife habitat have been proposed recently (Fulbright 1996; Brawn et al. 2001; Fuhlendorf and Engle 2001).

Disturbance patterns have been altered across much of the landscape in North America. Uniformity across the management unit has historically been a goal of rangeland and forest management. Most management strategies were designed to maximize uniformity because (1) it appeared easier to manage for single use objectives, and (2) an implied theory that economic returns from rangelands or forests are maximized by diverting energy flow from organisms such as wildlife and focusing it on a single product, for example, domestic livestock or timber. The common example of this is the active suppression of fire across much of North America. Fire suppression "preserved" forage and timber, but it resulted in lengthening of the fire return interval compared to pre-Columbian intervals on many rangelands and forests. Over time, plant communities in these areas started changing, and shrubs began to replace the original grasslands (Wright and Bailey 1982). Prescribed fire was reintroduced into many ecosystems during the late twentieth century to overcome this lack of disturbance and was applied at the management unit level.

When managers started reintroducing fire, prescriptions were developed to maximize the area burned to create a uniform vegetation response and reduce secondary disturbances. It was recognized that burned areas attracted grazing animals and overuse could occur if only a portion of a pasture or management unit was burned; it was also considered easier to manage a uniform pasture for grazing animals. The idea of maintaining uniformity was used in a variety of management strategies. Livestock water, for example, was uniformly distributed across a pasture to spread grazing disturbance evenly (Vallentine 2001). Uniform water distribution minimizes the spatial extent of areas of under- and overuse, again ensuring a predicted even vegetation response over the management unit.

The result of the uniform management philosophy was to push ecological conditions to be uniform across any particular pasture. The unintended consequence of these management practices was loss of heterogeneity, which in turn minimized ecological diversity of available habitat. As researchers and natural resource managers began examining the consequences of these strategies, it became apparent that in some cases new management approaches and paradigms were needed to increase heterogeneity.

As researchers and managers have realized the importance of diversity, new theories have been introduced to serve as the basis for maximizing heterogeneity in a management unit (Fuhlendorf and Engle 2001). Fuhlendorf and Engle (2001) proposed using multiple disturbances to increase grazing intensity in portions of a management unit to decrease uniformity. This consisted of burning small portions of a unit, and then, rather than deferring grazing animals as is traditionally done, allowing the animals to intensively graze these patches. Intensive grazing drives the vegetation to a higher proportion of forbs in the grasslands and increases landscape heterogeneity. Then other patches were burned in subsequent years to allow each patch a sufficient length of time between disturbances for recovery. The end result was a landscape mosaic consisting of patches of vegetation disturbed at varying intensities and in different stages of recovery from fire and grazing. Fuhlendorf and Engle (2001) proposed this in a relatively homogenous, mesic grassland ecosystem where disturbance would alter dominance of the forb and grass components of the system. Little work has been done in the arid and semi-arid areas.

Landscape homogeneity may reduce abundance of certain bird species associated with periodically disturbed habitats. Forty percent of North American bird species associated with habitats that are mediated by disturbance declined from 1966 to 1998 (Brawn et al. 2001). Different bird species reached peak densities at different periods of time following prescribed fires in the mixed grass prairie of the northern Great Plains (Johnson 1997). A landscape level management plan in which burning is temporally rotated among spatially separated patches to maintain a landscape mosaic of patches with varying time intervals since burning interspersed with unburned areas may maintain greater bird species diversity.

Maintaining coarse-scale variability in forage biomass across the landscape is important for winter survival of elk (Turner et al. 1997). Turner et al. (1997) designed a model to simulate burning effects in Yellowstone National Park. In the simulation, the effects of animals aggregating across the spatial heterogeneity created by fires were explored. When between-habitat heterogeneity within areas that elk were aggregating was removed from the model ungulate winter survival was reduced 30%.

In drier environments where deficient moisture often dictates vegetation composition, lack of moisture also can lengthen the recovery time needed between burns. In addition, the frequency of drought conditions must be viewed as an additional disturbance whose combined effects can leave intensive grazed areas and burned areas open and bare to excessive erosion and invasive plant species. These new communities could then be more homogeneous and resistant to change by grazing, burning, and other disturbances. While not a direct test of this idea, results following burning in Arizona indicate that grasslands dominated by invasive species can be very resistant to change following fire in an arid environment (Geiger and MacPherson 2005).

MAINTENANCE OF BIOLOGICAL DIVERSITY BY DISTURBANCE

Several hypotheses have been proposed to explain how biological diversity on landscapes is maintained by disturbance (Huston 1994). These include the intermediate disturbance hypothesis. A premise of the intermediate disturbance hypothesis is that disturbance is a natural and ongoing process in ecosystems. According to the hypothesis, different portions of the landscape may be in various stages of recovery from disturbance at any given time, and may be disturbed at different intensities by factors such as fire, floods, and grazing. Landscapes, then, are a mosaic of patches disturbed at varying intensities and in varying stages of recovery from disturbance. These patches support different plant communities depending on the intensity and frequency of disturbance. Patches that are intensely disturbed support a few species that are adapted to disturbed sites; those that are undisturbed are dominated by a few species adapted to undisturbed conditions.

Intermediately disturbed patches support the greatest diversity of plants according to the hypothesis because they support a mixture of disturbance-adapted species and species characteristic of undisturbed sites (Pickett and White 1985). Disturbance-adapted species dominate immediately after disturbance, but over time they decline in abundance and species adapted to undisturbed sites begin to colonize the patch. At some intermediate time following disturbance, the patch will support a mixture of these species and species richness will be at a maximum. Species richness declines as species adapted to undisturbed conditions begin to dominate the site and replace the disturbance-adapted species. The result of these dynamics is greater landscape diversity than would exist if the entire landscape were undisturbed (Petraitis et al. 1989).

PROBLEMS WITH USE OF DISTURBANCE TO MANIPULATE HABITATS

Several potential problems exist with respect to applying disturbance to manipulate habitat for wildlife in arid and semiarid environments: (1) the intermediate disturbance hypothesis has limited application in semiarid environments, (2) natural disturbances are more prone to increase diversity than human-imposed disturbances, (3) much of the landscape is already highly disturbed by humans and their livestock, (4) below a certain precipitation/site-productivity threshold patch dynamics are not driven by disturbance, (5) disturbance may facilitate ingress of invasive exotic grasses, and (6) simulating "natural" disturbance regimes may not have predictable outcomes on landscapes that have been altered by human activities and invasion by exotic plant species.

INTERMEDIATE DISTURBANCE HYPOTHESIS

The intermediate disturbance hypothesis may have limited application as a theoretical underpinning for management in arid and semiarid environments for several reasons. Most of the evidence supporting the intermediate disturbance hypothesis is from ecosystems in which rates of growth and of competitive displacement are high, such as in floodplains (Huston 1994; Brawn et al. 2001). In semiarid environments, extreme variation in rainfall in the form of prolonged drought or wet periods may have more pronounced effects on vegetation composition than attempts to create a human-imposed mosaic of disturbances. Rather than increasing diversity, periodic disturbances in patches interspersed in the landscape could result in reduced landscape diversity if these disturbances result in new stable states rather than stimulating directional succession. Finally, attempts to create a mosaic of patches differing in successional stages may be unsuccessful if (1) effects of human-imposed disturbances differ from those of natural disturbances, (2) the landscape is already disturbed by other factors such as livestock grazing, and (3) endogenous factors influence patch dynamics more than disturbance.

NATURAL VERSUS HUMAN-IMPOSED DISTURBANCES

Natural disturbances are 133% more likely to result in significant disturbance-species richness relationships than anthropogenic disturbances (Mackey and Currie 2001). In addition, effects of natural disturbances may differ considerably from those of human-imposed disturbances. For example, hurricanes primarily impact woody plant canopies, and recovery of vegetation may be rapid (Chazdon 2003). Bulldozing or long-term heavy grazing, in contrast, may cause soil erosion and loss of plant species, resulting in very slow recovery of vegetation.

PRIOR DISTURBANCE

Much of the landscape in semiarid areas is already highly disturbed from grazing. For example, about 27% of the rangeland in Texas in 1992 supported <26% climax vegetation (Mitchell 2000). Forty-eight percent of USDI Bureau of Land Management land in Arizona and 73% in New Mexico was in early or mid-seral stages of succession in 2005 (USDI Bureau of Land Management 2005). Imposing additional disturbance in these areas would be unlikely to increase diversity.

HYPOTHETICAL DIFFERENCES IN FUNCTIONS OF DISTURBANCE IN XERIC AND MESIC ENVIRONMENTS

Disturbance as a useful paradigm in habitat management may break down at some threshold along a gradient of decreasing precipitation or productivity. Beyond that threshold, minimizing disturbance may be a more viable paradigm. A premise underlying use of disturbance in managing wildlife habitat is that, following a disturbance event, patches undergo directional vegetation change leading back to climax. Disturbance may be less desirable in arid and semiarid environments wherein vegetation change is largely nondirectional, unless (1) it is known that the applied disturbance will be insufficiently severe to push the vegetation across a threshold to a new stable state, (2) the reversible state resulting from a disturbance is a desirable one for wildlife, or (3) the new stable state following a severe disturbance is predictable and is better for the wildlife species being managed. These conditions are difficult to define because thresholds and new stable states resulting from disturbance may be unknown.

Vegetation in arid and semiarid regions is often distributed in a pattern consisting of patches of dense vegetation interspersed in a matrix of bare ground or less-dense vegetation (Aguiar and Sala 1999). The most common forms of these clumps are somewhat linear bands of dense vegetation or more irregularly shaped clumps, often referred to as the "spotted" pattern. Patch dynamics in arid and semiarid systems that exhibit this type of vegetation patterning may be driven more by endogenous factors than by disturbance. This pattern results in redistribution of water and nutrients from the "sources," areas of bare ground or sparse vegetation, to "sinks," the clumps of dense vegetation. The result is the interspaces of sparse vegetation between the clumps of dense vegetation effectively become more xeric and less fertile, while the clumps of dense vegetation become more mesic and fertile. The end result is the overall system becomes more productive than would occur if vegetation were distributed more homogenously.

Maintaining patch structure by minimizing disturbance may be a better management strategy than introducing disturbance in some arid and semiarid landscapes where disturbance is not the major driver of patch dynamics (Aguiar and Sala 1999). The zone in the gradient between humid climatic zones and arid and semiarid zones in which the primary driver of patch dynamics switches from disturbance to endogenous factors is unclear; additional research is needed to determine where the threshold lies in different ecosystems. Hypothetically, perhaps there exists a threshold or transition zone between humid and semiarid bioclimatic regions in which disturbance drives patch dynamics in wet years or on fertile soils and water and nutrient redistribution drive patch dynamics in dry years or

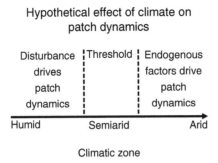

FIGURE 14.2   Hypothetical threshold between humid regions where patch dynamics are driven by disturbance and arid bioclimatic regions where patch dynamics are driven by endogenous factors; in the threshold region disturbance drives patch dynamics in wet years or on fertile soils and water and nutrient redistribution drive patch dynamics in dry years or on infertile soils.

on infertile soils (Figure 14.2). Minimizing disturbance and maintaining patch structure may become the aim of management beyond this threshold.

FACILITATION OF EXOTIC SPECIES INVASION

Different disturbance intensities caused by mechanical treatments, fire, grazing, or a combination of these factors may also provide optimum conditions for exotic invasive plant species to establish. Exotic species are organisms that evolved on different areas and were introduced, naturally or by humans, to new habitats (Westbrooks 1998). Disturbance does not appear to be required for exotic plant species to spread; however, disturbance does in many cases enhance invasion of exotic plants.

In some cases, exotic grasses were introduced to rehabilitate degraded grasslands and the most successful species appear to be those that tolerate drought, grazing, and fire (Geiger and McPherson 2005). Negative consequences have been associated with introductions of some of these plants, however, even though they may provide a high quantity of quality forage for domestic animals.

Exotic plant species are currently the second largest threat to the conservation of biodiversity, second only to loss of habitat (Zalba et al. 2000; Keane and Crawley 2002). Negative effects on ecological processes created by invasive plants include alteration of hydrological processes, energy flow, nutrient cycling, regeneration of native plants, and fire regimes (Masters and Shelley 2001). Areas dominated by the exotic C_4 grasses buffelgrass (*Pennisetum ciliare*) and Lehmann lovegrass (*Eragrostis lehmanniana*) support lower abundance and species richness of grassland birds and fewer insects than areas dominated by native grasses (Flanders et al. 2006).

GRAZING AND EXOTIC GRASSES

Grazing in most natural ecosystems is as much a part of the system as is the need for forage by grazing animals. Most native grazing lands evolved with the disturbance of domestic or wild animal grazing and plants tolerant to grazing (Vallentine 2001). Preference for native grasses by grazing herbivores, including domestic stock, may benefit exotic grasses by eliminating competition for sunlight, water, and nutrients. When herbivores consume primarily native grasses, introduced grasses have the potential to spread unchecked (D'Antonio and Vitousek 1992). In addition to animal selectivity, grazing intensity interactions with seasonal and annual variation of climatic variables may lead to overuse and overgrazing, which may create optimum conditions for exotic invaders, which in many cases are more tolerant to drought and grazing.

In humid environments dominated by mid-to-tall grasses, grazing has been recognized as a habitat management tool to reduce grass cover and increase forbs (Fulbright and Ortega-S. 2006). Cattle grazing, for example, increases species richness and abundance of native annual forbs in California coastal prairie (Hayes and Holl 2003). In another study, Harrison et al. (2003) reported that grazing increased native species richness on serpentine soils but not on nonserpentine soils in California. Although grazing may increase plant species richness and abundance, the effects of grazing can be positive or negative even in humid environments (Hayes and Holl 2003). If excessive grazing leads to increases in bare ground, invasive grass species more aggressive than the natives are likely to increase and become dominant. An example of this is invasions of King Ranch and Kleberg bluestem (*Bothriochloa ischaemum* and *Dichanthium annulatum*, respectively) occurring in native rangelands and introduced pastures ranging from subtropical areas of south Texas to the humid tropics in the state of Veracruz, Mexico along the Mexican Gulf Coast. Overgrazing in these areas of Mexico is one of the factors that facilitated the invasion because overgrazing is a primary factor affecting productivity of cattle operations in northern Mexico (Martinez et al. 1997).

The impact of grazing disturbance on habitat productivity and composition in arid and semi-arid ecosystems is as controversial as in humid environments, resulting partly from the difficulty of separating the effects of livestock from the interannual variation in climatic patterns. While some studies have found grazing to be the primary determinant of plant community composition (Conley et al. 1992; Fuhlendorf and Smeins 1997), others have reported climatic patterns to outweigh grazing effects (Herbel et al. 1972; Gibbens and Beck 1988). In addition, interactions between climate variation and grazing disturbances may exacerbate changes in grassland plant communities (Fuhlendorf and Smeins 1997). Examples of introduced grasses subsequently found to have negative effects upon native plant communities include Lehmann lovegrass and buffelgrass.

The available literature shows that, independent of the environment, grazing disturbance can have positive or negative effects on plant community composition and productivity and it is difficult to separate grazing effects from the climatic variation and interactions between these two factors. In this context, careful consideration must be taken when using grazing as a disturbance to manipulate plant communities because the proper conditions for invasion of exotic plant species may be created.

MECHANICAL BRUSH MANAGEMENT TREATMENTS AND EXOTIC GRASSES

Mechanical treatments were among the earliest brush control approaches developed and have been the most lasting for manipulating plant communities (Scifres and Hamilton 2003). Effects of woody plant control on plant community composition and ecological processes depend on various factors including treatment method, composition of the plant community before the treatments, soils, and rainfall before and after the treatment (Fulbright 1996). Increases in herbaceous vegetation after mechanical treatments have been documented. Roller chopping dense brush stands dominated by guajillo (*Acacia berlandieri*) and blackbrush acacia (*Acacia rigidula*) increases canopy cover of forbs preferred by white-tailed deer when rainfall is adequate (Bozzo et al. 1992). Soil disturbance caused by methods such as root plowing, discing, and roller chopping may cause increased abundance of unpalatable plants such as goldenweed (*Isocoma* spp.) and exotic grasses such as buffelgrass and Kleberg bluestem, forming dense stands that inhibit growth of forbs (Fulbright and Taylor 2001). Mechanical disturbances may be used to alter vegetation characteristics, however, negative impacts such as loss of plant diversity and exotic grasses invasion must be considered.

FIRE AND EXOTIC GRASSES

Burning is the oldest known practice used by humans as a disturbance to alter vegetation characteristics (Vallentine 1989). The initial use of fire focused on utilitarian purposes (Scifres and Hamilton 2003). Fire was used to manipulate vegetation to attract game for hunting, control insects, render the habitat suitable for living, and promote vegetation as a food source. Fire disturbance in combination with other factors such as drought and grazing, however, can have negative consequences on native plant communities by facilitating the invasion of exotic plants.

Establishment and growth of many African grasses introduced to North America is stimulated by fire, and they respond more positively to fire than native grasses do (Williams and Baruch 2000). Biomass production of buffelgrass, for example, is increased by fire (Scifres and Hamilton 1993). Lehmann lovegrass in Arizona appears to invade and replace native grasses because its germination strategies allow it to establish more rapidly following fire or drought than the natives (Angell and McClaran 2001). Rather than hastening the death of natives, the lovegrass simply replaces the natives when they die and continues to increase in density following fire or drought.

Cheatgrass (*Bromus tectorum*), an exotic C_3 grass, competes with native shrubs and grasses for soil water, reducing productivity and water status of the natives (Melgoza et al. 1990). Invasion by cheatgrass changes the fire frequency in the Great Basin region of the United States from a fire every 60–100 years to 5-year intervals (Whisenant 1990). More rapid fire return intervals result because cheatgrass forms a fine, continuous fuel (Allen 2004). Big sagebrush (*Artemisia tridentata*) is easily killed by fire, and cheatgrass invasion and subsequent fire has resulted in the loss of millions of hectares of big sagebrush. Conversion from sagebrush steppe to cheatgrass reduces abundance of certain passerine birds and sage grouse (*Centrocercus urophasianus*) (Brooks and Pyke 2001). In more mesic environments, cheatgrass is not as successful at competing for moisture with native grasses and is less likely to replace native plants (Grace et al. 2001).

Although fire stimulates many exotic grasses, fire has been suggested by some researchers as a tool to alter plant community composition by favoring native grasses and reducing exotic grasses and woody species (Geiger and McPherson 2005). Livestock were removed by Geiger and McPherson (2005), and fire was reintroduced in an attempt to restore herbaceous cover and diversity, lower cover of woody species, and to reduce exotic grasses. Results of this study indicate, however, that plant community structure in semi-desert grasslands dominated by introduced grasses has not markedly changed following the introduction of prescribed fires and removal of livestock. Changes in cover and diversity were associated with variability in precipitation. In contrast, in a study conducted in southern Texas during a 3-year wet cycle in a more humid environment, invasive guinea grass (*Panicum maximum* Jacq.) decreased with summer-prescribed fire and native plant species richness increased (Ramirez 2005). Native plant species richness in the summer burning and cattle grazing treatment increased from 1 to 4.4 species/0.25 m^2, which represented a 340% increase from the beginning to the end of the study. Ten important forbs used by deer and six used by quail were newly recorded or increased 1 year after the application of the burning treatments.

Grazing, mechanical treatments, and fire are disturbances that can be used to manipulate vegetation composition, however, the effects of these disturbances can be positive or negative, depending on the intensity of the disturbance, interactions among disturbances, and climatic variation. The risk of promoting invasion of exotic plants after these disturbances occur must be considered. Gleason (1926) suggested that, rather than following traditional directional models of succession, plant species composition after disturbance may be more a function of random chance in many situations. Availability of propagules near the disturbed site may be an important factor determining the new plant species composition of disturbed sites. For example, areas of native vegetation heavily disturbed by grazing, mechanical treatments, or fire close to patches dominated by Kleberg bluestem or buffelgrass may be invaded by these exotic grasses on soils where they are adapted if moisture conditions are adequate.

NATURAL DISTURBANCE REGIMES ON UNNATURAL LANDSCAPES

A central assumption underlying the use of disturbance to manipulate and improve habitat for wildlife is that native species have evolved under natural disturbance regimes (Angelstam 2003). This concept further assumes that the vegetation communities being manipulated are composed of native plants. On arid and semiarid lands in North America, most of the landscape is subjected to ongoing disturbance by livestock grazing, agriculture, recreational activities, and various other factors. Plant communities on most semiarid landscapes are in early- to mid-seral stages or are in new stable states. When the landscape is subject to ongoing disturbances resulting from human activities, or has been pushed to a new stable state, human-imposed disturbances may compound disturbance effects and make them more severe rather than simulating natural disturbance regimes. Further, much of the landscape has been transformed and extant plant communities differ considerably from the plant communities native wildlife evolved in, particularly where exotic plant species dominate the landscape. Results of simulated "natural disturbance regimes" may not have the predicted outcomes on these areas. In many cases, the most appropriate "management" paradigm may be to set aside areas of minimally disturbed land, rather than compounding disturbance effects with new disturbances.

CONCLUSIONS

Wildlife habitat management practices often cannot be applied in the same manner or with the same outcome in both mesic habitats and arid and semiarid environments. Using disturbances such as grazing in semiarid environments to cause retrogression of the plant community clearly will not benefit species such as bobwhites that are best adapted to later stages of succession in these environments. Disturbances to cause retrogression in patches will do little to increase landscape heterogeneity in landscapes where ongoing disturbances such as heavy livestock grazing maintain the landscape in mid- to early stages of succession. A better management approach to increase heterogeneity in such landscapes may be to reduce livestock stocking rates, or temporarily eliminate grazing, in portions of the landscape to allow succession to proceed to a later stage.

Wildlife and vegetation responses to habitat manipulation are difficult to predict in arid and semiarid environments compared to regions with higher precipitation that is seasonally distributed more evenly. Use of disturbance to manipulate wildlife habitat in arid and semiarid regions should be approached with caution because (1) long-term cycles of low precipitation may override vegetation changes and arrest succession, resulting in a new "stable state," and make the outcome of management practices unpredictable; (2) the extreme variation in precipitation in arid and semiarid environments may impact wildlife much more than even severe human-imposed disturbances including mechanical manipulations, fire, and livestock grazing; (3) in ecosystems where patch dynamics are driven by endogenous factors because of low rainfall or infertile soils, disturbances such as mechanical manipulation, fire, and grazing may have undesirable impacts; and (4) disturbance may facilitate invasion of exotic grasses and undesirable native plants.

An assumption underlying use of disturbance to manipulate and improve habitat for wildlife is that disturbance is a natural factor that has positive effects on wildlife and vegetation. Disturbance may disrupt ecological processes in ecosystems where patch dynamics are driven by endogenous rather than exogenous factors because of low rainfall or infertile soils. The assumption that disturbance will benefit wildlife, consequently, may be erroneous on many arid and semiarid landscapes.

Application of wildlife habitat management based on landscape ecology and patch dynamic theory, rather than simple theories of succession applied at the patch level, will become increasingly important as human populations continue to increase. Heretofore, managing habitats in a coordinated manner on extensive landscapes was virtually impossible. Geospatial technologies have provided the tools that enable researchers and managers to make the transition from the patch to the landscape scale. Future conservation of wildlife in natural habitats will depend in part on our ability to inhibit fragmentation and loss of critical habitats by applying theories of landscape ecology and visualizing the structure and organization of landscapes using geospatial technologies.

REFERENCES

Aguiar, M. R., and O. E. Sala. 1999. Patch structure, dynamics, and implications for the functioning of arid ecosystems. *Trends Ecol. Evol.* 14:273.

Angell, D. L., and M. P. McClaran. 2001. Long-term influences of livestock management and a non-native grass on grass dynamics in the Desert Grassland. *J. Arid Environ.* 49:507.

Allen, E. 2004. Restoration of Artemisia shrub lands invaded by exotic annual Bromus: A comparison between southern California and the Intermountain Region, in *Proc. Seed and Soil Dynamics in Shrubland Ecosystems*. Hild, A. L., N. L. Shaw, S. E. Meyer, D. T. Booth, and E. D. McArthur, compilers. Ft. Collins: U.S.D.A. For. Serv. Rocky Mnt. Res. Sta., RMRS-P-31.

Angelstam, P. 2003. Reconciling the linkages of land management with natural disturbance regimes to maintain forest biodiversity in Europe, in *Landscape Ecology and Resource Management*, J. A. Bissonette, and I. Storch (eds), Chap. 9. Washington DC: Island Press.

Bolen, E. G., and W. L. Robinson. 2003. *Wildlife Ecology and Management*, 5th edn. New Jersey: Prentice Hall.

Bozzo, J. A., S. L. Beasom, and T. E. Fulbright. 1992. Vegetation responses to 2 brush management practices in South Texas. *J. Range Manage.* 45:170.

Brawn, J. D., S. K. Robinson, and F. R. Thompson, III. 2001. The role of disturbance in the ecology and conservation of birds. *Ann. Rev. Ecol. Sys.* 32:251.

Brennan, L. A., J. M. Lee, E. Staller, S. Wellendorf, and R. S. Fuller. 2000. Effects of disking versus feed patch management on northern bobwhite brood habitat and hunting success, in *Quail IV: Proceedings of the Fourth National Quail Symposium*, Brennan, L. A., W. E. Palmer, L. W. Burger, Jr., and T. L. Pruden, (eds). Florida: The Tall Timbers Research Station, p. 59.

Briske, D. D., S. D. Fuhlendorf, and F. E. Smeins. 2005. State-and-transition models, thresholds, and rangeland health: a synthesis of ecological concepts and perspectives. *Rangeland Ecol. Manage.* 58:1.

Brooks, M. L., and D. A. Pyke. 2001. Invasive plants and fire in the deserts of North America, in *Proc. Invasive Species Workshop: The Role of Fire in the Control and Spread of Invasive Species, Fire Conference 2000: the First National Congress on Fire Ecology, Prevention, and Management, Misc. Pub. 11*, K. E. M. Galley, and T. P. Wilson (eds). Florida: Tall Timbers Research Station, p. 1.

Chazdon, R. L. 2003. Tropical forest recovery: legacies of human impact and natural disturbances. *Persp. Plant Ecol., Evol., and Syst.* 6:51.

Conley, W., M. R. Conley, and T. R. Karl. 1992. A computational study of episodic events and historical context in long-term ecological process: climate and grazing in the northern Chihuahuan dessert. *Coenoses* 7:55.

Connel, J. H., and R. O. Slatyer. 1977. Mechanisms of succession in natural communities and their roles in community stability and organization. *Am. Nat.* 111:1119.

D'Antonio, C. M., and P. M. Vitousek. 1992. Biological invasions by exotic grasses, the grass/fire cycle, and global change. *Ann. Rev. Ecol. Syst.* 23:63.

Doan-Crider, D. L. 2003. *Movements and Spaciotemporal Variation in Relation to Food Productivity and Distribution, and Population Dynamics of the Mexican Black Bear in the Serranias Del Burro, Coahuila, Mexico*, Ph.D. Dissertation. Kingsville: Texas A&M University — Kingsville and College Station: Texas A&M University.

Draeger, D. A. 1996. *Predicting seasonal flux in white-tailed deer carrying capacity in south Texas: root plowed versus undisturbed soil sites*, M.S. Thesis. Kingsville: Texas A&M University — Kingsville.

Egbert, S. L., S. Park, K. P. Price, R. Y. Lee, J. Wu, and M. D. Nellis. 2002. Using conservation reserve program maps derived from satellite imagery to characterize landscape structure. *Comput. Electr. Agric.* 37:141.

Flanders, A. A., W. P. Kuvlesky, Jr., D. C. Ruthren, III, R. E. Zaiglan, R. L. Bingham, T. E. Fulbright, F. Hernandez, and L. A. Brennan. 2006. Effects of invasive exotic grasses on south Texas rangeland breeding birds. *Auk.* 123:171.

Forman, R. T. T., and M. Godron. 1981. Patches and structural components for a landscape ecology. *Bioscience.* 31:733.

Friedel, M. H. 1991. Range condition assessment and the concept of thresholds: a viewpoint. *J. Range Manage.* 44:422.

Fuhlendorf, S. D., and F. E. Smeins. 1997. Long-term vegetation dynamics mediated by hervivores, weather, and fire in a *Juniperus-Quercus* savanna. *J. Veg. Sci.* 8:819.

Fuhlendorf, S. D., D. D. Briske, and F. E. Smeins. 2001. Herbaceous vegetation change in variable rangeland environments: the relative contribution of grazing and climatic variability. *Appl. Veg. Sci.* 4:177.

Fuhlendorf, S. D., and D. M. Engle. 2001. Restoring heterogeneity on rangelands: ecosystem management based on evolutionary grazing patterns. *BioScience.* 51:625.

Fulbright, T. E. 1996. Viewpoint: a theoretical basis for planning brush management to maintain species diversity. *J. Range Manage.* 49:554.

Fulbright, T. E. 1999. Response of white-tailed deer foods to discing in a semiarid habitat. *J. Range Manage.* 52:346.

Fulbright, T. E., and S. L. Beasom. 1987. Long-term effects of mechanical treatments on white-tailed deer browse. *Wildl. Soc. Bull.* 15:560.

Fulbright, T. E., and F. S. Guthery. 1996a. Long-term effects of mechanical brush management on shrub diversity, in *Proceedings of the Fifth International Rangeland Congress*, N. West (ed.). Salt Lake City, Utah, p. 166.

Fulbright, T. E., and F. S. Guthery. 1996b. Mechanical Manipulation of Plants, in *Rangeland Wildlife*, P. R. Krausman (ed.), Chap. 20. Denver: Society for Range Management.

Fulbright, T. E., and J. A. Ortega-S. 2006. *White-Tailed Deer Habitat: Ecology and Management on Rangelands*. College Station: Texas A&M University Press.

Fulbright, T. E., and R. B. Taylor. 2001. *Brush Management for White-Tailed Deer*. Kingsville: Caesar Kleberg Wildlife Research Institute and Texas Parks and Wildlife, Joint Publication.

Geiger, E. L., and G. R. McPherson. 2005. Response of semi-desert grasslands invaded by non-native grasses to altered disturbance regimes. *J. Biogeog.* 32:895.

Gibbens, R. P., and R. F. Beck. 1988. Changes in grass basal area and forbs density over a sixty four-year period on grassland types of the Jornada Experimental Range. *J. Range Manage.* 41:186.

Gleason, H. A. 1926. The individualistic concept of the plant association. *Bull. Torrey Bot. Club.* 53:1.

Grace, J. B., M. O. Smith, S. L. Grace, S. L. Collins, and T. J. Stohlgren. 2001. Interactions between fire and invasive plants in temperate grasslands of North America. In *Proc. Inv. Species Workshop: The Role of Fire in the Control and Spread of Invasive Species*, Fire Conference 2000: the First National Congress on Fire Ecology, Prevention, and Management, Misc. Pub. 11, K. E. M. Galley, and T. P. Wilson (eds). Florida: Tall Timbers Research Station, p. 40.

Harrison, S., B. D. Inouye, and H. D. Safford. 2003. Ecological heterogeneity in the effects of grazing and fire on grassland diversity. *Conserv. Biol.* 17:837.

Hayes, F. F., and K. D. Holl. 2003. Cattle grazing impacts on annual forbs and vegetation composition of mesic grasslands in California. *Conserv. Biol.* 17:1694.

Hellickson, M., and A. Radomski. 1999. *Bobwhites of the Wild Horse Desert: Status of Our Knowledge*. Kingsville: Caesar Kleberg Wildlife Research Institute, Texas A&M University — Kingsville.

Herbel, C. H., F. N. Ares, and R. A. Wright. 1972. Drought effects on a semidesert grassland range. *Ecology.* 53:1084.

Huston, M. A. 1994. *Biological Diversity: The Coexistence of Species on Challenging Landscapes*. Cambridge: Cambridge University Press.

Illius, A., and T. G. O'Connor. 1999. On the relevance of non-equilibrium concepts to arid and semi-arid grazing systems. *Ecol. Appl.* 9:798.

Johnson, D. H. 1997. Effects of fire on bird populations in mixed-grass prairie, in *Ecology and Conservation of Great Plains Vertebrates*, F. L. Knopf, and F. B. Samson (eds), Chap. 8. Ecological Studies Number 125, New York: Springer-Verlag.

Keane., R. M., and M. J. Crawley. 2002. Exotic plant invasions and the enemy release hypothesis. *Trends Ecol. Evol.* 17:164.

Keane, R. E., R. A. Parsons, and P. F. Hessburg. 2002. Estimating historical range and variation of landscape patch dynamics: limitations of the simulation approach. *Ecol. Modeling.* 151:29.

Mackey, R. L., and D. J. Currie. 2001. The diversity–disturbance relationship: is it generally strong and peaked? *Ecology.* 82:3479.

Martinez, M. A., V. Molina, F. S. Gonzalez, J. S. Marroquin, and J. Navar. 1997. Observations of white-tailed deer and cattle diets in Mexico. *J. Range Manage.* 50:253.

Masters, R. A., and R. L. Sheley. 2001. Principles and practices for managing rangeland invasive plants. *J. Range Manage.* 54:502.

Melgoza, G., R. S. Nowak, and R. J. Tausch. 1990. Soil water exploitation after fire: competition between *Bromus tectorum* (cheatgrass) and two native species. *Oecologia*. 83:7.

Mitchell, J. E. 2000. *Rangeland Resource Trends in the United States: A Technical Document Supporting the 2000 U.S.D.A. Forest Service RPA Assessment*, General Technical Report RMRS-GTR-68. Fort Collins, Colorado: U.S. Department of Agriculture, Forest Service, Rocky Mountain Research Station.

Mutz, J. L., C. J. Scifres, D. L. Drawe, T. W. Box, and R. E. Whitson. 1978. *Range Vegetation after Mechanical Brush Treatment on the Coastal Prairie*. Texas Agricultural Experiment Station Bulletin 1191, College Station: Texas A&M University.

Natural Resources Conservation Service. 2003. Ecological Site Information System. http://esis.sc.egov. usda.gov/, accessed 4 October 2006.

Nolte, K. L., T. M. Gabor, M. W. Hehman, M. A. Asleson, T. E. Fulbright, and J. C. Rutledge. 1994. Long-term effects of brush management on vegetation diversity in ephemeral drainages. *J. Range Manage.* 47:457.

Pearson, D. M. 2002. The application of local measures of spatial autocorrelation for describing pattern in north Australian landscapes. *J. Environ. Manage.* 64:85.

Peoples, A. D., R. L. Lochmiller, D. M. Leslie, Jr., and D. M. Eagle. 1994. Producing northern bobwhite food on sandy soils in semiarid mixed prairies. *Wild. Soc. Bull.* 22:204.

Petraitis, P. S., R. L. Latham, and R. A. Niesenbaum. 1989. The maintenance of species diversity by disturbance. *Quart. Rev. Biol.* 64:393.

Pickett, S. T. A., and P. S. White (eds). 1985. *The Ecology of Natural Disturbance and Patch Dynamics*. New York: Academic Press, Inc.

Pulliam, H. R. 1988. Sources, sinks, and population regulation. *Am. Nat.* 132:651.

Ramirez Y., L. E., 2005. *Prescribed Fire and Intensive Grazing to Control Invasive Guineagrass on Native Pastures in South Texas*, M.S. Thesis. Kingsville: Texas A&M University–Kingsville.

Rogers, J. O., T. E. Fulbright, and D. C. Ruthven, III. 2004. Vegetation and deer response to mechanical shrub clearing and burning. *J. Range Manage.* 57:41.

Ruthven, D. C., III, T. E. Fulbright, S. L. Beasom, and E. C. Hellgren. 1993. Long-term effects of root plowing on vegetation in the eastern South Texas Plains. *J. Range Manage.* 46:351.

Ruthven, D. C., III, E. C. Hellgren, and S. L. Beasom. 1994. Effects of root plowing on white-tailed deer condition, population status, and diet. *J. Wildl. Manage.* 58:59.

Saunders, D. A., R. J. Hobbs, and C. R. Margules. 1991. Biological consequences of ecosystem fragmentation: a review. *Cons. Biol.* 5:18.

Scifres, C. J., and W. T. Hamilton. 1993. *Prescribed Burning for Brushland Management: the South Texas Example*. College Station: Texas A&M University Press.

Scifres, C. J., and W. T. Hamilton. 2003. Range habitat management: the tools, In *Ranch Management: Integrating Cattle, Wildlife, and Range*, C. A. Forgason, F. C. Bryant, and P. C. Genho (eds). Kingsville: King Ranch.

Spears, G. S., F. S. Guthery, S. M. Rice, S. J. DeMaso, and B. Zaiglin. 1993. Optimum seral stage for northern bobwhites as influenced by site productivity. *J. Wildl. Manage.* 57:805.

Stewart, K. M., J. P. Bonner, G. R. Palmer, S. F. Pattern, and T. E. Fulbright. 1997. Shrub species richness beneath honey mesquite on root-plowed rangeland. *J. Range Manage.* 50:213.

Sullivan, S., and R. Rohde. 2002. On non-equilibrium in arid and semi-arid grazing systems. *J. Biog.* 29:1595.

Turner, M. G. 2005. Landscape ecology in North America: past, present, and future. *Ecology.* 86:1967.

Turner, M. G., S. M. Pearson, W. H. Romme, and L. L. Wallace. 1997. Landscape heterogeneity and ungulate dynamics: what spatial scales are important? in *Wildlife and Landscape Ecology: The Effect of Pattern and Scale*, J. A. Bissonette (ed.), Chap. 13. New York: Springer.

USDI, Bureau of Land Management. 2005. National Rangeland Inventory 2005. http://www.blm.gov/nstc/rangeland/pdf/Rangeland2005.pdf, accessed 18 July 2006.

Vallentine, J. F. 1989. *Range Developments and Improvements*, 3rd edn. San Diego: Academic Press.

Vallentine, J. F. 2001. *Grazing Management*, 2nd edn. San Diego: Academic Press, Inc.

Van der Maarel, E. 1988. Vegetation dynamics: patterns in time and space. *Vegetatio.* 77:7.

Weiers, S., M. Bock, M. Wissen, and G. Rossner. 2004. Mapping and indicator approaches for the assessment of habitats at different scales using remote sensing and GIS methods. *Landscape Urban Plan.* 67:43.

Wiens, J. A. 1984. On understanding a non-equilibrium world: myth and reality in community patterns and processes, in *Ecological Communities: Conceptual Issues and the Evidence*, D. R. Strong, Jr., D. Simberlong, Jr., K. G. Abele, and A. B. Thistle (eds). New Jersey: Princeton University Press, p. 439.

Westbrooks, R. 1998. *Invasive Plants, Changing the Landscape of America: Fact Book*, Washington, DC: Federal Interagency Committee for the Management of Noxious and Exotics Weeds.

Westoby, M. 1980. Elements of a theory of vegetation dynamics in arid rangelands. *Israel J. Bot.* 28:169.

Whisenant, S. G. 1990. Changing fire frequencies on Idaho's Snake River Plains: ecological and management implications, in *Proc. Symp. On Cheatgrass Invasion, Shrub Die-off, and Other Aspects of Shrub Biology and Management*. McArthur, E. D., E. M. Ranney, S. D. Smith, and P. T. Tueller. Compilers, Utah: U.S.D.A. For. Serv., Interm. Res. Sta. Gen. Tech. Rep. INT-276.

Wildlife Habitat Council 1999. *Northern Bobwhite (Colinus virginianus)*, U.S.D.A. Natural Resources Conservation Service, Fish and Wildlife Habitat Management Leaflet 9 (http://policy.nrcs.usda.gov/scripts/lpsiis.dll/TN/tn_b_6_a.pdf).

Williams, C. K., F. S. Guthery, R. D. Applegate, M. J. Peterson. 2004. The northern bobwhite decline: scaling our management for the twenty-first century. *Wildl. Soc. Bull.* 32:861.

Williams, D. G., and Z. Baruch. 2000. African grass invasion in the Americas: ecosystem consequences and the role of ecophysiology. *Biol. Invasions.* 2:123.

Wright, H. E., and A. W. Bailey. 1982. *Fire Ecology: United States and Southern Canada*. New York: John Wiley & Sons.

Wu, X. B., E. J. Redeker, and T. L. Thurow. 2001. Vegetation and water yield dynamics in an Edwards Plateau watershed. *J. Range Manage.* 50:98.

Zalba, S. M., M. I. Sonaglioni, C. A. Compagnoni, and C. J. Belenguer. 2000. Using a habitat model to assess the risk of invasion by an exotic plant. *Biol. Conserv.* 93:203.

Part IV

Animal Health and Genetics

15 The Introduction and Emergence of Wildlife Diseases in North America

Robert G. McLean

CONTENTS

Following a period of success in controlling infectious diseases with new vaccines, global vaccination programs (smallpox and polio), antibiotics, and advanced treatments, especially in the United States during the 1960s to the early 1980s, an era of invading, emerging, and reemerging diseases began. These diseases accelerated through the 1990s and early 2000s, resulting in new disease threats and outbreaks with increased human health risks and huge economic impacts [e.g., AIDS, Lyme disease (LD), West Nile (WN) virus, and severe acute respiratory syndrome-associated coronavirus (SARS-CoV)]. Of the 175 new human emerging diseases, 75% were caused by zoonotic disease agents transmitted between wild or domestic animals and humans (Cleaveland et al. 2001), and these emerging pathogens were predominantly viruses (Woolhouse and Gowtage-Sequeria 2005); for example, Hantavirus, WN virus, Monkeypox, SARS-CoV, and Nipah virus. Many of the newly emerging pathogens have seriously impacted the global public health and animal health infectious disease infrastructure, and some pathogens had the threat of producing pandemics, such as SARS-CoV and recently highly pathogenic avian influenza (HPAI) virus (Fauci et al. 2005). The causes and methods of dissemination of these invading and emerging diseases are as varied as the diseases themselves. Despite advances in medicine and technology, we have been unable to prevent their introduction, establishment, or spread. Recent developments in rapid detection and identification technology have greatly improved surveillance capabilities (Kuiken et al. 2003), but many of these diseases have wildlife as natural hosts and disseminators of the pathogens, and we have insufficient resources to effectively manage the diseases in wildlife populations. I discuss some of the major causes of invasive and emerging diseases and provide examples of wildlife diseases of public health and animal health importance that invaded or emerged in North America during the past few decades and some foreign animal diseases that threaten new invasions. I then discuss measures that are in place and those that could be improved to prevent, detect, and hopefully control these disease threats.

MAJOR CAUSES

The causes of the emergence, reemergence, and invasion of infectious diseases are varied and complex. Factors that are associated with and have contributed to emergence of pathogens include evolutionary changes in the pathogen (HPAI), changes in ecology of the host and pathogens (LD), and invasion of pathogens by movement of the infected host or vector species (WN virus) (Morse 1995; Wilson 1995; Lederberg 1998; Daszak et al. 2000; Cleaveland et al. 2001; Antia et al. 2003; Slingenbergh et al. 2004; Fauci et al. 2005; Gibbs 2005; Woolhouse and Gowtage-Sequeueria 2005). The frequency of new disease threats is increasing while the investment in public health and animal health infrastructure to deal with these challenges tries to keep up in the developed countries like the United States, but falls behind throughout the rest of the world.

One of the major causes of this increase of invasive and emerging diseases is unchecked human population growth. During the past 50 years, the world population increased more rapidly than ever before, and more rapidly than it will likely grow in the future. In 2000, the world population had reached 6.1 billion, and this number could rise to more than 9 billion in the next 50 years (Lutz and Qiang 2002). This human population growth continuously puts an enormous demand on undeveloped land for housing, agriculture, and production of goods, creates further urbanization of natural environments, and concentrates human populations, making them more exposed to and at higher risk for transmission of certain diseases. This human population growth also promotes further encroachment into wilderness habitats that are the natural niches of insect vectors and wildlife hosts and their shared pathogens, making humans more likely to be infected with exotic viruses such as Ebola, jungle yellow fever, Nipah, and HIV. The expanding demand for wood and agricultural products promotes the destruction of more and more tropical and temperate forests and exposes forest and agriculture workers and their families to disease pathogens such as Ebola in Africa, yellow fever and arenaviruses in South America, and Nipah virus in Malaysia.

Another cause of invading and emerging diseases is the increased frequency and rapidity of international travel that can transport people, animals, animal products, and pathogens worldwide within 1–2 days, well within the incubation period of most diseases. Travel associated with ecotourism, business, and leisure can move an individual exposed to a pathogen from one continent to another, arriving with an infectious disease that can be transmitted before symptoms appear and introduce an exotic disease such as Nipah or Rift Valley Fever (RFV), both of which have wildlife reservoirs and can severely affect domestic livestock. Luckily, most introductions are not successful, but some pathogenic microbes introduced into new areas can survive the introduction, infect susceptible hosts, cause disease, become established, and emerge into a major disease of public health or animal health importance (Wilson 1995). Emerging diseases are also caused by the global wildlife trade that rapidly transports wildlife through major international routes, mostly through uncontrolled or illegal networks, and involves millions of birds, mammals, reptiles, amphibians, and fish every year (Karesh et al. 2005). The intermixing of wildlife species from many parts of the world in crowded live-wildlife markets in China and other countries combined with close contact among domestic animals, such as poultry and pigs, and humans provides a great opportunity for disease transmission and the development of new emerging diseases such as SARS and HPAI strains. Once one of these diseases jumps to humans, then rapid international travel can disseminate the disease worldwide and cause major public health problems, resulting in enormous economic impact (SARS; Ksiazek et al. 2003). Legal trade of wildlife can also lead to emerging diseases. Wildlife hosts naturally infected in their native habitats where the disease does not cause clinical disease that are captured and transported to new environments or situations, again very rapidly, can transmit the disease to naïve wildlife hosts and cause disease and die-offs in these wild animals and expose associated humans (monkeypox).

Emergence of wildlife diseases can occur in wildlife populations when their natural sustaining habitats are destroyed or modified for human use, fragmented, or deteriorated (Friend et al. 2001). These negative ecological changes concentrate wildlife at high densities in inferior habitats, resulting in increased stress, reduced nutrition, and enhanced transmission of diseases. An important wetland region in central valley of California supported huge migratory and wintering waterbird populations, but >90% of these natural wetlands were drained or converted to support human population increase. This loss of critical habitat forced these birds to shift south to the Salton Sea in southern California, which is a 974-km^2 lake in the desert, and is an agriculture drainage reservoir of poor habitat quality with salinity (44 ppt) exceeding that of the ocean. The Salton Sea had high fish production that supported large waterbird populations, but habitat deterioration led to massive fish die-offs followed by significant disease outbreaks in waterbirds, particularly pelicans, caused by an unusual form of botulism involving fish (Nol et al. 2004).

SPECIFIC INVASIVE AND EMERGING WILDLIFE DISEASES

There have been a series of emerging and invasive wildlife diseases that affect humans and domestic animals during the past few decades. Some of these disease threats affected the entire North American continent and have become endemic, continuing to cause severe disease and mortality. A few of these diseases will be discussed in more detail (Table 15.1), including discussions about the reasons for the emergence and measures to control or prevent the disease (Table 15.2).

WEST NILE VIRUS

The most spectacular invading and emergent disease during the past 20 years was the West Nile virus (*Flavivirus*, Flaviviridae, and WNV). This mosquito-borne virus made it to New York City, probably via an infected mosquito, bird, or human from the Middle East during the spring or early summer 1999 (Lanciotti et al. 1999), and quickly amplified in local bird populations (Eidson et al. 2001a). By autumn 1999, a small human epidemic occurred (CDC 1999a), and the virus distribution

---

TABLE 15.1

Examples of Invading, Emerging, and Re-emerging Wildlife Diseases of Public Health and Animal Health Importance in the United States

Disease	First year reported in the United States	Pathogen type	Transmission method	Primary vertebrate host	Origin of disease agent
Raccoon rabies	1956	Virus	Animal bite	Carnivores	United States
Lyme disease	1982	Bacteria	Tick bite	Rodents	United States
Hantavirus	1993	Virus	Direct aerosol, animal contact	Deer mice	United States
West Nile virus	1999	Virus	Mosquito bite	Avian species	Middle East
Monkeypox	2003	Virus	Direct aerosol, animal bite, or contact	Rodents	West Africa

TABLE 15.2

Reasons for the Emergence and Re-emergence of Wildlife Diseases of Public Health and Animal Health Importance in the United States

Disease	First year	States started	States expanded	Reasons disease agent established/expanded
Raccoon rabies	1950s	FL	20	Slow gradual expansion of new strain northward from Florida to four states by early 1970s; translocation north in 1977 and rapid expansion in NE States in 1980–90s; west to Ohio in 1996 and north to Canada by 1999
Lyme disease	1982	CT	49 (reported)	Fragmented forest and suburban habitats supported high host and tick populations for expansion
Hantavirus	1993	NM, AZ, CO, and UT	30	Discovery of more infected locations — prevention/education, reduced risks and cases
West Nile virus	1999	NY	48	Ideal weather and susceptible host and vector populations to become established and virulence of virus-strain produced broad host/vector range to allow expansion to many new ecosystems in NA
Monkeypox	2003	TX	(8) (eliminated)	Rapid movement and mixing of infected rodents with native rodents in animal facilities — did not survive or become established

expanded outward from the introduction site in all directions to about a 160-km-diameter circle in three states surrounding New York City (Eidson et al. 2001b), as evidenced by WNV-positive dead birds. It became evident early that the WNV strain introduced was particularly virulent for native bird species and caused significant mortality, especially in Corvidae species (Bernard et al. 2001). This unique feature of mosquito-borne flaviviruses was utilized to detect and track the movement of WNV and became the primary tool for active surveillance by local and state public health departments for the first few years when only small numbers of human cases were occurring. This dead bird surveillance was accompanied by passive surveillance for human and equine cases and active testing of sentinel birds and mosquito collections in some states (CDC 2000).

The abundance of susceptible native avian species and optimum natural habitats for avian hosts and vector mosquitoes throughout NA supported the establishment and rapid expansion of WNV. The many millions of migratory birds moving south in the fall and north in the spring provided a means for the movement of the virus within NA and, subsequently, south to countries in the Caribbean and Central and South America. During the spread of WNV across the United States from 1999 to 2005, it caused 19,655 human cases, 23,117 equine cases, and was responsible for 53,268 dead birds from 308 species reported from all 48 U.S. states to the public health surveillance network (Farnon 2006), with an estimated mortality in the millions of birds. Few other zoonotic diseases have been as successful in becoming established and in disseminating so rapidly and extensively as WNV. This successful invader is now endemic throughout most of NA, with WNV activity in 2005 reported in all 48 of the continental states of the United States, and its virulence has apparently not changed during the past 7 years. It is currently invading South America (Mattar et al. 2005) and could spread throughout that continent in the near future.

The national WNV surveillance infrastructure that was quickly established in the eastern states and expanded throughout the United States included a national database (ArboNet), rapid testing and weekly reporting of surveillance data by states, a weekly updated national surveillance map displaying the continuing detection and spread of the virus (Marfin et al. 2001), and an annual surveillance conference to modify and improve the surveillance network was an excellent model for dealing with invasive and emerging diseases. Funding provided by the Centers for Disease Control and Prevention (CDC) through congressional appropriations to directly support surveillance and targeted research, and by the National Institutes of Health and CDC to fund research through grants, was effective in dealing with this massive disease threat. However, the public and media are now complacent about this disease that made big news for the past 6 years and the public may let down their guard to keep using the best protective measures, such as vaccination of their horses, eliminating mosquito breeding sites on their properties, and personal protection against mosquito bites, which could result in a resurgence of clinical disease.

HANTAVIRUS

A novel hantavirus, Sin Nombre (SN) virus, Bunyaviridae, was discovered in 1993 during an outbreak of acute cardiopulmonary disease in humans living in a rural area in the Four Corners states of the southwestern United States (Nichol et al. 1993). This human epidemic followed a significant El Nino southern oscillation event resulting in an unusually wet winter and spring in this normally dry environment, and this increased precipitation promoted vegetation growth and subsequently produced very high populations of rodents during the summer months (Hjelle and Glass 2000). The high rodent populations amplified virus transmission and increased human contact with rodents, particularly following invasion and infestations of houses and outside buildings, and expanded exposure to infected rodents. Humans were exposed to the SN virus by the aerosol route through inhalation of virus-contaminated excreta from the natural reservoir host, the deer mouse (*Peromyscus maniculatus*). Deer mice are not affected by SN virus infections and excrete the virus in their urine, feces, and saliva, but a severe disease known as hantavirus pulmonary syndrome (HPS) with high mortality occurs in infected humans (Zeitz et al. 1995). This initially appeared to be a new emerging disease; however, it later became evident that HPS has been endemic in the United States for more than three decades with human cases recognized as early as 1959 (Frampton et al. 1995). The disease has been reported sporadically in humans throughout the range of the deer mouse in the Western and Midwestern states during the past 13 years following the initial outbreak, further indicating the broad endemicity of this virus and emphasizing the single host species and single virus relationships of this group of viruses. There have been 384 cases of HPS reported in the United States from 1993 to 2004. Other single host–virus combinations have been discovered throughout North and South America (Schmaljohn and Hjelle 1997). Control of the disease and prevention of human cases is targeted at reducing contact with infected rodents. Humans contract the disease mostly in and around their permanent or seasonal residences; therefore, the primary strategy for reducing exposure and infection with HPS is rodent prevention and control in and around the home (CDC 2006).

LYME DISEASE

Lyme disease is caused by the spirochete *Borrelia burgdorferi* and is transmitted through the bite of *Ixodes* spp. ticks. The natural history of the disease in the eastern United States includes rodents (primarily white-footed mice, *Peromyscus leucopus*, and eastern chipmunk, *Tamias striatus*) as the primary host species for the spirochete and for immature stages of the vector deer tick (*Ixodes scapularis*) and the white-tailed deer (*Odocoileus virginianus*) as the primary host maintaining the adult ticks (Lane et al. 1991; Steere 2001). LD was identified as a clinical syndrome of juvenile rheumatoid arthritis in children in Lyme, Connecticut, in 1976 (Steere et al. 1977) and the causative spirochete of LD was discovered in 1981 (Burgdorfer et al. 1982). Retrospective analysis of human cases found LD had occurred in Cape Cod in the 1960s and PCR analysis of museum specimens of ticks and rodents from Long Island found evidence of *B. burgdorferi* DNA from the late 1800s and early 1900s. However, few cases were reported before the national surveillance in the United States was started by the CDC in 1982 and LD was not designated as a nationally notifiable disease until 1991. LD began to emerge as the number of reported cases increased steadily since 1982 and LD distribution expanded in the northeastern and north central United States until it is now the most commonly reported arthropod-borne illness in the United States and Europe, with about 20,000 cases reported annually in the United States alone (CDC 2002).

The emergence of LD during the past 20 years was facilitated in the northeastern United States by the improving conditions for the ecology of LD. Before the disease emergence, this region was predominately farmland as a result of the clearing of the extensive woodlands during early colonization by Europeans. At the same time, deer populations were decimated by hunting. Farming declined in this region during the past 40–50 years and farmland gradually reverted to meadows, shrubs, and secondary growth woodlands that provided food and shelter for increasing populations of deer and rodents. These habitat changes combined with a rapid expansion of human development in the region that converted the rural woodlands into wooded suburbs with grass yards and backyard woods where rodents and deer proliferated allowing the deer tick populations to thrive and expand. These suburban regions also have restrictions on hunting deer that have contributed to an even greater abundance of deer. This increase in the host populations for the spirochete and for the ticks enhanced transmission of the spirochete within the extensive suburbs and exposed the associated human populations to LD in their own yards and in recreational areas. The origin and progression of the LD region in the north-central states was different and likely started in the late 1970s in central Wisconsin (Davis et al. 1984). The distribution of the tick and LD gradually expanded westward through western Wisconsin and into Minnesota in habitats conducive for the survival of the tick and the LD spirochete. This emergence was supported by the natural ecology of the region and represented a slow dispersal of the vector tick species and LD via the movement of its more mobile vertebrate hosts of deer and birds (McLean et al. 1993).

Because of the predominance of domicile transmission, prevention and control of LD has concentrated primarily on insecticide treatment of backyard habitats, acaricide treatment of mice to reduce tick abundance, or landscape changes to discourage use by rodents and deer. Advances have occurred with various control methods to reduce risk; nevertheless, the methods have generally been ineffective in significantly reducing transmission, although education for the use of personal protection measures may help. The number of reported human cases in the United States has remained at about 20,000 cases per year for the past few years.

MONKEYPOX

Monkeypox virus belongs to the Orthopoxvirus group of viruses that include variola (smallpox), vaccinia (used in smallpox vaccine), and cowpox viruses (Nalca et al. 2005). It is a rare viral disease in Africa that includes clinical signs and symptoms resembling those of smallpox, but which are usually milder. Humans are exposed to monkeypox from an infected animal through a bite, direct contact with fluids, or aerosols and person to person transmission can occur through the respiratory route, but less efficiently. Human outbreaks have been reported from areas in Central and West Africa with a fatality rate of 1–10% of cases. Wild mammal involvement with monkeypox virus in Africa is known mostly from serology and monkeys are thought to be incidental hosts similar to humans; whereas, multiple species of rodents are the likely reservoirs (Khodakevich et al. 1988). The only confirmed virus isolation was from a rope squirrel (*Funisciuris anerythrus*) from Zaire (Khodakevich et al. 1986).

Monkeypox was unknown in the western hemisphere until the virus was introduced into United States in a legal shipment of 762 African rodents, including some infected rodents, imported from Ghana, West Africa, by an exotic pet dealer in Texas (CDC 2003a). Most of these exotic mammals were subsequently shipped to an animal dealer in Iowa, although 178 of the African rodents could not be traced beyond the point of entry in Texas because records were not available. Some of the infected African rodents were then shipped from Iowa with other animals to a dealer in Illinois who housed these animals in the same room with 200 native prairie dogs (*Cynomys* sp.). Over half (110) of the prairie dogs that were exposed to infected African rodents were later shipped to animal dealers in multiple states and were sold to the public as pets before 15 became sick or died. Of the

15 ill prairie dogs, 10 died rapidly, and 5 exhibited anorexia, wasting, sneezing, coughing, swollen eyelids, and ocular discharge. Infection and pathologic studies of infected prairie dogs showed the animals had bronchopneumonia, conjunctivitis, and tongue ulceration (Guarner et al. 2004). Active viral replication was observed in the lungs and tongue indicating that both respiratory and direct mucocutaneous exposure are potentially important routes of transmission of monkeypox virus between rodents and to humans. The remaining prairie dogs that could be located were destroyed. The high susceptibility of native prairie dogs to monkeypox was unexpected and was responsible for most of the human cases. Also, some of the African rodents became ill and died after arriving in the United States and were PCR positive for Monkeypox virus (CDC 2003b), including three dormice (*Graphiurus* sp.), two rope squirrels, and one Gambian giant pouched rat (*Cricetomys* sp.).

There were 71 reported cases of monkeypox in humans in the United States associated with the infected rodents, primarily as a result of contact with infected prairie dogs that had acquired monkeypox from diseased African rodents, and 35 cases were laboratory-confirmed in Illinois, Indiana, Kansas, Missouri, and Wisconsin (CDC 2003c; Sejvar et al. 2004; Kile et al. 2005). Most patients had mild, self-limited febrile rash illness; however, 18 were hospitalized (some for isolation purposes). Two of the hospitalized cases were children who required intensive care, one for severe monkeypox-associated encephalitis, and one with profound painful cervical and tonsillar adenopathy and diffuse pox lesions (Huhn et al. 2005). Both children recovered from their illness.

Non-native animal species, such as the African rodents, have become popular pets in the United States, but they can create serious public health and animal health problems when they introduce a new disease, such as monkeypox, to the native animal and human populations. The transportation, sale, or distribution of infected animals or the release of infected animals into the environment can result in the further spread of diseases to other animal species and to humans (CDC 2003c). Certain aspects of the importation and movement of exotic animals into and within the United States are under the jurisdiction and regulation of different federal and state agencies. As this disease situation progressed, it became clear that the state regulations were limited to their respective jurisdictions. Regulations differed among states in the types of animals and response actions that were covered and state rules expired on specific dates, all of which hampered efforts to manage and control the movement of the animals and the disease. Communicable diseases that are not confined by State borders, however, may require Federal action to help prevent their spread. The CDC and the Food and Drug Agency issued a joint order (DHHS 2003) to place a temporary embargo on the importation of all rodents from Africa and also banned the sale, movement, or release of prairie dogs into the environment to halt the dissemination of the monkeypox outbreak. Improvements in the regulation and control of the trade of wildlife exotic pets into and within the United States are needed to prevent future disease invasions. Human infections with monkeypox virus may be prevented by vaccination with vaccinia virus (the smallpox vaccine); even up to 14 days after exposure, but there are no licensed antiviral drugs available for post-exposure therapy (Nalca et al. 2005).

RACCOON RABIES

Rabies is an acute fatal encephalitis caused by neurotropic viruses in the genus *Lyssavirus*, family *Rhabdoviridae*. Rabies is a preventable disease of mammals that is transmitted primarily by the bite of a rabid animal. Preventable measures include pre-exposure vaccination and post-exposure treatment. Dog rabies was the predominant form of rabies from 1938 when national data on the incidence of rabies were first compiled until the 1950s. Rabies in wildlife was virtually unknown, but eventually became evident, and reporting began to increase as domestic animal rabies was drastically reduced and came under control through nationwide mandatory dog vaccination programs in the 1950s and then attention shifted to the underlying problem of wildlife rabies (McLean 1970). Rabies in raccoons (*Procyon lotor*) appeared in southern Florida in 1955–56 where it was unknown previously and raccoon rabies began to emerge as a new disease (Kappus et al. 1970). By the late 1960s and early 1970s, epizootics of raccoon rabies were occurring throughout Florida (Bigler et al. 1973), and raccoon rabies began to spread northward through Florida to Georgia (McLean 1971).

Although the existence of distinct genetic variants of rabies viruses was not documented until the late 1970s, the rabies virus in Florida raccoons was apparently a new variant, and raccoons began to emerge as an important new rabies host (Smith et al. 1984). Rabies in raccoons was spreading slowly northward to South Carolina, but its northward movement was assisted by humans with the translocation of infected raccoons from Florida to the Virginia/West Virginia border in 1977 for hunting purposes (Nettles et al. 1979). This introduction started a new focus of raccoon rabies that emerged rapidly and spread northward throughout the northeastern United States to Canada, southward to join the expanding front in South Carolina, and eventually westward to include all of the states east of the Appalachian Mountains and Ohio, Tennessee, and Alabama (Slate et al. 2005).

Small, targeted vaccination efforts to control raccoon rabies began in the mid-1990s utilizing a vaccinia-rabies glycoprotein recombinant (VRG) vaccine in a fishmeal bait (Hanlon et al. 1998). To expand the vaccination efforts, a coordinated oral rabies vaccination (ORV) program was implemented in 1998 by Wildlife Services, APHIS, USDA, to halt the westward spread of the raccoon rabies variant and to eventually eliminate this variant from the eastern United States (Slate et al. 2005). Millions of VRG vaccine baits are distributed, mostly by aircraft, each year in habitats that support raccoons to create immune buffer zones to stop the spread of raccoon rabies. In 2003, 4.23 million baits were dropped to target raccoons in states containing the Appalachian Ridge covering a 64,122-km^2 area in six states at a cost of about $96/$km^2$(Slate et al. 2005). Benefits from this vaccination program are in the expected savings in reduced costs for treatment of humans exposed to rabid or potentially rabid animals and reduced costs of public health programs for rabies detection, testing, prevention, and control in the United States, which has been estimated to be over $300 million/year (Krebs et al. 1998). A similar vaccination program in South Texas contained the northward spread from Mexico of a canine strain of rabies adapted to coyotes (*Canis latrans*) and subsequently eliminated coyote rabies from the state (Fearneyhough et al. 1998). A vaccination buffer is maintained along the Texas–Mexico border to prevent the reentry of coyote rabies. Immediate goals of the National ORV Program are to prevent specific strains of the rabies virus in the raccoon, gray fox, and coyote from spreading to new, uninfected areas. The long-range goal is to eliminate these strains.

FOREIGN WILDLIFE DISEASES THAT COULD INVADE THE UNITED STATES

There are a number of wildlife diseases from throughout the globe that could invade NA under specific conditions, and many could become established. A few diseases will be presented as examples of the types of pathogens, the variety of vertebrate hosts involved, and the potential routes of entry into NA (Table 15.3).

HIGHLY PATHOGENIC AVIAN INFLUENZA

The most likely new invasive disease for NA is the HPAI strain of H5N1 subtype, type A virus. Aquatic birds, particularly Anseriformes (ducks, geese, and swans) and Charadriiformes (gulls, terns, and shorebirds or waders) are infected with a variety of subtypes of influenza A (AI) viruses and are likely the natural reservoirs (Krauss et al. 2004). Nearly all of the subtypes of AI viruses are endemic in and circulate in wild bird populations, predominantly in waterfowl species (Webster et al. 2006). Low-pathogenic avian influenza (LPAI) viruses have been isolated from more than 100 wild bird species and all of the AI virus subtypes have been detected in wild bird reservoirs and poultry (Olsen et al. 2006). Many strains of AI virus can infect a variety of domestic birds, such as chickens, turkeys, pheasants, quail, ducks, geese, and guinea fowl, and cause varying amounts of clinical illness. The pathogenicity of AI viruses are based on the severity of the disease they

TABLE 15.3

Examples of Foreign Wildlife Diseases of Public Health and Animal Health Importance That Could Invade the United States

Disease	Method of potential introduction	Pathogen type	Transmission method	Primary vertebrate hosts	Origin of pathogen
HP H5N1 Asian avian influenza	Migratory waterfowl, poultry, humans	Virus	Direct/ aerosol-ingestion	Waterfowl, poultry	SE Asia
Rift Valley Fever	Infected mosquito, rodent import or human	Virus	Mosquito bite, direct	Rodents, sheep, cattle	Africa, Arabian Peninsula
Japanese encephalitis	Infected mosquito or human	Virus	Mosquito bite	Waterbirds, pigs	SE Asia
Nipah virus	Infected bat or human	Virus	Direct	Fruit bats, pigs	Australia, Malaysia

cause, and most of these subtypes are LPAI forms that cause little or no disease although some strains are capable of mutating under field conditions or passage in chickens into HPAI viruses. HPAI viruses are an extremely infectious and fatal form of the disease that, once established, can spread rapidly among chickens and from flock to flock. Influenza viruses are unstable and specific mutations and evolution of these viruses occur with unpredictable frequency through the constant mingling of multiple subtypes in wild waterfowl populations and the frequent exchange of genetic material (Webster et al. 1992).

A HPAI virus strain, H5N1 subtype, evolved in China and was originally detected in 1996 when it caused mortality in wild geese at Qinghai Lake, China (Liu et al. 2005), which was unusual because AI subtypes do not usually cause disease in the natural hosts. This goose virus acquired other gene segments from quail and ducks and became the dominant genotype being transmitted in live poultry markets in Hong Kong in 1997 (Webster et al. 2006), causing extensive mortality in poultry and in 6 of 18 infected humans (de Jong et al. 1997). This genotype disappeared when all domestic poultry in Hong Kong were culled, but other reassortants from duck and goose reservoirs appeared with similar characteristics. These H5N1 viruses continued to develop until a single genotype in 2002 killed most of the wild and domestic waterfowl in Hong Kong (Sturm-Ramirez et al. 2004) and spread to humans. This 2002 genotype was the precursor of the Z genotype that later became the dominant genotype that spread from China quickly south to Vietnam, Cambodia, Thailand, Laos, and Indonesia where it has caused numerous outbreaks in poultry and many human cases associated with sick or dead poultry. As of April 3, 2006, 165 human cases with 94 deaths (57%) from HPAI H5N1 infections have been reported in China and Southeast Asia (WHO 2006). The H5N1 genotype subsequently spread west from SE Asia to Russia, Europe, the Middle East, and Africa causing outbreaks in poultry, some wild birds and scattered human cases (25 cases in four countries, with 13 deaths). Nearly all of the human cases were confirmed to have resulted directly from interactions with poultry.

The geographical spread of the virus was a result of a combination of factors, many of which can be attributed to humans. Local spread is likely achieved by human movement of poultry and poultry products to and from markets and commercial and backyard flocks, movement and interchange of fighting cocks, and local intermingling of domestic ducks (Webster et al. 2006). Longer-distance spread, particularly within a region, can be accomplished by commercial trade of poultry and poultry products, disseminating ducks and other aquatic birds that move seasonally through harvested rice fields, and migratory birds. The role of migratory birds in spreading AI viruses, especially LPAI, is well known and the Anseriformes and Charadriiformes are the major natural reservoirs for these viruses (Olsen et al. 2006). Millions of migratory birds move within and between large continents along major routes or flyways where bird populations connect with each other after sharing either common breeding areas, staging areas, or wintering grounds. Infected birds can transmit their viruses to susceptible birds that in turn can move the viruses to new areas. For example, migratory birds moving within the West Pacific and the East Asian-Australasian flyways overlap with each other and with birds in Alaska where some of them share common breeding areas with NA birds (Webster et al. 2006).

Serious concerns have been raised about the potential impact of HPAI H5N1 virus on domestic poultry, wild bird populations, and humans in the event that it is introduced into the United States. Potential routes of introduction of H5N1 into the United States could be through the illegal import-ation of domestic, pet, or wild birds (legal importation of birds is restricted or the birds must undergo 30-day quarantine), contaminated poultry products, infected human travelers (although there is no evidence yet of human to human sustainable transmission), bioterrorism event, or migra-tion of infected wild birds through Alaska and the Pacific flyway or through eastern Canada and the Atlantic flyway. In response to these concerns, the U.S. government developed a "National Strategy for Pandemic Influenza" that outlines the responsibilities that local, state, and federal government departments and industry and individuals have in preparing for and responding to an influenza pandemic. Funding was provided to translate the national strategy into an Implement-ation Plan that provides guidance for the development of individual plans, identifies actions for Federal departments and agencies, sets clear expectations for local and state governments and for nongovernment entities, and provides guidance for individuals and families to prepare for a pan-demic (http://www.pandemicflu.gov/plan/tab1.html). Three major components of the strategy are preparedness and communication, surveillance and detection, and response and containment.

The federal government's role in the surveillance and detection part of the "National Strategy for Pandemic Influenza" was to develop an interagency strategic plan for an early detection system for highly pathogenic H5N1 avian influenza in wild migratory birds (USDA 2006a). The plan outlines five major surveillance strategies for detecting H5N1: (1) investigation of morbidity/mortality events, (2) surveillance of live wild birds, (3) surveillance of hunter-killed birds, (4) sentinel species, and (5) environmental sampling. The National Wildlife Research Center was designated to conduct the environmental sampling strategy and will be testing fecal and water samples collected from high-risk waterfowl habitats across the United States. Surveillance will initially be focused in Alaska, where H5N1 is likely to be introduced from Asia, and secondarily on the Atlantic coast, where HPAI could be introduced with migratory birds that cross over the Atlantic Ocean from Europe to Canada and the eastern coast of the United States. Special attention will also be given to locations along major flyways, particularly the Pacific and Mississippi flyways, that migratory waterfowl use when moving south from Alaska during the fall and winter in the southern United States and farther south into the Caribbean and Latin America.

RIFT VALLEY FEVER

Rift Valley Fever (RVF) is a vector-transmitted, viral zoonoses of domestic livestock and other mam-mals in sub-Saharan Africa. The ecology of RFV is unique (Wilson 1994), because the virus can survive in arid grasslands where it persists in the eggs of multiple species of *Aedes* mosquitoes, the primary vectors and reservoirs of the virus, that hatch in natural depressions in the grasslands when they are flooded following periodic heavy rains (Davies et al. 1985). The adult mosquitoes emerge already infected and feed on nearby mammals, particularly domestic ungulates, and initiate local virus transmission. The cattle, sheep, and goats that become infected circulate high amounts of virus and infect mosquitoes and other arthropods. This circulation of the virus amplifies transmission that leads to periodic epizootics in domestic animals, causing abortions and death in susceptible animals. Human infections occur through vector transmission, aerosols, or direct contact with infected anim-als, and epidemics of about 27,000 cases with 170 deaths have been reported in Kenya, East Africa

(Woods et al. 2002). The RVF virus escaped from the African Continent for the first time in 2000 through the transport of infected livestock to Saudi Arabia and Yemen where it caused an epizootic in livestock and a subsequent human epidemic (Madani et al. 2003). Other means by which this virus could be transported out of Africa and into the United States are through infected wild mammals (rodents in the pet trade), humans as airline passengers, infected adult mosquitoes, or infected *Aedes* mosquito eggs transported on a passenger plane or ship. Once the virus arrives there are numerous *Aedes* and other mosquito species in the United States that are competent vectors, abundant wildlife species, especially rodents, as natural reservoirs (Gora et al. 2000), and enormous populations of susceptible livestock throughout the country that would serve as amplifying hosts. This virus could easily become established as WNV did and would have a huge impact on the sheep and cattle industry.

Nipah Virus

Nipah virus is an emerging zoonotic virus in the new genus Henipavirus within the family Paramyxoviridae that also includes Hendra virus. Nipah is a highly pathogenic virus that emerged from fruit bats in Malaysia in 1998 largely due to shifts in livestock production and alterations to reservoir host habitat. The virus caused outbreaks of fatal disease in domestic pigs and humans with substantial economic loss to the local pig industry (CDC 1999b). The disease in pigs showed respiratory and neurological signs that spread to humans causing severe febrile encephalitis resulting in death in 40–75% of cases (Chua et al. 1999). Fruit bats in the genus *Pteropus* are the natural reservoir host of Nipah virus (Johara et al. 2001) and these bat populations have been substantially reduced in Southeast Asia during the past two decades because of extensive deforestation of their natural habitats and climatic effects (Chua et al. 2002). In 1997/98, slash-and-burn of forests led to agricultural expansion and intensification in the modified areas, including the development of piggeries located in cultivated fruit orchards. The deforestation also reduced the availability of fruiting forest trees for forging fruit bats forcing them to encroach upon cultivated fruit orchards. These changes allowed the juxtaposition of the natural bat host with a highly susceptible domestic pig in the fruit orchards that allowed transmission of a novel virus from its reservoir host to the domestic pig and subsequently to the farmers attending the pigs. The virus distribution expanded to Australia and Singapore and caused five subsequent outbreaks between 2001 and 2005 in Bangladesh. During these outbreaks, the virus appears to have been transmitted directly from bats to humans and person-to-person transmission possibly occurred, suggesting an increased public health risk (Epstein et al. 2006).

Fruit bats became a popular exotic animal introduced into the pet trade in the United States in the early 1990s, but the threat of introducing a pathogenic virus resulted in a complete embargo on importation. This regulatory action reduced the possibility of an introduction of Nipah-virus-infected bats similar to the introduction of monkeypox with infected rodents, but with a much more lethal virus. However, some risks of introduction still exist from illegal importation of bats, movement of infected pigs or pig products, infected human travelers, and bioterrorism (Lam 2003). The introduction of Nipah virus into NA could have severe consequences for domestic and wild pigs and associated humans.

Japanese Encephalitis

Japanese encephalitis (JE) is a common but serious human disease in 16 countries of eastern and southern Asia and is the leading cause of viral encephalitis in these countries with 30,000–50,000 cases reported annually. Case-fatality rates vary from 0.3 to 60% (Endy and Nisalak 2002). Severe clinical disease and death from JE is age related, with most cases occurring in the very young and elderly. The majority of cases are mild infections, but severe infections can progress from acute encephalitis with high fever, disorientation, tremors, convulsions (especially in infants), spastic paralysis, coma, and death. Japanese encephalitis is a mosquito-borne virus (Flavivirus, Flaviviridae) closely related to WNV and St. Louis encephalitis virus that has birds as the natural hosts. Rice-field-breeding mosquitoes (*Culex tritaeniorhynchus* group) are the primary enzootic vector, and waterbirds (herons and egrets) are the major avian amplifying host species in SE Asia. Black-crowned night herons (*Nycticorax nycticorax*), egrets (*Egretta* sp.) and European starlings (*Sturnus vulgaris*) were shown experimentally to be competent hosts to infect *Culex* mosquitoes (Soman et al. 1977). Periodic epidemics occur when the virus is brought into peridomestic environments by birds and mosquito bridge vectors are infected by feeding on the birds and transmitting the virus to domestic pigs that serve as additional amplifying hosts for other mosquitoes to pass the virus to humans (Burke and Leake 1988). Some countries that have had major epidemics are controlling the disease through human vaccination programs. The virus has spread recently to northern Australia, probably by migratory birds (Hanna et al. 1996). The risks and routes of introduction of JE virus into the United States are similar to what they were for WNV. If JE virus is introduced, it could become as easily established and disseminated as WNV because of the availability of abundant avian hosts and competent mosquito vectors, but possibly more confined to areas around bodies of water that contain more waterbirds. Many regions in the southern states also have large populations of free-ranging feral pigs mixed among the wetland habitats containing many species of waterbirds that could support intense amplification and outbreaks of the disease. There may be some cross protection of infections between the closely related JE and WNV viruses (Tesh et al. 2002).

PREVENTION, DETECTION, AND CONTROL OF INVADING AND EMERGING DISEASES

There are other foreign animal diseases (FAD), besides monkeypox and WN viruses, that could be introduced with infected wild animals [rodents (RVF, Lassa Fever), birds (JE, HPAI), bats (Nipah), primates (yellow fever)], domestic animals (Foot and Mouth Disease, African swine fever, Venezuelan equine encephalitis, RVF), or humans (Ebola, Lassa fever, and Nipah). Other emerging zoonotic diseases need to be monitored such as bovine TB, leptospirosis, tularemia, and *Escherichia coli* 0157:H7 and other pathogenic bacteria. Theoretically, the most effective prevention and control methods are obviously to prevent the introduction or emergence of wildlife diseases of public health or animal health importance rather than attempting to control them after they have become a problem. Prevention of introduction is a daunting task because of the many sources and routes of introduction into the United States, and early detection is one of the keys to prevention and rapid control or containment. Research advances have helped to manage and mitigate some of the effects of invasive and emerging infectious diseases such as improved worldwide surveillance, improved diagnostics methods, and the development of new vaccines and antiviral agents (Kuiken et al. 2003). A number of preventative methods are in place, although many can be improved, intensified, or broadened. A variety of FADs have been identified throughout the world that could potentially be introduced, and information is available and has been obtained on many diseases; however, a more systematic and thorough collection of detailed and comprehensive information on the natural hosts, vectors, and disease manifestations in native hosts and in potential hosts in the United States needs to be completed. To this end, experimental studies of susceptibility of native wildlife species to FADs could be conducted before the potential disease invasion and provide valuable planning information to improve detection and surveillance (Tesh et al. 2004). There are many information sources that can be used, such as

World Organization for Animal Health (Organization International Epizootics, OIE)
World Health Organization (WHO)
Pan American Health Organization (PAHO)
Food and Agriculture Organization (FAO)

Individual governments and their agencies such as

United States Department of Agriculture (USDA)
Centers for Disease Control and Prevention (CDC)
Department of Homeland Security (DHS)
Department of Defense (DOD)
Department of Interior (DOI) in the United States
Universities
Nongovernment organizations

Potential routes of FAD entry into the United States must be carefully analyzed and all possible scenarios examined. This information is needed to identify gaps in regulatory authority for specific wildlife or vector species and holes in observations and testing procedures during importation and quarantine that could miss another wildlife disease entry like monkeypox. Certain aspects of the importation and movement of exotic animals into and within the United States are under the jurisdiction and regulation of different federal and state agencies. Increased coordination and communication among agencies would improve chances of detecting an invasive disease. Changes in restrictions on importation of high-risk animals or products and increases in quarantine time periods and species to be quarantined are available and are used when necessary. However, screening of incoming wild animals and animal products at ports of entry for potential infectious diseases are inadequate. Technological advances have made rapid, sensitive, and accurate detection equipment and procedures available (Kuiken et al. 2003), but they must be deployed to test high-risk wildlife species at first entry for specific diseases.

In addition to the established ports of entry to prevent introduction, a nationwide passive and active surveillance system of wildlife diseases and nationwide information network for disease reporting and evaluation are needed for early detection and reporting of those wildlife diseases of concern that successfully became established before we are surprised by a disease outbreak in wildlife, humans, or domestic animals. Beyond the initial detection of a disease pathogen, a plan to integrate nationwide resources for a rapid and adequate response to contain, mitigate, and control high-risk disease outbreaks in wildlife is essential; for example, a discovery-to-control continuum process (Murphy 1998). The National Strategy for Pandemic Influenza plan for the early detection of HPAI H5N1 virus in the United States and for the implementation of a nationwide response is a good example of being prepared for the potential invasion of a high impact disease.

The real challenge is the management and elimination of the diseases once they have been introduced or emerged. The immediate containment and eradication of the WNV introduction into New York City, United States, in 1999 would have saved thousands of human and equine cases, millions of birds that died, and hundreds of millions of dollars to deal with this exotic disease that got a foothold in a small area and subsequently spread throughout the western hemisphere. The transmission cycles of most zoonotic, wildlife diseases are known and include disease agents and hosts for diseases transmitted directly between hosts like rabies and include a vector(s) and possibly multiple hosts for diseases transmitted indirectly between hosts such as WNV. The components of transmission cycles that are theoretically the most critical and vulnerable to manipulation (weakest links) are the ones targeted for intervention to interrupt or stop transmission. Control of direct transmitted diseases like wildlife rabies is focused on reducing contact between an infected and a susceptible animal by either population suppression to reduce the probability of contact or vaccination of susceptible animals. Vaccinated animals are dead-ends for the virus and thus limit transmission and allow the disease to burn itself out (Slate et al. 2005). The weakest link in most vector-transmitted diseases is the vector and not the vertebrate host and vector control is accomplished best through an integrated pest management approach in community wide programs (CDC 2000). Vector-transmitted diseases of mammals such as LD are easier to attack, because the vertebrate host species and tick vectors are relatively sedentary and thus transmission is more predictable. Mosquito-transmitted viral diseases of birds such as WNV are more difficult to predict where virus transmission is or will be occurring, because the avian host species are not sedentary and the diseases are spatially and temporally dynamic. This uncertainty makes it problematical in controlling mosquito populations to reduce transmission; therefore, disease control methods utilize chemical and biological compounds to reduce larval production through early control in historical problem-breeding sites and population reduction of adult mosquitoes over larger areas when active transmission is elevated (Moore et al. 1993). The availability and use of equine vaccines have greatly reduced the number of equine cases from some high impact, mosquito-borne diseases such as WNV and eastern equine encephalitis (USDA 2006b) although cases still occur in areas where the vaccines are not used regularly or effectively.

REFERENCES

Antia, R., .R. R. Regoes, J. C. Koella, and C. T. Bergstrom. 2003. The role of evolution in the emergence of infectious diseases. *Nature* 426:658.

Bernard, K. A., J. G. Maffei, S. A. Jones, E. B. Kauffman, G. D. Ebel, A. P. Dupuis II, K. A. Ngo, D. C. Nicholas, D. M. Young, P.- Y. Shi, V. L. Kulasekera, M. Eidson, D. J. White, W. B. Stone, NY State West Nile Virus Surveillance Team, and L. D. Kramer. 2001. West Nile virus infection in birds and mosquitoes, New York State, 2000. *Emerg. Infect. Dis.* 7:679.

Bigler, W. J., R. G. McLean, and H. A. Trevino. 1973. Epizootiologic aspects of raccoon rabies in Florida. *Am. J. Epidemiol.* 98:326.

Burgdorfer, W., A. G. Barbour, S. F. Hayes, J. L. Benach, E. Grunwaldt, and J. P. Davis. 1982. Lyme disease — a tick-borne spirochetosis?, *Science* 216:1317.

Burke, D. S., and C. J. Leake. 1988. Japanese encephalitis, in *The Arboviruses: Epidemiology and Ecology*, T. P. Monath (ed.), vol. 3. Boca Raton: CRC Press, p. 63.

CDC, Centers for Disease Control and Prevention. 1999a. Outbreak of West Nile-like viral encephalitis — New York, 1999. *Morb. Mortal. Wkly Rep.* 48:845.

CDC, Centers for Disease Control and Prevention. 1999b. Outbreak of Hendra-like virus — Malaysia and Singapore, 1998–1999. *Morb. Mortal. Wkly Rep.* 48:265.

CDC, Centers for Disease Control and Prevention. 2000. Guidelines for Surveillance, Prevention, and Control of West Nile virus infection — United States. *Morb. Mortal. Wkly Rep.* 49:25.

CDC, Centers for Disease Control and Prevention. 2002. Lyme disease: United States. *Morb. Mortal. Wkly Rep.* 51:29.

CDC, Centers for Disease Control and Prevention. 2003a. Multistate outbreak of monkeypox — Illinois, Indiana, and Wisconsin, 2003. *Morb. Mortal. Wkly Rep.* 52:537

CDC, Centers for Disease Control and Prevention. 2003b. Update: multistate outbreak of monkeypox — Illinois, Indiana, Kansas, Missouri, Ohio, and Wisconsin, 2003. *Morb. Mort. Wkly Rep.* 52:616.

CDC, Centers for Disease Control and Prevention. 2003c. Update: multistate outbreak of monkeypox — Illinois, Indiana, Kansas, Missouri, Ohio, and Wisconsin, 2003. *Morb. Mort. Wkly Rep.* 52:642.

CDC, Centers for Disease Control and Prevention. 2006. Prevent rodent infestations, All about Hantavirus, http://www.cdc.gov/rodents/prevent_rodents/index.htm.

Chua, K. B., K. J. Goh, K. T. Wong, A. Kamarulzaman, P. S. Tan, T. G. Ksiazek, S. R. Zaki, G. Paul, S. K. Lam, and C. T. Tan. 1999. Fatal encephalitis due to Nipah virus among pig-farmers in Malaysia. *Lancet* 354:1257.

Chua, K. B., B. H. Chua, and C. W. Wang. 2002. Anthropogenic deforestation, El Nino and the emergence of Nipah virus in Malaysia. *Malays. J. Pathol.* 24:15.

Cleaveland, S., M. K. Laurenson, and L. H. Taylor. 2001. Diseases of humans and their domestic mammals: pathogen characteristics, host range and the risk of emergence. *Trans. R. Soc. Lond. B. Biol. Sci.* 356:991.

Daszak, P., A. A. Cunningham, and A. D. Hyatt. 2000. Emerging infectious diseases of wildlife — threats to biodiversity and human health. *Science* 287:443.

Davies, F. G., K. J. Linthicum, and A. D. James. 1985. Rainfall and epizootic Rift Valley fever. *Bull. World Health Organ.* 63:941.

Davis, J. P., W. L. Schell, T. E. Amundson, M. S. Godsey, A. Spielman, W. Burgdorfer, A. G. Barbour, M. LaVenture, and R. A. Kaslow. 1984. Lyme disease in Wisconsin: epidemiologic, clinical, serologic, and entomologic findings. *Yale J. Biol. Med.* 57:685.

de Jong, J. C., E. C. Class, A. D. Osterhaus, R. G. Webster, and W. L. Lim. 1997. A pandemic warning? *Nature* 389:554.

DHHS, Department of Health and Human Services. 2003. Control of Communicable Diseases; Restrictions on African Rodents, Prairie Dogs, and Certain Other Animals, Doc 03-2755. Federal Register 68:62353.

Eidson, M., N. Komar, F. Sorhage, R. Nelson, T. Talbot, F. Mostashari, R. McLean, and West Nile Virus Avian Mortality Surveillance Group. 2001a. Crow deaths as a sentinel surveillance system for West Nile virus in the North-eastern United States, 1999. *Emerg. Infect. Dis.* 7:615.

Eidson, M., L. Kramer, W. Stone, Y. Hagiwara, K. Schmit, and West Nile Virus Avian Mortality Surveillance Group. 2001b. Dead bird surveillance as an early warning system for West Nile virus. *Emerg. Infect. Dis.* 7:631.

Endy, T. P., and A. Nisalak. 2002. Japanese Encephalitis Virus: Ecology and epidemiology, in *Japanese encephalitis and West Nile viruses*. Current Topics in Microbiology and Immunology, J. S. Mackenzie, A. D. T. Barrett, and V. Deubel (eds) Vol. 267. Berlin: Springer-Verlag, p. 11.

Epstein, J. H., H. E. Field, S. Luby, J. R. Pulliam, and P. Daszak. 2006. Nipah virus: impact, origins, and causes of emergence. *Curr. Infect. Dis. Rep.* 8:59.

Farnon, E. 2006. Summary of West Nile Virus Activity, United States 2005. *Seventh National Conference on West Nile Virus in the United States.* Centers for Disease Control and Prevention, http://www.cdc.gov/ncidod/dvbid/westnile/conf/February_2006.htm.

Fauci, A. S., N. A. Touchette, and G. K. Folkers. 2005. Emerging infectious diseases: A 10-year perspective from the National Institute of Allergy and Infectious Diseases. *Emerg. Infect. Dis.* 11:519.

Fearneyhough, M. G., P. J. Wilson, K. A. Clark, D. R. Smith, D. H. Johnston, B. N. Hicks, and G. M. Moore. 1998. Results of an oral rabies vaccination program for coyotes. *J. Am. Vet. Med. Assoc.* 212:498.

Frampton, J. W., S. Lanser, and C. R. Nichols. 1995. Sin Nombre virus infection in 1959. *Lancet* 346:781.

Friend, M., R. G. McLean, and F. J. Dein. 2001. Disease emergence in birds: challenges for the twenty-first century. *Auk* 118:290.

Gibbs, E. P. J. 2005. Emerging zoonotic epidemics in the interconnected global community. *Vet. Rec.* 157:673.

Gora, D., T. Yaya, T. Jocelyn, F. Didier, D. Maoulouth, S. Amadou, T. D. Ruel, and J. P. Gonzalez. 2000. The potential role of rodents in the enzootic cycle of Rift Valley fever in Senegal. *Microbes Infect.* 2:343.

Guarner, J., B. J. Johnson, C. D. Paddock, W. J. Shieh, C. S. Goldsmith, M. G. Reynolds, I. K. Damon, R. L. Regnery, S. R. Zaki: Veterinary Monkeypox Virus Working Group. 2004. Monkeypox transmission and pathogenesis in prairie dogs. *Emerg. Infec. Dis.* 10:426.

Hanlon, C. A., M. Niezgoda, A. N. Hamir, C. Schumacher, H. Koprowski, and C. E. Rupprecht. 1998. First North American field release of a vaccinia-rabies glycoprotein recombinant virus. *J. Wildl. Dis.* 34:228.

Hanna, J. N., S. A. Ritchie, D. A. Phillips, J. Shield, M. C. Bailey, J. S. Mackenzie, M. Poidinger, B. J. McCall, and P. J. Mills. 1996. An outbreak of Japanese encephalitis in the Torres Strait, Australia, 1995. *Med. J. Aust.* 165:256.

Hjelle, B., and G. E. Glass. 2000. Outbreak of hantavirus infection in the four corners region of the United States in the wake of the 1997–1998 El Nino — southern oscillation. *J. Infect. Dis.* 181:1569.

Huhn, G. D., A. M. Bauer, K. Yorita, M. B. Graham, J. Sejvar, A. Likos, I. K. Damon, M. G. Reynolds, and M. J. Kuehnert. 2005. Clinical characteristics of human monkeypox, and risk factors for severe disease. *Clin. Infect. Dis.* 41:1742.

Johara, M. Y., H. Field, A. M. Rashdi, C. Morrissy, B. van der Heide, P. Rota, A. bin Adzhar, J. White, P. Daniels, A. Jamaluddin, and T. Ksiazek. 2001. Nipah virus infection in bats (order Chiroptera) in peninsular Malaysia. *Emerg. Infect. Dis.* 7:439.

Kappus, K. D., W. J. Bigler, R. G. McLean, and H. A. Trevino. 1970. The raccoon, an emerging rabies host. *J. Wildl. Dis.* 6:507.

Karesh, W. B., R. A. Cook, E. L. Bennett, and J. Newcomb. 2005. Wildlife trade and global disease emergence. *Emerg. Infect. Dis.* 11:1000.

Khodakevich, L., Z. Jezek, and K. Kinzanzka. 1986. Isolation of monkeypox virus from wild squirrel infected in nature. *Lancet* 1:98.

Khodakevich, L., Z. Jezek, and D. Messinger. 1988. Monkeypox virus: ecology and public health significance. *Bull. World. Health Organ.* 66:747.

Kile, J. C., A. T. Fleischauer, B. Beard, M. J. Kuehnert, R. S. Kanwal, P. Pontones, H. J. Messersmith, R. Teclaw, K. L. Karem, Z. H. Braden, I. Damon, A. S. Khan, and M. Fischer. 2005. Transmission of monkeypox among persons exposed to infected prairie dogs in Indiana in 2003. *Arch. Pediatr. Adolesc. Med.* 159:1022.

Krauss, S., D. Walker, S. P. Pryor, L. Niles, L. Chenghong, V. W. Hinshaw, and R. G. Webster. 2004. Influenza A viruses of migrating wild aquatic birds in North America. *Vector-Borne Zoonotic Dis.* 4:177.

Krebs, J. W., S. C. Long-Martin, and J. E. Childs. 1998. Causes, costs, and estimates of rabies postexposure prophylaxis treatments in the United States. *J. Public Health Manage. Pract.* 4:56.

Ksiazek, T. G., D. Erdman, C. S. Goldsmith, S. R. Zaki, T. Peret, S. Emery, S. Tong, C. Urbani, J. A. Comer, W. Lim, P. E. Rollin, S. F. Dowell, A. E. Ling, C. D. Humphrey, W. J. Shieh, J. Guamer, C. D. Paddock, P. Rota, B. Fields, J. DeRisi, J. Y. Yang, N. Cox, J. M. Hughes, J. W. LeDuc, W. J. Bellini, L. J. Anderson; SARS Working Group. 2003. A novel coronavirus associated with severe acute respiratory syndrome. *N. Engl. J. Med.* 348:1967.

Kuiken, T., R. Fouchier, G. Rimelzwann, and A. Osterhaus. 2003. Emerging viral infectious in a rapidly changing world. *Curr. Opin. Biotech.* 14:641.

Lam, S. K. 2003. Nipah virus — a potential agent of bioterrorism? *Antiviral Res.* 57:113.

Lanciotti, R. S., J. T. Roehrig, V. Deubel, J. Smith, M. Parker, K. Steele, B. Crise, K. E. Volpe, M. B. Crabtree, J. H. Scherret, R. A. Hall, J. S. MacKenzie, C. B. Cropp, B. Panigrahy, E. Ostlund, B. Schmitt, M. Malkinson, C. Banet, J. Weissman, N. Komar, H. M. Savage, W. Stone, T. McNamara, and D. J. Gubler. 1999. Origin of the West Nile virus responsible for an outbreak of encephalitis in the northeastern United States. *Science* 286:2333.

Lane, R. S., J. Piesman, and W. Burgdorfer. 1991. Lyme borreliosis: relation of its causative agents to its vector and hosts in North America and Europe. *Annu. Rev. Entomol.* 36:587.

Lederberg, J. 1998. Emerging infections: An evolutionary perspective. *Emerg. Infect. Dis.* 4:366.

Liu, J., H. Xiao, F. Lei, Q. Zhu, K. Qin, X. W. Zhang, X. L. Zhang, D. Zhao, G. Wang, Y. Feng, J. Ma, W. Liu, J. Wang, and G. F. Gao. 2005. Highly pathogenic H5N1 influenza virus infection in migratory birds. *Science* 309:1206.

Lutz, W., and R. Qiang. 2002. Determinants of human population growth. *Philos. Trans. R. Soc. Lond. B. Biol. Sci.* 357:1197.

Madani, T. A., Y. Y. Al-Mazrou, M. H. Al-Jeffri, A. A. Mishkhas, A. M. Al-Rabeah, A. M. Tukistani, M. O. Al-Sayed, A. A. Abodahish, A. S. Khan, T. G. Ksiazek, and O. Shobokshi. 2003. Rift Valley fever epidemic in Saudi Arabia: epidemiological, clinical, and laboratory characteristics. *Clin. Infect. Dis.* 37:1084.

Marfin A. A., L. R. Peterson, M. Eidson, J. Miller, J. Hadler, C. Farello, B. Werner, G. L. Campbell, M. Layton, P. Smith, E. Bresnitz, M. Cartter, J. Scaletta, G. Obiri, M. Bunning, R. C. Craven, J. T. Roehrig, K. G. Julian, S. R. Hinten, D. J. Gubler, and the ArboNET Cooperative Surveillance Group. 2001. Widespread West Nile Virus Activity, Eastern United States, 2000. *Emerg. Infect. Dis.* 7:730.

Mattar, S., E. Edwards, J. Laguado, M. Gonzalez, J. Alvarez, and N. Komar. 2005. West Nile virus antibodies in Columbian horses. *Emerg. Infect. Dis.* 11:1497.

McLean, R. G. 1970. Wildlife rabies in the United States: recent history and current concepts. *J. Wildl. Dis.* 6:229.

McLean, R. G. 1971. Raccoon rabies in the southeastern United States. *J. Infect. Dis.* 123:680.

McLean, R. G., S. R. Ubico, C. A. Hughes, S. M. Engstrom, and S. M. Johnson. 1993. Isolation and characterization of *Borrelia burgdorferi* from blood of a bird captured in the Saint Croix River Valley. *J. Clin. Microbiol.* 31:2038.

Moore, C. G., R. G. McLean, C. J. Mitchell, R. S. Nasci, T. F. Tsai, C. H. Calisher, A. A. Marfin, P. S. Moore, and D. J. Gubler. 1993. *Guidelines for arbovirus surveillance programs in the United States.* Fort Collins: Centers for Disease Control and Prevention, 83 pp.

Morse, S. S. 1995. Factors in the emergence of infectious diseases. *Emerg. Infect. Dis.* 1:7.

Murphy, F. A. 1998. Emerging Zoonoses. *Emerg. Infect. Dis.* 4:429.

Nalca, A., A. W. Rimoin, S. Bavari, and C. S. Whitehouse. 2005. Reemergence of Monkeypox: prevalence, diagnostics, and countermeasures. *Clin. Infect. Dis.* 41:1765.

Nettles, W. F., J. H. Shaddock, R. K. Sikes, and C. R. Reyes. 1979. Rabies in translocated raccoons. *Am. J. Public Health* 69:601.

Nichol, S. T., C. F. Spiropoulou, S. Morzunov, P. E. Rollin, T. G. Ksiazek, H. Feldmann, A. Sanchez, J. Childs, S. Zaki, and C. J. Peters. 1993. Genetic identification of a hantavirus associated with an outbreak of acute respiratory illness. *Science* 262:914.

Nol, P., T. E. Rocke, K. Gross, and T. M. Yuill. 2004. Prevalence of neurotoxic Clostridium botulinum type C in the gastrointestinal tracts of tilapia (Oreochromis mossambicus) in the Salton Sea. *J. Wildl. Dis.* 40414.

Olsen, B., V. J. Munster, A. Wallensten, J. Waldenstrom, A. D. Osterhaus, and R. A. Fouchier. 2006. Global patterns of influenza A virus in wild birds. *Science* 312:384.

Schmaljohn, C., and B. Hjelle. 1997. Hantaviruses: a global disease problem. *Emerg. Infect. Dis.* 3:95.

Sejvar, J. J., Y. Chowdary, M. Schomogyi, J. Stevens, J. Patel, K. Karem, M. Fischer, M. J. Kuehnert, S. R. Zaki, C. D. Paddock, J. Guamer, W. J. Shieh, J. L. Patton, N. Bernard, Y. Li, V. S. Olson, R. L. Kline, V. N. Loparev, D. S. Schmid, B. Beard, R. R. Regnery, and I. K. Damon. 2004. Human Monkeypox infection: a family cluster in the Midwestern United States. *J. Infect. Dis.* 190:1833.

Slate, D., C. E. Rupprecht, J. A. Rooney, D. Donovan, D. H. Lein, and R. B. Chipman. 2005. Status of oral rabies vaccination in wild carnivores in the United States. *Virus Res.* 111:68.

Slingenbergh, J., M. Gilbert, K. I. de Balogh, and W. Wint. 2004. Ecological sources of zoonotic diseases. *Rev. Sci. Tech. Off. Int. Epiz.* 23:467.

Smith, J. S., J. W. Sumner, L. F. Roumillat, G. M. Baer, and W. G. Winkler. 1984. Antigenic characteristics of isolates associated with a new epizootic of raccoon rabies in the United States *J. Infect. Dis.* 149:769.

Soman, R. S., F. M. Rodrigues, S. N. Guttikar, and P. Y. Guru. 1977. Experimental viraemia and transmission of Japanese encephalitis virus by mosquitoes and ardeid birds. *Ind. J. Med. Res.* 66:709.

Steere, A. C. 2001. Lyme disease. *N. Engl. J. Med.* 345:115.

Steere, A. C., S. E. Malawista, D. R. Snydman, R. E. Shope, W. A. Andiman, M. R. Ross, and F. M. Steele. 1977. Lyme arthritis: an epidemic of oligoarticular arthritis in children and adults in three Connecticut communities. *Arthritis Rheum.* 20:7.

Sturm-Ramirez, K. M., T. Ellis, B. Bousfield, L. Bissett, K. Dyrting, J. E. Rehg, L. Poon, Y. Guan, M. Peiris, and R. B. Webster. 2004. Reemerging H5N1 influenza viruses in Hong Kong in 2002 are highly pathogenic to ducks. *J. Virol.* 78:4892.

Tesh, R. B., A. P. Travassos da Rosa, H. Guzman, T. P. Araujo, and S. Y. Xiao. 2002. Immunization with heterologous flaviviruses protective against fatal West Nile encephalitis. *Emerg. Infect. Dis.* 8:245.

Tesh R. B., D. M. Watts, E. Sbrana, M. Siirin, V. L. Popov, and S. Y. Xiao. 2004. Experimental infection of ground squirrels (*Spermophilus tridecemlineatus*) with monkeypox virus. *Emerg. Infect. Dis.* 10:1563.

WHO, World Health Organization website. 2006. Avian Influenza, http://www.who.int/csr/disease/avian_influenza.html

USDA, United States Department of Agriculture. 2006a. Wild Bird Plan: An early detection system for highly pathogenic H5N1 avian influenza in wild migratory birds: U.S. Interagency Strategic Plan. News Release No. 0094.06. Web: http://www.usda.gov

USDA, United States Department of Agriculture. 2006b. West Nile virus, http:// www.aphis.usda.gov/lpa/pubs/fsheet_faq_notice/fs_ahwnv.html

Webster, R. G., W. J. Bean, O. T. Gorman, T. M. Chambers, and Y. Kawaoka. 1992. Evolution and ecology of influenza A viruses, *Microbiol. Rev.* 56:152.

Webster, R. G., M. Peiris, H. Chen, and Y. Guan. 2006. H5H1 outbreaks and enzootic influenza. *J. Infect. Dis.* 12:3.

Wilson M. E., Travel and the emergence of infectious diseases. 1995. *Emerg. Infect. Dis.* 1:39.

Wilson, M. L. 1994. Rift Valley fever ecology and the epidemiology of disease emergence. *Ann. N. Y. Acad. Sci.* 740:169.

Woods C. W., A. M. Karpati, T. Grein, N. McCarthy, P. Gaturuku, E. Muchiri, L. Dunster, A. Henderson, A. S. Khan, R. Swanepoel, I. Bonmarin, L. Martin, P. Mann, B. L. Smoak, M. Ryan, T. G. Ksiazck, R. R. Arthur, A. Ndikuyeze, N. N. Agata, C. J. Peters: World Health Organization Hemorrhagic Fever Task Force. 2002. An Outbreak of Rift Valley Fever in Northeastern Kenya, 1997–98. *Emerg. Infect. Dis.* 8:138.

Woolhouse, M. E. J., and S. Gowtage-Sequeria. 2005. Host range and emerging and re-emerging pathogens, *Emerg. Infect. Dis.* 11:1842.

Zeitz, P. S., J. C. Butler, J. E. Cheek, M. C. Samuel, J. E. Childs, L. A. Shands, R. E. Voorhees, J. Sarisky, P. E. Rollin, T. G. Ksiazek, L. Chapman, S. E. Reef, K. K. Komatsu, C. Dalton, J. W. Krebs, G. O. Maupin, K. Gage, C. M. Sewell, R. F. Brelman, and C. J. Peters. 1995. A case–control study of hantavirus pulmonary syndrome during an outbreak in the southwestern United States. *J. Infect. Dis.* 171:864.

16 Wildlife Disease Management: An Insurmountable Challenge?

Scott E. Henke, Alan M. Fedynich, and Tyler A. Campbell

CONTENTS

Wildlife biologists are increasingly being thrust onto the frontlines of wildlife disease management and surveillance in the United States. For example, biologists now routinely engage in oral vaccination of wildlife [e.g., Oral Rabies Vaccination Programs (ORVPs)] and collect specimens for both diagnostic purposes and disease surveillance or monitoring (e.g., avian influenza in migratory birds). For many, these responsibilities are novel compared to the more traditional roles of population estimation, habitat manipulation, and formulation of harvest recommendations. Wildlife disease investigation and management, in reality, are in their infancy compared with disciplines of human and domestic animal disease management. The recent focus on wildlife diseases is attributable to (1) the emergence of zoonoses that have a clear wildlife component (e.g., Lyme disease), (2) the recognition that wildlife can serve as reservoirs for diseases important to domestic animals (e.g., pseudorabies), (3) the increase in game farming and the associated risk of disease transmission to free-living wildlife (e.g., chronic wasting disease), (4) recognition of risks associated with the translocation of wildlife, and (5) intensified management for species at risk of extinction (Wobeser 2002). Clearly, interest in wildlife diseases and their management will continue to expand as humans, livestock, and wildlife come into closer contact as a direct or indirect result of the burgeoning human population.

In the context of *"Linking Ecological Theory and Management Applications,"* we believe it appropriate and timely to examine historic and current wildlife disease management activities aimed at preventing, controlling, or eradicating a particular disease-causing agent. Although some of these management activities have been highly successful at achieving their objectives, others have been replete with challenges. Nevertheless, we do not believe wildlife disease management is an insurmountable challenge. Successful wildlife disease management is possible when scientists (e.g., ecologists, biologists, modelers, pathologists, virologists, and toxicologists) and practitioners (e.g., managers, agriculturalists, and veterinarians) participate together in formulating and implementing management plans with clearly stated goals and objectives (Wobeser 2006), and with the benefit of sufficient resources. We conclude this chapter with how we think the profession of wildlife disease management is evolving.

DISEASE MANAGEMENT: PRE-1900

Generally, few substantive regulations were passed during the eras of wildlife abundance (1600–1849) and overexploitation (1850–99) aimed at conserving wildlife (Taber and Payne 2003), and notions of wildlife disease management were largely nonexistent. Early concepts pertaining to infectious disease agents were not yet developed, primarily because of uncertainty of what actually caused disease. Not until the invention of a sufficiently powerful microscope in 1674 by Anton van Leeuwenhoek were scientists able to observe organisms as small as protozoans. The ability to observe microbes and microscopically examine diseased tissues dramatically advanced concepts pertaining to human disease agents and provided the foundation for advancing science-based human disease management and epidemiological theory. Advances in understanding human diseases provided the foundation from which domestic animal and wildlife disease management would emerge. This allowed for progress to be made in describing and classifying disease agents, assigning clinical signs to the disease, developing concepts pertaining to disease agent transmission, and characterizing mortality events within the context of causative agents.

During the infancy of domestic animal and wildlife disease management, scientists and practitioners began to recognize the ability and need to manage several diseases important to livestock. In 1884, the Bureau of Animal Industry (BAI) was established specifically to eradicate the infectious cattle disease, contagious bovine pleuropneumonia (CBPP) (Walton 2000). The sweeping success of the CBPP eradication program (i.e., accomplished in only 8 years through quarantine of contact animals, slaughter of infected animals, and disinfection of infected premises) ensured the continuation of this new, yet important Bureau within the United States Department of Agriculture (USDA). The BAI was the forerunner of the USDA Animal and Plant Health Inspection Service (APHIS), the agency currently charged with "Protecting America's Agriculture," including wildlife damage and disease-related issues. Though not the intended purpose, many USDA APHIS activities in managing selected livestock diseases (Table 16.1) altered the dynamics of diseases as they relate to wildlife hosts as well.

DISEASE MANAGEMENT: 1900–1949

Wildlife management during the first half of the twentieth century was characterized by eras of protection (1900–1929) and game management (1930–65) (Taber and Payne 2003). Aldo Leopold's (1933) landmark text *Game Management* shaped wildlife management theory and practices through the later years of this period and into the present. Concurrently, scientists began foundational investigations involving diseases of wildlife. Some of these early studies included determining the vulnerability of small mammals to plague (McCoy 1911), describing tularemia in rodents (McCoy and Chapin 1912), and characterizing avian botulism in waterfowl (Kalmbach and Gunderson 1934).

Establishment of The Wildlife Society and its affiliated scientific publication *Journal of Wildlife Management* in 1937 marked the beginning of formalized study in wildlife management. The importance of wildlife diseases was recognized and emphasized from the journal's inception. For example, in Volume 1 there was a call for cooperation between parasitologists and wildlife biologists (Van Cleave 1937); Volume 2 included articles concerning leucocytozoonosis in ruffed grouse

TABLE 16.1
Partial List of Events, Occurrences, Milestones, or Accomplishments of USDA's Animal and Plant Health Inspection Service (APHIS)

Disease	Year	Event, occurrence, milestone, or accomplishment
Classical swine fever (CSF)	1833	First reported in southern Ohio and along the Wabash River in Indiana
	1903	Discovered that CSF is caused by a virus
	1906	Serum-virus method of immunizing swine developed
	1913	First license issued for production of anti-CSF serum
	1951	First use of modified live virus vaccines for CSF
	1961	Congress authorizes CSF eradication program
	1965	All states are enrolled in a four-phase CSF eradication program
	1969	Government bans the use of modified live virus CSF vaccines
	1970	Task force approach first used to eradicate a CSF outbreak
	1972	Secretary of Agriculture declares CSF emergency, providing additional funding
	1973	First CSF-free month in more than 100 years
	1978	United States officially declared "CSF free"; program costs $140 million versus $1.12 billion estimated disease costs without a program
Foot-and-mouth disease (FMD)	1870	First known outbreak of FMD in the United States
	1914	Largest outbreak of FMD occurs in the United States; more than 3,500 livestock herds are infected
	1924	United States Army kills 22,000 deer in containing FMD outbreak
	1929	Last of 9 FMD outbreaks occurs in the United States (1870, 1880, 1884, 1902, 1908, 1914, 1924 [2], and 1929) is eradicated
	1930	Tariff Act of 1930 prohibits imports of animals or animal products from countries infected with FMD
	1979	APHIS establishes FMD "vaccine bank" with Canada and Mexico
Bovine tuberculosis (BTB)	1892	First tuberculin test of a herd was made in the United States; of the 79 animals tested, 30 were reactors
	1900	Tuberculin test is required for all imported cattle
	1917	Congress appropriates $75,000 to begin efforts against bovine TB; 5% of the nation's cattle are infected
	1936	Bovine TB reactor rate drops to new low of 1%
	1940	All states are Modified Accredited Bovine TB areas (cattle infection rate <0.5%)
	1974	Bovine TB reactor rate drops to new low of 0.03%
	1984	Bovine TB reactor rate drops to new low of 0.003%
	1991	Bovine TB recognized as a serious problem in deer and elk
Brucellosis	1905	Brucellosis discovered in imported goats
	1910	*Brucella abortus* first isolated from cattle in the United States
	1916	Committee on Contagious Abortion formed by the United States Livestock Sanitary Association to encourage control of brucellosis
	1918	Discovery of the relationship between brucellosis organisms that cause the disease in cattle, swine, and goats
	1930	Discovery of Strain 19 vaccine for brucellosis
	1934	Efforts against bovine brucellosis begin as part of a "cattle reduction plan" caused by severe drought conditions
	1936	National brucellosis herd infection rate estimated at 14–15%
	1940	Vaccination with Strain 19 becomes part of the brucellosis control program
	1942	North Carolina becomes the first Modified-Certified Brucellosis Area (infection in <5% of herds and 1% of cattle)
	1947	First Uniform Methods and Rules (UM&R) are adopted for brucellosis program
	1952	Brucellosis milk ring test adopted as a surveillance tool in dairy herds

Continued

TABLE 16.1
Continued

Disease	Year	Event, occurrence, milestone, or accomplishment
Brucellosis (Continued)	1954	Accelerated brucellosis eradication program begins; an estimated 124,000 herds are infected nationwide
	1960	Market cattle testing program adopted as a surveillance tool for brucellosis in beef cattle
	1961	Swine brucellosis eradication program begins
	1978	Brucellosis Technical Commission reports its findings that "control leading to eradication is biologically feasible"
	1979	New standards adopted for brucellosis eradication program
	1982	New state classifications become effective for brucellosis eradication program
	1988	*Ad hoc* committee on brucellosis in Yellowstone National Park is formed
	1989	APHIS begins "Rapid Completion Plan" to finish its brucellosis eradication program
	1990	Number of cattle herds quarantined for brucellosis drops below 1,000 for the first time
	1992	Number of cattle herds quarantined for brucellosis drops below 500 for the first time
	1995	Number of cattle herds quarantined for brucellosis drops below 100 for the first time
Screwworm	1938	The theory of using laboratory-reared sterilized insects to control and eradicate pest populations is conceived
	1951	Procedures developed for sterilizing screwworms, enabling control of this pest and others with sterile insect technology
	1954	Screwworms successfully eradicated from the Island of Curacao after 4 months using the sterile insect technique
	1958	Screwworm eradication begins in the southeastern United States using the sterile insect technique; eradication is accomplished in 1959 and a quarantine line is established at the Mississippi River to prevent reinfestation
	1962	Screwworm eradication program begins in the southwestern United States
	1966	Overwintering screwworm populations are eradicated in the United States
	1981	APHIS closes screwworm production plant in Mission, Texas
Psuedorabies virus (PRV)	1983	Five states (Iowa, Illinois, North Carolina, Pennsylvania, and Wisconsin) partake in pilot projects against PRV
	1986	First genetically engineered vaccine is licensed by APHIS (for PRV in swine)
	1989	APHIS adopts standards for 5-stage State-Federal-industry program to eradicate PRV in swine
	1995	All states in stage 2 or higher; 14 states have "free" status
Rabies	1991	Field trial successfully completed on Parramore Island, Virginia, for the first genetically engineered oral raccoon rabies vaccine
	1995	APHIS launches campaign against coyote rabies in Texas using genetically engineered oral vaccine

Note: Includes predecessor organizations in managing selected livestock diseases important to wildlife from 1833 to 1995 (modified from http://permanent.access.gpo.gov/lps3025/history.html)

(*Bonasa umbellus*) (Clarke 1938), waterfowl parasites (Gower 1938), and coccidiosis in muskrats (*Ondatra zibethica*) (Shillinger 1938); and Volume 3 contains papers involving blood parasites of deer (Dougherty 1939; Whitlock 1939a), *Echinococcus* infections in wildlife (Riley 1939), larval tapeworms in cottontail rabbits (*Sylvilagus floridanus*) (Whitlock 1939b), *Plasmodium* infections in wild birds (Wetmore 1939), and a description of "a freak deer head" with cutaneous fibromas (Honess 1939). This interest in wildlife pathogens and diseases set the stage for other emerging professional organizations such as the Wildlife Disease Association, which was formed in 1951 and consists mostly of wildlife disease specialists. Volume 1 of the Bulletin of the Wildlife Disease Association was published in 1965 and became *Journal of Wildlife Diseases* in 1970.

A vast majority of the scientific studies conducted from 1900 to 1950 involving wildlife diseases were descriptive in nature, aimed at identifying the disease-causing agent, suitable hosts, and disease presence. Our understanding of most wildlife diseases had not yet advanced to the point where management recommendations could be made with a healthy degree of scientific rigor. In attempts to eradicate the large liver fluke (*Fascioloides magna*), for example, broadcast application of copper sulfate was used to kill aquatic snails, the fluke's intermediate host (Swales 1935). This management practice was successful in eradicating the liver fluke, but at the expense of beneficial snails (Pybus 1990). The unintended consequences of such disease management activities pointed to the need to develop a more thorough ecosystem-based theoretical approach to disease management.

PRESENT DISEASE MANAGEMENT

This period has seen movement away from preoccupation with large mortality events often focused upon by biologists in earlier time periods and toward adoption of a theoretical based approach to disease and disease management. Emphasis is being directed on the role that diseases actually play in wildlife populations, understanding the disease cycle to find the weakest link, and identifying the underlying mechanisms that promote stable and unstable host–pathogen systems (Anderson and May 1978; May and Anderson 1978).

There are two general or overarching strategies employed in present disease management. The first option is the "hands-off" approach, which is applicable to infectious disease agents. Its foundation is established in ecological theory, in which all organisms are components of the ecosystem and the concept that ecological systems evolve toward equilibrium states. In this view, the host and the pathogen are in a continuous battle of survival: The pathogen seeks to maintain a viable population within a susceptible host without "damaging" the host population to the extent that it causes the pathogen's own extinction. Simultaneously, selective pressures on the host result in behavioral modifications to avoid the pathogen or enhancements in immunological resistance. When the pathogen causes a significant disruption in the host population, a locally unstable situation arises until the equilibrium state is re-established, thereby permitting coexistence of host and pathogen. In situations where there are large, healthy wildlife populations, human intervention is often considered ineffective and economically costly, because an equilibrium state will establish itself naturally.

Equilibrium states that occur between pathogens and hosts have been demonstrated in several long-term field studies. The seminal study by van Riper et al. (1986) and van Riper (1991) on the impact of the introduced malarial parasite *Plasmodium relictum* on Hawaiian avifauna found that natural selection eliminated individuals that slept with their head and legs exposed to mosquito vectors. They also found that the daily bird movement patterns in some species were modified to avoid mosquito vectors. Ultimately, some avian species adapted in response to the exotic pathogen, whereas others did not and their populations declined. Thus, the host–pathogen system reached equilibrium, but host species composition and abundance was substantially different from before the introduction of the pathogen (regionally unstable equilibrium state). In another study, Pence and Windberg (1994) followed a disease cycle involving an outbreak of sarcoptic mange in a coyote (*Canis latrans*) population in southern Texas. Although there was about 80% mortality in the coyote population at the peak of the epizootic, at the end of the 10-year cycle, coyote abundance was approximately the same as prior to the epizootic, which suggested both locally and regionally stable equilibriums for this host–pathogen system.

Although the hands-off approach may apply to wildlife populations that are large and widely distributed, it does not seem applicable in three situations: (1) species that are threatened or endangered, for example, an outbreak of avian cholera in the Nebraska Rainwater Basin during peak migration of whooping cranes (*Grus americana*) would be more risky to the cranes than to waterfowl co-occurring in the same area; (2) game species that reach a new equilibrium with a pathogen in which host density is substantially lower would not be viewed favorably by hunters; and (3) host–pathogen systems that involve zoonotic potential.

If the hands-off approach is not applicable, intervention can be an option. Some diseases can be prevented or reduced. This option can be particularly effective for certain noninfectious and infectious diseases that are caused or facilitated by human activities. Development of disease transmission theory is essential for the success of this approach, as it is necessary to determine where the weakest link is in the disease cycle and focus efforts there to have the highest probability of success.

Instances in which humans have aided in exposing wildlife to disease agents are numerous, including selenium poisoning of waterfowl and shore birds in California because of irrigation runoff (Ohlendorf et al. 1988), lead poisoning in waterfowl and upland game birds from the use of lead shot for hunting (Deuel 1985), introduction of avian malaria to Hawaii (van Riper et al. 1986), and translocation of raccoons (*Procyon lotor*), and subsequently rabies, from Florida to mid-Atlantic coast states (Nettles et al. 1979). The first two examples highlight human activities — irrigation for agriculture and hunting, which resulted in environmental contamination and wildlife diseases. The latter two examples involved translocation of animals for human purposes, which inadvertently translocated the animals' disease agents (i.e., virons, bacteria, helminths, etc.). In the case of avian malaria, the mosquito vector was brought to Hawaii through ballast water that contained larvae (Warner 1968). Malarial parasites (*P. relictum*) were later introduced to Hawaii with importation of exotic birds (van Riper et al. 1982). Now, avian malaria is well established in Hawaii, and it has negatively affected the native avifauna. Raccoons were transported and released by hunting clubs who wished to increase raccoon density, but by doing so, also introduced protozoa, helminths (Schaffer et al. 1978), and rabies (Nettles et al. 1979) to the eastern United States. In addition, urban and suburban sprawl has lead to habitat loss, resulting in the crowding of wildlife into smaller tracts of land, thereby promoting disease transmission. In each case, human activities have created or exacerbated a disease scenario, and thus, it is human responsibility to circumvent wildlife disease problems.

There are three reasons why humans consider intervention wildlife disease management: (1) disease can be a health risk to humans, (2) disease can be deleterious to domestic livestock, and (3) disease can negatively affect wildlife considered beneficial to humans. It is rare for a disease management program to evolve and not to be justified by one or more of these reasons. For example, the oral rabies vaccine program, which has placed millions of vaccine-laden baits for raccoons along the eastern coast of the United States (Hanlon et al. 1989) and for coyotes and gray foxes (*Urocyon cinereoargenteus*) in Texas (Texas Department of Health 1994; Farry et al. 1998a), has a central objective to reduce the risk of rabies to humans. Without this focus on human health, it is likely, little effort and financial resources would be expended on eradicating rabies from wildlife populations.

EXAMPLES OF DISEASE MANAGEMENT TO AID LIVESTOCK INDUSTRY

The following are examples in which concepts pertaining to disease transmission theory were used to identify the weakest link and focus control efforts to break the disease cycle.

Swine Brucellosis: Swine brucellosis is a good example of disease management to aid the livestock industry. The brucellosis eradication program primarily is concerned with bovine brucellosis (*Brucella abortus*). At present, Idaho, Wyoming, and Texas are Class A (not disease-free) for *B. abortus*. For swine brucellosis, only Texas remains at Stage 2 (not disease-free). Swine brucellosis is caused by *Brucella suis*, a small Gram-negative bacterium. Infected animals may be asymptomatic, or have chronic clinical signs including abortion, fetal reabsorption, infertility in sows, orchitis (inflammation of the testes) in boars, lameness, and a high-mortality rate in piglets (Tessaro 1990; Davidson and Nettles 1997). Transmission occurs by oral and venereal routes, and the bacteria localize in lymph nodes with an incubation period from 2 weeks to several months (Davidson and Nettles 1997; Conger et al. 1999). A fully effective vaccine has not yet been developed, and there is no known cure for the disease. In the United States, brucellosis has been found in Alabama (Davidson and Nettles 1997), Arkansas (Zygmont et al. 1982), California (Sweitzer et al. 1996; Davidson and Nettles 1997), Florida (Zygmont et al. 1982; Belden 1993; van der Leek et al. 1993a; Davidson and Nettles 1997), Georgia (Hanson and Karstad 1950; Zygmont et al. 1982; Davidson and Nettles 1997), Hawaii (Davidson and Nettles 1997), Louisiana (Zygmont et al. 1982; Davidson and Nettles 1997), Oklahoma (Davidson and Nettles 1997), South Carolina (Wood et al. 1976; Zygmont et al. 1982; Davidson and Nettles 1997; Gresham et al. 2002), and Texas (Randhawa et al. 1977; Corn et al. 1986; Davidson and Nettles 1997). Prevalence in feral swine populations can range from 0 to 44% (Dees 1999). Due to the potential spread of brucellosis from feral hogs to domestic pigs, it is recommended that domestic swine facilities be double fenced to reduce the chances of direct contact between feral and domestic swine. In addition, periodic disease testing of domestic swineherds is advisable.

Pseudorabies: Pseudorabies virus (PRV, Aujeszky's disease, Mad Itch) is an alphaherpes virus (suid herpesvirus 1) that occurs in swine, but can be lethal to nonswine species that contract the virus (Kocan 1990). When infection occurs, the virus travels along peripheral sensory nerves toward neurons in ganglia, where the virus maintains its latent status until reactivated during periods of host stress (Romero et al. 2003). In swine, the disease ranges from asymptomatic to fatal in young animals, and depends on strain of the disease and age of the infected animal (Davidson and Nettles 1997). Clinical signs include fever, respiratory infection, loss of coordination, abortion, mummified fetuses, stunted growth, and high mortality in piglets less than four weeks old (Kocan 1990; Davidson and Nettles 1997). A current theory is that modes of transmission differ in feral pigs versus domestic pigs due to different ganglial sites of latency. The virus settles in the sacral (most common in feral pigs) and trigeminal ganglia (most common in domestic pigs) of the nervous system tissues, and can be isolated from the tonsil (Romero et al. 2003). In feral hogs, because the virus is in sacral ganglia, venereal transmission has the highest frequency (Romero et al. 1997, 2001, 2003), unlike that in domestic pigs where the virus is predominantly transmitted through exchange of oral and nasal fluids. However, PRV has occurred by aerosol transmission (Schoenbaum et al. 1990; Christensen et al. 1993), infected meat, and contaminated food and water (Kocan 1990; Hahn et al. 1997; Kluge et al. 1999). The wild-type of PRV, found in feral swine, appears to be attenuated, with lower pathogenicity than those found in domestic herds. Therefore, it may not manifest similar symptoms, making it difficult to recognize the virus in domestic herds (Romero et al. 1997).

Feral hogs, as a disease reservoir, can be economically significant. The U.S. pork industry is valued at $30 billion annually, employs over 600,000 people, and produces 10% of the world's pork supply (APHIS 2003; Witmer et al. 2003), giving the industry valid concern when it comes to disease management. Since 1989, the domestic pork industry has participated in a USDA-coordinated national campaign to eradicate PRV. PRV alone costs the national pork industry an estimated $40 million annually, not including loss of market opportunity internationally (NIAA: www.animalagriculture.com). The PRV program has five stages: stage I is preparation, stage II is control, stage III is mandatory cleanup of all pseudorabies-infected herds, stage IV is surveillance to verify no infection remains, and stage V status is when all herds are pseudorabies-free for 1 year or more. As of late 2004, all states, Puerto Rico, and the Virgin Islands were at stage V (NIAA: www.animalagriculture.com). The threat of reintroduction of these diseases to uninfected domestic herds by diseased feral populations has been considered in the scientific literature, and only recently has a disease management program been initiated (Wyckoff et al. 2005).

Pseudorabies appears to be well established in feral populations throughout the United States, and persists in populations through time (Gresham et al. 2002; Corn et al. 2004). Infected populations have been found in Florida (van der Leek et al. 1993a,b), Georgia (Pirtle et al. 1989), Oklahoma (Davidson and Nettles 1997), South Carolina (Wood et al. 1992; Gresham et al. 2002), Texas (Corn et al. 1986), and in 12 unlisted states (Miller 1993). Rates of infection have varied from not present to 70% (Pritle et al. 1989; van der Leek et al. 1993a; Sweitzer et al. 1996; Hahn et al. 1999; Gresham et al. 2002; Corn et al. 2004). Rates of infection seemingly depend on location, season of sampling, and age structure of the sampled population (Romero et al. 1997). Similar management recommendations as suggested for brucellosis have been offered for pseudorabies.

Screwworm Eradication: Control of the blowfly *Callitroga hominivorax*, implemented under the federal Screwworm Eradication Program, provides an example for a disease management strategy to aid wildlife considered beneficial to humans, although the initial reason for the program was to aid the livestock industry. This fly lays its eggs in homeotherms, and historically cattle and white-tailed deer (*Odocoileus virginianus*) were preferred hosts. The larvae feed on living tissue, debilitating the host. In Texas, mortality of white-tailed deer fawns from screwworm infections reached 80% before control programs were implemented (Wobeser 1994). Control of this parasitic infection was based on behavioral biology of the fly; the fly breeds only once each year. Massive numbers of irradiated, and thus sterile, adult male flies were released. Female flies mated with sterile males and viable eggs were not produced, thus reducing parasitic infection in cattle and deer. Suppression resulted for several years, but occasional outbreaks have occurred (Richardson et al. 1982). Efforts are now underway to eradicate the screwworm throughout Americas.

PRESENT STRATEGIES OF WILDLIFE DISEASE MANAGEMENT

Disease management can take three forms: prevention, control, and eradication. Most disease management programs involve components of prevention, control, and eradication.

Prevention is designed to keep a disease from entering an unaffected area. Prevention typically involves restrictions on importation or translocation of certain animals. Prevention often is the easiest and most economic method of disease management. Consequences from accidental introduction of a disease can be disastrous. Zebu cattle from India were introduced into Africa and, consequently, rinderpest swept across the continent. Mortality rates for wild ruminants exceeded 90% for some species (Henderson 1982).

Disease control is designed to reduce the frequency of disease to some tolerable level. Control implies that the disease will persist in the host population and its environment, but at a level that will produce negligible effects to humans or human interests. The ORVP is a control program to reduce the infection rate of rabies in wild animals [i.e., raccoons, coyotes, gray fox, and striped skunks (*Mephitis mephitis*)]. The ORVP involved many stages, from vaccine development (Rupprecht et al. 1988), bait development (Farry et al. 1998b; Steelman et al. 1998), baiting strategies (Farry et al. 1998a), and program assessment.

Based on disease transmission theory, control can be attempted by manipulating four basic factors: the disease agent, the host population, the environment, and human activities, or by a combination of these factors. These factors have been incorporated into various theoretical models developed by R. M. Anderson and R. M. May. Management of infectious disease agents is complicated by replication of the disease agent and by transmission to other susceptible individuals in the host population (Anderson and May 1979). Reproduction rate of a disease, R, is the average number of secondary infections caused by a single infected individual that was introduced into a completely susceptible population (Fine et al. 1982). In other words, an intrinsic reproductive rate of 8 means that, on average, each infected individual resulted in the infection of eight susceptible individuals. Anderson (1982) defined the reproductive rate (R) of a disease as the density of susceptible individuals in the population (X) divided by the threshold density for disease persistence (N_T), or $R = X/-N_T$. The proportion of susceptible individuals in a population to a disease is reduced by immunization; therefore, immunization reduces R. If $R = 1$, then the disease agent is just able to maintain itself in a population. If $R < 1$, then the disease agent cannot sustain itself and will eventually become extinct from the population. Only when $R > 1$ will a disease agent become obvious in a host population. Therefore, the objective of many disease control programs is to depress the reproductive rate of a disease agent below unity. However, a disease agent with a high rate of reproduction is more difficult to control than one with a low rate (Anderson 1982). The proportion of a population (p) that must be immunized to eradicate a disease must exceed $1 - 1/R$ (Anderson 1982). For example, for reproductive rates greater than 5, more than 80% of the susceptible population must be vaccinated to eradicate the disease. It becomes obvious that diseases with high reproductive rates are extremely difficult to manage through immunization because of the large proportion of the population that must be immunized. Consider a population with a threshold density of 1 animal/km^2 and a population density of 2, 4, and 8 animals/km^2. Under these conditions, the proportion of the population that must be immunized to eliminate a disease would be 50, 75, and 88%, respectively. Another obstacle with immunization that must be considered is the average age of a susceptible host when it is exposed to the disease. For an immunization program to be effective, animals must be immunized prior to their exposure to disease. Therefore, animals that are exposed to disease at an early age are more difficult to control by immunization.

Disease eradication involves complete elimination of a disease from an area for an indefinite period. Such programs typically are large in scale and require a large investment of time and money. When foot-and-mouth disease was accidentally introduced into California in 1923, a deer eradication program was successfully initiated (Brooksby 1968).

The most direct way to manage a disease is by manipulating the disease agent. It is often easier to manipulate a noninfectious disease agent, such as a toxicant, than an infectious disease, such as a parasite. A biological or synthesized toxin eventually degrades with time, so its effects lessen as long as no new toxin is added to the environment. However, infectious agents can replicate themselves without new additions. The pesticide DDT (dichlorodiphenyltrichloroethane) was once utilized throughout the world. DDT and its metabolites DDE (dichlorodiphenyldichloroethylene) and DDD (dichlorodiphenyldichloroethane) have high stability and persistence within the environment. This organochlorine pesticide was implicated as the causative agent for eggshell thinning of birds, and thus the reason for reproductive failure in many bird species (Spitzer et al. 1978). Once highlighted as a mortality problem, efforts were made to stop the use of DDT in the United States. Upon cessation, reproductive success of birds improved, but not as dramatically as might have been expected. This pesticide and its metabolites are still being found in tissues of birds (Mora 1995; Wainwright et al. 2001). Screwworm (previously discussed) is a parasitic infection that plagued cattle and white-tailed deer. However, manipulation of the disease agent, the blowfly *C. hominivorax*, reduced the incidence of disease in wild cervids.

Disease management through manipulation of the host population has been attempted. Dispersion of animals to manage a disease can be useful if there is no chance of spreading the disease agent to a new area. Manipulation of host populations has included culling diseased animals, test and slaughter programs, and reducing population density. Selective culling can only work if infected animals are easily identified and if the disease is slow to spread through a population. Test and slaughter programs are of limited use for wildlife because of the difficulty of capturing, holding in captive facilities, and testing all individuals of a population. Reducing host population density on a local or regional scale has been attempted, but the effort to do so is typically intensive, and the results are temporary because of animal repopulation of the control area (Henke and Bryant 1999; Henke et al. 2002). Reduction of host population density through reproductive control of the host has been suggested, but does not appear to have been attempted till date. Creation of a barrier (e.g., geographic area in which animals were vaccinated to the disease) to restrict disease spread has been attempted for striped skunks to reduce the spread of rabies in Alberta (Gunson et al. 1978), for red fox (*Vulpes vulpes*) to reduce rabies spread in Europe (Wandeler et al. 1974), and for African buffalo (*Syncerus caffer*) to reduce the spread of rinderpest in Uganda (Anonymous 1953). Depopulation has been attempted for American coots (*Fulica americana*) in Virginia to control avian cholera (Purseglove et al. 1976), for ground squirrels (*Spermophilus* sp.) in Colorado to control plague (Waltermire 1982), and for European badgers (*Meles meles*) in England to control tuberculosis (Henderson 1982).

Immunization of the host population has potential as a disease control manipulation because immunization reduces the proportion of susceptible individuals in a population, thus reducing R. Vaccines to immunize animals are generally available. The problem typically lies with how to deliver the vaccine to wild animals, and whether the methodology to do so is feasible and economical. The ORVP (previously described) for coyotes in Texas provides an example of a successful immunization program to control disease in a wild animal population (Texas Department of Health 1994).

Disease management through manipulation of the environment can be used to reduce the causative agent, the host population, populations of other species involved in the disease, and other factors involved in disease occurrence. Environmental manipulation typically does not provide quick results in disease management, but it does usually provide long-lasting results. Avian botulism in waterfowl provides a good example of habitat manipulation (Wobeser 1994). The bacterium, *Clostridium botulinum* type C, is the causative agent of avian botulism. This organism is found as a resistant spore within the soils of wetlands. Under anaerobic conditions, the bacteria produce a toxin. Waterfowl become poisoned when they consume vegetation or invertebrates containing the toxin. The bacteria use decaying animal matter, more so than decaying vegetative matter, as a substrate upon which to grow. The concept here is to reduce the amount of substrate, which reduces the number of bacterial spores, and thus reduces the level of toxin. Invertebrates that die because of changing water depths can provide suitable substrate for the bacteria, and thus the toxin. Occurrence of botulism can be reduced by maintaining consistent water depths via water control devices.

Wildlife disease management usually involves people management. Public support of a disease management program is necessary to get compliance with the program and financial support to conduct the program. Education programs for the public are aimed at acquiring this support. The public must understand the biology of the disease (i.e., risk to people from zoonoses), justification for disease management, and how the program will be conducted. Often, human activities must be altered to reduce risk of disease to humans and to wildlife. A rabies education program was initiated in southern Texas during the ORVP for coyotes when it was discovered that most of the public did not vaccinate their family pets against rabies (Kresta and Henke 2000). An education program explained how rabies was transmitted, the risk of exposure to humans, the effects of the disease, and how to reduce risk of exposure.

Lead poisoning of waterfowl illustrates another example of altering human activities. The effect of ingesting spent shotgun pellets was described by Westmore (1919) early in the history of wildlife management. Zwank et al. (1985) searching two lakes in Louisiana during 3 months found 783 of 1171 sick or dead ducks (67%) caused by lead poisoning. Other management techniques, such as hazing birds from areas of heavy pellet deposition and plowing areas in an attempt to bury lead shot deep in the soil, have been attempted (Wobeser 1994).

Use of nontoxic shot, beginning with first generation steel shot, was promoted as the solution to reduce lead poisoning in waterfowl. To force this change from lead shot to nontoxic shot, the U.S. Fish and Wildlife Service initiated regulations that systematically eliminated lead shot for hunting waterfowl beginning with hot-spots and eventually ending with a U.S.-wide ban in 1991 (USFWS 2004). However, the perception of additional federal regulation of hunting activity, coupled with poor ballistic performance of steel shot (compared with lead), and the much higher price per shell were not looked upon favorably by the waterfowl hunting community. With advances in nontoxic shot alternatives (bismuth–tin, tungsten–bronze, tungsten–iron, tungsten–matrix, tungsten–nickel–iron, tungsten–polymer, tungsten–tin–bismuth, and tungsten–tin–iron–nickel) allowed by the U.S. Fish and Wildlife Service during the 2005 waterfowl hunting season (http://www.wildlifedepartment.com/approvednontoxic.htm) more ballistically effective loads are now available for hunting waterfowl and upland game birds. The conversion from lead shot to non-toxic shot has been a slow process, and education efforts were not the most efficient. Such difficulties in education and acceptance point to the need for focusing on information transfer to stakeholders and for effective public relations personnel in state and federal wildlife agencies.

FUTURE DIRECTIONS

Clearly, the wildlife disease profession has come a long way from its earliest endeavors of monitoring and assessing disease impacts in the early part of the century to the present where intervention strategies are being adopted and implemented. Wobeser (2002) suggested that the desire to actively manage infectious disease in wild animals is a relatively recent phenomenon, compared to disease management strategies for humans and domestic animals. Therefore, with greater scientist and general public awareness of the effect of wildlife disease agents, what does wildlife disease management hold for the future? At least four areas of focus will likely emerge. These include (1) increased global cooperation in surveillance and management of wildlife diseases, (2) increased communication with the public and people management, (3) developing sophisticated tools for solving complex disease issues, and (4) continued focus on basic and applied research.

INCREASED GLOBAL COOPERATION

Based on the recent past, emerging and reemerging diseases will likely be center-stage, particularly those infectious agents that have pandemic potential and can rapidly spread among contients. For example, the foot-and-mouth disease virus was responsible for an explosive pandemic affecting Asia, Africa, and Europe during 1998–2001 (Knowles et al. 2005). West Nile virus and avian flu are other recent examples of diseases that have spread across continents rapidly. Additionally, there is the possibility of bioterrorism using militarized infectious agents and exotic zoonotic agents. Militarized zoonotic disease agents such as anthrax have recently captured the attention of the media, public, and U.S. Homeland Security Agency. Other agents such as cattle plague, foot-and-mouth disease, African swine fever, classical swine fever (hog cholera), and avian influenza have been identified as potential agents for agricultural terrorism (Committee on Confronting Terrorism in Russia 2002). Such disease agents have the potential to disrupt many globally linked economies. For example, control measures and economic sanctions imposed during the 2001 foot-and-mouth disease outbreak in the United Kingdom had a direct loss to agriculture and its associated food chain of approximately $3.1 billion (Thompson et al. 2002). Consequently, there clearly is a need for significant collaboration among global stakeholders. Governments will increasingly realize the need to form cooperative linkages to provide better disease surveillance, sharing of information, and database management. Benefits of such cooperation will include advanced disease detection, which should aid in control and prevention before agents reach epizootic or panzootic proportions. Both the United States and Canada have recognized the need for collaborative activities and incorporated them into their disease management strategies (USGS 2004; Ministry of Natural Resources 2005). However, increased knowledge from sharing information about country-specific disease agents will likely lead to increased controls on importation, exportation, and movement of wildlife among countries and, for zoonotic agents, potential restrictions on human travel.

INCREASED COMMUNICATION WITH THE PUBLIC

To implement an effective wildlife disease management plan, it is critical to inform the public. Many past problems in disease management were not necessarily related to the plan of action involving control or eradication, but to the ability to convince the public that the plan was appropriate (e.g., chronic wasting disease in Wisconsin, fowl plague in California). Such a lack of support can often be traced to lack of public awareness about the problem or its effect on wildlife, domestic animals, and humans.

Consequently, there will be an increasing need to have highly trained public relations personnel who can bridge information gaps between researchers and the general public and "sell" the disease management plan. Additionally, Daszak et al. (2000) suggested that it might become increasingly important to include the potential of wildlife disease impacts in Environmental Impact Statements so that the public is aware that disease issues are being considered.

As multimedia technology advances, new communication tools can be used. These tools may include Internet newsgroups, which provide rapid dissemination of credible information (Daszak et al. 2000), and pod-casts and other rapidly advancing technological outlets that can be used for quick and accurate information transfer by public relations personnel in state and federal agencies. Unfortunately, with such advances in communication technology, various interest groups can and will present biased information to advance their agendas. Consequently, the public could be swayed by inaccurate information. The importance of providing reliable and trustworthy information will become paramount.

Developing New Scientific Tools

Innovative and cutting-edge science will be needed to control increasingly complex disease issues. Greater emphasis is needed in developing improved modeling and quantitative tools. Development of such capability will provide insights regarding transmission and spread of diseases and aid in evaluating management actions used in specific control or eradication programs (USGS 2003). Additionally, there is a need to develop models of the economics of managing infectious wildlife diseases (Horan and Wolf 2005).

Advanced technologies, such as Geographic Information Systems (GIS) and Global Positioning Systems (GPS), which incorporate satellite imaging, weather maps, habitat and land use features, and geographic distributions of disease carriers, reservoirs, and vectors, will become increasingly important in monitoring enzootic diseases and spotting early stages of emerging diseases. Studies at the Caesar Kleberg Wildlife Research Institute are incorporating GIS and GPS technology to monitor feral pig movements across the landscape to develop better control strategies. Another study is using GIS to examine relationships between *Baylisascaris procyonis,* raccoons, and human populations. Use of such technological resources is clearly part of future efforts to manage wildlife diseases.

Emphasis on More Research

Emerging zoonotic diseases have recently been the primary focus of disease management, particularly chronic wasting disease, avian flu, and West Nile virus. There will be an increasing need to understand the dynamics of emerging diseases to develop effective management strategies (USGS 2003). Basic information will be required for monitoring and surveillance, which will provide information on which types of diseases are present, incidence of disease, patterns within susceptible populations, and associated risks to humans and livestock (Duff 2003). Additionally, more sophisticated assessments of wildlife diseases will be needed including disease dynamics, risk analysis, and development of more effective sampling techniques (USGS 2004). Focus of disease research will increasingly be on ecology, pathology, and population biology of host–pathogen systems approached from individual, population, and environmental perspectives. Thus, disease research will necessitate a multidisciplinary approach to identify causes of disease outbreaks and develop effective control measures (Daszak et al. 2000). Additionally, there is a need for studies focusing on economic costs of disease management strategies. Wobeser in his Carlton Herman Founders Fund Lecture at the 2006 Wildlife Disease Association meeting stressed the need to move from observational and descriptive studies to experimental studies in which biotic and abiotic factors are manipulated to assess impacts of disease agents on hosts under experimental scenarios. He also indicated that the large, long-term disease and mortality datasets being compiled by various state and federal agencies need to be examined using epizootiological approaches to understand population level dynamics of disease agents.

Looking to the future, it seems that epizootiologists will encounter increased complexity. However, with multinational cooperation, greater public awareness, and the development of advanced scientific tools, wildlife disease managers can continue their efforts in moving from the theoretical realm to the applied in bettering the management of wildlife populations well into the twenty-first century.

REFERENCES

Anderson, R. M. 1982. Transmission dynamics of indirectly transmitted disease agents: The vector component. In *Population Biology of Infectious Diseases*, R. M. Anderson, and R. M. May (eds). Berlin: Springer-Verlag, p. 149.

Anderson, R. M., and R. M. May. 1978. Regulation and stability of host–parasite population interactions. I. Regulatory processes. *J. Anim. Ecol.* 47:219.

Anderson, R. M., and R. M. May. 1979. Population biology of infectious disease. *Nature* 280:361.

Anonymous. 1953. Rinderpest-buffalo free zone, Uganda. *Bull. Epiz. Dis. Afr.* 1:46.

APHIS. 2003. Pseudorabies, www.aphis.usda.gov/vs/nahps/pseudorabies/q-a.html (accessed December 19, 2003).

Belden, R. C. 1993. Feral hogs: The Florida experience. In *Proceedings on Feral Swine: A Compendium for Resource Managers*, D. Rollins (ed.). Austin: Texas Animal Health Commission, p. 101.

Brooksby, J. B. 1968. Wild animals and the epizootiology of foot-and-mouth disease. *Symp. Zool. Soc. Lond.* 24:1.

Christensen, L. S., S. Mortensen, A. Botner, B. S. Strandbygaard, L. Ronsholt, C. A. Henriksen, and J. B. Andersen. 1993. Further evidence of long distance airborne transmission of Aujeszky's disease (pseudorabies) virus. *Vet. Rec.* 132:317.

Clarke, C. H. D. 1938. Organisms of a malarial type in ruffed grouse, with a description of the schizogony of *Leucocytozoon bonasae*. *J. Wildl. Manage.* 2:146.

Committee on Confronting Terrorism in Russia. 2002. High-impact terrorism: Proceedings of a Russian–American Workshop, National Academy Press. http://darwin.nap.edu/books/0309082706/html/207.html (accessed April 12, 2006).

Conger, T. H., E. Young, and R. A. Heckmann. 1999. *Brucella suis* in feral swine. *Proc. First Nat. Feral Swine Conf.* 1:98.

Corn, J. L., P. K. Swideren, B. O. Blackburn, G. A. Erickson, A. B. Thierman, and V. F. Nettles. 1986. Survey of selected disease in wild swine in Texas. *J. Amer. Vet. Med. Assoc.* 189:1029.

Corn, J. L., D. E. Stallknecht, N. M. Mechlin, M. P. Luttrell, and J. R. Fischer. 2004. Persistence of pseudorabies virus in feral swine populations. *J. Wildl. Dis.* 40:307.

Daszak, P., A. A. Cunningham, and A. D. Hyatt. 2000. Emerging infectious diseases of wildlife — Threats to biodiversity and human health. *Science* 287:443.

Davidson, W. R., and V. F. Nettles. 1997. *Field Manual of Wildlife Diseases in the Southeastern United States*, 2nd edn. Southeastern Cooperative Wildlife Disease Study, Athens: University of Georgia.

Dees, T. 1999. Feral/wild swine surveillance for foreign animal diseases and some field study projects on brucellosis and pseudorabies in the southeastern USA. *Proc. First Nat. Feral Swine Conf.* 1:121.

Deuel, B. 1985. Experimental lead dosing of northern pintails in California. *Calif. Fish Game* 71:125.

Dougherty, R. W. 1939. Sickle cells in the blood of western deer. *J. Wildl. Manage.* 3:17.

Duff, P. 2003. Wildlife disease surveillance by the veterinary laboratories agency. *Micro. Today* 30:157.

Farry, S. C., S. E. Henke, A. M. Anderson, and M. G. Fearneyhough. 1998a. Efficacy of bait distributional strategies to deliver canine rabies vaccines to coyotes in southern Texas. *J. Wildl. Dis.* 34:23.

Farry, S. C., S. E. Henke, A. M. Anderson, and M. G. Fearneyhough. 1998b. Responses of captive and free-ranging coyotes to simulated oral rabies vaccine baits. *J. Wildl. Dis.* 34:13.

Fine, P. E. M., J. L. Aron, J. Berger, D. J. Bradley, H. J. Burger, E. G. Knox, H. P. R. Seeliger, C. E. G. Smith, K. W. Ulm, and P. Yekutiel. 1982. The control of infectious disease group report. In *Population Biology of Infectious Diseases*, R. M. Anderson, and R. M. May (eds). Berlin: Springer-Verlag, p. 121.

Gower, W. C. 1938. Seasonal abundance of some parasites of wild ducks. *J. Wildl. Manage.* 2:223.

Gresham, C. S., C. A. Gresham, M. J. Duffy, C. T. Faulkner, and S. Patton. 2002. Increased prevalence of *Brucella suis* and pseudorabies virus antibodies in adults of an isolated feral swine population in coastal South Carolina. *J. Wildl. Dis.* 38:653.

Gunson, J. R., W. J. Dorward, and D. B. Schowalter. 1978. An evaluation of rabies control in skunks in Alberta. *Can. Vet. J.* 19:214.

Hahn E. C., G. R. Page, P. S. Hahn, K. D. Gillis, C. Romero, J. A. Annelli, and E. P. J. Gibbs. 1997. Mechanisms of transmission of Aujeszky's disease virus originating from feral swine in the USA. *Vet. Microbiol.* 55:123.

Hahn, N., C. Hsu, and B. Paszkiet. 1999. Research on PRV in feral swine: Past, present and future direction. *Proc. First Nat. Feral Swine Conf.* 1:75.

Hanlon, C. L., D. E. Hayes, A. N. Hamir, D. E. Snyder, S. Jenkins, C. P. Hable, and C. E. Rupprecht. 1989. Proposed field evaluation of a rabies recombinant vaccine for raccoons (*Procyon lotor*): Site selection, target species characteristics, and placebo baiting trials. *J. Wildl. Dis.* 25:555.

Hanson, R. P., and L. Karstad. 1950. Feral swine in the southeastern United States. *J. Wildl. Manage.* 23:64.

Henderson, W. M. 1982. The control of disease in wildlife when a threat to man and farm livestock. In *Animal Disease in Relation to Animal Conservation*, M. A. Edwards, and U. McDonell (eds). London: Academic Press, p. 287.

Henke, S. E., and F. C. Bryant. 1999. Effects of coyote removal on the faunal community in western Texas. *J. Wildl. Manage.* 63:1066.

Henke, S. E., D. B. Pence, and F. C. Bryant. 2002. Effect of short-term coyote removal on populations of coyote helminths. *J. Wildl. Dis.* 38:54.

Honess, R. F. 1939. A freak deer head. *J. Wildl. Manage.* 3:360.

Horan, R. D., and C. A. Wolf. 2005. The economics of managing infectious wildlife disease. *Amer. J. Agr. Econ.* 87:537.

Kalmbach, E. R., and M. F. Gunderson. 1934. Western duck sickness — A form of botulism. Technical Bulletin, No. 411, United States Department of Agriculture, Washington, DC.

Kluge, J. P., G. W. Beran, H. T. Hill, and K. B. Platt. 1999. Pseudorabies (Aujeszky's Disease). In *Diseases of Swine*, B. E. Straw, et al. (eds). Ames: Iowa State University Press, p. 233.

Knowles, N. J., A. R. Samuel, P. R. Davies, R. J. Midgley, and J. Valarcher. 2005. Pandemic strain of foot-and-mouth disease virus serotype O. *Emerg. Inf. Dis.* 11:1887.

Kocan, A. 1990. Pseudorabies. In *Review of Wildlife Disease Status in Game Animals in North America*, R. Lind (ed.). Saskatchewan, Canada: ADF–Saskatchewan Agriculture Development Fund, p. 43.

Kresta, A. E., and S. E. Henke. 2000. Attitudes towards rabies in southern Texas: A need for public education. In *Proceedings of the Nineteenth Vertebrate Pest Conference*, D. A. Whisson, and R. M. Timm (eds). San Diego, p. 113.

Leopold, A. 1933. *Game Management*. New York: Charles Scribner's Sons.

May, R. M., and R. M. Anderson. 1978. Regulation and stability of host–parasite population interactions. II. Destabilizing processes. *J. Anim. Ecol.* 47:249.

McCoy, G. W. 1911. The susceptibility to plague of the weasel, the chipmunk, and the pocket gopher. *J. Infect. Dis.* 8:42.

McCoy, G. W., and C. W. Chapin. 1912. Bacterium tularense the cause of a plague-like disease of rodents. *U.S. Public Health Marine Hosp. Bull.* 53:17.

Miller, J. E. 1993. A national perspective on feral swine. In *Proceedings of Feral Swine: A Compendium for Resource Managers*, D. Rollins (ed.). Austin: Texas Animal Health Commission, p. 9.

Ministry of Natural Resources. 2005. *Our Sustainable Future*. Ontario: Ontario Ministry of Natural Resources Strategic Directions.

Mora, M. A. 1995. Residues and trends of organochlorine pesticide and polychlorinated biphenyls in birds from Texas, 1965–88. U.S. Department Interior, National Biological Service, Fish Wildlife Research 14, Washington, DC.

Nettles, V. F., J. H. Shaddock, R. K. Sikes, and C. R. Reyes. 1979. Rabies in translocated raccoons. *Am. J. Publ. Health* 69:601.

Ohlendorf, H. M., A. W. Kilness, J. L. Simmons, R. K. Stroud, D. J. Hoffman, and J. F. Moore. 1988. Selenium toxicosis in wild aquatic birds. *J. Toxicol. Environ. Health* 24:67.

Pence, D. B., and L. A. Windberg. 1994. Impact of a sarcoptic mange epizootic on a coyote population. *J. Wildl. Manage.* 58:624.

Pirtle, E. C., J. M. Sacks, V. F. Nettles, and E. A. Rollor, III. 1989. Prevalence and transmission of pseudorabies virus in an isolated population of feral swine. *J. Wildl. Dis.* 25:605.

Purseglove, S. R., Jr., D. F. Holland, F. H. Settle, and D. G. Gnegy. 1976. Control of a fowl cholera outbreak among coots in Virginia. *Proc. Ann. Conf. S. E. Assoc. Fish Wildl. Agencies* 30:602.

Pybus, M. J. 1990. Survey of hepatic and pulmonary helminths of wild cervids in Alberta, Canada. *J. Wildl. Dis.* 26:453.

Randhawa, A. S., V. P. Kelly, and E. F. Baker, Jr. 1977. Agglutinins to *Coxiella burnetii* and *Brucella* spp., with particular reference to *Brucella canis* in wild animals of southern Texas. *J. Amer. Vet. Med. Assoc.* 171:939.

Richardson, R. H., J. R. Ellison, and W. W. Averhoff. 1982. Autocidal control of screwworms in North America. *Science* 215:361.

Riley, W. A. 1939. The need for data relative to the occurrence of hydatids and of *Echinococcus granulosus* in wildlife. *J. Wildl. Manage.* 3:255.

Romero, C. H., P. Meade, J. Santagata, K. Gillis, G. Lollis, E. C. Hahn, and E. P. J. Gibbs. 1997. Genital infection and transmission of pseudorabies virus in feral swine in Florida, USA. *Vet. Microbiol.* 55:131.

Romero, C. H., P. N. Meade, J. E. Shultz, H. Y. Chung, E. P. Gibbs, E. C. Hahn, and G. Lollis. 2001. Venereal transmission of pseudorabies viruses indigenous to feral swine. *J. Wildl. Dis.* 37:289.

Romero, C. H., P. N. Meade, B. L. Homer, J. E. Shultz, and G. Lollis. 2003. Potential sites of virus latency associated with indigenous pseudorabies viruses in feral swine. *J. Wildl. Dis.* 39:567.

Rupprecht, C. E., A. N. Hamir, D. H. Johnston, and H. Koprowski. 1988. Efficacy of vaccinia rabies glycoprotein recombinant virus vaccine in raccoons (*Procyon lotor*). *Rev. Infect. Dis.* 10:S803.

Schaffer, G. D., W. L. Hanson, W. R. Davidson, and V. F. Nettles. 1978. Hematotropic parasites of translocated raccoons in the southeast. *J. Am. Med. Assoc.* 173:1148.

Schoenbaum, M. A., J. J. Zimmerman, G. W. Beran, and D. P. Murphy. 1990. Survival of pseudorabies virus in aerosol. *Amer. J. Vet. Res.* 51:331.

Shillinger, J. E. 1938. Coccidiosis in muskrats influenced by water levels. *J. Wildl. Manage.* 2:233.

Spitzer, P. R., R. W. Risebrough, W. Walker, R. Henderson, A. Poole, D. Puleston, and I. C. T. Nisbet. 1978. Productivity of ospreys in Connecticut-Long Island increases in DDE residues decline. *Science* 202:333.

Steelman, H. G., S. E. Henke, and G. M. Moore. 1998. Gray fox response to baits and attractants for oral rabies vaccination. *J. Wildl. Dis.* 34:764.

Swales, W. E. 1935. The life cycle of *Fascioloides magna* (Bassi, 1875), the large liver fluke of ruminants, in Canada. *Can. J. Res.* 12:177.

Sweitzer, S., I. A. Gardener, B. J. Gonzales, D. Van Uren, and W. M. Boyce. 1996. Population densities and disease surveys of wild pigs in the coast ranges of central and northern California. *Proc. Vertebrate Pest Conf.* 7:75.

Taber, R. D., and N. F. Payne. 2003. *Wildlife, Conservation, and Human Welfare: A United States and Canadian Perspective*. Malabar: Krieger Publishing.

Tessaro, S. V. 1990. Brucellosis caused by *Brucella suis*. In *Review of Wildlife Disease Status in Game Animals in North America*, R. Lind (ed.). Saskatchewan, Canada: Saskatchewan Agriculture Development Fund, p. 105.

Texas Department of Health. 1994. Gray fox rabies in Texas: A status report. 20 June 1994, Texas Department of Health, Austin.

Thompson, D., P. Muriel, D. Russell, P. Osborne, A. Bromley, M. Rowland, S. Creigh-Tyte, and C. Brown. 2002. Economic costs of the foot-and-mouth disease outbreak in the United Kingdom in 2001. *Rev. Sci. Tech.* 21:675.

USFWS. 2004. *Waterfowl Population Status, 2004*. Washington, DC: U.S. Dep. Inter.

USGS. 2003. Helping to combat chronic wasting disease. U.S. Geological Survey — National Wildlife Health Center Information Sheet.

USGS. 2004. Wildlife health: Thirty years of science. U.S. Department of the Interior, U.S. Geological Survey.

Van Cleave, H. J. 1937. Worm parasites in their relations to wildlife investigations. *J. Wildl. Manage.* 1:21.

van der Leek, M. L., H. N. Becker, E. C. Pirtle, P. Humphrey, C. L. Adams, B. P. All, G. A. Erickson, R. C. Belden, W. B. Frankenberger, and E. P. J. Gibbs. 1993a. Prevalence of pseudorabies (Aujeszky's Disease) virus antibodies in feral swine in Florida. *J. Wildl. Dis.* 29:403.

van der Leek, M. L., H. N. Becker, P. Humphrey, C. L. Adams, R. C. Belden, W. B. Frankenberger, and P. L. Nicoletti. 1993b. Prevalence of *Brucella* sp. antibodies in feral swine in Florida. *J. Wildl. Dis.* 29:410.

van Riper, C., III. 1991. The impact of introduced vectors and avian malaria on insular passeriform bird populations in Hawaii. *Bull. Soc. Vector Ecol.* 16:59.

van Riper, C., III, S. G. van Riper, M. L. Goff, and M. Laird. 1982. The impact of malaria on birds in Hawaiian Volcanoes National Park. Technical Report 47, CPSU/UH 022, University of Hawaii at Manoa, Honolulu.

van Riper, C., III, S. G. van Riper, M. L. Goff, and M. Laird. 1986. The epizootiology and ecological significance of malaria in Hawaiian land birds. *Ecol. Monogr.* 56:327.

Wainwright, S. E., M. A. Mora, J. L. Sericano, and P. Thomas. 2001. Chlorinated hydrocarbons and biomarkers of exposure in wading birds and fish of the Lower Rio Grande Valley, Texas. *Arch. Environ. Contam. Toxicol.* 40:101.

Waltermire, R. G. 1982. Analysis of small mammal control in a Colorado campground. MS Thesis, Colorado State University, Ft Collins.

Walton, T. E. 2000. The impact of diseases on the importation of animals and animal products. *Ann. New York Acad. Sci.* 916:36.

Wandeler, A. J. Muller, G. Wachendorfer, W. Schale, U. Forster, and F. Steck. 1974. Rabies in wild carnivores in central Europe: Ecology and biology of the fox in relation to control operations. *Z. Vet. Med. B* 21:765.

Warner, R. E. 1968. The role of introduced diseases in the extinction of the endemic Hawaiian avifauna. *Condor* 70:101.

Westmore, A. 1919. Lead poisoning in waterfowl. U.S. Dept. Agric. Bull. No. 793, Washington, DC.

Wetmore, P. W. 1939. A species of *Plasmodium* for the sharp-tailed grouse infected to other birds. *J. Wildl. Manage.* 3:361.

Whitlock, S. C. 1939a. Studies of the blood of white-tailed deer. *J. Wildl. Manage.* 3:14.

Whitlock, S. C. 1939b. Infection of cottontail rabbits by *Cysticerus pisiformis* (*Taenia pisiformis*). *J. Wildl. Manage.* 3:258.

Witmer, G. W., R. B. Sanders, and A. C. Taft. 2003. Feral swine — Are they a disease threat to livestock in the United States. In *Proceedings of the 10th Wildlife Damage Management Conference*, K. A. Fagerstone, and G. W. Witmer (eds). Fort Collins: Wildlife Damage Management Working Group of The Wildlife Society, p. 316.

Wobeser, G. A. 1994. *Investigation and Management of Disease in Wild Animals*. New York: Plenum Press, p. 265.

Wobeser, G. 2002. Disease management strategies for wildlife. *Rev. Sci. Off. Int. Epiz.* 21:159.

Wobeser, G. 2006. *Essentials of Disease in Wild Animals*, Chap. 13. Ames: Blackwell.

Wood, G. W., J. B. Hendricks, and D. E. Goodman. 1976. Brucellosis in feral swine. *J. Wildl. Dis.* 12:579.

Wood, G. W., L. A. Woodward, D. C. Mathews, and J. R. Sweeney. 1992. Feral hog control efforts on a coastal South Carolina plantation. *Ann. Conf. Southeast. Assoc. Fish Wildl. Agencies* 45:167.

Wyckoff, A. C., S. E. Henke, T. Campbell, D. G. Hewitt, and K. VerCauteren. 2005. Preliminary serologic survey of selected diseases and movements of feral swine in Texas. In *Proceedings of the 11th Wildlife Damage Management Conference*, K. A. Fagerstone (ed.). Ft Collins: Wildlife Damage Management Working Group of The Wildlife Society.

Zwank, P. J., V. L. Wright, P. M. Shealy, and J. D. Newsom. 1985. Lead toxicosis in waterfowl on two major wintering areas in Louisiana. *Wildl. Soc. Bull.* 13:17.

Zygmont, S. M., V. F. Nettles, E. B. Shotts Jr., W. A. Carment, and B. O. Blackburn. 1982. Brucellosis in wild swine: A serologic and bacteriologic survey in the southeastern United States and Hawaii. *J. Am. Vet. Med. Assoc.* 181:1285.

17 Conservation Genetics of Marine Turtles — 10 Years Later

John C. Avise

CONTENTS

Marine turtles are model subjects for genetic evaluations in a conservation context (Figure 17.1). All seven to eight extant species (depending on taxonomic opinion) are considered threatened or endangered under various formal listings such as the U.S. Endangered Species Act and the World Conservation Union's Red List. These charismatic animals are "high profile" in public as well as scientific arenas. Finally, sea turtles have long generation lengths, oceanic habits, and protracted

FIGURE 17.1   Marine turtles are model subjects for genetic evaluations in a conservation context. Photograph Copyright Timothy E. Fulbright.

migrations that can make direct field studies difficult, yet numerous behavioral, ecological, and demographic features (often of potential relevance to conservation) can be illuminated by molecular genetic assays.

Ten years ago, Bowen and Avise (1996) summarized pioneering genetic studies of marine turtles based on molecular markers (mostly mitochondrial DNA restriction sites and sequences). Since then, the number of publications on the behavioral, ecological, and evolutionary genetics of marine turtles has grown steadily from the 28 papers in our original review to more than 100 papers available today (Figure 17.2). Here, I review these latter studies and comment on where they may have confirmed, extended, or perhaps contradicted (but no examples of the latter were found) the earlier genetic findings.

An assumption of this review is that readers are familiar with the general biology of sea turtles, including their life cycles and conservation challenges. These topics have been addressed at length in several excellent treatments of traditional (mostly nonmolecular) biological information (e.g., Bjorndal 1995; Klemens 2000; Bolton and Witherington 2003; Lutz et al. 2003; Spotila 2004).

GENETIC PATERNITY AND MATING SYSTEMS

GENERAL BACKGROUND

Most reports on the mating behaviors of sea turtles have been anecdotal, but a few intensive observational efforts have given better indications of where and when mating typically occurs. Notably, field studies on several populations of green turtle (*Chelonia mydas*) indicate that courtship and copulation in this species often take place near nesting beaches, just before the nesting season and immediately after adults of both sexes have migrated to the region from foraging grounds that may be hundreds or even thousands of kilometers distant (Booth and Peters 1972; Broderick and Godley 1997; Godley et al. 2001). A female's successive migration/nesting episodes are typically 2 or 3 years apart; the interval for males may be shorter (Limpus 1993; Miller 1997). Males, who do not maintain territories, often seem to be relentless in their attempts to copulate with females, suggesting that the behavioral mating system could be described as scramble polygamy. However, the apparent or social mating system of any species can differ from the actual or genetic mating system (as realized by

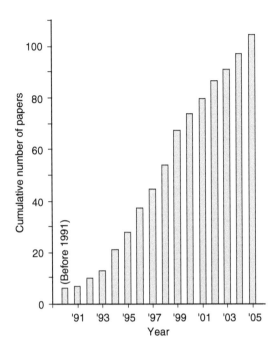

FIGURE 17.2 Cumulative numbers of scientific publications and theses in which molecular markers were employed to address ecological or evolutionary issues in marine turtles. Listings before 1996 were papers referenced in the original review by Bowen and Avise (1996). Tallies for 1996 and beyond were compiled primarily from "Recent Publications" reported quarterly in issues of the *Marine Turtle Newsletter* (ISSN 0839-7708).

progeny production); the latter requires knowledge on genetic paternity that is unobtainable from field observations alone.

Synopsis of Pre-1996 Genetic Findings

By comparing allozyme markers in female loggerhead turtles (*Caretta caretta*) and in progeny from their respective nests, Harry and Briscoe (1988) deduced which alleles in each clutch were of paternal origin. From direct counts of such paternal alleles, the authors estimated that at least 30% of the 45 surveyed broods must have had more than one sire. Apart from that landmark report documenting multiple paternities within clutches, no other substantial analyses of genetic parentage had been conducted on marine turtles before the mid-1990s.

Updated Genetic Discoveries

Since 1996, hypervariable microsatellite markers (Goldstein and Schlötterer 1999) have been employed in about a dozen genetic studies to assess paternity in at least five marine turtle species (Table 17.1). Results generally confirm that hatchlings within a nest often represent mixtures of full-sibs and maternal half-sibs with different fathers. The reported frequencies of multiple paternity in Table 17.1 are provisional due to uncertainties that could bias the estimates either downward (limited resolving power of the markers employed, incomplete sampling of clutches) or upward (*de novo* mutations as a confounding source of "extra" paternal alleles in some families). Nonetheless, it is now clear that multiple paternity in marine turtles is a common phenomenon (albeit variable in frequency among populations and species). Indeed, statistical and model-based re-analyses of the

TABLE 17.1

Summary of Microsatellite-Based Findings on Genetic Paternity and Female Sperm Storage in Marine Turtles

Species	No. of clutches assayed	Mean no. of hatchlings assayed per clutch	No. of loci scored	% Clutches with multiple paternity[a]	Female sperm storage inferred across clutches?	References
Green turtle	3	11	—[b]	33	Not examined	Parker et al. (1996)
Green turtle	22	41	5	9	Yes, within season	FitzSimmons (1998)
Green turtle	3	15	2	100	Not examined	Ireland et al. (2003)
Green turtle	18	16	5	61	Not examined	Lee and Hays (2004)
Loggerhead	3	21	2	33	Not examined	Bollmer et al. (1999)
Loggerhead	70	10	4	31	Not examined	Moore and Ball (2002)
Kemp's ridley	26	9	3	>31[c]	Not examined	Kichler et al. (1999)
Olive ridley	10	70	2	20	Not examined	Hoekert et al. (2002)
Leatherback	17	10	6	0	Yes, within season	Dutton et al. (2000)
Leatherback	50	20	3	16	Yes, within season	Crim et al. (2002)

[a] See text for qualifications about the accuracy of these estimates.

[b] Multilocus "DNA fingerprinting" (rather than by accumulation of data from discrete microsatellite loci).

[c] 70–81%, according to a statistical reanalysis of these same data by Neff et al. (2002).

genetic data suggest that the true frequency of clutches with multiple sires may approach 100% in extreme cases (Kichler et al. 1999; Neff et al. 2002).

Additional discoveries about mating activities have emerged from the microsatellite surveys in Table 17.1. First, at face value, the genetic markers typically distinguished only two fathers for each documented half-sib nest, but the actual number of sires might often be higher (because of limited resolving power in some of the genetic studies, as mentioned above). Indeed, a few genetic studies on marine turtles have firmly documented instances of 3–5 sires for particular nests (Crim et al. 2002; Moore and Ball 2002; Lee and Hays 2004). Second, the genetic analyses have shown that the relative contributions of different fathers to a clutch can range from highly skewed (e.g., 90% and 10%) to nearly equal (examples in Hoekert et al. 2002; Lee and Hays 2004). Third, several studies examined successive clutches laid within a nesting season by a given female, and some of these clutches proved to have had the same sire. Unless a given female had remated exclusively with one male at different times during the season (which seems unlikely given the biology of the species), these genetic findings imply that females can store and utilize sperm across multiple nesting excursions. Longer-term sperm storage by females (i.e., across years) may be possible as well — as has been documented in some terrestrial and freshwater turtle species (Palmer et al. 1998; Pearse et al. 2001; review in Pearse and Avise 2001) — but the phenomenon has not yet been well investigated in marine turtles.

Multiple paternity within a clutch evidences polygamous behavior by a female. Given the availability of suitable markers, such polyandry is easy to detect when it exists, because each dam and her known progeny can be jointly assayed, such that paternal alleles in a surveyed clutch are straightforward to deduce. Do males show polygamous behavior also? The documentation of successful polygyny via molecular markers is more challenging, because it requires that multiple clutches sired by a focal male have been included in the genetic survey and that the markers display adequate variation to genetically exclude nearly all other candidate sires. These conditions have seldom been met in available analyses, but Crim et al. (2002) were able to infer two separate instances in which a male leatherback turtle (*Dermochelys coriacea*) apparently had sired progeny in the nests of at least two females. Given the known proclivities of male turtles to seek copulations rather indiscriminately, this genetic documentation of polygyny is probably only the tip of the iceberg of successful multiple mating by males.

BIOLOGICAL AND CONSERVATION RELEVANCE

In most animal species, the potential benefits of polygynous matings to males (increased fitness) are obvious, whereas any reproductive advantages of polyandrous matings to females are harder to understand (especially when males do not contribute directly to offspring care). Conventional speculation is that a female, by mating with more than one male, may obtain any of several potential fitness benefits including fertilization insurance, better paternal genes for her progeny, more options in sperm usage, and clutches with higher genetic variety. Any or all of these possibilities seem plausible (but remain unproven) for marine turtles.

An entirely different view is that male coercion accounts for polyandry. Under this hypothesis, a female concedes to superfluous mating only to escape incessant male harassment. A recent genetic study of genetic paternity in green turtles provided some empirical support for this idea. In comparisons of single-sired versus multiple-sired clutches on Ascension Island, Lee and Hays (2004) found no significant association of polyandry with any monitored estimator of reproductive success (clutch size, proportion of eggs fertilized, hatching success, or hatchling survival). The authors interpreted these data as consistent with the notion that females gain no reproductive advantage from multiple mating but instead "make the best of a bad job" by acquiescing to aggressive male courtship as a personal damage-control tactic.

With regard to conservation relevance, polygamy in principle can have several consequences relevant to population viability, especially in low-density circumstances. For example, populations of sea turtles with highly polygynous tendencies should be somewhat buffered against major reproductive failures that otherwise might arise from local shortages of males. The mating system can also impact a population's genetic effective size (N_e), which in turn has a crucial influence on inbreeding and rates of change in gene diversity (Sugg and Chesser 1994). For example, when some males are excluded from reproduction, or when the variance in reproductive success among males is very high (as can occur in polygynous species with high female fecundities), N_e can fall well below the adult census size (N) of a population. On the other hand, polyandrous matings by highly fecund females could engage many more males in fatherhood than might otherwise be the case, and perhaps thereby decrease the variance in male reproductive success. In such situations, N_e can be much larger than would be true for a monogamous or polygynous population of same adult census size.

Although such speculations are common in the scientific literature, the fact remains that genetic studies of mating systems in marine turtles have not yet, to my knowledge, been translated into practical population management plans, nor is their immediate relevance to conservation efforts particularly clear.

POPULATION STRUCTURE, INTER-ROOKERY GENE FLOW, AND NATAL HOMING

GENERAL BACKGROUND

Decades of tagging efforts on nesting beaches, most notably for green turtles, loggerheads, and hawksbills (*Eretmochelys imbricata*), had demonstrated that most adult females, despite their high vagility and migratory nature, return faithfully to nest at a particular rookery location, often on a 2–3-year interval. What remained unknown from these tagging studies (due to the logistic difficulty of attaching physical tags to hatchlings for possible recovery more than a decade later) is whether the particular rookery to which a nesting adult is fidelic was also her natal site. Information on this key issue awaited assays of natural molecular tags, most importantly from maternally inherited mitochondrial (mt) DNA.

SYNOPSIS OF PRE-1996 GENETIC FINDINGS

Landmark population genetic studies of green turtles, loggerheads, and to some extent hawksbills showed that conspecific rookeries within an ocean basin often display large differences in the frequencies (but seldom in the magnitudes of sequence divergence) of mtDNA haplotypes (review in Bowen and Avise 1996). From these data, genetically based estimates of female-mediated inter-colony gene flow per generation (the Nm parameter as inferred from mtDNA markers) were often close to zero, especially for rookeries situated in different regions within an ocean. These discoveries of dramatic but evolutionarily shallow spatial structure of "matrilineal surnames" (Avise 1989) implied that nesting females have strong but imperfect natal-homing tendencies. If, instead, females routinely nested at non-natal locales, or if they frequently switched nesting areas during their lifetimes, then mtDNA haplotypes should be shared extensively among rookeries under most demographic scenarios. Conversely, if natal homing was consistently perfect or nearly so across evolutionary time, then conspecific rookeries within an ocean basin should show genetic separations far deeper than those observed.

A few early reports also addressed possible male-mediated genetic exchange between rookeries. These analyses involved assessing population structures using bi-parentally inherited molecular markers (allozymes or restriction fragment length polymorphisms in nuclear DNA), and then comparing results to those from mtDNA. For green and loggerhead turtles, these studies revealed only occasional differences in nuclear allele frequencies between pairs of conspecific rookeries, and the resulting Nm values generally tended to be much higher than those inferred from mtDNA. These genetic findings were interpreted to imply that breeding males may have modest tendencies for natal-homing (as opposed to randomly mating with females from any source), but that males also mediate appreciable gene flow between nesting populations via inter-rookery copulations that result in "foreign" paternity.

UPDATED GENETIC DISCOVERIES

Since 1996, molecular surveys addressing conspecific population structure and gene flow have been extended to faster evolving mtDNA sequences (notably the control region of the molecule), to more highly polymorphic nuclear markers (especially microsatellites), to multiple life history stages of marine turtles, and to many additional rookeries and species (Table 17.2). Most of the results nonetheless remain qualitatively similar to those of the earlier surveys.

For example, as summarized by Roberts et al. (2004), mtDNA assays have identified significant genetic differences in 86% of 43 pair-wise comparisons of green turtle rookeries within an ocean basin (Atlantic or Pacific), and yielded a mean Nm value (the inferred number of females exchanged between pairs of rookeries per generation) of only 0.25; whereas, microsatellites have shown statistically significant genetic distinctions in only about 12% of 57 such pair-wise comparisons, and have indicated a mean Nm value (in this case the estimated number of inter-rookery migrants of unspecified sex per generation) of about 6.0. The higher (but still modest) Nm estimates from nuclear markers as compared to mitochondrial markers probably reflect, in part, appreciably higher levels of male-mediated than female-mediated genetic exchange between nesting sites (but see the following sections). Similarly for loggerhead turtles, nesting colonies along the southeastern U.S. coast display highly significant population structure in mtDNA haplotypes but not so in microsatellite alleles, a finding that Bowen et al. (2005) attributed to natal homing by females coupled with inter-rookery nuclear gene flow due to opportunistic mating of males with females traveling along migration corridors.

TABLE 17.2
Primary Studies since 1996 of Population Genetic Structure and Inferred Gene Flow among Conspecific Marine Turtle Rookeries and/or Their Adjoining Mating Areas[a] (See Also Table 17.3)

Species	No. of rookeries	No. of specimens	Nesting areas showing moderate or strong genetic differences	References
Green turtle	9	147	Most surveyed sites in the N. Atlantic and Mediterranean	Encalada et al. (1996)
Green turtle[b]	9	ca. 275	Sites in northern versus eastern Australia	FitzSimmons et al. (1997a,b)
Green turtle[c]	16	337	Atlantic versus Pacific sites, primarily	Roberts et al. (2004)
Loggerhead	12	249	Most surveyed sites in the N. Atlantic and Mediterranean	Encalada et al. (1998)
Loggerhead	4	259	Several Japanese sites	Hatase et al. (2002)
Loggerhead[c]	9	23–123	Primarily southeastern United States	Bowen et al. (2005)
Olive ridley	7	80	Major Indo-Pacific regions, and these versus the Atlantic	Bowen et al. (1997)
Olive ridley	4	81	Sites in eastern India versus Sri Lanka	Shanker et al. (2004)
Olive ridley	6	149	Primarily Mexican sites in the eastern Pacific	López-Castro and Rocha-Olivares (2005)
Hawksbill	7	103	Several sites in the Caribbean and western Atlantic	Bass et al. (1996)
Hawksbill	4	127	Seychelles, Chagos Islands, and the Arabian Peninsula	Mortimer and Broderick (1999)
Hawksbill	3[d]	136	Indo-Pacific sites, and these versus the Caribbean	Okayama et al. (1999)
Hawksbill	3	143	Cuba, Puerto Rico, and Mexico	Díez-Fernández et al. (1999)
Leatherback	11	175	Several sites within and among the world's oceans	Dutton et al. (1999)

[a] Except where otherwise noted, all assessments were based on mtDNA control region sequences.
[b] Assessments based on anonymous single-copy nuclear loci as well as mtDNA control region sequences.
[c] Assessments based on nuclear microsatellite loci (in addition to mtDNA control region sequences in the loggerhead study).
[d] Plus seven foraging areas.

Important qualifications apply to any such inferences about gene flow based on the indirect evidence of population genetic structure (see Waples 1998; Whitlock and McCauley 1999; Balloux and Lugon-Moulin 2002). For example, homoplasy — which is known to be especially common at microsatellite loci (Avise 2004) — can produce inflated estimates of nuclear genetic exchange when alleles identical by state but not by descent arise in different populations by convergent mutation. A more general problem is the inherent difficulty of teasing apart the population effects of recent gene flow (or lack thereof) from population effects due to genetic drift and historical lineage sorting. For example, large populations that currently exchange few or no migrants could nonetheless share many alleles due to the retention of ancestral markers; and conversely, small populations that frequently exchanged migrants in the past could nonetheless share few nuclear or mitochondrial alleles at present due to the chance effects of recent bottlenecks or genetic drift. Another potential complication in sexual species is that mtDNA, by virtue of its uniparental and haploid inheritance mode, has a fourfold-lower effective population size (all else being equal) than autosomal loci. This itself can cause, via genetic drift and more rapid lineage sorting, the appearance of higher population structure and diminished gene flow in cytoplasmic compared with nuclear markers. Such complications notwithstanding, the fact remains that spatial structure, especially in mtDNA, is typically pronounced among marine turtle rookeries. Furthermore, in the few comparisons attempted to date, no appreciable temporal shifts in mtDNA composition have been documented at rookeries resampled across different time periods, including different years (e.g., Bowen and Bass 1997).

Thus, many of the recent molecular surveys have sought to clarify which particular rookeries within an ocean basin differ genetically, and, more generally, at what average spatial scales. An emerging consensus is that regional aggregations of conspecific nesting populations often differ sharply from others in mtDNA haplotype frequencies, but that rookeries within such regions sometimes do and sometimes do not show clear matrilineal distinctions from one another.

For example, in green turtles of the Atlantic and Mediterranean, almost no overlap in mtDNA haplotypes exists between four nesting areas (Encalada et al. 1996; Lahanas et al. 1998): a western set of rookeries in Costa Rica, Mexico, and Florida; a central set involving Aves Island and Surinam; a southeastern set in Brazil, Ascension Island, and Africa's Guinea Bissau; and an eastern site (Cyprus) in the Mediterranean. In contrast, rookeries within each of these areas tend to share common and sometimes even rare mtDNA haplotypes (albeit not always in identical frequencies). In the Indo-Pacific also, green turtles show strong regional differences in matriline frequencies (see Bowen and Avise 1996), but four nesting locales in Michoacan, Mexico, displayed no detectable mtDNA population genetic substructure (Chassin-Noria et al. 2004). Similarly, in the olive ridley (*Lepidochelys olivacea*) mtDNA genotypes clearly distinguish three regional populations (East Pacific, Indian-West Pacific, and Atlantic; Bowen et al. 1997), but specimens from four nesting areas along India's east coast appear to belong to one well-mixed population (Shanker et al. 2004) and several nesting populations along the Pacific beaches of Mexico show only moderate (albeit statistically significant) mtDNA population structure (López-Castro and Rocha-Olivares 2005).

Qualitatively similar patterns of mtDNA spatial structure have been reported for loggerheads, hawksbills (review in Bass 1999), and leatherback turtles (Table 17.2). In considering the totality of available genetic evidence, Bowen and Bass (1997) concluded, "In general, sea turtle nesting sites separate by a few hundred km are distinguishable with mtDNA data, and nesting sites separated by less than about 100 km are not. Exceptions to this rule exist and are an expected byproduct of colonization events." This statement continues to hold today, to a first approximation, based on subsequent population surveys of mtDNA sequences in marine turtles. Reece et al. (2005) recently applied nested clade analysis (Templeton 1998) to previously published mtDNA control region data for green turtles, hawksbills, and loggerheads in the Atlantic/Mediterranean, and these authors similarly concluded that metapopulations of these species have had idiosyncratic histories involving "complex patterns of historical population subdivision, long-distance dispersal, and restricted gene flow."

Lee et al. (2007) recently took quite a different population genetic approach to address the possibility of even finer-scale natal homing in green turtles. Using a battery of nuclear microsatellite loci to assay specimens from three proximal nesting beaches on Ascension Island, these researchers showed that adult females sampled from a particular beach had significantly higher probabilities of genetic assignment to a specific beach (based on molecular data from other specimens) than to other nesting beaches only a few kilometers away. The same was not true for surveyed males (actually, for sires whose genotypes had been inferred from genetic paternity analyses, because the adult males were seldom captured). Lee and colleagues interpreted these and associated genetic observations as indicative of fine-scale beach discrimination by females for nesting, but not by males for mating. These findings raise the possibility, which needs further study, that at least some female sea turtles may show natal homing tendencies at far finer spatial scales than formerly might have been supposed.

Another extension of earlier work has involved characterizing population genetic structures at different life history stages. In probably the largest of these treatments, Bowen et al. (2005) reported that magnitudes of mtDNA population structure in loggerhead turtles increase through successive life stages as follows: pelagic post-hatchlings (which show no significant structure across the northern Atlantic); sub-adults (which show low but statistically significant population genetic structure in the southeastern United States); and adults nesting along this same coastline (which show strong mtDNA population differentiation). These results probably reflect the following, respectively: extensive intermingling of hatchlings from different rookeries during the pelagic phase; a tendency for subadults to recruit to neritic habitats somewhat near their natal sites; and a strong proclivity of adult females to nest at their respective natal locations.

BIOLOGICAL AND CONSERVATION RELEVANCE

As pointed out by Avise (1995) and since reiterated by many researchers, the primary conservation relevance of salient mtDNA population structure resides in a special genetics–demography connection for matrilines. Namely, when the per-generation number of breeding females exchanged between particular rookeries or regional sets of rookeries is low (as generally appears to be true for nesting marine turtles based on mtDNA markers), then those rookeries at the present time must be demographically independent of one another with regard to reproductive output. So, for example, if a nesting colony is extirpated by humans or natural causes, it will unlikely be reestablished (at least in the near term) by natal-homing females hatched elsewhere. This conclusion holds regardless of the extent of male-mediated gene flow. Even if rookeries were panmictic (and thus would show absolutely no population structure in neutral nuclear markers because of extensive inter-rookery mating), each colony would be demographically autonomous in offspring production provided that nesting females remain philopatric to their natal sites. Thus, with regard to reproduction, each distinguishable rookery would be a potential "management unit" or MU (see Avise 2004: 205–6).

It is also important to appreciate an inherent asymmetry in this type of reasoning: Whereas conspicuous mtDNA divergence implies that rookeries are demographically autonomous or quasi-autonomous in reproductive output, and hence perhaps worthy of separate management or protection, lack of mtDNA divergence does not necessarily imply that rookeries are demographically unified today. The evolutionary timeframe of detectable mtDNA divergence may be centuries or millennia, whereas the ecological timeframe of demographic parameters relevant to population management is normally years or decades. Thus, demography-based conservation plans should incorporate, in addition to mtDNA data, modern field observations on contemporary behavioral and life history characteristics.

Natal homing by female turtles cannot be absolute, of course; rookeries exist at many locations, so wandering females do establish new colonies at least occasionally. Considering the mtDNA findings together with general thought about sea turtle biology, the following scenario for intra-ocean rookery dynamics seems probable. Nesting beaches are evolutionarily ephemeral because of physical and biotic changes in the environment, and female natal homing is strong but imperfect. Thus, matrilineal meta-populations within an ocean are periodically restructured via rookery extinctions and recolonizations. Across evolutionary time, this kaleidoscopic population genetic process can produce a "concerted" pattern of mtDNA evolution: at any time horizon (such as the present) rookeries may appear tightly connected genealogically (i.e., their coalescent was rather recent), yet they display strong matrilineal spatial structure because of the inherent normal tendency for female natal homing. Superimposed on this evolutionary operation is "routine" ongoing gene flow mediated disproportionately (but not exclusively) by males, who via the mating system provide an additional avenue for the effective transfer of nuclear markers between rookeries.

In the scenario described above, genetic diversities within rookeries should in theory be lower and the distinctions between rookeries should be sharper if local effective population sizes were small and if each colonization event involved only a small number of individuals (in the extreme, a single gravid waif). Although molecular data on these points are inconclusive, they do seem to eliminate the possibility that effective population sizes at particular marine turtle rookeries are normally huge. For example, many surveyed rookeries (including some that are exceptionally large today, such as the green turtle colony on Ascension Island) show few mtDNA haplotypes and low (near zero) mtDNA nucleotide diversities. On the other hand, some nesting sites display several to many mtDNA haplotypes and at least modest nucleotide diversities consistent with effective population sizes perhaps in the thousands (contingent on several problematic assumptions about population boundaries, current and past gene flow, and mtDNA evolutionary rates; see Avise et al. 1988).

The finding that different life history phases in marine turtles can have different population genetic structures also has conservation ramifications. For example, in the case of loggerheads in the North Atlantic (Bowen et al. 2005), the genetic inference that post-hatchlings from different rookeries are well mixed during the pelagic phase implies that any salient positive or negative impacts on the survival of these juveniles can have diffuse demographic effects on multiple rookeries; whereas, the lesser mixing inferred from the greater mtDNA population structures of subadults and nesting adults suggests that any such demographic impacts at these life stages will tend to be focused or even pinpointed on the corresponding breeding populations.

In summarizing this section, the most salient genetic finding has been that conspecific rookeries within an ocean basin often show pronounced yet shallow matrilineal distinctions from one another, thus implying strong but imperfect natal homing tendencies by females. In turn, the primary conservation relevance of this finding is that for all practical purposes, each rookery should be managed as if it is demographically autonomous in terms of its current reproductive output.

ROOKERY ORIGINS OF INDIVIDUALS AT NON-NESTING PHASES OF THE LIFE CYCLE

GENERAL BACKGROUND

Marine turtles have long generation lengths — age at first nesting is usually 10 years or more, and life spans extend several decades. Sea turtles also move great distances both as hatchlings (who may drift entire ocean basins) and adults (who may swim hundreds or thousands of kilometers on each migratory circuit between foraging and reproductive areas). The discovery that particular rookeries or regional aggregations of nesting sites often carry diagnostic mtDNA haplotypes raised early hopes that these molecular tags would be effective for tracking the precise rookery origins of marine turtles at other phases of the life cycle, and at spatial/temporal scales that can defy conventional field observations (Avise and Bowen 1994; Bowen 1995).

SYNOPSIS OF PRE-1996 GENETIC FINDINGS

At the time of our 1996 review, these hopes had only begun to be realized. For loggerhead turtles, mtDNA markers had been provisionally employed to identify the regional rookery sources of juveniles on foraging grounds in South Carolina, those inadvertently captured in the Spanish long-line fishery (a majority of which proved to have come from the Western Atlantic), and those captured in the north-central Pacific or along the coast of Baja California (most of which proved to have originated in Japan). Similarly for hawksbills, mtDNA haplotypes were employed to address the rookery origins of forging individuals in northern Australia and in Puerto Rico.

UPDATED GENETIC DISCOVERIES

The past 10 years have been witness to many additional studies that have capitalized on forensic applications for genetic markers in sea turtles. Also, statistical approaches including maximum-likelihood, Bayesian approaches, and Markov chain Monte Carlo have been improved for quantifying rookery contributions to "mixed stocks" when molecular markers in source populations overlap in frequency (Pella and Masuda 2001; Bolker et al. 2003). Some genetic stock assessments on marine turtles have also incorporated demographic data (e.g., population sizes of potential source rookeries) or ecological information (e.g., distances to possible rookery sources) into the statistical analyses (e.g., Okuyama and Bolker 2005).

Table 17.3 summarizes results from studies that used mtDNA markers to address the natal origins of marine turtles at other stages of life. The following generalities emerged. First, the genetic data have bolstered other evidence that juveniles can traverse great distances after leaving their natal sites

TABLE 17.3
Genetic Studies since 1996 That Utilized mtDNA Markers in a Forensic Context to Identify the Probably Rookery Origins of Marine Turtles at Other Life History Phases

Species	No. of specimens in the diagnoses	Geographic setting and general biological conclusion	References
Green turtle	80	Juveniles in a foraging area near Grand Inagua, Bahamas, originated from multiple rookeries mostly in the western Atlantic	Lahanas et al. (1998)
Green turtle	61	Juveniles along the east Florida coast originated from rookeries in the United States/Mexico, Costa Rica, and Aves Island/Surinam	Bass and Witzell (2000)
Green turtle	60	Juveniles in a foraging area near Barbados, West Indies, originated from multiple rookeries mostly in the western Atlantic	Luke et al. (2003)
Green turtle	60	Adults "sleeping" off the northeastern coast of Nicaragua originated mostly from Costa Rica, but some also from Aves Island or Surinam	Bass et al. (1998)
Loggerhead	207	Bycatch specimens in various Mediterranean fisheries originated from western Atlantic as well as Mediterranean nesting areas	Laurent et al. (1998)
Loggerhead	65	Juveniles inadvertently caught in an Italian fishery originated from multiple western Atlantic and perhaps Mediterranean nesting areas	Casale et al. (2002)
Loggerhead	45	Juveniles in a foraging area in Panama originated from nesting beaches in southern Florida and Mexico	Engstrom et al. (2002)
Loggerhead	82	Specimens stranded along beaches in the northeastern United States originated from three nesting regions the southern United States and Mexico	Rankin-Baransky et al. (2001)
Loggerhead	121	Juveniles in pelagic realms of the eastern Atlantic originated from nesting sites in the southeast United States and Yucatan	Bolten et al. (1998)
Loggerhead	106	Juveniles in a foraging area along eastern Florida originated from multiple nesting areas in the United States and Mexico	Witzell et al. (2002)
Loggerhead	100	Most specimens in a North Carolina foraging area originated from Florida, but smaller numbers also came from Mexico and other nesting sites	Bass et al. (2004)
Loggerhead	216	Sub-adults on offshore feeding grounds along the U.S. east coast came disproportionately from nearby nesting beaches	Roberts et al. (2005)
Loggerhead	1437	Juveniles in their westward trans-Atlantic migrations to the eastern United States tend to return to foraging areas near their natal sites	Bowen et al. (2004)
Loggerhead	106	Juveniles in foraging areas along the southern Italian coastline originated mostly from nesting sites within the Mediterranean	Maffucci et al. (2006)
Hawksbill	41	Adults at a Puerto Rican foraging area probably originated from multiple nesting sites across the Caribbean	Bowen et al. (1996)
Hawksbill	345	Populations of foragers near Cuba, Puerto Rico, and Mexico consisted of mixtures of individuals from different Caribbean nesting areas	Díez-Fernández et al. (1999)
Hawksbill	241	Juveniles foraging in the Seychelles and Chagos Islands probably originated from rookeries in these same Indian Ocean archipelagos	Mortimer and Broderick (1999)

(including entire ocean basins such as the North Pacific or North Atlantic). Second, the molecular data have helped to elucidate specific migration routes and destinations of both juveniles and adults by provisionally identifying the rookery origins of turtles on foraging grounds or elsewhere. Third, the genetic data have shown that particular foraging areas often are co-utilized by turtles from multiple nesting regions, and these data have permitted estimates of the relative rookery contributions to these mixed stocks. Finally, the molecular data have permitted quantitative appraisals of which reproductive populations are impacted when turtles die as bycatch in oft-distant (and sometimes devastating; see Lewison et al. 2004) fisheries operations.

Such studies are not without limitations (see Mrosovsky 1997). First, available mtDNA markers cannot distinguish all nesting areas, so pinpointing the exact (as opposed to regional) natal origin of an individual turtle is seldom feasible. Second, many of the available estimates of haplotype frequencies at nesting locations have been based on small or moderate sample sizes, and this can translate into wide confidence limits surrounding quantitative appraisals of rookery contributions to mixed stocks. Third, not all nesting areas have been genetically surveyed, so the possibility remains that specimens taken on foraging grounds or elsewhere came from unsampled rookeries rather than those to which they were provisionally assigned by mtDNA. Notwithstanding such limitations, the genetic approach to studying turtle movements in the open ocean has greatly advanced our scientific understanding of these animals' behavioral patterns.

BIOLOGICAL AND CONSERVATION RELEVANCE

The biological significance of identifying sea turtles' natal origins is obvious — it helps reveal where those animals have traveled on their lengthy sojourns. Almost no such information was available from field observations or traditional tagging studies due to the inherent difficulty of monitoring seafaring individuals, beginning as hatchlings, across multiple years. The genetic approach is novel and has yielded many interesting discoveries. For example, loggerhead nesting is unknown in the eastern and central Pacific, but juveniles occur in large numbers near Baja California and also are captured in a north-central Pacific driftnet fishery. Nearly all specimens sampled from these two distinct areas have proved to carry Japanese mtDNA genotypes and, hence, probably represent trans-Pacific migrants.

In terms of conservation relevance, an important discovery is that many foraging populations of marine turtles contain mixtures of individuals from multiple natal regions. For example, juvenile green turtles along the east coast of Florida appear to represent rookeries in the southern United States/Mexico (ca. 42%), Costa Rica (53%), and Aves Island/Surinam (4%) (Bass and Witzell 2000); loggerheads stranded in the northeastern United States came from nesting areas in northeast Florida/North Carolina (25%), southern Florida (59%), and Quintana Roo, Mexico (16%) (Rankin-Baransky et al. 2001); and hawksbills foraging off southeastern Cuba appear to have come from nesting areas in Cuba (70%), Puerto Rico (12%), and Mexico (7%) (Díez-Fernández et al. 1999). Such genetic findings imply that any source of turtle mortality on foraging grounds can have a demographic impact on multiple rookeries, including some that are far distant.

Thus, although most rookeries are apparently quite independent of one another with respect to reproductive output by natal-homing females, those same rookeries may be demographically interconnected with respect to sources of mortality at other stages of life. This can be good or bad news for conservation efforts. For example, excessive human harvesting of turtles at a given foraging area can be simultaneously detrimental to several nesting populations, but decreasing that harvest [as Cuba appears to have done for hawksbills in the 1990s; see Mrosovsky (2000)] can be simultaneously beneficial to those same rookeries. Of course, the demographic impact of feeding-ground mortality on any specific rookery will also depend on additional factors such as the size of that rookery and the fraction of its potential breeders that inhabit a particular foraging region.

Genetic assays have also enabled appraisals of which nesting areas are impacted when turtles are harvested as bycatch (or sometimes purposefully) in various marine fisheries. For example, Laurent et al. (1998) concluded from mtDNA markers that about one-half of the loggerheads taken as bycatch in a long-line fishery in the pelagic Mediterranean had come from Mediterranean (and the other half from Atlantic) nesting areas, whereas nearly all of the loggerhead bycatch in bottom trawl fisheries in Tunisia, Egypt, and Turkey had originated from Mediterranean nesting locales. From this evidence, the authors concluded that turtle mortality from commercial Mediterranean fisheries was of greater concern to Mediterranean loggerhead populations than formerly supposed. Thus, with regard to practical management, the pronounced intra-ocean population-genetic structure in marine turtles has many ramifications for forensic analyses, ranging from the reconstruction of migration routes to assessing which rookeries are impacted by various sources of mortality at non-nesting phases of the life cycle.

INTRASPECIFIC PHYLOGEOGRAPHY AT GREATER SPATIAL AND TEMPORAL SCALES

GENERAL BACKGROUND

As discussed above, matriline *frequencies* as estimated by mtDNA assays often differ sharply across nesting areas. However, in terms of *magnitudes of sequence differences*, those same mtDNA haplotypes can range from barely distinguishable in some inter-rookery comparisons, to quite highly divergent (up to 1% or more sequence divergence) in others. In marine turtles, the largest intraspecific genetic distances typically (but not invariably) are observed in inter-ocean comparisons. These presumably reflect more ancient matrilineal separations (i.e., deeper coalescent times) than those typically characteristic of turtles within an ocean basin or nesting sub-region.

SYNOPSIS OF PRE-1996 GENETIC FINDINGS

At the time of our earlier review, available phylogeographic data had come mostly from mtDNA restriction-site assays. The largest of these genetic surveys canvassed green turtle and loggerhead rookeries globally. For green turtles, the deepest split in the mtDNA gene tree separated all rookeries in the Atlantic and Mediterranean from all surveyed rookeries in the Indian and Pacific Oceans. These findings were interpreted to evidence an ancient vicariant event (probably the rise of the Central American land bridge approximately 3 million years ago) that sundered this circum-tropical species into two fundamental phylogeographic units. Another finding from these same assays was that the "black turtle" of the eastern Pacific (traditionally recognized by some authors as a distinct species, *Chelonia agassizi*) was essentially indistinguishable in mtDNA sequences from adjacent samples of the Pacific green turtle (review in Karl and Bowen 1999).

A basal bifurcation of similar magnitude was observed in the mtDNA gene tree for loggerheads, but the two primary mtDNA clades were well represented in both the Atlantic and Indo-Pacific basins. Results again were interpreted to suggest an ancient vicariant event (as for green turtles), but in this case followed by secondary and perhaps ongoing mtDNA lineage transfers between the Atlantic and Indo-Pacific via occasional loggerhead movements around Africa's southern coastline. This interpretation seems at least plausible biologically, because loggerheads in the Indian Ocean currently nest within 1000 km of the Cape of Good Hope (in Natal, South Africa), and, in general, this species is better adapted to cool temperate regimes than is the more tropic-confined green turtle.

UPDATED GENETIC DISCOVERIES

Phylogeographic efforts since 1996 have extended such global surveys to three more sea turtle species or species complexes (Tables 17.2 and 17.3): hawksbills, olive and Kemp's ridleys (*L. olivacea* and *L. kempi*), and leatherbacks. Most of these analyses entailed direct examination of rapidly evolving mtDNA control region (CR) sequences.

For the hawksbill turtle, the deepest bifurcation in the mtDNA gene tree (3.5% CR sequence divergence) separated all surveyed animals in the Indo-Pacific region (Japan, Indonesia, Philippines, Solomon Islands, Fiji, Maldives, and Seychelles) from those taken at several locations (Mexico, Puerto Rico, and Cuba) in the Caribbean (Okayama et al. 1999). Within each ocean basin, some shallower, but well-supported mtDNA clades were also present, and they tended to be spatially localized (Bass 1999; Okayama et al. 1999).

For the ridley complex, the deepest bifurcation (approximately 6.0% CR sequence divergence) distinguished Kemp's ridleys, which nest only in the western Gulf of Mexico, from olive ridleys sampled across the Indo-Pacific as well as the South Atlantic (Bowen et al. 1997). Within olive ridleys, another deep split (ca. 3.0% CR sequence divergence) separated nearly all specimens from the eastern coast of India from those elsewhere in the Indo-Pacific and Atlantic, all of which were related much more closely to one another. Indeed, common mtDNA haplotypes in the Atlantic differed by only a few mutation steps from common haplotypes throughout most of the Indo-Pacific (Bowen et al. 1997; Shanker et al. 2004). Overall, the genetic findings were interpreted to suggest a relatively ancient vicariant separation between Kemp's and olive ridley, again perhaps stemming from the rise of the Central American land bridge (Bowen et al. 1997); and, for the olive ridley, a probable ancestral Indo-West Pacific source for later range expansions (Bowen et al. 1997; Shanker et al. 2004), including a secondary invasion of the Atlantic within the past 300,000 years [as Pritchard (1969) had also inferred from nonmolecular lines of evidence].

For leatherback turtles, a global phylogeographic survey of nesting areas by Dutton et al. (1999) uncovered 11 mtDNA haplotypes, all of which were closely related (0.6% mean sequence divergence in the CR). Frequencies of these haplotypes often differed significantly from colony to colony (suggesting a surprisingly high level of natal homing by females in this highly migratory species), but the shallow global coalescent nonetheless indicates that no extant matrilines trace to long-term vicariant (or other) population separations. This latter outcome differs rather dramatically from the deep splits observed in the mtDNA genealogies of most other marine turtle species surveyed to date, and it may evidence the unusually high mobility (including high-latitude occupancy) of leatherback turtles.

Quite a different approach to assessing sea turtle phylogeography was carried out by Rawson et al. (2003) who assayed mtDNA sequences from coronulid barnacles (*Chelonibia testudinaria*) carried on the shells of loggerhead turtles. By sequencing a cytochrome oxidase gene in 79 barnacles from five regions of the world's oceans, the researchers reported clean genetic distinctions between all surveyed specimens in the eastern Pacific, western Pacific, and western-Atlantic/Mediterranean. Although these barnacles probably have high dispersal capabilities via their larvae (which are suspected, but not yet proven, to be planktonic in this species), the authors concluded that the phylogeographic patterns also were influenced by host-mediated dispersal. For example, the close genetic similarity of barnacles in the western Atlantic and Mediterranean could result in whole or part from a recent colonization of the Mediterranean by Atlantic loggerheads carrying adult *C. testudinaria* on their shells.

BIOLOGICAL AND CONSERVATION RELEVANCE

The deep phylogeographic separations observed within most globally surveyed species of marine turtles strongly imply long-standing isolations of matrilines in different regions of the world. Furthermore, the particular placements and inferred evolutionary depths of these mtDNA separations often appear to make considerable sense in terms of known or suspected historical vicariant events (operating in conjunction with each species' inherent dispersal capability, habitat preference, and ecological tolerance).

Deep phylogeographic structure has relevance to conservation and management efforts, because each long-separated set of populations carries a significant and unique fraction of a species' total genetic variation or hereditary legacy. Recognizing this fact, biologists often refer to such genetically distinctive population arrays as evolutionarily significant units or ESUs (see Avise 2004: 205–6). However, current genetic inferences about ESUs in marine turtles come with a qualifier: they have thus far been based primarily or exclusively on mtDNA evidence from nesting females and hatchlings. Male turtles, via their movements and mating practices, could in principle have mediated extensive nuclear gene flow between even highly divergent matrilineal phylogroups, in which case ESUs provisionally identified by mtDNA sequences might instead be merely MUs. In other words, such populations would be largely independent of one another with regard to female reproduction but not necessarily so with regard to the evolutionary histories of most of their nuclear genes.

For marine turtles, however, such caveats should not be particularly worrisome for at least two reasons. First, the provisional ESUs identified by mtDNA sequences usually have proved to reside in separate ocean basins or distant regions, between which extensive and ongoing exchanges of males seems biologically implausible. Second, at least some of the phylogeographic surveys included juvenile or adult males in the assayed samples, yet the mtDNA haplotypes carried by these individuals typically belonged to their respective indigenous phylogroups rather than to foreign or far-distant matrilineal clades. Nonetheless, future phylogeographic surveys on sea turtles should be extended to a variety of nuclear loci as well. In summarizing this section, the deep phylogeographic splits observed within each of several marine turtle species identify potential intraspecific ESUs that should be recognized and included in any comprehensive conservation and management plan.

SYSTEMATICS AND DEEPER PHYLOGENETICS

GENERAL BACKGROUND

Issues regarding intraspecific ESUs often translate into concerns about where to draw biological and taxonomic species boundaries, which in turn can translate into concerns about where to place conservation priorities. Sea turtle biologists have not been immune from controversial issues of this sort. Higher-level phylogenetic relationships are also relevant to taxonomy and systematics, and sometimes to setting conservation priorities (e.g., Avise 2005; Purvis et al. 2005). These issues too have been addressed in marine turtles using molecular genetic data.

SYNOPSIS OF PRE-1996 GENETIC FINDINGS

The genetic discoveries mentioned above about global phylogeographic patterns in the green/black (*mydas/agassizi*) complex of *Chelonia* turtles and in the olive/Kemp's (*olivacea/kempi*) complex of *Lepidochelys* turtles illustrate both sides of the species-taxonomy coin. In the case of Pacific green and Pacific black turtles, a conventional and presumably secure species-level distinction gained no support from the newer mitochondrial and nuclear DNA data, thus leading some authors to challenge (Grady and Quattro 1999; Karl and Bowen 1999) but others to defend (Pritchard 1999) the traditional morphology-based taxonomy. Conversely, in the case of ridley turtles, mtDNA data bolstered what had been a problematic species-level distinction between *olivacea* and *kempi*. Previously, these species were separable only on the basis of somewhat questionable morphological evidence (Bowen et al. 1991).

With respect to higher-echelon relationships, Bowen et al. (1993) used mtDNA sequences from cytochrome *b* to generate the first molecular phylogenetic estimate for all living sea turtles. Results included the following: the leatherback lineage branched off earliest in the marine turtle tree; Australia's flatback turtle (*Natator depressus*) is quite distinct from all other extant species, perhaps supporting its removal from *Chelonia* and resurrection of the genus *Natator* (Limpus et al. 1988); and the hawksbill is genetically allied most closely with the loggerhead/ridley complex.

UPDATED GENETIC DISCOVERIES

Since 1996, molecular surveys generally have supported the earlier genetic findings and have produced no further major surprises about marine turtle relationships. Thus, even after additional sampling of nesting areas and mtDNA sequences, the basal matrilineal divergence between olive and Kemp's ridleys is still evident (Bowen et al. 1997; Shanker et al. 2004), as is the lack of appreciable matrilineal distinction between Pacific black and green turtles (Dutton et al. 1996; Karl and Bowen 1999). Furthermore, a phylogenetic analysis of all extant species of marine turtle species based on additional mtDNA loci (Dutton et al. 1996) yielded consensus and total evidence phylogenies that were closely similar in structure to those originally presented by Bowen et al. (1993).

BIOLOGICAL AND CONSERVATION RELEVANCE

Proper taxonomies are crucial in conservation efforts, because they inevitably influence human perceptions about how the biological world is genetically partitioned, and because preserving genetic diversity is conservation's ultimate goal. By providing direct genetic evidence on genealogical relationships that formerly could only be surmised indirectly from phenotypes, molecular data can help systematists erect proper taxonomies and classifications. The conclusions discussed above about genetic relationships between the green and black turtles, and between Kemp's and olive ridleys, are prime examples of how molecular input can help clarify problematic taxonomies upon which conservation programs have been built.

At higher taxonomic echelons also, phylogenetic uniqueness has sometimes been promoted as a criterion for prioritizing extant species for conservation efforts. A typical argument goes as follows: Species that are phylogenetically most distinctive should have the highest conservation priority, because they contribute disproportionately to overall genetic diversity within a clade. In other words, distinctive genomes that carry much "independent evolutionary history" (IEH) should be valued more highly than those that have mostly shared histories with close living relatives (Krajewski 1994; May 1994; Humphries et al. 1995; Crozier 1997). By the IEH criterion, the leatherback should rank highest among all extant sea turtle species for preservation efforts, because its lineage is demonstrably the most ancient and distinctive.

However, many other ranking criteria (including rarity, geographic distribution, ecological importance, management feasibility, etc.) should also be factored into any conservation plans (Avise 2005). The bottom line is that all sea turtle species are inherently valuable, because, among other reasons, they are (or were) important biological components of marine ecosystems, and also, because they are aesthetically as well as scientifically compelling creatures.

CONCLUSIONS AND PERSPECTIVE

The post-1995 literature summarized in this review evidences an ongoing maturation of ecological– and evolutionary–genetic analyses of marine turtles. Several key biological conclusions from earlier genetic studies have been bolstered by the newer information, and additional details about sea turtle behaviors and life histories have come to light from numerous focused appraisals using molecular markers. My hope is that the recent molecular–genetic analyses summarized here, like their intellectual and technological predecessors from the prior decade, will contribute to our understanding and appreciation of marine turtles in ways that can promote efforts to protect these magnificent creatures.

However, scientific findings *per se* are only a part of any future equation for turtle conservation; societal and political actions will undoubtedly be paramount in determining whether these ancient mariners will someday thrive again or be lost. In commenting on the 25th Annual Symposium on Sea Turtle Biology and Conservation (held in Savannah, Georgia, in January 2005), the economist Paul Ferrero (2005) was struck by an odd juxtaposition: "the methodological sophistication that accompanied the analyses of sea turtle biology …and the absence of sophistication that accompanied the analyses of sea turtle conservation efforts." The molecular–genetic sciences as applied to marine turtles have indeed become quite refined. Whether or not societal attitudes toward sea turtles become enlightened, let us at least hope that future conservation efforts will be based on sound science and most of all that they will be effective.

REFERENCES

Avise, J. C. 1989. Nature's family archives. *Nat. Hist.* 3:24.
Avise, J. C. 1995. Mitochondrial DNA polymorphism and a connection between genetics and demography of relevance to conservation. *Conserv. Biol.* 9:686.
Avise, J. C. 2004. *Molecular Markers, Natural History, and Evolution*, 2nd edn. Sunderland, MA: Sinauer.
Avise, J. C. 2005. Phylogenetic units and currencies above and below the species level. In *Phylogeny and conservation*, A. Purvis, T. Brooks, and J. Gittleman (eds), Chap. 4. Cambridge: Cambridge University Press.
Avise, J. C., R. M. Ball, Jr., and J. Arnold. 1988. Current versus historical population sizes in vertebrate species with high gene flow: A comparison based on mitochondrial lineages and inbreeding theory for neutral mutations. *Mol. Biol. Evol.* 5:331.
Avise, J. C., and B. W. Bowen. 1994. Investigating sea turtle migration using DNA markers. *Curr. Biol.* 4:882.
Balloux, F., and N. Lugon-Moulin. 2002. The estimation of population differentiation with microsatellite markers. *Mol. Ecol.* 11:155.
Bass, A. L. 1999. Genetic analysis to elucidate the natural history and behavior of hawksbill turtles (*Eretmochelys imbricata*) in the wider Caribbean: A review and re-analysis. *Chelonian Conserv. Biol.* 3:195.
Bass, A. L., D. A. Good, K. A. Bjorndal, J. I. Richardson, Z. -M. Hillis, A. Horrocks, and B. W. Bowen. 1996. Testing models of female reproductive migratory behaviour and population structure in the Caribbean hawksbill turtle, *Eretmochelys imbricata*, with mtDNA sequences. *Mol. Ecol.* 5:321.
Bass, A. L., S. P. Epperlt, and J. Braun-McNeill. 2004. Multi-year analysis of stock composition of a loggerhead turtle (*Caretta caretta*) foraging habitat using maximum likelihood and Bayesian methods. *Conserv. Genet.* 5:783.
Bass, A. L., C. J. Lagueux, and B. W. Bowen. 1998. Origin of green turtles, *Chelonia mydas*, at "Sleeping Rocks" off the northeast coast of Nicaragua. *Copeia* 1998:1064.
Bass, A. L., and W. N. Witzell. 2000. Demographic composition of immature green turtles (*Chelonia mydas*) from the east central Florida coast: Evidence from mtDNA markers. *Herpetologica* 56:357.
Bjorndal, K. A. (ed.). 1995. *Biology and Conservation of Sea Turtles*. Washington, DC: Smithsonian Institution Press.
Bolker, B., T. Okuyama, K. Bjorndal, and A. Bolten. 2003. Sea turtle stock estimation using genetic markers: Accounting for sampling error of rare genotypes. *Ecol. Appl.* 13:763.
Bollmer, J. L., M. E. Irwin, J. P. Reider, and P. G. Parker. 1999. Multiple paternity in loggerhead turtle clutches. *Copeia* 1999:475.
Bolten, A. B., K. A. Bjorndal, H. R. Martins, T. Dellinger, M. J. Biscoito, S. E. Encalada, and B. W. Bowen. 1998. Trans-Atlantic developmental migrations of loggerhead sea turtles demonstrated by mtDNA sequence analysis. *Ecol. Appl.* 8:1.
Bolton, A. B., and B. E. Witherington (eds). 2003. *Loggerhead Sea Turtles*. Washington, DC: Smithsonian Institution Press.
Booth, J., and J. A. Peters. 1972. Behavioural studies on the green turtle (*Chelonia mydas*) in the sea. *Anim. Behav.* 20:808.
Bowen, B. W. 1995. Tracking marine turtles with genetic markers. *BioScience* 45:528.
Bowen, B. W., and J. C. Avise. 1996. Conservation genetics of marine turtles. In *Conservation Genetics: Case Histories from Nature*, J. C. Avise, and J. L. Hamrick (eds), Chap. 7. New York: Chapman and Hall.
Bowen, B. W., and A. L. Bass, 1997. Movement of hawksbill turtles: What scale is relevant to conservation, and what scale is resolvable with mtDNA data? *Chelonian Conserv. Biol.* 2:440.
Bowen, B. W., A. M. Clark, F. A. Abreu-Grobois, A. Chaves, H. A. Reichart, and R. J. Ferl. 1997. Global phylogeography of the ridley sea turtles (*Lepidochelys* spp.) as inferred from mitochondrial DNA sequences. *Genetica* 101:179.

Bowen, B. W., A. L. Bass, L. Soares, and R. J. Toonen. 2005. Conservation implications of complex population structure: Lessons from the loggerhead turtle (*Caretta caretta*). *Mol. Ecol.* 14:2389.

Bowen, B. W., A. B. Meylan, and J. C. Avise. 1991. Evolutionary distinctiveness of the endangered Kemp's ridley sea turtle. *Nature* 352:709.

Bowen, B. W., W. S. Nelson, and J. C. Avise. 1993. A molecular phylogeny for marine turtles: Trait mapping, rate assessment, and conservation relevance. *Proc. Natl Acad. Sci. USA* 90:5574.

Bowen, B. W., A. L. Bass, A. Garcia-Rodriguez, C. E. Diez, R. van Dam, A. Bolten, K. A. Bjorndal, M. M. Miyamoto, and R. J. Ferl. 1996. Origin of hawksbill turtles in a Caribbean feeding area as indicated by genetic markers. *Ecol. Appl.* 6:566.

Bowen, B. W., A. L. Bass, S. -M. Chow, M. Bostrom, K. A. Bjorndal, A. B. Bolten, T. Okuyama, B. M. Bolker, S. Epperly, E. Lacasella, D. Shaver, M. Dodd, S. R. Hopkins-Murphy, J. A. Musick, M. Swingle, K. Rankin-Baransky, W. Teas, W. N. Witzell, and P. H. Dutton. 2004. Natal homing in juvenile loggerhead turtles (*Caretta caretta*). *Mol. Ecol.* 13:3797.

Broderick, A. C., and B. J. Godley. 1997. Observations of reproductive behavior of male green turtles (*Chelonia mydas*) at a nesting beach in Cyprus. *Chelonian Conserv. Biol.* 2:615.

Casale, P., L. Laurent, G. Gerosa, and R. Argano. 2002. Molecular evidence of male-biased dispersal in loggerhead turtle juveniles. *J. Exp. Mar. Biol. Ecol.* 267:139.

Chassin-Noria, O., A. Abreu-Grobois, P. H. Dutton, and K. Oyama. 2004. Conservation genetics of the East Pacific green turtle (*Chelonia mydas*) in Michoacan, Mexico. *Genetica* 121:195.

Crim, J. L., L. D. Spotila, J. R. Spotila, M. O'Connor, R. Reina, C. J. Williams, and F. V. Paladino. 2002. The leatherback turtle, *Dermochelys coriacea*, exhibits both polyandry and polygyny. *Mol. Ecol.* 11:2097.

Crozier, R. H. 1997. Preserving the information content of species: Genetic diversity, phylogeny, and conservation worth. *Annu. Rev. Ecol. Syst.* 28:243.

Díez-Fernández, R., T. Okayama, T. Uchiyama, E. Carrillo, G. Espinosa, R. Márquez, C. Diez, and H. Koike. 1999. Genetic sourcing for the hawksbill turtle, *Eretmochelys imbricata*, in the northern Caribbean region. *Chelonian Conserv. Biol.* 3:296.

Dutton, P., E. Bixby, and S. K. Davis. 2000. Tendency toward single paternity in leatherbacks detected with microsatellites. In *Proceedings of the 18th International Symposium on Sea Turtle Biology and Conservation*, F. A. Abreu-Grobois, R. Briseño-Dueñas, R. Márquez, and L. Sarti (eds). Washington, DC: NOAA Technical Memorandum, p. 156.

Dutton, P. H., B. W. Bowen, D. W. Owens, A. Barragan, and S. K. Davis. 1999. Global phylogeography of the leatherback turtle (*Dermochelys coriacea*). *J. Zool. Lond.* 248:397.

Dutton, P. H., S. K. Davis, T. Guerra, and D. Owens. 1996. Molecular phylogeny for marine turtles based on sequences of the ND4-leucine tRNA and control regions of mitochondrial DNA. *Mol. Phylogenet. Evol.* 5:511.

Encalada, S. E., P. N. Lahanas, K. A. Bjorndal, A. B. Bolten, M. M. Miyamoto, and B. W. Bowen. 1996. Phylogeography and population structure of the Atlantic and Mediterranean green turtle *Chelonia mydas*: A mitochondrial DNA control region sequence assessment. *Mol. Ecol.* 5:473.

Encalada, S. E., K. A. Bjorndal, A. B. Bolten, J. C. Zurita, B. Schroeder, E. Possardt, C. J. Sears, and B. W. Bowen. 1998. Population structure of loggerhead turtle (*Caretta caretta*) nesting colonies in the Atlantic and Mediterranean regions as inferred from mtDNA control region sequences. *Mar. Biol.* 130:567.

Engstrom, T. N., P. A. Meylan, and A. B. Meylan. 2002. Origin of juvenile loggerhead turtles (*Caretta caretta*) in a tropical developmental habitat in Caribbean Panama. *Anim. Conserv.* 5:125.

Ferrero, P. J. 2005. An economist's reflections on the 25th Annual Symposium on Sea Turtle Biology and Conservation: Empirical program evaluation and direct payments for sea turtle conservation. *Mar. Turtle Newslett.* 109:2.

FitzSimmons, N. N. 1998. Single paternity of clutches and sperm storage in the promiscuous green turtle (*Chelonia mydas*). *Mol. Ecol.* 7:575.

FitzSimmons, N. N., C. J. Limpus, J. A. Norman, A. R. Goldizen, J. D. Miller, and C. Moritz. 1997a. Philopatry of male marine turtles inferred from mitochondrial DNA markers. *Proc. Natl Acad. Sci. USA* 94:8912.

FitzSimmons, N. N., C. Moritz, C. J. Limpus, L. Pope, and R. Prince. 1997b. Geographic structure of mitochondrial and nuclear gene polymorphisms in Australian green turtle populations and male-biased gene flow. *Genetics* 147:1843.

Godley, B. J., A. C. Broderick, and G. C. Hays. 2001. Nesting of green turtles *Chelonia mydas* at Ascension Island, South Atlantic. *Biol. Conserv.* 97:151.

Goldstein, D. B., and C. Schlötterer. 1999. *Microsatellites: Evolution and Applications.* Oxford: Oxford University Press.

Grady, J. M., and J. M. Quattro. 1999. Using character concordance to define taxonomic and conservation units. *Conserv. Biol.* 13:1004.

Harry, J. L., and D. A. Briscoe. 1988. Multiple paternity in the loggerhead turtle (*Caretta caretta*). *J. Hered.* 79:96.

Hatase, H., M. Kinoshita, T. Bando, N. Kamezaki, K. Sato, Y. Matsuzawa, K. Goto, K. Omuta, Y. Nakashima, H. Takeshita, and W. Sakamoto. 2002. Population structure of loggerhead turtles, *Caretta caretta*, nesting in Japan: Bottlenecks on the Pacific population. *Mar. Biol.* 141:299.

Hoekert, W. E. J., N. Neuféglise, A. D. Schouten, and S. B. J. Menken. 2002. Multiple paternity and female-biased mutation at a microsatelite locus in the olive ridley sea turtle (*Lepidochelys olivacea*). *Heredity* 89:107.

Humphries, C. J., P. H. Williams, and R. I. Vane-Wright. 1995. Measuring biodiversity for conservation. *Annu. Rev. Ecol. Syst.* 26:93.

Ireland, J. S., A. C. Broderick, F. Glen, B. J. Godley, G. C. Hays, P. L. M. Lee, and D. O. F. Skibinski. 2003. Multiple paternity assessed using microsatellite markers, in green turtles *Chelonia mydas* (Linnaeus, 1758) of Ascension Island, South Atlantic. *J. Exp. Mar. Biol. Ecol.* 291:149.

Karl, S. A., and B. W. Bowen. 1999. Evolutionary significant units versus geopolitical taxonomy: Molecular systematics of an endangered sea turtle (genus *Chelonia*). *Conserv. Biol.* 13:990.

Kichler, K., M. T. Holder, S. K. Davis, R. Márquez, and D. W. Owens. 1999. Detection of multiple paternity in the Kemp's ridley sea turtle with limited sampling. *Mol. Ecol.* 8:819.

Klemens, M. W. (ed.). 2000. *Turtle Conservation.* Washington, DC: Smithsonian Institution Press.

Krajewski, C. 1994. Phylogenetic measures of biodiversity: A comparison and critique. *Biol. Conserv.* 69:33.

Lahanas, P. N., K. A. Bjorndal, A. B. Bolten, S. E. Encalada, M. M. Miyamoto, R. A. Valverde, and B. W. Bowen. 1998. Genetic composition of a green turtle (*Chelonia mydas*) feeding ground population: Evidence for multiple origins. *Mar. Biol.* 130:345.

Laurent, L., P. Casale, M. N. Bradai, B. J. Godley, G. Gerosa, A. C. Broderick, W. Schroth, B. Schierwater, A. M. Levy, D. Freggi, E. M. Abd El-Mawla, D. A. Hadoud, H. E. Gomati, M. Domingo, M. Hadjchristophorou, L. Kornaraky, F. Demirayak, and C. Gautier. 1998. Molecular resolution of marine turtle stock composition in fishery bycatch: A case study in the Mediterranean. *Mol. Ecol.* 7:1529.

Lee, P. L. M., and G. C. Hays. 2004. Polyandry in a marine turtle: Females make the best of a bad job. *Proc. Natl Acad. Sci. USA* 101:6530.

Lee, P. L. M., P. Luschi, and G. C. Hays. 2007. Detecting female precise natal site philopatry in green turtles using assignment methods. *Mol. Ecol.* 16:61.

Lewison, R. L., S. A. Freeman, and L. B. Crowder. 2004. Quantifying the effects of fisheries on threatened species: The impact of pelagic long lines on loggerhead and leatherback sea turtles. *Ecol. Lett.* 7:221.

Limpus, C. J. 1993. The green turtle, *Chelonia mydas*. In Queensland: Breeding males in the southern Great Barrier Reef. *Wildl. Res.* 20:513.

Limpus, C. J., E. Gyuris, and J. D. Miller. 1988. Reassessment of the taxonomic status of the sea turtle genus *Natator* McCullock, 1908, with redescription of the genus and species. *Trans. R. Soc. S. Australia* 112:1.

López-Castro, M. C., and A. Rocha-Olivares. 2005. The panmixia paradigm of eastern Pacific olive ridley turtles revised: Consequences for their conservation and evolutionary biology. *Mol. Ecol.* 14:3325.

Luke, K., J. A. Horrocks, R. A. LeRoux, and P. H. Dutton. 2003. Origins of green turtle (*Chelonia mydas*) feeding aggregations around Barbados, West Indies. *Mar. Biol.* 144:799.

Lutz, P. L., J. A. Musick, and J. Wyneken (eds). 2003. *The Biology of Sea Turtles*, Vol. 2. Boca Raton, FL: CRC Press.

Maffucci, F., W. H. C. F. Kooistra, and F. Bentivegna. 2006. Natal origin of loggerhead turtles, *Caretta caretta*, in the neritic habitat off the Italian coasts, central Mediterranean. *Biol. Conserv.* 127:183.

May, R. M. 1994. Conceptual aspects of the quantification of the extent of biological diversity. *Philos. Trans. R. Soc. Lond. B* B. 345:13.

Miller, J. D. 1997. Reproduction in sea turtles. In *The Biology of Sea Turtles*, P. L. Lutz, and J. A. Musick (eds), Chap. 3. Boca Raton, FL: CRC Press.

Moore, M. K., and R. M. Ball, Jr. 2002. Multiple paternity in loggerhead turtle (*Caretta caretta*) nests on Melbourne Beach, Florida: A microsatellite analysis. *Mol. Ecol.* 11:281.

Mortimer, J. A., and D. Broderick. 1999. Population genetic structure and developmental migrations of sea turtles in the Chagos Archipelago and adjacent regions inferred from mtDNA sequence variation. In *Ecology of the Chagos Archipelago*, C. R. C. Sheppard, and M. R. D. Seaward (eds). *Linnean Soc. Occas. Publ. 2*, p. 14.

Mrosovsky, N. 1997. Movement of hawksbill turtles — A different perspective on the DNA data. *Chelonian Conserv. Biol.* 2:438.

Mrosovsky, N. 2000. *Sustainable Use of Hawksbill Turtles: Contemporary Issues in Conservation.* Issues in Management No. 1, Key Centre for Tropical Wildlife Management, Darwin Australia, 1.

Neff, B. D., T. W. Pitcher, and J. A. Repka. 2002. Bayesian model for assessing the frequency of multiple mating in nature. *J. Hered.* 93:406.

Okayama, T., R. Díaz-Fernández, Y. Baba, M. Halim, O. Abe, N. Azeno, and H. Koikel. 1999. Genetic diversity of the hawksbill turtle in the Indo-Pacific and Caribbean regions. *Chelonian Conserv. Biol.* 3:362.

Okuyama, T., and B. M. Bolker. 2005. Combining genetic and ecological data to estimate sea turtle origins. *Ecol. Appl.* 15:315.

Palmer, K. S., D. C. Rostal, J. S. Grumbles, and M. Mulvey. 1998. Long-term sperm storage in the desert tortoise (*Gopherus agassizi*). *Copeia* 1998:702.

Parker, P. G., T. A. Waite, and T. Peare. 1996. Paternity studies in animal populations. In *Molecular Genetic Approaches in Conservation*, T. B. Smith, and R. K. Wayne (eds). New York: Oxford University Press, p. 413.

Pearse, D. E., and J. C. Avise. 2001. Turtle mating systems: Behavior, sperm storage, and genetic paternity. *J. Hered.* 92:206.

Pearse, D. E., F. J. Janzen, and J. C. Avise. 2001. Genetic markers substantiate long-term storage and utilization of sperm by female painted turtles. *Heredity* 86:378.

Pella, J. J., and M. Masuda. 2001. Bayesian methods for analysis of stock mixtures from genetic characters. *Fish. Bull.* 99:151.

Pritchard, P. C. H. 1969. Studies of the systematics and reproductive cycles of the genus *Lepidochelys*. PhD Thesis, University of Florida, Gainesville.

Pritchard, P. C. H. 1999. Status of the black turtle. *Conserv. Biol.* 13:1000.

Purvis, A., T. Brooks, and J. Gittleman (eds). 2005. *Phylogeny and Conservation.* Cambridge: Cambridge University Press.

Rankin-Baransky, K., C. J. Williams, A. L. Bass, B. W. Bowen, and J. R. Spotila. 2001. Origin of logger-head turtles stranded in the northeastern United States as determined by mitochondrial DNA analysis. *J. Herpetol.* 35:638.

Rawson, P. D., R. MacNamee, M. G. Frick, and K. L. Williams. 2003. Phylogeography of the coronulid barnacle, *Chelonibia testudinaria*, from loggerhead sea turtles, *Caretta caretta*. *Mol. Ecol.* 12:2697.

Reece, J. S., T. A. Castoe, and C. L. Parkinson. 2005. Historical perspectives on population genetics and conservation of three marine turtle species. *Conserv. Genet.* 6:235.

Roberts, M. C., C. J. Anderson, B. Stender, A. Segars, J. D. Whittaker, J. M. Grady, and J. M. Quattro. 2005. Estimated contribution of Atlantic coastal loggerhead turtle nesting populations to offshore feeding aggregations. *Conserv. Genet.* 6:133.

Roberts, M. A., T. S. Schwartz, and S. A. Karl. 2004. Global population genetic structure and male-mediated gene flow in the green sea turtle (*Chelonia mydas*): Analysis of microsatellite loci. *Genetics* 166:1857.

Shanker, K., J. Ramadevi, B. C. Choudhury, L. Singh, and R. K. Aggarwal. 2004. Phylogeography of olive ridley turtles (*Lepidochelys olivacea*) on the east coast of India: Implications for conservation theory. *Mol. Ecol.* 13:1899.

Spotila, J. R. 2004. *Sea Turtles.* Baltimore, MD: Johns Hopkins University Press.

Sugg, D. W., and R. K. Chesser. 1994. Effective population sizes with multiple paternity. *Genetics* 137:1147.

Templeton, A. R. 1998. Nested clade analysis of phylogeographic data: Testing hypotheses about gene flow and population history. *Mol. Ecol.* 7:381.

Waples, R. S. 1998. Separating the wheat from the chaff: Patterns of genetic differentiation in high gene flow species. *J. Hered.* 89:438.

Whitlock, M. C., and D. E. McCauley. 1999. Indirect measures of gene flow and migration: $F_{ST} \neq 1/(4Nm+1)$. *Heredity* 82:117.

Witzell, W. N., A. L. Bass, M. J. Bresette, D. A. Singewald, and J. C. Gorham. 2002. Origin of immature loggerhead sea turtles (*Caretta caretta*) at Hutchinson Island, Florida: Evidence from mtDNA markers. *Fish. Bull.* 100:624.

18 Genetics and Applied Management: Using Genetic Methods to Solve Emerging Wildlife Management Problems

Randy W. DeYoung

CONTENTS

The science and profession of wildlife management were born during the early twentieth century as the need for a sound knowledge base and a corps of professionals to gather and implement the knowledge (e.g., biologists, managers, and wildlife scientists) became apparent (Mackie 2000). By this time, many wildlife species had declined in number or were locally extirpated in the United States due to overexploitation and loss of habitat. Accordingly, early wildlife management and research efforts in the United States were heavily influenced by a mandate of preservation and recovery. By the mid-to-late twentieth century, many charismatic species [e.g., deer, elk (*Cervus elaphus*), turkey (*Meleagris gallopavo*), and many species of waterfowl, wading birds, and raptors] were beginning to recover. The restoration of these species is a major conservation success story; so successful in fact that few outside of the wildlife realm are aware just how severe the declines were a few decades before. As game species recovered, a portion of wildlife research and management efforts shifted to focus on the sustainable use of these recovered species, developing harvest theory and refining survey methods. At the same time, many rare or lesser-known threatened and endangered species began to receive increased attention.

Today, wildlife managers are increasingly faced with a different set of problems. While conservation and the sustainable use of natural resources remain important, issues involving disease concerns, animal damage, and invasive species are becoming increasingly common. Each new wildlife management challenge requires reliable knowledge of animal behavior and population attributes upon which to base management decisions. In many cases, traditional approaches to wildlife research (e.g., tagging, banding, radiotelemetry) are inefficient (e.g., limited by cost, resources) or inadequate to provide this knowledge. Furthermore, contemporary wildlife management issues often involve multiple spatial scales, necessitating a transition from population-level management to management at the scale of landscapes or at least to the geographic extent of the population. Wildlife scientists and managers must be flexible enough to adjust their focus and change their scientific and management toolkits to confront the management issues looming on the horizon. The ability to recognize impending challenges and to efficiently use all available tools will be paramount. One set of tools, genetic methods, essentially form a "molecular toolbox" that have thus far received little attention in the realm of applied ecology and wildlife management (DeYoung and Honeycutt 2005).

BRIEF HISTORY OF GENETIC TECHNIQUES

Genetic tools first became available for use in wildlife in the form of a class of genetic markers termed allozymes. Pioneered by Lewontin and Hubby (1966) and Harris (1966), allozyme markers involved the detection of alternative forms of proteins and enzymes among individuals, populations, and species (Avise 2004). Before this time, a large body of theoretical genetic research existed, but was limited in practice because the ability to index genetic variation below the level of quantitative traits was limited (Hedrick 2000). Identification of species, populations, demes, and individuals requires the presence of genetic variation as a basis for decision. For many decades, the only means of detecting population genetic variation was by quantitative characters (e.g., differences in color, morphology), chromosomal variants, or blood antigen groups, all of which face severe limitations in the amount and type of genetic variation available for study (Hedrick 2000; Avise 2004). Allozymes became the first method for assessing genetic variation at the molecular level, allowing the application of population genetics theory to empirical data; the intellectual legacy of giants in the field of theoretical population genetics, such as Sewall Wright, Theodosius Dobzhansky, Ronald A. Fisher, J. B. S. Haldane, and many others, could now be tested, refined, and used to make inferences about populations (Table 18.1).

Allozyme markers, which are easy to use and require relatively little in terms of specialized equipment, fostered important advances in understanding the partitioning of genetic variation within and among populations. However, allozymes underestimate the amount of genetic variation present because only mutations that affect the net charge of proteins, and thus their rate of migration through a gel medium when exposed to electric current, are detected (Avise 2004). Allozymes also require relatively large samples of tissue, often necessitating euthanasia of the organism. Advances in DNA sequencing technology (Sanger et al. 1977) during the 1970s and 1980s have permitted the detection and characterization of genetic variation at the DNA sequence level. The description of the polymerase chain reaction and the discovery of thermostable DNA polymerase in the 1980s (Saiki et al. 1988) allowed the *in vitro* amplification of minute quantities of DNA (as little as one molecule) and rendered the thermal cycling process amenable to automation. Thus, nondestructive sampling, including noninvasive sampling, became possible, and a wider range of species could be studied. However, use of the new technology required considerable technical expertise, was time consuming and limited in terms of throughput, and could be costly in terms of instrumentation and reagents. As a result, genetic studies of wildlife species were largely limited to threatened and endangered species or to questions of higher-level taxonomy.

TABLE 18.1

Pioneers in the Field of Theoretical Population Genetics

Theoretician	Contribution
Sewall Wright (1889–1988)	Provided the theoretical basis that underpins much of modern population genetics, including inbreeding, genetic drift, and population size and structure. Wright's 1968, 1969, 1977, and 1978 volumes provide a thorough and extensive overview of population genetic theory
Theodosius Dobzhansky (1900–1975)	Extensive influence on diverse fields of biological science; several of his students became prominent scientists; *Genetics and Origin of Species* (Dobzhansky 1941) was a key synthesis of modern evolutionary theory
Ronald A. Fisher (1890–1962)	A prominent statistician, Fisher also made major contributions linking population genetics and evolutionary theory, theoretical aspects of selection and estimation of genetic parameters; *The Genetical Theory of Natural Selection* (1930) unified natural selection and population genetics
J. B. S. Haldane (1882–1964)	Haldane's contributions, together with Wright and Fisher, arguably provide the foundation of population genetic theory. Haldane's mathematical approach provided insights into the interaction of selection and mutation, and to understanding the dynamics of allelic polymorphism

Source: Information from Hedrick, P. W. 2000. *Genetics of Populations*, 2nd edn. Sudbury, MA: Jones and Bartlett.

Analyses based on DNA sequence data represent the most accurate method of detecting genetic variation at the nucleotide level, and the ease with which DNA sequences can be obtained has increased markedly in recent years (Avise 2004). Continuing advances in the number and type of genetic markers available have revolutionized genetic approaches to population biology (Honeycutt 2000; DeYoung and Honeycutt 2005). The discovery that simple sequence repeats are widely distributed throughout the genome and could be used as a source of highly variable genetic markers was especially important for population genetics. One class of genetic markers, DNA microsatellites, has proven particularly useful. Microsatellites are short (10–100 bases), highly repetitive sequences (Weber and May 1989) occurring in the form of 2–5 base-pair repeats (e.g., $[AC]_n$ or $[CAG]_n$). Microsatellite loci have higher mutation rates than most other DNA sequences (Hancock 1999), resulting in a large number of alleles per locus. This variability makes microsatellite loci particularly valuable genetic markers for studies of wildlife populations, especially studies that focus on gene flow and dispersal, social and geographic structuring, and recent population history (Beaumont and Bruford 1999).

The availability of highly variable genetic markers and the development of automated DNA sequencing instrumentation have made large-scale genetic studies of wildlife populations attainable (Honeycutt 2000; DeYoung and Honeycutt 2005). Although genetic analyses are not cheap, the cost per sample is decreasing, as increased automation multiplies the number of samples that can be processed and reduces labor cost and time investment. Importantly, the ability to rapidly and efficiently generate large genetic datasets has spurred the development of new analytical methods that take advantage of continuing increases in desktop computing power, making possible the use of the large body of genetic theory.

Thus, a suite of technical and theoretical advances has enabled analyses and applications that were too expensive, too difficult, or in some cases impossible, only a short while earlier. The combination of demographic information, spatial data, and molecular techniques can be extremely valuable for better understanding the social biology, population structure, and population dynamics of wildlife (Hampton et al. 2004b; DeYoung and Honeycutt 2005). In turn, these parameters are important in formulating and implementing effective management plans for issues ranging from wildlife disease, wildlife damage, and invasive species. Genetic tools have been used in a conservation context for many years and have recently become highly popular for investigating animal behavior and population-level questions (Hedrick and Miller 1992; Hughes 1998; Avise 2004). In fact, the use of genetic markers to investigate animal ecology and behavior is now widely considered a discipline itself, termed *molecular ecology* (Burke 1994; Palsböll 1999). Although genetic tools have received little use in an applied management context to date, this may be part of a natural progression from specialized use to more widespread application as the technology and analytical methods are refined and more labs focus on the use of genetic methods in wildlife species. This chapter is focused on what I perceive to be some current and future challenges in the applied ecology and management of wildlife species, and how genetic tools can help surmount these challenges.

DISEASE, DAMAGE, AND INVASIVE SPECIES: NEW CHALLENGES IN WILDLIFE MANAGEMENT

Historically, human–wildlife conflicts revolved mainly around the take of livestock by predators (e.g., Ballard and Gipson 2000). In the social and political climate of the time, the solution was fairly simple: eradicate all predators that affected livestock. Today's wildlife professionals face new and potentially devastating challenges involving disease, damage, and invasive species (Table 18.2). Some of the specific challenges raised can be illustrated by the following three examples. These examples illustrate how genetic methods could be applied to improve the effectiveness of management.

CASE STUDY 1: WHITE-TAILED DEER OVERABUNDANCE, DAMAGE, AND DISEASE

It is ironic that some new management challenges are a direct result of the success of past management actions and serve to emphatically illustrate the transition from historic to current management challenges during the past few decades; white-tailed deer (*Odocoileus virginianus*) are a prime example. White-tailed deer were nearly extirpated in the southeastern United States by the early 1900s because of overexploitation. Deer recovered due to the passage and enforcement of game laws, establishment of refuges, and vigorous trapping and transplanting programs (Blackard 1971). In fact, deer in the southeastern United States and elsewhere have recovered to the extent that they are considered overabundant in many areas (McShea et al. 1997).

The population recovery and overabundance of white-tailed deer has led to several management problems. High densities of deer typically result in damage to natural habitat to the extent of changing plant communities and plant successional trends and affecting other wildlife species (Waller and Alverson 1997; Côté et al. 2004; Gordon et al. 2004). Agricultural crops and ornamental plants in urban neighborhoods also suffer damage from overbrowsing (Waller and Alverson 1997). Second,

TABLE 18.2
Emerging Wildlife Management Challenges

Issue	Challenge	Examples
Overabundance	Preserve habitat quality, minimize human–wildlife conflicts	White-tailed deer, feral pigs
Disease	Manage endemic pathogens, contain foreign pathogens	Chronic wasting disease, rabies, foot and mouth
Invasive species	Potential to limit population expansion or reduce damage	Feral pigs, Norway rat, fire ant
Scale	Manage at population scale, not local scale or by arbitrary units	Populations with continuous distribution, highly vagile species

where high densities of deer occur in proximity to roadways, collisions with automobiles increase, resulting in property damage and the potential for human injury (Conover et al. 1995). Third, high densities of deer result in the spread of pathogens that affect humans, livestock, and other cervids. Examples of these pathogens include bovine tuberculosis, ticks that carry Lyme disease, chronic wasting disease, and a type of brainworm that white-tailed deer tolerate but is deadly to elk and moose (*Alces alces*) (Conover 1997; Davidson and Doster 1997). Finally, white-tailed deer populations are expanding into areas of the western United States where they have not historically occurred, hybridizing with mule deer (*Odocoileus hemionus*) (Cathey et al. 1998).

These management problems are not simple to solve. In many cases, hunting alone will not suffice because harvest pressure will not increase sufficiently, even if bag limits are raised, due to hunter saturation; each hunter or family can only process and consume a certain amount of deer meat, and many hunters cease to harvest after their individual needs are met (Riley et al. 2003). Reduction of deer density in local areas through removal or sterilization has been recommended for disease and damage control (Muller et al. 1997). However, deer are distributed continuously in many areas, making it difficult to define the geographic area to target or to predict and interrupt disease transmission. Approaches based on social behavior of female white-tailed deer have been recommended (Porter et al. 1991; McNulty et al. 1997), but it is not certain if these approaches will apply in all deer populations, especially in high-density populations or where high rates of female dispersal occur due to limited availability of cover during parts of the year (e.g., Nixon et al. 1991).

CASE STUDY 2: FERAL SWINE, AN EXOTIC INVASIVE THAT POSES RISKS FROM DAMAGE AND DISEASE

Feral swine (*Sus scrofa*) are an exotic invasive pest species that were first introduced into the United States as early as the 1400s when Europeans were exploring and settling in North America (Mayer and Brisbin 1991). Since this time, many accidental and intentional introductions consisting of domestic and wild stock have occurred. Although some feral swine have been present in the United States for >200 years, the number and distribution of feral swine have increased dramatically in recent decades. For instance, the Southeastern Cooperative Wildlife Disease Study (2004) recently reported feral swine occurring in 28 states, spanning the United States from California to Virginia. The United States population is estimated at 4 million individuals (Nettles 1997; Pimentel et al. 2000), with as many as half occurring in Texas (Mapston 2004).

Increased damage to agriculture, natural ecosystems, and the environment has been coincident with the explosion in feral swine. Feral swine consume most types of agricultural crops produced in the United States (Donkin 1985; Sweeney et al. 2003). Furthermore, feral swine wallowing behavior can cause sedimentation of livestock ponds and tanks (Mapston 2004), resulting in algae blooms, oxygen depletion, bank erosion, and soured water (Sweeney et al. 2003). Feral swine cause livestock losses by depredating on sheep (Moule 1954; Rowley 1970; Pavlov et al. 1981; Choquenot et al. 1997), goats, and newborn cattle. Feral swine also cause extensive damage to native plant communities by rooting, or using their snout to dig for food items (Bratton 1975; Wood and Barrett 1979; Stone and Keith 1987). Swine consume a variety of wildlife, including earthworms, grasshoppers, beetles, salamanders, frogs, snakes, rodents, eggs and chicks of ground-nesting birds, and white-tailed deer fawns (Wood and Roark 1980; Howe et al. 1981; Baber and Coblentz 1987; Hellgren 1993).

Furthermore, there are serious concerns regarding the potential of large populations of feral swine to act as a reservoir for disease. Feral swine harbor numerous viral and bacterial diseases (Williams and Barker 2001) and are susceptible to many internal and external parasites, such as nematodes, roundworms, flukes, lice, and ticks (Samuel et al. 2001). Many of these diseases and parasites also affect livestock, other wildlife, and humans. Of particular concern are pseudorabies, swine brucellosis, bovine tuberculosis, vesicular stomatitis, and leptospirosis. There is also concern that feral swine could play a significant role in the spread of an exotic animal disease, such as foot and mouth, rinderpest, African swine fever, or classical swine fever (Witmer et al. 2003).

Attempts to control feral swine populations have traditionally used both lethal and nonlethal methods. Nonlethal methods include exclusion by fencing and habitat modification (Littauer 1993; Mapston 2004). Lethal methods for feral swine control include snares, cage traps, hunting, and aerial shooting (Littauer 1993). Fencing, however, requires considerable maintenance [in the form of vegetation control; Littauer (1993)] and may not permanently control feral swine (Mapston 2004), functioning primarily by shifting the problem to adjacent areas. Removal methods also have limitations and drawbacks, including high manpower and decreased effectiveness over time (trapping), low population impact (snares), high cost, and limited area of effectiveness (aerial shooting).

Eradication of feral swine is not feasible in most situations. An integrated approach, using a variety of lethal methods complemented by the best available information on population dynamics and structure, is often recommended to temporarily control feral swine to alleviate seasonal damage (Kammermeyer et al. 2003). However, managed areas are often quickly recolonized, and thus damage becomes a chronic, recurring problem.

CASE STUDY 3: GRAY FOX AND RABIES IN THE SOUTHWESTERN UNITED STATES

In the United States, animal rabies generally occurs in free-ranging species of mammals, often small carnivores such as raccoons (*Procyon lotor*), skunks, foxes, and bats, where genetically distinct rabies strains are present in distinct geographical areas. For instance, ~92% of reported United States rabies cases in 2004 were in wild animals (Krebs et al. 2005). The transmission of rabies in wild populations occurs primarily among conspecifics and in defined geographic regions, with a low rate of interspecific infection. Within these regions, rabies outbreaks can be highly persistent, lasting decades, and perhaps longer once established (Real et al. 2005). The geographic area harboring infected animals may be temporally variable and appears to be affected by population processes, terrain features that influence animal movements, and population density (Childs et al. 2000, 2001).

In central Texas, a distinct gray fox (*Urocyon cinereoargenteus*) rabies strain is maintained, posing a significant threat to human and animal health. To combat this threat, the Texas Department of State Health Services and Texas Wildlife Services began an oral rabies vaccine (ORV) program in 1996. The aim of the program is to aerially disperse edible baits containing a rabies vaccine throughout the geographic area of infection. Animals consuming the baits become immunized; when a sufficient portion of the population is immune, the enzootic is disrupted. The current gray fox ORV zone in Texas extends from the Mexican border to west–central Texas, requiring the release of two million ORV baits in 2003, a considerable expense in terms of cost and manpower. During the course of the ORV program, it has become apparent that more information is needed regarding gray fox movements and dispersal. For instance, breaks in the ORV zone (e.g., rabid foxes outside the present vaccination zone) appear to occur only in select geographical locations. It is suspected that these are located in areas where terrain features promote dispersal or long-distance movements, but this is difficult to verify through traditional means, such as radiotelemetry or recovery of marked animals.

COMMON THEMES IN APPLIED MANAGEMENT CASE STUDIES

A thorough understanding of population biology, social behavior and social structure, and animal movements at multiple scales is needed to provide effective disease containment and damage-management strategies. Perhaps most important is the need to increase the efficiency and effectiveness of existing control methods so that management goals are achievable in a timely fashion with a minimal impact on animal and human welfare. For instance, predictions of disease transmission for nonvector-borne diseases are most reliable when informed by detailed data on contact rates among individuals and populations. Contact rates are influenced by a variety of factors, including dispersal distances, habitat, and social structure (Alitzer et al. 2003). Contact rates among individuals in social groups may be estimated by visual observation if individuals occupy open habitats. However, rates of cryptic or infrequent contact, such as sexual contact, among individuals in wild populations may be difficult to estimate through visual observation, even where individuals appear to be highly visible. Consider the high rates of promiscuity in many species of birds, which were thought to be monogamous before the advent of genetic parentage testing (Petrie and Kempenaers 1998) and the finding that social dominance may not equate to reproductive success in species of large mammals (Coltman et al. 1999; Worthington Wilmer et al. 1999; Gemmel et al. 2001). These and many other similar observations are prime examples of the inadequacy of visual observations to track true patterns of behavior. Unfortunately, the lack of knowledge of animal behavior may severely affect accuracy and conclusions of epidemiologic models. For example, the validity of modeling efforts aimed at predicting the spread of chronic wasting disease in deer and elk has been criticized because transmission modes and rates of contact among individuals are poorly known (Schauber and Woolf 2003).

Management units may be defined as populations or groups of populations that exchange few or no individuals such that they are functionally independent of one another, yet are not so different as to be phylogenetically unique (Moritz 1994). Management units may be relatively easy to define in species that are habitat specialists simply by delineating habitat boundaries. The issue becomes more complicated for species with a high capacity for dispersal, species that display migratory behavior, or species that are apparently continuously distributed. Thus, in the absence of prior knowledge about population structure, it may be very difficult to define boundaries for some populations. Management units are often defined arbitrarily, such as along property or political boundaries (e.g., county, state, national borders). However, animal movements and dispersal are not random across the landscape, but are influenced by a variety of environmental (e.g., habitat, terrain) and social (e.g., dispersal, social structure) factors.

The uninformed definition of management units often results in negative or ineffective outcomes for management actions. For instance, elimination of threats from animal disease or damage may require removal of individuals through trapping or euthanasia to reduce population size (and thus the amount of damage) or to decrease the probability of contact among individuals. Population reduction may be inefficient in terms of manpower and resources, especially for highly vagile species, which can quickly recolonize managed areas. For populations that are continuously distributed, the problem is how to define the target area when there are no obvious breaks or population boundaries. In these situations, long-term control requires a twofold action: definition of a target area for management and preventing recolonization of the managed area. Conservative definition of management units increases cost of control methods while a focused approach may not affect the entire local population. Therefore, one must (1) manage at the scale of local populations and (2) identify and target dispersal corridors. Management decisions informed by population structure, including natural population boundaries and dispersal corridors (rivers, streams, etc.) could dramatically increase success of management actions. In this manner, management efforts could be concentrated at specific sites, thus increasing efficiency and effectiveness of management actions. Furthermore, managers could take advantage of habitat features or animal behavior. For instance, population boundaries could be used in a "divide-and-conquer" strategy, rather than focus removal efforts over a vast area.

Ingenuity and innovation in new management strategies may help clear some hurdles. However, solutions for many of these new management challenges clearly lie in application of old-fashioned applied wildlife management. The missing ingredient is often knowledge of specific population parameters or behaviors, and the interaction of these attributes with biotic and abiotic features of the environment. How, then are we to achieve this knowledge so that we may surpass management

TABLE 18.3

Genetic Approaches to Wildlife Management Problems

Issue	Challenge	Approach
Emerging infectious disease or pathogen	Predict transmission rate or prevalence	Dispersal, parentage, relatedness, landscape genetic methods
	Containment	Management units, dispersal, assignment, landscape genetics
	Assess effectiveness of control	Genetic bottleneck, effective population size, STAR
	Increase efficiency of control	Management units, dispersal, assignment, landscape genetics
Animal damage	Containment	Management units, dispersal, landscape genetics
	Predict future occurrence	Dispersal, landscape genetics
	Assess effectiveness of control	Genetic bottleneck, effective population size, STAR
	Increase efficiency of control	Population structure, landscape genetic methods
Invasive species	Containment	Management units, dispersal, landscape genetics
	Identify population of origin or source of invasion	Assignment methods
	Hybridization	Assignment methods

Specific methods are described in text.

obstacles? Despite the fact that these generalized impending management challenges arrive from diverse fronts, there are commonalities in that management solutions rely on knowledge of basic animal behavior and population attributes, including

1. Population boundaries, management units, or neighborhood size.
2. Population connectivity, interrelation between population dynamics, dispersal, and habitat continuity.
3. Identification of immigrant and resident individuals.
4. Identification of landscape features affecting animal movements and dispersal.

Thus, recognition and application of new tools aimed at securing reliable knowledge to inform conventional management approaches should be a priority. Genetic approaches offer a great deal of promise for applied ecology and management in that genetic approaches have been explicitly developed for the study of animal behavior and population attributes. Now that suitable markers are available which permit acquisition of data, the large and well-developed body of population genetic theory can be applied to nearly any management challenge (Table 18.3).

THEORETICAL FOUNDATIONS OF POPULATION GENETICS

Differences in mating system, social behavior, dispersal, population size, habitat variables, and so forth, may contribute to the structuring of populations into subpopulations or demes (Chesser 1991a,b; Sugg et al. 1996; Tiedemann et al. 2000), some of which may occur at very fine scales even in highly vagile organisms (e.g., Purdue et al. 2000; Nussey et al. 2005). Thus, estimation of population structure and exploration of factors causing structure have long been of fundamental interest and importance in population genetics. An important early contribution to population genetics, especially to detecting the influence of demographic and other processes on patterns of genetic variation, was the concept of describing populations in terms of allele frequencies rather than genotype frequencies

(Hedrick 2000). This led to the development of the Hardy–Weinberg (HW) principle, independently conceived by G. H. Hardy and W. Weinberg in 1908, which states that in an idealized population characterized by random mating and absence of gene flow, selection, and mutation, allele frequencies will remain unchanged among generations (Hedrick 2000). Departure of allele frequencies from HW expectations therefore indicates that one or more assumptions of the ideal population are violated. For instance, Wahlund (1928) showed that the grouping of samples from populations differing in allele frequencies results in a departure from HW proportions in the form of an excess of homozygotes, even if the separate populations are themselves in equilibrium. The detection of a "Wahlund effect" thus indirectly indicates the presence of population structure.

Wright (1951, 1965) developed the first formal means of describing population structure. Wright's method involves correlation coefficients termed "F-statistics" that partition genetic variation over the total population, among population subdivisions, and among individuals within populations. The coefficients are commonly used in population genetics, where F_{ST} represents the amount of genetic differentiation among subpopulations, F_{IT} the deviation from HW expectations in the total population, and F_{IS} the deviation from HW expectations within subpopulations. Wright's basic approach has been modified and extended (e.g., Weir and Cockerham 1984; Nei 1987), and in some ways superseded by newer approaches, but remains important as a theoretical basis for assessing relative degrees of population differentiation and gene flow (Neigel 2002).

Several conceptual models of population structure have been developed which can be extended to assess gene flow and migration rates (Neigel 1997; Hedrick 2000). Wright's continent–island model (Wright 1940), where some individuals from a large "continent" population disperse to several "island" populations each generation, was one of the first attempts to understand the effect of gene flow and population size on genetic similarity and diversity. For the case of populations that are continuously distributed, demes may become differentiated if dispersal distance is limited through isolation by distance (Wright 1938, 1940). For this case, Wright (1943) proposed the term "neighborhood," an area defined by the standard deviation of the per-generation gene flow (V), where the size of the neighborhood circle is $4\pi V$ (Hedrick 2000). This approximates the geographic distance beyond which subpopulations are effectively independent. Models have been developed and extended to consider more complex population structures, including the stepping-stone model, where migration occurs only among geographically proximate populations (Maruyama 1970), and metapopulation models, where more complex migration, extinction, and colonization events are considered (Hastings and Harrison 1994; Harrison and Hastings 1996). Other approaches for assessing population structure include analysis of molecular variance (AMOVA), an approach akin to an analysis of variance on allele frequency data (Cockerham 1969, 1973; Excoffier et al. 1992; Weir and Cockerham 1984; Weir 1996). The AMOVA approach allows population structure to be examined in a hierarchical fashion. For example, genetic variation may be partitioned among groups, among populations within group, among individuals, and within individuals (Weir 1996).

POPULATION STRUCTURE: SOCIAL STRUCTURE, MANAGEMENT UNITS, AND FACTORS AFFECTING POPULATION DISTRIBUTION AND EXCHANGE

Theoretical models are an important foundation for understanding population structure. However, some theoretical approaches are limited in a management context because the spatial location of discontinuities is not explicitly addressed. Furthermore, assumptions of simple population models, such as the continent–island model, are not realistic for many natural populations, especially when populations have been admixed or have different demographic histories (Hedrick 1999; Nei and Kumar 2000). Therefore, indirect estimates of gene flow derived using these simple population models are often unrealistic (Whitlock and McCauley 1999). Finally, it can be difficult to avoid the arbitrary definition of population boundaries or sampling areas, which may not capture true population parameters.

Recently, there has been increased emphasis on addressing the spatial genetic structure of populations in a more explicit manner, especially identification of geographic features influencing population distribution and exchange (Holderegger and Wagner 2006). Geographic variation in gene frequencies can be used to explore how ecological characteristics of populations and landscape features (or changes in features) lead to nonrandom spatial associations (Sokal et al. 1997; Epperson 2003). A variety of approaches to define population boundaries or the location of genetic discontinuities have been proposed or refined (reviewed in Manel et al. 2003; Scribner et al. 2005). Essentially, these landscape genetic approaches involve integration of two or more data sets composed of genetic and ecological or geographical information (Manel et al. 2003; Scribner et al. 2005). The choice of methods may depend on the amount, extent, and type of genetic data that can be collected. Often, two or more approaches are used in concert to provide a greater strength of evidence. The combination of spatial and genetic data, and especially the integration of genetic and GIS technology, bears perhaps the greatest promise for applied management.

Two relatively straightforward methods for assessing population boundaries detect the presence of structure or dispersal barriers indirectly. Estimating the correlation between genetic and geographic distances allows the detection of a pattern of isolation by distance, expected where dispersal is limited in distance compared to the extent of sampling. Correlations between matrices of genetic and geographic distances among sampling sites are performed using Mantel or partial Mantel methods (Mantel 1967). Discontinuities in allele frequencies among sampling sites (indicative of barriers to dispersal or exchange among populations) can be indirectly detected by noting changes in the correlation among sampling sites on either side of putative barriers. The weakness of this method is that hidden or cryptic barriers may be difficult to detect and the spatial extent of the relationship is not defined (Diniz-Filho and Telles 2002).

Similarly, one may indirectly detect the presence of genetic discontinuities caused by barriers to dispersal through the serial pooling of data. One obtains samples from a number of sites spanning regular intervals of geographic distance, for instance in a linear fashion. Standard measures of population subdivision, such as F_{ST} are used first to test for the presence of population structure. If significant structure is present, then one can assess the scale of structure by systematically pooling samples in order of geographic proximity, calculating F_{IS} at each pooling step. An increase in F_{IS} between two pooling steps is evidence that the pooled sample includes more than one genetically distinct unit in terms of allele frequencies (e.g., Goudet et al. 1994).

Spatial autocorrelation is a statistical approach that describes the autocorrelation of allele frequencies between individuals or populations as a function of spatial distance, thus allowing an estimate of nonrandom patterns of genetic variation arising from family or social structure, incomplete dispersal, and so forth. Moran's I is often used as the autocorrelation statistic, and provides an estimator of Wright's coefficient of relationship when computed from individual allele frequencies (Hardy and Vekemans 1999). The approach is to calculate pairwise values of Moran's I between all individuals in sets of arbitrary distance classes to determine the mean value within each distance class. The resulting correlogram can indicate the geographic distance over which samples are effectively independent (neighborhood size), read as the last distance class for which the autocorrelation statistic is significantly different from a null or permuted value (Figure 18.1; Diniz-Filho and Telles 2002). The shape of the correlogram itself is also informative, indicating whether autocorrelation arises from factors such as limited dispersal distance or local structure (Diniz-Filho and Telles 2002). Spatial autocorrelation methods represent improvement over indirect methods because individuals can be used as the basis for comparison and the spatial extent of the correlation can be identified, but do not allow the precise location of barriers or boundaries (Manel et al. 2003). Other approaches include space–time autoregressive (STAR), a method for the joint consideration of temporal and spatial processes affecting nonrandom association of alleles (Scribner et al. 2005). Empirical examples of autocorrelation and STAR methods in applied management are described in Scribner et al. (2005).

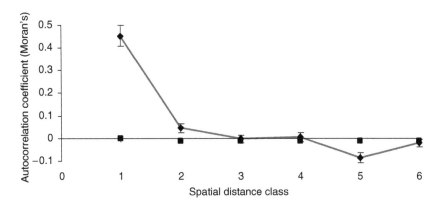

FIGURE 18.1 Correlogram derived from spatial autocorrelation analysis of allele frequency data in deer (R. W. DeYoung, unpublished data). The autocorrelation is significant for distance classes 1 and 2 and becomes nonsignificant by class 3. The squares represent null expected values derived through permutation; bars are ±1SE.

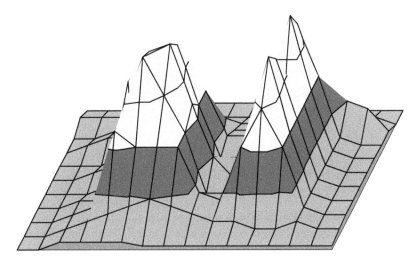

FIGURE 18.2 Hypothetical interpolation map of principal component scores derived from genetic marker data. In this manner, the spatial location of genetic discontinuities within a sampling area can be visualized. This analysis suggests two genetically distinct groups.

Concurrent with advances in genetic methods has been the introduction and proliferation of geographic information systems (GIS), providing the means to conduct detailed analyses of spatial patterns of variation in a geo-referenced environment. Thus, the combination of genetic marker data and GIS allows a detailed study of environmental features affecting population boundaries and population connectivity. Detection of changes in allele frequencies over short geographic distances can indicate barriers to gene flow. One approach involves collection of genetic data at several sampling sites spanning the area of interest followed by a principal components analysis on the allele frequency data. Principal component scores of each sampling site are interpolated and barriers are identified as zones of maximum slope following the contours of PC scores and overlaid on landcover maps (e.g., Cavalli-Sforza et al. 1994; Piertney et al. 1998). Thus, geographic features acting as barriers may be identified and visualized (Figure 18.2; Manel et al. 2003). A Bayesian approach attempts to group individuals into putative populations, emphasizing minimal departure from HW expectations or linkage equilibrium (e.g., Pritchard et al. 2000). Clusters of individuals that meet the criteria for population membership can then be plotted on a map of the area to visualize the population distribution and boundaries. Alternatively, a relatively new method employs a Bayesian Markov chain Monte Carlo (MCMC) method to explicitly identify the location of population boundaries by modeling the global set of sampled individuals as a spatial mixture of panmictic populations (Guillot et al. 2005).

ASSIGNMENT METHODS: DIRECT IDENTIFICATION OF INDIVIDUALS, MIGRANTS, AND POPULATIONS

Assignment methods, commonly referred to as "assignment tests," are a collection of related methods that seek to identify individuals or populations based on allele frequency data (e.g., Paetkau et al. 1995; Rannala and Mountain 1997; Cornuet et al. 1999; Pritchard et al. 2000). Individuals are assigned to their most likely population of origin based on the probability of their genotype occurring in a population (Manel et al. 2005). As discussed previously, estimation of migration rates between two or more populations may be unrealistic if conditions of the underlying theoretical model are violated, which occurs in most real populations (Whitlock and McCauley 1999). In contrast, assignment methods attempt direct, explicit identification of migrants or individuals with migrant ancestors, where the confidence in the results can be explicitly stated in terms of probability (Manel et al. 2002). First developed to identify dispersers and hybrids, assignment tests may be used in a variety of applied contexts, including verifying population of origin for disease-positive individuals, verifying illegal releases or transfers, identifying population of origin for introduced or invasive species, and indexing population structure (Rannala and Mountain 1997; Paetkau et al. 1998; Pritchard et al. 2000; Blanchong et al. 2002; Manel et al. 2002; Berry et al. 2004).

GENETIC BOTTLENECKS AND EFFECTIVE SIZE: ASSESSING DEMOGRAPHIC HISTORY AND EFFECTIVENESS OF CONTROL METHODS

The principle of adaptive management strikes a balance between uncertainty (lack of knowledge) and the urgent need for management action. The goal is to proceed with management based on the best available knowledge, then modify management actions based upon their success or failure and incorporate new knowledge as it becomes available, thus improving the effectiveness of management over time. Clearly, gauging the need to adapt relies on the ability to generate information on effectiveness of management actions. For instance, there is often uncertainty regarding the recent and historical demographic history of many populations. Managers may need to know if current management problems are the result of recent or historical increases in population size or geographic extent of populations. Furthermore, the effectiveness of removal or population reduction methods may be difficult to gauge because suitable survey methods are lacking for many species; thus, there is uncertainty as to the extent that management actions have actually affected the target population.

Demographic histories of populations may be estimated from genetic marker data in single or temporally spaced samples. Tests for genetic bottlenecks rely on the theoretical prediction that alleles are lost before heterozygosity declines during a drastic reduction in population size. Heterozygosity may remain relatively high for several generations (depending on the effective population size) until a balance is reestablished between the number of alleles and average heterozygosity (Cornuet and Luikart 1996). Tests designed to detect this temporary heterozygosity excess appear to perform well on simulated and real data (Cornuet and Luikart 1996; Luikart et al. 1998). For recently founded or invasive populations, the number of founders determines the number of alleles in the population, while average heterozygosity is influenced mainly by the population's growth rate (Nei et al. 1975; Hedrick 2000). Populations that increase rapidly retain more neutral genetic variation because any losses of heterozygosity occur over a shorter period of time (Hedrick 2000), thus providing a means of assessing the population trajectories of introduced species.

The principle of HW equilibrium states that allele frequencies remain relatively constant in an idealized population under certain conditions (e.g., large, closed, random mating, absence of mutation or selection). Therefore, the degree to which allele frequencies vary between samples taken at different time periods indicates the amount of genetic drift that has occurred, forming the basis for "temporal variance" methods (reviewed in Spencer et al. 2000; Berthier et al. 2002; Leberg 2005). Effective size of a population, loosely termed the number of breeding individuals, is inversely proportional to the amount of genetic drift expected, and thus temporal variance, between the samples. Other methods based on DNA sequence data, often from maternally inherited mitochondrial DNA, allow inference of historical changes in population size and geographical distribution (Templeton 1998; Emerson et al. 2001; Strimmer and Pybus 2001; Templeton 2004). Empirical uses of effective size or tests for genetic bottlenecks include testing hypotheses pertaining to control efforts and distribution of feral pigs in Australia (Hampton et al. 2004a,b) and *Anopheles* mosquitoes in Africa (Lehmann et al. 1998).

PARENTAGE AND RELATEDNESS: INFERENCES INTO ANIMAL BEHAVIOR

Before the advent of genetic methods for parentage assignment, parentage and relatedness were estimated through visual observations. Observation-based estimates were straightforward and appeared to work well for species or populations that could be sighted regularly, where individuals could be recognized, or where males provided parental care. However, studies of parentage and breeding success for cryptic or rare species were problematic, and were restricted to females of species in which males provided no parental care. Recently, genetic methods of parentage determination have revolutionized the study of mating success (Hughes 1998). It is now clear that the social and genetic mating system of a species or population may be quite different from expected (Fleischer 1996). In retrospect, observation-based studies of parentage are often inaccurate or misleading even under the best of conditions because not all of the individuals can be observed continuously.

Genetic data can reveal alternative mating tactics, (e.g., Hogg and Forbes 1997) and rates of female promiscuity (Figure 18.3; Petrie and Kempenaers 1998). Effects of changes to habitat, population density, and distribution of resources on mating strategies and success of individuals can be quantified (Langbein and Thirgood 1989; Clutton-Brock et al. 1997; Komers et al. 1997; Rose et al. 1998; Coltman et al. 1999; Hoelzel et al. 1999; Pemberton et al. 1999). Estimates of interpopulation relatedness can document fine-scale genetic structure and detect sex-biased dispersal patterns (Ohnishi et al. 2000) and kin structure (Richard et al. 1996). Thus, genetic studies of parentage and relatedness can provide direct estimates of contact rates among individuals, rates of hybridization, and effects of social structure and dispersal on disease transmission.

Genetic markers, such as DNA microsatellites, are highly variable and are inherited in a known (Mendelian) manner; in diploid ($2N$) species individuals have two copies of an allele at each locus and the offspring receives one allele at random from each parent. Therefore, one can identify individuals and estimate relationships among individuals, including parentage. If genetic data can be obtained from all of the parents, then parentage determination is a matter of simply excluding all potential parent–offspring pairs who do not share at least one allele at each locus, provided that a sufficient number of variable markers are typed such that there is no more than one nonexcluded sire or dam. In most real-world situations, however, a complete sample of parents is not available. Furthermore, the occurrence of mutations, null or nonamplifying alleles, or errors in the data set may result in the false exclusion of a true parent (Jones and Ardren 2003).

A fractional allocation approach has been used when the primary question of parentage is the age class of parents and not specific individuals. A fraction ($1/n$) of parentage is allocated to all

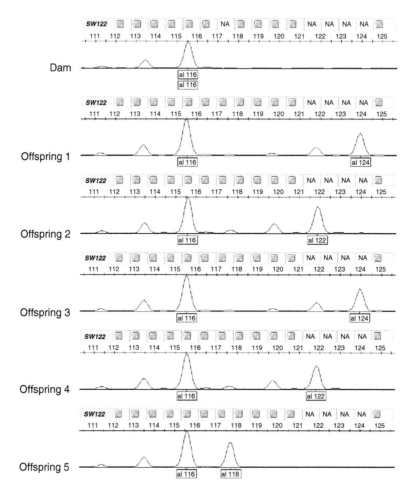

FIGURE 18.3 Microsatellite electropherogram depicting evidence for multiple paternity within a litter of feral pigs; numbers indicate allele size in base-pairs (R. W. DeYoung, unpublished data). Since pigs are diploid, each parent contributes one allele at each genetic locus. The dam's genotype is known, so the identification of more than two paternal alleles at multiple loci is evidence that more than one male sired this litter.

nonexcluded individuals or as a weighted proportion if behavioral data indicate one individual or age class is more likely to produce offspring but cannot be separated from other candidates based on genetic data. Fractional allocation is obviously limited in inferential power and undesirable when estimates of individual breeding and fitness are desired. Therefore, parentage assignment based on likelihood ratios have been developed to circumvent the shortcomings of exclusion and fractional allocation of parents (reviewed in Jones and Ardren 2003; DeWoody 2005). Typically, simulations are performed on the genetic data set and used to estimate the confidence in parentage assignments in the presence of incomplete sampling of parents, missing genetic data for some individuals, genotyping errors, or mutations.

Allele frequency data can also be used to estimate relationships among individuals. Individuals share alleles in direct proportion to the degree of cosanguity, and relatedness estimators incorporate the degree of allele sharing into a measure of identity by descent, the probability that alleles are inherited from a common ancestor (Blouin 2003). For diploid individuals, expected relationship coefficients for individuals related at the level of parent–offspring or full siblings are 0.5 (corresponding to 50% similarity); expected values for half siblings, first cousin, and unrelated are 0.25, 0.125, and 0.0, respectively. Estimators of relatedness have a high variance and cannot always determine the exact relationship between individuals unless a large number (ca. 30–40) of genetic loci are used. However, relatedness estimators can be very useful for comparison among groups (Van De Casteele et al. 2001; Blouin 2003) and for the spatial autocorrelation analyses described in the population structure section.

MANAGEMENT IMPLICATIONS

Viable long-term solutions to wildlife disease, damage, and invasive species problems clearly must place a greater emphasis on animal behavior and population structure than has been previously considered. Analyses based on genetic data can provide an objective means of defining population boundaries and estimating rates of dispersal through a landscape. Genetic data present a distinct advantage over traditional management approaches in that genetic markers are heritable in a known fashion, thus permitting identification of lineages and relationships among individuals and populations (Avise 2004). Genetic data also offer a means for the independent evaluation of data derived through traditional methods (Honeycutt 2000). Genetic approaches are attractive in that explicit consideration of factors such as dispersal and population and social structure are integral to and deeply rooted in population genetic theory and can be addressed within a single conceptual framework. This approach assures that research is focused in a process-oriented and scale-appropriate manner, with emphasis on interactions among populations. Therefore, genetic approaches offer a great deal of promise for applied management now that suitable markers are available which permit acquisition of data and application of the large and well-developed body of population genetic theory (Table 18.3). Overall, genetic methods offer powerful inferential tools that have thus far been vastly underutilized in applied wildlife biology and management. All that is needed is creativity and vision in their application.

REFERENCES

Alitzer, S., C. L. Nunn, P. H. Thrall, J. L. Gittleman, J. Antonovics, A. A. Cunningham, A. P. Dobson, V. Ezenwa, K. E. Jones, A. B. Pedersen, M. Poss, and J. R. C. Pulliam. 2003. Social organization and parasite risk in mammals: Integrating theory and empirical studies. *Ann. Rev. Ecol. Evol. Syst.* 34:517.

Avise, J. A. 2004. *Molecular Markers, Natural History, and Evolution*, 2nd edn. Sunderland, MA: Sinauer Associates.

Baber, D. W., and B. E. Coblentz. 1987. Diet, nutrition, and conception in the feral pigs on Santa Catalina Island. *J. Wildl. Manage.* 51:306.

Ballard, W. B., and P. S. Gipson. 2000. Wolf. In *Ecology and Management of Large Mammals in North America*, S. Demarais, and P. R. Krausman (eds), Chap. 16. Upper Saddle River, NJ: Prentice Hall.

Beaumont, M. W., and Bruford, M. W. 1999. Microsatellites in conservation genetics. In *Microsatellites: Evolution and Applications*, D. B. Goldstein, and C. Schlötterer (eds), Chap. 13. New York: Oxford University Press.

Berry, O., M. D. Tocherand, and S. D. Sarre. 2004. Can assignment tests measure dispersal? *Mol. Ecol.* 13:551.

Berthier, P., M. A. Beaumont, J. M. Cornuet, and G. Luikart. 2002. Likelihood-based estimation of the effective population size using temporal changes in allele frequencies: A genealogical approach. *Genetics* 160:741.

Blackard, J. J. 1971. Restoration of the white-tailed deer to the southeastern United States. MS Thesis, Louisiana State University, Baton Rouge.

Blanchong, J. A., K. T. Scribner, and S. R. Winterstein. 2002. Assignment of individuals to populations: Bayesian methods and multi-locus genotypes. *J. Wildl. Manage.* 66:321.

Blouin, M. S. 2003. DNA-based methods for pedigree reconstruction and kinship analysis in natural populations. *Trends Ecol. Evol.* 18:503.

Bratton, S. P. 1975. The effects of the European wild boar, *Sus scrofa*, on gray beech forest in the Great Smoky Mountains. *Ecology* 56:1356.

Burke, T. 1994. Spots before the eyes: Molecular ecology. *Trends Ecol. Evol.* 9:355.

Cathey, J. C., J. W. Bickham, and J. C. Patton. 1998. Introgressive hybridization and nonconcordant evolutionary history of maternal and paternal lineages in North American deer. *Evolution* 52:1224.

Cavalli-Sforza, L. L., L. P. Menozzi, and A. Piazza. 1994. *The History and Geography of Human Genes.* Princeton, NJ: Princeton University Press.

Chesser, R. K. 1991a. Gene diversity and female philopatry. *Genetics* 127:437.

Chesser, R. K. 1991b. Influence of gene flow and breeding tactics on gene diversity within populations. *Genetics* 129:573.

Childs, J. E., A. T. Curns, M. E. Dey, L. A. Real, L. Feinstein, O. N. Bjørnstad, and J. W. Krebs. 2000. Predicting the local dynamics of epizootic rabies among raccoons in the United States. *Proc. Natl Acad. Sci. USA* 97(13):666.

Childs, J. E., A. T. Curns, M. E. Dey, L. A. Real, C. E. Rupprecht, and J. W. Krebs. 2001. Rabies epizootics among raccoons vary along a North–South gradient in the Eastern United States. *Vec. Borne Zool. Dis.* 1:253.

Choquenot, D., B. Lukins, and G. Curan. 1997. Assessing lamb predation by feral pigs in Australia's semi-arid rangelands. *J. Appl. Ecol.* 34:1445.

Clutton-Brock, T. H., K. E. Rose, and F. E. Guinness. 1997. Density-related changes in sexual selection in red deer. *Proc. R. Soc. Lond. B* 264:1509.

Cockerham, C. C. 1969. Variance of gene frequencies. *Evolution* 23:72.

Cockerham, C. C. 1973. Analysis of gene frequencies. *Genetics* 74:679.

Coltman, D. W., D. R. Bancroft, A. Robertson, J. A. Smith, T. H. Clutton-Brock, and J. M. Pemberton. 1999. Male reproductive success in a promiscuous mammal: Behavioral estimates compared with genetic paternity. *Mol. Ecol.* 8:1199.

Conover, M. R. 1997. Monetary and intangible valuation of deer in the United States. *Wildl. Soc. Bull.* 25:298.

Conover, M. R., W. C. Pitt, K. K. Kessler, T. J. DuBow, and W. A. Sanborn. 1995. Review of human injuries, illnesses, and economic losses caused by wildlife in the United States. *Wildl. Soc. Bull.* 23:407.

Cornuet, J. M., and G. Luikart. 1996. Description and power analysis of two tests for detecting recent population bottlenecks from allele frequency data. *Genetics* 144:2001.

Cornuet, J. M., S. Piry, G. Luikart, A. Estoup, and M. Solignac. 1999. New methods employing multilocus genotypes to select or exclude populations as origins of individuals. *Genetics* 153:1989.

Côté, S. D., T.P. Rooney, J.-P. Tremblay, C. Dussault, and D. M. Waller. 2004. Ecological impacts of deer abundance. *Ann. Rev. Ecol. Evol. Syst.* 35:113.

Davidson, W. R., and G. L. Doster. 1997. Health characteristics and population density in the southeastern United States. In *The Science of Overabundance: Deer Ecology and Population Management*, W. J. McShea, H. B. Underwood, and J. H. Rappole (eds). Washington, DC: Smithsonian Institution Press.

DeWoody, J. A. 2005. Molecular approaches to the study of parentage, relatedness, and fitness: Practical applications for wild animals. *J. Wildl. Manage.* 69:1400.

DeYoung, R. W., and R. L. Honeycutt. 2005. The molecular toolbox: Genetic techniques in wildlife ecology and management. *J. Wildl. Manage.* 69:1362.

Diniz-Filho, J. A. F., and M. P. D. Telles. 2002. Spatial autocorrelation analysis and the identification of operational units for conservation in continuous populations. *Conserv. Biol.* 16:924.

Dobzhansky, T. 1941. *Genetics and the Origin of Species.* New York: Columbia University Press.

Donkin, R. A. 1985. The peccary — With observations on the introduction of pigs to the New World. *Trans. Am. Philos. Soc.* 75:1.

Emerson, B. C., E. Paradis, and C. Thébaud. 2001. Revealing the demographic histories of species using DNA sequences. *Trends Ecol. Evol.* 16:707.

Epperson, B. K. 2003. *Geographical Genetics.* Princeton, NJ: Princeton University Press.

Excoffier, L., P. Smouse, and J. Quattro. 1992. Analysis of molecular variance inferred from metric distances among DNA haplotypes: Application to human mitochondrial DNA restriction data. *Genetics* 131:479.

Fisher, R. A. 1930. *The Genetic Theory of Natural Selection.* New York: Dover.

Fleischer, R. C. 1996. Application of molecular methods to the assessment of genetic mating systems in vertebrates. In *Molecular Zoology: Advances, Strategies, and Protocols*, J. D. Ferraris, and S. R. Palumbi (eds), Chap. 7. New York: Wiley-Liss.

Gemmel, N. J., T. M. Burg, I. L. Boyd, and W. Amos. 2001. Low reproductive success in territorial male Antarctic fur seals (*Arctocephalus gazella*) suggests the existence of alternative mating strategies. *Mol. Ecol.* 10:451.

Gordon, I. J., A. J. Hester, and M. Festa-Bianchet. 2004. The management of wild large herbivores to meet economic, conservation and environmental objectives. *J. Appl. Ecol.* 41:621.

Goudet, J., T. DeMeeüs, A. J. Day, and C. J. Gliddon. 1994. Gene flow and population structure. In *Genetics and Evolution of Aquatic Organisms*, M. A. Beaumont (ed.), Chap. 2. London: Chapman & Hall.

Guillot, G., A. Estoup, F. Mortier, and J. F. Cossonet. 2005. A spatial statistical model for landscape genetics. *Genetics* 170:1261.

Hampton, J. O., J. R. Pluske, and P. B. S. Spencer. 2004a. A preliminary genetic study of the social biology of feral pigs in south-western Australia and the implications for management. *Wildl. Res.* 31:375.

Hampton, J. O., P. B. S. Spencer, D. L. Alpers, L. E. Twigg, A. P. Woolnough, J. Doust, T. Higgs, and J. Pluske. 2004b. Molecular techniques, wildlife management and the importance of genetic population structure and dispersal: A case study with feral pigs. *J. Appl. Ecol.* 41:735.

Hancock, J. M. 1999. Microsatellites and other simple sequences: Genomic context and mutational mechanisms. In *Microsatellites: Evolution and Applications*, D. B. Goldstein, and C. Schlötterer (eds), Chap. 1. New York: Oxford University Press.

Hardy, O. J., and X. Vekemans. 1999. Isolation by distance in a continuous population: Reconciliation between spatial autocorrelation analysis and population genetics models. *Heredity* 83:145.

Harris, H. 1966. Enzyme polymorphisms in man. *Proc. Roy. Soc. Lond. B* 164:298.

Harrison, S., and A. Hastings. 1996. Genetic and evolutionary consequences of metapopulation structure. *Trends Ecol. Evol.* 11:180.

Hastings, A., and S. Harrison. 1994. Metapopulation dynamics and genetics. *Ann. Rev. Ecol. Syst.* 25:167.

Hedrick, P. W. 1999. Perspective: Highly variable loci and their interpretation in evolution and conservation. *Evolution* 53:313.

Hedrick, P. W. 2000. *Genetics of Populations*, 2nd edn. Sudbury, MA: Jones and Bartlett.

Hedrick, P. W., and P. S. Miller. 1992. Conservation genetics: Techniques and fundamentals. *Ecol. Appl.* 2:30.

Hellgren, E. C. 1993. Biology of feral Hogs (*Sus scrofa*) in Texas. In *Proc. Conf. Feral Swine: A Compendium for Resource Managers*, C. W. Hanselka, and J. F. Cadenhead (eds). College Station, TX: Tex. Ag. Ext. Serv., p. 50.

Hoelzel, A. R., B. J. Le Boeuf, J. Reiter, and L. Campangna. 1999. Alpha-male paternity in elephant seals. *Behav. Ecol. Sociobiol.* 46:298.

Hogg, J. T., and S. H. Forbes. 1997. Mating in bighorn sheep: Frequent male reproduction via a high-risk "unconventional" tactic. *Behav. Ecol. Sociobiol.* 41:33.

Holderegger, R., and H. H. Wagner. 2006. A brief guide to landscape genetics. *Landscape Ecol.* 21:793.

Honeycutt, R. L. 2000. Genetic applications for large mammals. In *Ecology and Management of Large Mammals in North America*, S. Demarais, and P. R. Krausman (eds), Chap. 12. Upper Saddle River, NJ: Prentice Hall.

Howe, T. D., F. J. Singer, and B. B. Ackerman. 1981. Foraging relationships of European wild boar invading northern hardwood forest. *J. Wildl. Manage.* 45:748.

Hughes, C. 1998. Integrating molecular techniques with field methods in studies of social behavior: A revolution results. *Ecology* 79:383.

Jones, A. G., and W. R. Ardren. 2003. Methods of parentage analysis in natural populations. *Mol. Ecol.* 12:2511.

Kammermeyer, K., J. Bowers, B. Cooper, D. Forster, K. Grahl, T. Holbrook, C. Martin, S. McDonald, N. Nicholson, M. Van Brackle, and G. Waters. 2003. Feral hogs in Georgia: Disease, damage and control. Georgia Dep. Nat. Res., Wildl. Res. Div., Social Circle, Georgia.

Komers, P. E., C. Pelabon, and D. Stenstrom. 1997. Age at first reproduction in male fallow deer: Age-specific versus dominance-specific behaviors. *Behav. Ecol.* 8:456.

Krebs, J. W., E. J. Mandel, D. L. Swerdlow, and C. E. Rupprecht. 2005. Rabies surveillance in the United States during 2004. *J. Am. Vet. Med. Assoc.* 227:1912.

Langbein, J., and S. J. Thirgood. 1989. Variation in mating systems of fallow deer (*Dama dama*) in relation to ecology. *Ethology* 83:195.

Leberg, P. L. 2005. Genetic approaches for estimating the effective size of populations. *J. Wildl. Manage.* 69:1385.

Lehmann, T. 1998. The effective population size of *Anopheles gambiae* in Kenya: Implications for population structure. *Mol. Biol. Evol.* 15:264.

Lewontin, R. C., and J. L. Hubby. 1966. A molecular approach to the study of genic heterozygosity in natural populations. II. Amount of variation and degree of heterozygosity in natural populations of *Drosophila pseudoobscura*. *Genetics* 54:595.

Littauer, G. A. 1993. Control techniques for feral hogs. In *Proc. Conf. Feral Swine: A Compendium for Resource Managers*, C. W. Hanselka, and J. F. Cadenhead (eds). College Station, TX: Tex. Ag. Ext. Serv., p. 139.

Luikart, G., W. A. Hawley, H. Grebert, and F. H. Collins. 1998. Distortion of allele frequency distributions provides a test for recent population bottlenecks. *J. Hered.* 89:238.

Mackie, R. J. 2000. History of management of large mammals in North America. In *Ecology and Management of Large Mammals in North America*, S. Demarais, and P. R. Krausman (eds), Chap. 15. Upper Saddle River, NJ: Prentice Hall.

Manel, S., P. Berthier, and G. Luikart. 2002. Detecting wildlife poaching: Identifying the origin of individuals with Bayesian assignment tests and multilocus genotypes. *Conserv. Biol.* 16:650.

Manel, S., M. K. Schwartz, G. Luikart, and P. Taberlet. 2003. Landscape genetics: Combining landscape ecology and population genetics. *Trends Ecol. Evol.* 18:189.

Manel, S., O. E. Gaggiotti and R. S. Waples. 2005. Assignment methods: Matching biological questions with appropriate techniques. *Trends Ecol. Evol.* 20:136.

Mantel, N. 1967. The detection of disease clustering and a generalized regression approach. *Cancer Res.* 27:209.

Mapston, M. E. 2004. *Feral Hogs in Texas*. Tex. Coop. Ext. Serv., College Station, TX: Texas A&M University Press.

Maruyama, T. 1970. On the rate of decrease of heterozygosity in circular stepping stone models of populations. *Theoret. Popul. Biol.* 1:101.

Mayer, J. J., and I. L. Brisbin. 1991. *Wild Pigs in the United States*. Athens, GA: University of Georgia Press.

McNulty, S. A., W. F. Porter, N. E. Mathews, and J. A. Hill 1997. Localized management for reducing white-tailed deer populations. *Wildl. Soc. Bull.* 25:265.

McShea, W. J., H. B. Underwood, and J. H. Rappole (eds). 1997. *The Science of Overabundance: Deer Ecology and Population Management*. Washington, DC: Smithsonian Institution Press.

Moritz, C. 1994. Defining 'evolutionarily significant units' for conservation. *Trends Ecol. Evol.* 9:373.

Moule, G. R. 1954. Observations on mortality amongst lambs in Queensland. *Aust. Vet. J.* 30:153.

Muller, L. I., R. J. Warren, and D. L. Evans. 1997. Theory and practice of immunocontraception in wild mammals. *Wildl. Soc. Bull.* 25:504.

Nei, M. 1987. *Molecular Evolutionary Genetics*. New York: Columbia University Press.

Nei, M., and S. Kumar. 2000. *Molecular Evolution and Phylogenetics*. New York: Oxford University Press.

Nei, M., T. Maruyama, and R. Chakraborty. 1975. The bottleneck effect and genetic variability in populations. *Evolution* 29:1.

Neigel, J. E. 1997. A comparison of alternative strategies for estimating gene flow from genetic markers. *Ann. Rev. Ecol. Syst.* 28:105.

Neigel, J. E. 2002. Is F_{ST} obsolete? *Cons. Gen.* 3:167.

Nettles, V. F. 1997. Feral swine: Where we've been, where we're going. In *Proc. National Feral Swine Symposium*, K. L. Schmitz (ed.). USDA, Anim. Plant Health Insp. Serv., p. 1.1

Nixon, C. M., L. P. Hansen, P. A. Brewer, and J. E. Chelsvig. 1991. Ecology of white-tailed deer in an intensively farmed region of Illinois. *Wildl. Monogr.* 118.

Nussey, D. H., D. W. Coltman, T. Coulson, L. E. B. Kruuk, A. Donald, S. J. Morris, T. H. Clutton-Brock, and J. Pemberton. 2005. Rapidly declining fine-scale spatial genetic structure in female red deer. *Mol. Ecol.* 14:3395.

Ohnishi, N., T. Saitoh, and Y. Ishibashi. 2000. Spatial genetic relationships in a population of the Japanese wood mouse *Apodemus argenteus*. *Ecol. Res.* 15:285.

Paetkau, D., W. Calvert, I. Stirling, and C. Strobek. 1995. Microsatellite analysis of population structure in Canadian polar bears. *Mol. Ecol.* 4:347.

Paetkau, D., G. F. Shields, and C. Strobek. 1998. Gene flow between insular, coastal, and interior populations of brown bears in Alaska. *Mol. Ecol.* 7:1283.

Palsböll, P. 1999. Genetic tagging: Contemporary molecular ecology. *Biol. J. Linn. Soc.* 68:3.

Pavlov, P. M., R. J. Kilgour, and H. Pederson. 1981. Predation by feral pigs on merino lambs at Nyngan, New South Wales. *Aust. J. Exp. Ag. Anim. Husb.* 21:570.

Pemberton, J. A., D. W. Coltman, J. A. Smith, and J. G. Pilkington. 1999. Molecular analysis of a promiscuous, fluctuating mating system. *Biol. J. Linn. Soc.* 68:289.

Petrie, M., and B. Kempenaers. 1998. Extra-pair paternity in birds: Explaining variation between species and populations. *Trends Ecol. Evol.* 13:52.

Piertney, S. B., A. D. C. MacColl, P. J. Bacon, and J. F. Dallas. 1998. Local genetic structure in red grouse (*Lagopus lagopus scoticus*): Evidence from microsatellite DNA markers. *Mol. Ecol.* 7:1645.

Pimentel, D., L. Lach, F. Zuniga, and D. Morrison. 2000. Environmental and economic costs of nonindigenous species in the United States. *Bioscience* 50:53.

Porter, W. F., N. E. Mathews, H. B. Underwood, R. W. Sage, and D. F. Behrend. 1991. Social organization in deer: Implications for localized management. *Environ. Manage.* 15:809.

Pritchard, J. K., M. Stephens, and P. Donnelly. 2000. Inference of population structure using multilocus genotype data. *Genetics* 155:945.

Purdue, J. R., M. H. Smith, and J. C. Patton. 2000. Female philopatry and extreme spatial genetic heterogeneity in white-tailed deer. *J. Mammal.* 81:179.

Rannala, B., and J. L. Mountain. 1997. Detecting immigration by using multilocus genotypes. *Proc. Natl Acad. Sci. USA* 94:9197.

Real, L. A., C. Russell, L. Waller, D. Smith, and J. Childs. 2005. Spatial dynamics and molecular ecology of North American rabies. *J. Hered.* 96:1.

Richard, K. R., M. C. Dillon, H. Whitehead, and J. M. Wright. 1996. Patterns of kinship in groups of free-living sperm whales (*Physeter macrocephalus*) revealed by multiple molecular genetic analyses. *Proc. Natl Acad. Sci. USA* 93:8792.

Riley, S. J., D. J. Decker, J. W. Enck, P. D. Curtis, T. B. Lauber, and T. L. Brown. 2003. Deer populations up, hunter populations down: Implications of interdependence of deer and hunter population dynamics on management. *Ecoscience* 10:455.

Rose, K. E., T. H. Clutton-Brock, and F. E. Guinness. 1998. Cohort variation in male survival and lifetime breeding success in red deer. *J. Anim. Ecol.* 67:979.

Rowley, I. 1970. Lamb predation in Australia: Incidence, predisposing conditions and identification of wounds. *CSIRO Wildl. Res.* 15:79.

Saiki, R. K., D. H. Gelfand, S. Stoffel, S. J. Scharf, R. Higuchi, G. T. Horn, K. B. Mullis, and H. A. Erlich. 1988. Primer-directed enzymatic amplification of DNA with a thermostable DNA polymerase. *Science* 239:487.

Samuel, W. M., M. J. Pybus, and A. A. Kocan. 2001. *Parasitic Diseases of Wild Mammals.* Ames, IA: Iowa State University Press.

Sanger, F., S. Nicklen, and A. R. Coulson. 1977. DNA sequencing with chain-terminating inhibitors. *Proc. Natl Acad. Sci. USA* 74:5463.

Schauber, E. M., and A. Woolf. 2003. Chronic wasting disease in deer and elk: A critique of current models and their application. *Wildl. Soc. Bull.* 31:610.

Scribner, K. T., J. A. Blanchong, D. J. Bruggeman, B. K. Epperson, C. Y. Lee, Y. W. Pan, R. I. Shorey, H. H. Prince, S. R. Winterstein, and D. R. Luukkonen. 2005. Geographical genetics: Conceptual foundations and empirical applications of spatial genetic data in wildlife management. *J. Wildl. Manage.* 69:1434.

Sokal, R. R., N. L. Oden, and B. A. Thomson. 1997. A simulation study of microevolutionary inferences by spatial autocorrelation analysis. *Biol. J. Linn. Soc.* 60:73.

Spencer, C. C., J. E. Neigel, and P. L. Leberg. 2000. Experimental evaluation of the usefulness of microsatellite DNA for detecting demographic bottlenecks. *Mol. Ecol.* 9:1517.

Stone, C. P., and J. O. Keith. 1987. Control of feral ungulates and small mammals in Hawaii's National Parks: Research and management strategies. In *Control of Mammal Pests*, C. G. H. Richards, and T. Y. Ku (eds). London: Taylor and Francis.

Strimmer, K., and O. G. Pybus. 2001. Exploring the demographic history of DNA sequences using the generalised skyline plot. *Mol. Biol. Evol.* 18:2298.

Sugg, D. W., R. K. Chesser, F. S. Dobson, and J. L. Hoogland. 1996. Population genetics meets behavioral ecology. *Trends Ecol. Evol.* 11:338.

Sweeney, J. R., J. M. Sweeney, and S. W. Sweeney. 2003. Feral Hog: *Sus scrofa*. In *Wild Mammals of North America*, G. A. Feldhamer, B. C. Thompson, and J. A. Chapman (eds). Baltimore, MD: John Hopkins University Press.

Templeton, A. R. 1998. Nested clade analyses of phylogeographic data: Testing hypotheses about gene flow and population history. *Mol. Ecol.* 7:381.

Templeton, A. R. 2004. Statistical phylogeography: Methods of evaluating and minimizing inference errors. *Mol. Ecol.* 13:789.

Tiedemann, R., O. Hardy, X. Vekemans, and M. C. Milinkovitch. 2000. Higher impact of female than male migration on population structure in large mammals. *Mol. Ecol.* 9:1159.

Van De Casteele, T., P. Galbusera, and E. Matthysen. 2001. A comparison of microsatellite-based pair wise relatedness estimators. *Mol. Ecol.* 10:1539.

Wahlund, S. 1928. Zusammensetzung von populationen und correlation-serscheinungen von standpunkt der verebungslehre aus betrachtet. *Hereditas* 11:65.

Waller, D. M., and W. S. Alverson. 1997. The white-tailed deer: A keystone herbivore. *Wildl. Soc. Bull.* 25:217.

Weber, J. L., and P. E. May. 1989. Abundant class of human DNA polymorphisms which can be typed using the polymerase chain reaction. *Am. J. Hum. Gen.* 44:388.

Weir, B. S. 1996. *Genetic Data Analysis II: Methods for Discrete Population Genetic Data*. Sunderland, MA: Sinauer Associates.

Weir, B. S., and C. C. Cockerham. 1984. Estimating F-statistics for the analysis of population structure. *Evolution* 38:1358.

Whitlock, M. C., and D. E. McCauley. 1999. Indirect measures of gene flow and migration: F_{ST}? $1/(4N_m + 1)$. *Heredity* 82:117.

Williams, E. S., and I. K. Barker. 2001. *Infectious Diseases of Wild Animals*. Ames, IA: Iowa State University Press.

Witmer, G. W., R. B. Sanders, and A. C. Taft. 2003. Feral swine — Are they a disease threat to livestock in the United States? *Proc. Wildl. Damage Manage. Conf.* 10:316.

Wood, G. W., and R. H. Barrett. 1979. Status of wild pigs in the United States. *Wildl. Soc. Bull.* 7:237.

Wood, G. W., and D. N. Roark. 1980. Food habits of feral hogs in coastal South Carolina. *J. Wildl. Manage.* 44:506.

Worthington Wilmer, J., P. J. Allen, P. Pomeroy, S. D. Twiss, and W. Amos. 1999. Where have all the fathers gone? An extensive microsatellite analysis of paternity in the grey seal (*Halichoerus grypus*). *Mol. Ecol.* 8:1417.

Wright, S. 1938. The distribution of gene frequencies under irreversible mutation. *Proc. Natl. Acad. Sci. USA* 245:253.

Wright, S. 1940. Breeding structure of populations in relation to speciation. *Am. Nat.* 74:232.

Wright, S. 1943. Isolation by distance. *Genetics* 28:114.

Wright, S. 1951. The genetical structure of populations. *Ann. Eugen.* 15:323.

Wright, S. 1965. The interpretation of population structure by F-statistics with special regard to systems of mating. *Evolution* 19:395.

Wright, S. 1968. *Evolution and the Genetics of Populations*, Vol. 1. Chicago: University of Chicago Press.

Wright, S. 1969. *Evolution and the Genetics of Populations*, Vol. 2. Chicago: University of Chicago Press.

Wright, S. 1977. *Evolution and the Genetics of Populations*, Vol. 3. Chicago: University of Chicago Press.

Wright, S. 1978. *Evolution and the Genetics of Populations*, Vol. 4. Chicago: University of Chicago Press.

Part V

Economic and Social Issues Affecting Wildlife Science

19 Society, Science, and the Economy: Exploring the Emerging New Order in Wildlife Conservation

Shane P. Mahoney and Jackie N. Weir

CONTENTS

Historical perspectives of the wildlife conservation movement reveal the complex interplay between evolving states of knowledge and evolving societal values and expectations. From its first awakening in the early-to-mid 1800s, and through the formative years of the late nineteenth and early twentieth centuries, the movement to safeguard North American wildlife reflected the great concerns for population and species depletions that lay strewn in the wake of unbridled slaughter and industrial expansion. Emergent policies and paradigms confronted this excess with a focus on protection and recovery, wilderness set-asides, forestry reform, and game laws. The underpinnings of natural history served these initiatives reasonably well, until moderate successes in recovering some populations but persistent declines in others revealed the vacancy between knowledge and applied policy (Trefethen 1975).

Wildlife science emerged from this tension as a prerequisite to all formalized conservation efforts (Geist et al. 2001; Mahoney 2004), eventually incorporating an explosion of knowledge about species dynamics, habitat requirements, and predator–prey interactions into harvest regimes and refuge designs. Some of this information derived from wildlife studies per se, but much was also borrowed from the broader reach of ecology, the discipline that most purposefully sought understanding of how natural systems behaved and were regulated. This engagement between a science focused on defining the biological imperatives of species important to wildlife management, and that which concerned itself with questions of natural system engineering and persistence, was to prove a long and fruitful one. Continuously, as ecology applied models and quantitative methods to analyze the natural world and predict its response to perturbations, wildlife science would incorporate these insights, improving its ability to more precisely integrate both human harvest and protection policies within the capacities of natural systems. In its turn, wildlife science contributed its detailed expositions of the life history and landscape requirements of certain managed species as baselines to assess the accuracy of ecology's more conceptual approach.

However, it was not only the ratcheting of scientific inquiry that steered the course of wildlife conservation. From its inception, the movement had been influenced, founded even, by two recognizably distinct societal views. These focused on either a utilitarian philosophy of nature's worth, or on a belief in the inherent value of the natural world, often represented as the anthropocentric versus biocentric rationale for conserving nature (Hendee and Stankey 1973; Paterson 2006). In truth, these value streams may not be entirely distinct, but certainly, they remain the most conceptually instructive dichotomy in conservation focus, and are reflected vividly in the historical legacy of wildlife policy, law, and management (Reiger 1975; Meine 1988). Wildlife science was of value to either approach of course, as both sought to preserve the natural world and required knowledge to do so.

However, the interplay between these philosophical contours and science was, and remains, far more complex than this, for obviously societal emphases help direct the focus of science, and coerce and enjoin its financial support. For these concrete reasons, and for many that are far more subtle, science and society do blend, fraying the lines of demarcation between social and conservation policy. And social policy, of course, is heavily defined by economics, ensuring that conservation approaches and the science attendant to them will never lie beyond the influence of society's valuations of what wildlife is worth. These relationships persist through the cyclic phases of precedence that the intrinsic versus utilitarian conservation agendas assume (Mulder and Coppolillo 2005).

Thus, even as the accretion of ecological knowledge continued through the mid-twentieth century, new and powerful social ideals were to erupt that would alter the course of wildlife conservation and help restructure both the social-scientific and the economic perspectives that had hitherto guided its approach. The environmental awakening of the 1960s, for example, repositioned humans as both dependant entity and custodian within conservation's purpose, while at the same time forecasting the inevitable consequences for human health and economic opportunity that ecological impoverishment would derive. These realities drove a new social awareness that demanded, in its turn, new and improved science that could offer alternative approaches to resource use and extraction, and would identify new bench marks such as biodiversity on the one hand, and endangered species on the other hand, as primary indicators of wildlife conservation success. Within this context, wildlife science could no longer be concerned with just ensuring game species were in ready supply. It needed to turn its attention to ecosystems and their dynamics, intellectual arenas ecology called home.

Somewhere in all of this, "game management" (Leopold 1933), the initial driver and raison d'etre for wildlife science (but not ecology), struggled for position and profile, and the economics of conservation in the broadest sense forced the question of who was to pay for the new ideals. In addition to the environmental movement, other structural changes in North American society helped drive this debate and impinged on the nature and focus of wildlife conservation. Many of these social alterations freighted important economic challenges. Thus, the economies surrounding nonhunting engagements with wildlife, such as wildlife viewing and bird feeding, escalated to unheard of proportions (Kellert and Smith 2000; Bolen and Robinson 2003), slowly reformulating valuations for wildlife in general. These trends were coupled with social patterns of increased urbanization, and decreased familiarity with nature in any practical sense. These, in turn, changed societal expectations for wildlife conservation, and increasingly demanded a science that embraced issues of animal over-abundance and wildlife-to-human disease transmission, but still, somehow, maintained its broader ecological focus. Farming and ranching of wild animals has recently led to other complex debates, and placed further demands on wildlife science to expand (Geist 1989; Rasker et al. 1992).

It is important to recognize that the trade between economic/social installations and wildlife science is not a one-way street. The new developments and insights revealed by science exerted their own influence on social awareness and priority setting, ensuring a level of symbiotic entanglement between these seemingly independent human endeavors. Furthermore, these tensions between wildlife and other resource interests are destined to intensify as human populations and resource demands rise (Klein 2000). Under such circumstances, wildlife science and ecology will both be called to attest their worth in light of new trends in social priorities, and traffic in similar ideas to remain relevant to their respective audiences. It is in this manner that the ties between ecology, wildlife science, and economy are repeatedly reinforced over time; and while new paradigms continuously emerge, this pattern of concept introgression and the hybridizing of the science around wildlife and ecosystems remains consistent. As the following discussion illustrates, this integration has led to major advances in wildlife science and helped prepare it to embrace the emerging conservation order.

ECOLOGICAL THEORY AND WILDLIFE SCIENCE

GENERAL BACKGROUND

While ecology had originally been considered a subdiscipline of physiology, its evolution toward a community-focused science was well in hand by the later part of the nineteenth century. By this time, the broad principles of how natural systems functioned were identified, and new insights and discoveries surrounding the complexity of ecological communities could be incorporated into a systematic understanding of how species interacted with one another and their physical environment. Over time, many of these advances (below) were to influence wildlife conservation and science, and critical transformations that would link economics and ecological theory also emerged. The latter have been especially influential in recent decades, when many concepts and theories central to ecology have been revised.

These shifts in thinking have had important implications for resource management and utilization (Pimm 1991; Pickett et al. 1992; Fiedler et al. 1997; Wallington et al. 2005), and have forced significant innovation in wildlife science approaches. In this regard, the most fundamental paradigm shift has been a change in perception of ecosystems as rather static and predictable, to entities that are complex, dynamic, and unpredictable across time and space (Holling 1986; Fiedler et al. 1997; Scoones 1999; Wallington et al. 2005). It is now generally recognized that disturbances (natural and human-caused) are among the most important factors shaping ecosystem health and performance, and that change, rather than being exceptional, is very much the one constant in nature's economy. Furthermore, classical ecology viewed humans as vagrants in ecological systems, and there was the general belief that natural systems would balance themselves if the influence of humans was removed. Current ecological approaches recognize that humans are an integral component of most ecosystems, and the world's growing human population is now considered the principal threat to biological diversity and persistence (McKee et al. 2003).

This emergent view of ecological systems has made the job of managing and protecting wildlife far more challenging. Recognition that disturbances are integral components of natural systems means that efforts to manage and conserve wildlife and their habitats must include consideration of disturbance processes, and not just their effects (Hobbs and Huenneke 1992). This new approach increasingly requires "active" management rather than the historical approach, which allowed nature to "take care of itself" (Wallington et al. 2005). Human demands and aspirations are now considered implicit to all wildlife management and conservation approaches; and human societies are not just end users of the resource, but shapers and drivers of ecological processes themselves.

These conceptual changes have led to a growing recognition that it is the responsibility of the society to collaboratively choose possible management and conservation options (Bradshaw and Bekoff 2001; Robertson and Hull 2001). What they have not altered is the responsibility scientists have to ensure that societal decisions around wildlife are based on the best available information. While a growing task for wildlife managers and policy makers is to determine how to include societal values in decision-making processes, integrating ecological theory has been, and must remain, a consistent and well-attended priority for wildlife science practitioners. In the midst of increasingly rapid change, science capacity becomes ever more relevant, and the ability to efficiently integrate new findings from disparate sources, ever more important.

ECOLOGICAL CONCEPTS AND THE WILDLIFE SCIENCE HORIZON

Ecology is a broad science, and one that has increasingly developed an appreciation of the effects of scale, both in the functioning of natural systems, and as a powerful conceptual lens for elucidating patterns and predictive models. Thus, its hierarchical subfields — behavioral, population, community, and landscape — have all contributed to the maturation and substance of understandings regarding wildlife fluctuations, and to our management prescriptions for wildlife conservation. Ecological subfields have also transferred such insights to the role of humans in the world's ecology, helping stimulate a new calibration of the human–nature equation.

BEHAVIORAL ECOLOGY

Behavioral ecology emerged from the field of ethology, which focused mainly on the description of innate and fixed-action patterns of animal behavior. Following the pioneering work of Niko Tinbergen and Konrad Lorenz, there was a focus on understanding the proximate and ultimate causes and functions of individual behaviors. Behavioral ecology expanded on ethology by focusing on both the ecological and evolutionary basis for animal activity, and the role of behavior in enabling an animal to adapt to its environment. For some time, behavioral ecologists have argued that an understanding of individual and group behavior is fundamental to successful wildlife management and conservation efforts. While there is uncertainly as to the degree to which this knowledge is being applied (Harcourt 1999), there is a growing recognition that behavioral ecology has a great deal to offer (Curio 1996; Lima and Zollner 1996; Sutherland 1998; Caro 1999; Anthony and Blumstein 2000). For example, optimal foraging theory is used to predict why and how individuals move through the landscape, and this knowledge is critical to our understanding of habitat selection for species generally and for identifying critical habitat necessary for the protection of species at risk. The theory's predictive capacity also contributes significantly to the development of policies that address the effects of landscape and habitat alteration.

Knowledge of species' reproductive behavior is also practically applied, being vital for both field and captive breeding programs associated with species recovery efforts. It also provides valuable information for predicting demographic and behavioral impacts of ecosystem exploitation (Caro 1999). Models link these specific behavioral responses to population effects, and are thus used to predict the far reaching consequences of resource extraction on conservation efforts (Sutherland and Gill 2001). Furthermore, enhanced understanding of species' social structures and avoidance behavior assists in developing accurate environmental assessment reviews and mitigation efforts for industrial undertakings (Mahoney and Schaefer 2002), and for effective predator–prey management strategies (NRC 1997; Vucetich et al. 1997).

POPULATION AND COMMUNITY ECOLOGY

Population and community ecology have provided insights into how ecological communities and components (individuals, species, and populations) are structured, and how they interact with their environment. Both have provided key insights that have become well established in wildlife science, and in conservation practices around the world. Both have long pedigrees, with Thomas Malthus, it may be said, launching the former with his 1798 *Essay on the Principle of Population*; and community ecology gradually emerging from the great European tradition of plant sociology that flourished throughout the nineteenth century. The theory of island biogeography (MacArthur and

Wilson 1967), and associated concepts of food webs and predator–prey dynamics, all emerged from these trails of inquiry and greatly improved our understanding of species interactions and the mechanisms influencing the distribution and abundance of species within an ecological community. Their combined influence on conservation has been of enormous significance, and many of our widely accepted approaches to wildlife exploitation and management are derived from or built directly upon their constructs.

By illustration, the equilibrium theory of island biogeography has had many far-reaching applications. The theory proposes that the number of species on any island reflects an equilibrium between the rate at which new species colonize it and the rate at which populations of established species become extinct. These two processes, in turn, are controlled by the size of the island, and the island's distance from the nearest mainland; smaller islands have larger extinction rates and islands closer to the mainland receive more immigrants. The deep simplicity of island biogeography theory and its immediate accessibility to a wide range of figurative "island" circumstances saw it applied to many kinds of problems, including selecting the minimum area required for nature reserves, selecting wetlands for protection, and predicting changes in the distribution and abundance of wildlife caused by habitat fragmentation (Bolen and Robinson 2003). It also contributed to wildlife science in a still more general sense, by firmly reinforcing the practical importance of considering scale as an independent variable of high resolution when calibrating ecological problems.

Like island biogeography, the concept of food chain (Elton 1927) has also, since its inception, played a critical role in our understanding of how ecological communities function; and has been generally applied to a wide range of ecological and wildlife conservation issues. We now know, of course, that ecosystems are organized in food webs, a theory that extends the food chain concept from a simple linear pathway to a complex network of interactions. Wildlife practitioners have increasingly recognized the importance of food-web structure and dynamics in understanding how ecosystems function, and how scale- and system-based approaches are required to predict and monitor the response of ecosystems to anthropogenic disturbances such as climate change, overfishing, pollution, and the introduction of invasive species. Wildlife science has not only logically borrowed from this concept in its studies of predator–prey relationships but has extended the concept to help design and execute broad synecological investigations, culminating in studies such as the Kluane Boreal Forest Ecosystem Project (Krebs et al. 2001), an elaborate experimental and multiscale study of species interactions in the Canadian north.

Predator–prey theory has contributed to wildlife science and management since the development of the Lotka–Volterra Predator–Prey Model (Lotka 1925; Volterra 1926). The model assumed a certain potential rate of increase for the predator population when the prey population was abundant, and an increase in the prey population when the predator population was low or absent. Like many theoretical constructs, the Lotka–Volterra Model's initial derivation was overly simplistic and assumed that prey populations would continue to increase as long as predators were absent or were removed from the system. However, the decline of numerous game species in North America during the early twentieth century, despite heavy predator control practices, challenged the simple cause-and-effect association between few predators and abundant prey, and led to the development of the concept of carrying capacity (Leopold 1933) and density dependence (Andrewartha and Birch 1954), two theories that became central to wildlife management and drove decades of fruitful wildlife research. Collectively, these ideas positioned, for the first time in wildlife circles, the notion of overabundance and its inevitable corollaries of disease, death, and decline.

Significantly, it was this science centered on wildlife populations themselves that provided, through broadly applied predator control programs, the long-term and large-scale "experiments" of predator–prey dynamics that ecological research *per se* was not positioned to undertake. Of course, these Malthusian principles came to wildlife science through hard and practical lessons, like the now classic irruption and decline of mule deer on the Kaibab Plateau, Arizona (Leopold 1943, cited in Bolen and Robinson 2003). After 20 years of predator control and no-shooting regulations, the mule deer population stomped and chewed its way to forage depletion and 60% of the animals died over two successive winters, eventually collapsing from 100,000 animals to no more than 10,000. Through this and other similar observations, wildlife ecologists realized that prey populations were regulated by factors other than predators, such as competition for food and space, and that predators play an important and valuable role in ecosystem functioning. For some, like Aldo Leopold himself, this insight was an epiphany (Meine 1988), changing fundamentally and forever how predators were viewed and their ecological profession assessed.

Eventually such allowances were incorporated in wildlife and ecological research, and predator–prey models were developed which incorporated logistic self-limitation or carrying-capacity components (Rosenzweig and MacArthur 1963). Since then, numerous predator–prey models have been developed to predict the effect of the functional and numerical response of predators on the abundance of prey, and outcomes from these models have been used to support always expensive and often controversial predator reintroduction and predator control management strategies. These innovations and the empirical studies that forced their development inevitably focused wildlife science on the broader question of how all populations, predators and prey alike, are regulated; and how human extraction can be managed within such ecological imperatives.

Certainly, concepts of population regulation, such as logistic growth and carrying capacity, continue to have significant implications for wildlife conservation. These have been deeply and variously integrated in the management and use of wildlife populations through the application of Maximum Sustainable Yield (MSY) theory. The goal of MSY is to hold population size at a constant level by harvesting the individuals that would be normally added to the population, and, by doing so, avoid driving a population to extinction. MSY is obtained at a harvest rate, which is roughly half the carrying capacity. Below this level yield is limited, because there are only a few individuals reproducing, and above it, density-dependent factors limit breeding until carrying capacity is reached and there are no surplus individuals to be harvested. Therefore, medium-sized populations with a high potential for growth produce the highest yields. Although MSY has been applied extensively in wildlife management, its utility has been criticized extensively, especially following the collapse of numerous fisheries worldwide managed under the MSY approach (reviewed in Ludwig et al. 1993). However, MSY still plays an important role in wildlife science and management, and its general principles remain relevant. Indeed MSY has re-emerged in the goals and objectives of Sustainable Use and Development, newly emergent paradigms now guiding the international conservation community (see below).

LANDSCAPE ECOLOGY

Developed almost as a hybrid subdiscipline of ecology and geography, landscape ecology was first described by the German geographer Carl Troll, who developed many of the formative concepts for the discipline while applying interpretations of aerial photographs to human-altered landscapes in Europe (Troll 1939). Until the emergence of landscape ecology, the influence of spatial scale and pattern on ecological processes was often neglected in wildlife investigations. While other theories also contributed (e.g., island biogeography), landscape ecology exerted unique influence, by emphasizing the importance of landscape diversity at *multiple* scales as a primary factor for predicting and assessing resistance to and recovery from disturbance. Thus, while classical ecology focused on homogeneity in landscapes, landscape ecology emphasized heterogeneity; and while classical ecology stressed and sought to elucidate nonanthropogenic influences as drivers of ecological processes, landscape ecology explicitly included, indeed focused upon, human factors as primary imperatives in real-world ecology.

Through its focus on the role of humans as part of the landscape rather than as a force external to it, landscape ecology provided a means to understand impacts of human disturbance on landscape structure and organism abundance (Naveh and Lieberman 1984) and highlighted the importance of considering fragmentation and scale in development of all wildlife conservation or habitat restoration strategies. Furthermore, the expansive geography which landscape ecology embraced allowed it to integrate a wide range of ecological theory arising from other applications. Thus, meta-population theory (Levins 1969), while becoming central to landscape ecology, originated from the theory of island biogeography. Because meta-population theory stressed the importance of habitat connectivity and corridors for persistence of wildlife populations in fragmented habitats, its relevance to landscape ecology was predetermined. In a similar fashion, these ecological insights are collectively poised to contribute increasingly to conservation as their arena of disrupted and discontinuous landscapes can only expand as human influence on earth's ecology grows.

This new emphasis on managing and protecting landscapes at multiple scales, and the necessity for connectivity between landscapes fractured by anthropogenic forces, has influenced numerous wildlife conservation initiatives and has recently spurred efforts at unprecedented ecosystem scales. For example, the Boreal Forest Conservation Framework (Canadian Boreal Initiative 2003) is a conservation approach that seeks to sustain ecological and cultural integrity of the entire Canadian boreal region (1.4 billion ha, 58% of Canada's land mass) by protecting at least 50% of the region in a network of large, interconnected protected areas, and by managing the remaining landscape through an ecosystem-based resource management approach (see below).

For endangered wildlife species, such as woodland caribou (*Rangifer tarandus* spp.), and for large predators generally (Peters 1983), there can be no substitute for such initiatives; space and ecological runway (the response opportunity afforded by habitat heterogeneity) are essential to the behavioral ecology of many large mammals. Large expanses of land will promote biodiversity protection in general, and ensure that ecosystem functions are not impaired (Gilbert et al. 1998; Harrison and Bruna 1999). Flow is essential to nature; as restriction is essential to zoos. In executing such a biome- and continent-wide application of ecological theory, we realize how much wildlife conservation, and the science it has engendered and depended upon, have borrowed and benefited from ecology's ever-broadening reach. Inevitably, this very science would itself influence the emergence of new management paradigms that, in their turn, would require not only additional science for wildlife conservation, but also, arguably, new kinds of science — moving from the reductionist to the integrative, and from the linear to the synthetic.

ECOSYSTEM MANAGEMENT AND THE ECONOMICS OF ECOLOGY

Just as the evolving character of wildlife science in the early-to-mid twentieth century was to reshape how we understood the role of predators in ecosystems, so attempts to understand the ecological conditions required to husband wider communities of animals and plants would inevitably lead conservation efforts to embrace ever-expanding hierarchies (Mulder and Coppolillo 2005). Thus, as ecology developed and our understanding of the functioning and complexity of ecosystems grew, there emerged a pervading recognition that human beings and their effects on natural systems needed to be considered integral, rather than just disruptive, if the science was to serve conservation in any practical sense. In the applied science that centered on wildlife populations and management, this was an easy assumption to integrate, it being at the heart of the discipline's origins; but for ecology, it was a more significant re-evaluation. Nevertheless, it has come to increasingly affect both disciplines, and stimulated a new approach to natural resource management and conservation that, by definition, incorporates human dimensions and demands new science ventures in turn.

Termed *Ecosystem Management*, the goal of this new approach is to ensure productive, healthy ecosystems by incorporating social, economic, physical, and biological values in management decisions. Unlike many other management approaches, Ecosystem Management is focused on long-term sustainability of resources rather than maximizing short-term yield; and economic gain is not the sole valuation on which management practices are constructed (Christensen et al. 1996). Although the International Union for the Conservation of Nature (IUCN) has developed, through its

Convention on Biological Diversity, a set of principles that help define the ecosystem management approach, and despite its widespread reference in the conservation literature, critics have suggested that the approach lacks clear, measurable objectives (Sedjo 1996). While this may be true, Ecosystem Management's explicit recognition of diverse values inherent to natural resources and systems, and its codifying of what these values are, represent a conceptual advance over previous approaches that focused on single resource values for specific ecosystems. Furthermore, Ecosystem Management has done a great deal to re-emphasize the inherent reliance of human life and society on the very processes that control and regulate ecosystems themselves, thus bringing wildlife, the habitats they require, and our own human existence within one holistic framework. To paraphrase Sir Francis Bacon's famous quip concerning justice, Ecosystem Management makes the case that if we maintain natural systems, they will maintain us.

To make its philosophy accessible, the Ecosystem Management movement adopted an economics frame of reference that not only evaluated resources, such as timber and wildlife, in the classic manner, but also evaluated the economics of ecosystem processes themselves. In the new lexicon, these processes were collectively termed *Ecosystem Services*, thus making humans ecosystem clients. Not poetic, certainly, but at least bringing some humility to our position in the natural scheme of things, and forcefully challenging the man-outside-of-nature syndrome. Specifically, these Ecosystem Services are defined as "the conditions and processes through which natural ecosystems, and the species that make them up, sustain and fulfill human life" (Daily 1997). In addition to provisioning of goods (food, freshwater, fuel, wood and fiber, and medicines), ecosystems also provide a variety of supporting (nutrient cycling, soil formation, primary production, and provision of habitat), regulating (climate regulation, flood control, pollination, and water purification), and cultural (spiritual, recreation, aesthetic, and educational) services that directly affect human well-being (security, health, shelter, and good social relations) (Daily et al. 1997; Millennium Ecosystem Assessment 2005; Pereira and Cooper 2006). Ecosystem Services are valuable to humans in that they support our lives, are cheap, and cannot easily be replaced with human-engineered alternatives (Cork 2001).

Explicit recognition that ecosystem functions have value beyond their inherent worth has prompted unprecedented attention from scientists and economists around the world. Their shared focus has been on describing, measuring, and valuating the entire range of ecosystem goods and services (Constanza et al. 1997; Daily 1997; Pimm 1997; Allen and Loomis 2006; Christie et al. 2006), a process sufficiently robust to be now recognized as a distinct field called *ecological economics* (Constanza 1989). This endeavor is reminiscent of efforts in the eighteenth and nineteenth centuries aimed at cataloguing species themselves, and indicates the heightened influence that ecological awareness is exerting within mainstream economics. It also provides clear evidence of how economic rationalizations are being used to buttress arguments in support of conservation. Increasingly, international and continental appraisals of the earth's ecological health are persuaded by this highly utilitarian perspective. Thus, while estimates and valuation techniques have met with some criticism (Pimm 1997; Ludwig 2000), the release of the United Nations sponsored Millennium Ecosystem Assessment (2005) has yet again emphasized the dependence of human well-being on ecosystems, the negative state of the world's ecosystems, and the urgent need to better value (ecologically, culturally, and economically) the goods and services they provide.

These new amalgams of ecology and finance are now influencing regional and local conservation initiatives worldwide (Czech 2000), as communities and individuals come to understand the interconnectedness of these phenomena, and the increasingly rapid pace of resource depletion and conflict. Indeed, the curve of knowledge is beginning to bend on itself as increasing numbers of people recognize that ecosystem services are declining because of a loss of biological diversity, which itself is a direct consequence of human actions (World Resources Institute 2000). This, of course, was the very worry that launched North American wildlife (and forestry) science in the first instance; although long before we understood in any detail how ecosystems actually worked. In a fascinating conceptual evolution, we have been forcefully returned to earlier fears of anthropogenic impacts by vastly improved knowledge, gathered largely through ecological studies that were focused on "natural" processes.

Wildlife science marched along through this process, keeping an eye on ecology and borrowing from its achievements, while at the same time improving its own capacity to better serve wildlife management. And the latter always maintained human requirements and valuations as central to its mission. Might we say that ecology has, in Ecosystem Management and its attendant science, converged on the focus wildlife science never abandoned? And in seeking an incentive-driven paradigm to convince ecological conscience and initiate conservation action, have we not returned to Leopold's "conservation economics" of the 1930s and to George Perkins Marsh's (1864) forewarnings in *Man and Nature*, the root philosophical treatise that may be said to have launched conservation in America?

No matter. We are going to need the best of both, indeed of all our disciplines. Over the past few centuries, humans have increased the extinction rates of species for all taxonomic groups by as much as 1000 times the historical rates, and future extinction rates are projected to be more than 10 times higher than the current rate. The most important drivers of biodiversity loss are habitat and climate change, invasive species, overexploitation, and pollution. The impact of these factors, all of which are associated with human activity, is predicted to remain constant or increase rapidly (Millennium Assessment 2005). Most experts take the latter view.

Little wonder then, that the global degradation of ecosystem services, and the recognition of the economic and intrinsic value of these services to humans, led to a heightened focus among international conservation agencies, and to development of an international protocol for the sustainable use and development of our natural resources. In many ways, this protocol reaffirms the principles articulated in the North American wildlife management approach; namely, that vested self interest and regulated harvest can be critical to long-term conservation efforts. While far broader than "wildlife" in its focus, this new international protocol has implications for wildlife science. Predictably, it will encourage wildlife research to become more integrative and multidisciplinary; it may also lead wildlife research further from its cherished history of ever more detailed studies of animal ecology. In all these regards, it will move wildlife science in many of the same directions promoted by landscape ecology, but this time with the combined force of political and social agendas that are coordinated by some of the world's most powerful organizations. In terms of its North American domain, wildlife science will become influenced by a world order now rapidly defining conservation agendas in terms reminiscent of North America's own first awakenings; incentive-based conservation has gone global, and the science required to sustain it, including wildlife science, will become more globalized in turn.

SUSTAINABLE USE AND CONSERVATION: THE NEW ORDER

At the international level, conservation had for many reasons become more preservationist oriented as the mid-twentieth century approached (Mulder and Coppolillo 2005). However, in time, many social and scientific influences came to challenge this largely protectionist approach, especially as it became clear that policies such as land protection and no development zones were entirely insufficient for preventing further declines in environmental standards. Changing lifestyles and societal aspirations in the developing world also made it clear that a more comprehensive approach to all lands and resources was required. It was UNESCO's (the United Nations Educational, Scientific, and Cultural Organization) Man and Biosphere program in the 1960s that may have first heralded a new international pragmatism. By proposing that human demands be included within international conservation policy, a new vision for conservation was launched, one in which resource consumption might actually help safeguard, rather than inevitably deplete, natural abundance (Batisse 1982). This was to move us conceptually beyond simply accounting for human influence in the environment, a central premise of Ecosystem Management, to viewing human activities as an important source for conservation itself. Meanwhile, global recognition that the environment was endangered by human activity, while escalating in scientific circles for decades, was cemented politically at the United Nations Conference on the Human Environment held in Stockholm, Sweden, in 1972, again making it clear that some new strategy for conserving the planet's environment and resources was desperately required.

Thus, the groundwork was laid to advance a new agenda, and both the motivations and general directions were obvious; however, as with so many conceptual leaps, no knowledge undercarriage existed to launch this new order. The international community was clear in identifying the need for greater understanding of the linkages between the environment and socioeconomic forces, and thus identified the knowledge gap. But the looming question was how to close the gap. As with Ecosystem Management (and its Ecosystem Services repertoire), a terminology was required for this new conservation strategy, a common language the world could agree upon when discussing the interface of environmental, social, and economic issues.

To address this need, the concept of *Sustainable Development* was derived, and gained near-immediate acceptance with the publication in 1987 of *Our Common Future*, a United Nations sponsored report by the World Commission on Environment and Development. Defined as "development that meets the needs of the present without compromising the ability of future generations to meet their own needs," the concept was quickly adopted at the 1992 United Nations Conference on Environment and Development held in Rio de Janeiro, Brazil. At this *Rio Earth Summit*, Sustainable Development was formally identified as the guiding vision for the development efforts of all countries, and led to the adoption of *Agenda 21*, a wide-ranging blueprint for action to achieve Sustainable Development worldwide. Ten years later, this plan of action was reviewed at the World Summit on Sustainable Development in Johannesburg, South Africa, with disappointing results. Nevertheless, Sustainable Development remains a significant guiding principle, one with sufficient international presence to see governments worldwide agree to a wide array of commitments and cooperative programs under its aegis. Many of these have implications for wildlife conservation and for wildlife science.

For example, one of the agreements adopted at the 1992 Rio de Janeiro meeting was the *Convention on Biological Diversity* (CBD), which concerns itself with conservation of the world's biological diversity and the sustainable use of its components. The CBD defines sustainable use as "the use of components of biological diversity in a way and at a rate that does not lead to long-term decline of biological diversity, thereby maintaining its potential to meet the needs and aspirations of present and future generations." Under this convention, sustainable use of the world's natural diversity was identified as essential to achieving the broader goal of Sustainable Development, leading, as a consequence, to the *Conservation through Sustainable Use Strategy* (CSU), introduced at the 2000 IUCN 2nd World Conservation Congress in Amman, Jordan.

The CSU, in its turn, proposed that wildlife conservation could sometimes be enhanced through harvest of wildlife, and recognized that economic benefit is critical for the success of many conservation plans. The IUCN supports use of resources only when it is sustainable and proceeds in a manner that minimizes losses to biodiversity. Accordingly, an adaptive management approach is considered critical for sustainable use to be effective as a conservation tool. This approach requires continuous monitoring and the ability to modify management practices to take account of risk and uncertainty. Thus, it requires a complementary stream of relevant science.

In addition to the CBD and the IUCN, the World Wildlife Fund (WWF) International, and the *Convention on the International Trade in Endangered Species of Wild Fauna and Flora* (CITIES) also support the CSU strategy, making it now possibly the most influential conservation paradigm in the world. This international pedigree is an important distinction, as there are numerous conservation approaches around the world that much earlier expressed similar points of view. The century-old North American Wildlife Conservation Model, for example, has forever incorporated similar philosophies of wise use and vested interest as its foundations for biodiversity protection (Mahoney 2004). However, it, like other conservation models around the world, was more regional in its application and effect. Through the CSU, many principles of the North American Model have gone global, most crucially the notion of vested interest or incentive-driven conservation as a compelling force for biodiversity protection.

However, despite the growing acceptance of CSU by the international community, not everyone in the field of wildlife conservation appreciates and supports the sustainable use approach. One of the main objections stems from the numerous historical examples where use of wildlife led to overexploitation and extinction (Ludwig et al. 1993; Lavigne et al. 1996; Ludwig 2001), in contrast to the few well-documented examples where extractive use has been shown to have enhanced conservation effects, especially for vulnerable species and populations (Webb 2002). Others view sustainable use as a part of a "new world deception" that has empowered charlatan promoters of the paradigm (Willers 1994), those who would seek, under its aegis and cloak of respectability, to advance pro-development regimes no more ecologically sensitive than previous. Other criticisms address the range of "uses" considered acceptable for conservation, and the lack of clarity in the definition of "sustainability" itself (Taylor and Dunstone 1996; Hutton and Leader-Williams 2003). Of course, some proponents of conservation, as well as animal rights and welfare, are opposed to any use of wildlife purely for ethical reasons, and oppose sustainable use, regardless of whether it does or does not lead to conservation benefits (Webb 2002).

Regardless of such reservations, sustainable use is the new conservation agenda: A first-time global acceptance of vested interest and utilitarianism, although contextualized certainly by reference to a broad range of human values and aspirations. For wildlife science and its game management framework (later expanded to include all biota) as evolved in North America, the international approach is remarkably familiar. A combination of practical conservation and idealistic activism, focused upon ecological diversity, integrity, and continuance, is a long tradition in the shared agenda of Canada and the United States (Mahoney 2004). And, at the base of it all, a vital community of restrained self-interest financing supports the very infrastructure of conservation, including the wildlife science programs essential to success. The ark of conservation progress seemingly bends toward such principles and the paradigms they encompass.

CONCLUSIONS

It is clear that the world's biota and natural systems are under ever-increasing strain. Seemingly, whatever advances we have attained in knowledge have been insufficient in themselves to halt biodiversity loss and environmental degradation. Even as ecology and wildlife science have advanced, so too have the scale and diversity of the ecological problems we face. We need a new approach. Sustainable-use paradigms reflect this reality, and seek to integrate ecological, social, and economic demands. This international approach represents the most recent realignment in our thinking and is viewed by many as crucial to successful conservation initiatives in the twenty-first century. In and of itself, however, sustainable use can only provide conceptual guidance; to be effective, it will require much of wildlife science and ecology. These will continue to be the beasts of burden, provisioning the new zeitgeist and enabling it to evaluate approaches and opportunities.

From a North American perspective, wildlife science emerged from a cradle of unbridled resource slaughter. It was nourished from its earliest years by citizen activism, which increasingly saw conservation of the continent's wild diversity as a near measure of nationhood and progress itself. This was true of both Canada and the United States, and led eventually to a cultural tradition in both nations for a practical inquiry that would support the continued use of wildlife, yet ensure its valuation as an intergenerational equity. Furthermore, the context for this science embraced a range of human activities and values, including the aesthetic and spiritual. Nevertheless, it was and remains a science in service of a utilitarian ethic, a rationalized insurance of continued supply simultaneous with continued demand, each being promoted by a social and political agenda that sees wildlife as crucial to our lives.

As the historical review shows, this utilitarianism was not a recipe for closeted inquiry. Wildlife science had its focus, of course; and in its earlier years, this was primarily the understanding of a specific suite of organisms, those large mammals and game birds sought by the recreational hunter. However, the sinuous path of knowledge acquisition itself and ecology's neighboring status coerced wildlife science, expanding its horizons and *raison d'etre*. Increasingly, ecosystem dynamics and the full range of biodiversity became the arena of wildlife investigations and the conservation of these its objectives. It could leave to ecology the task of refining theoretical models, but its own work was done with one eye to the intellectual culture; ecology was advancing, and its results were to be evaluated by a dual relevance, to wildlife in the practical sense and to validating and improving ecological insights themselves. In reaching this position, it helped clarify other scientific niches, and in concert with the looming biodiversity crisis, helped launch the discipline of conservation biology.

Wildlife science has matured greatly since its humble beginnings, but like many other disciplines finds difficulty in striking a balance between its historical focus and the knowledge requirements imposed by current social and ecological realities. Nevertheless, to maintain its relevance in the twenty-first century our discipline must find this balance. Doing so will mean positioning itself to more effectively engage in interdisciplinary research, especially in the socioeconomic and human dimensions arenas; and to more efficient communication with policy makers on issues of sustainability. It will also mean increasing its profile in the international arena and working much harder to inform the public of its contributions to conservation. Even more importantly, however, wildlife science must recommit to its responsibility for educating the public concerning the true nature of ecosystems, and for offering practical, humanity-relevant agendas for conservation. At the same time, it must recognize and promote the irreplaceable value of practical experience, and the collective intelligence of public discourse in addressing conservation challenges. By doing so, and by exploring the magic and mystery of wildlife, and our shared and inalienable responsibility for its continuance, our science will continue to inform the greatest of human challenges: how to maintain a glorious diversity of life and a civilized future for humankind.

REFERENCES

Allen, B. P., and J. B. Loomis. 2006. Deriving values for the ecological support function of wildlife: An indirect valuation approach. *Ecol. Econom.* 56:49.
Andrewartha, H. G., and L. C. Birth. 1954. *The Distribution and Abundance of Animals*. Chicago: University of Chicago Press.
Anthony, L. L., and D. T. Blumstein. 2000. Integrating behaviour into wildlife conservation: The multiple ways that behaviour can reduce N_e. *Biol. Conserv.* 95:303.
Batisse, M. 1982. The biosphere reserve: A tool for environmental conservation and management. *Environ. Conserv.* 9:101.
Bolen, E. G., and W. L. Robinson. 2003. *Wildlife Ecology and Management*, 5th edn. Upper Saddle River, NJ: Prentice Hall.
Bradshaw, G. A., and M. Bekoff. 2001. Ecology and social responsibility: The re-embodiment of science. *Trends Ecol. Evol.* 16:460.
Canadian Boreal Initiative. 2003. *The Boreal Forest Conservation Framework 2003*. Canadian Boreal Initiative homepage, www.borealcanada.ca, Accessed March 31, 2006.
Caro, T. 1999. The behavior–conservation interface. *Trends Ecol. Evol.* 14:366.
Christensen, N. L., A. M. Bartaska, J. H. Brown, S. Carpanter, C. D'Antonio, R. Francis, J. F. Franklin, J. A. McMahon, R. F. Noss, D. Z. Parsons, C. H. Peterson, M. G. Turmer, R. G. Woodmansee. 1996. The report of the Ecological Society of America committee on the basis for ecosystem management. *Ecol. Appl.* 6:665.
Christie, M., N. Hanley, J. Wamen, K. Murphy, and R. Wright. 2006. Valuing the diversity of biodiversity. *Ecol. Econom.* 58:304.
Constanza, R. 1989. What is ecological economics? *Ecol. Econom.* 1:1.

Constanza, R., R. D'Arge, R. deGroot, S. Farber, M. Grasso, B. Hannon, K. Limburg, S. Naeem, R. V. O'Neill, J. Paruelo, R. G. Raskin, P. Sutton, and M. van den Belt. 1997. The value of the world's ecosystem services and natural capital. *Nature* 387:253.

Cork, S. 2001. Ecosystem services: The many ways in which biodiversity sustains and fulfills human life. Presented at *Healthy People and a Healthy Planet 2001 Internet Conference*, organized by the Nature and Society Forum. http://conference.natsoc.org.au, Accessed February 20, 2006.

Curio, E. 1996. Conservation needs ethology. *Trends Ecol. Evol.* 11:260.

Czech, B. 2000. The importance of ecological economics to wildlife conservation. *Wildl. Soc. Bull.* 28:2.

Daily, G. (ed.). 1997. *Nature's Services: Societal Dependence on Natural Ecosystems.* Washington, DC: Island Press.

Daily, G., S. Alexander, P. R. Erlich, L. Goulder, J. Lobchenco, P. A. Matson, H. A. Mooney, S. Postel, S. H. Schneider, D. Tilman, and G. M. Woodwell. 1997. Ecosystem services: Benefits supplied to human societies by natural ecosystems. *Issues Ecol.* 2:1.

Elton, C. S. 1927. *Animal Ecology.* London: Sidgwick and Jackson.

Fiedler, P. L., P. S. White, and R. A. Leidy. 1997. The paradigm shift in ecology and its implication for conservation. In *The Ecological Basis of Conservation: Heterogeneity, Ecosystems, and Biodiversity*, S. T. A. Pickett, R. S. Ostfeld, M. Shachak, and G. E. Likens (eds). New York: Chapman and Hall, p. 83.

Geist, V. 1989. Game ranching: Threat to wildlife conservation in North America. *Wildl. Soc. Bull.* 13:594.

Geist, V., S. P. Mahoney, and J. F. Organ. 2001. Why hunting has defined the North American model of wildlife conservation. *Transactions of the 66th North American Wildlife and Natural Resources Conference*, p. 175.

Gilbert, F., A. Gonzalez, and I. Evans-Freke. 1998. Corridors maintain species richness in the fragmented landscapes of a microecosystem. *Proc. R. Soc. Lond. B* 265:577.

Harcourt, A. H. 1999. The behavior–conservation interface. *Trends Ecol. Evol.* 14:490.

Harrison, S., and E. Bruna. 1999. Habitat fragmentation and large scale conservation: What do we know for sure? *Ecography* 22:225.

Hendee, J. C., and G. H. Stankey. 1973. Biocentricity in wilderness management. *BioScience* 23:535.

Hobbs, R. J., and L. F. Huenneke. 1992. Disturbance, diversity, and invasion: Implications for conservation. *Conserv. Biol.* 6:324.

Holling, C. 1986. The resilience of terrestrial ecosystems: Local surprise and global change. In *Sustainable Development of the Biosphere*, W. Clark, and R. Munn (eds). Cambridge: Cambridge University Press, p. 292.

Hutton, J. M., and N. Leader-Williams. 2003. Sustainable use and incentive-driven conservation: Realigning human and conservation interests. *Oryx* 37:215.

Kellert, S. R., and C. P. Smith. 2000. Human values toward large mammals. In *Ecology and Management of Large Mammals in North America*, S. Demarais, and P. Krausman (eds). Upper Saddle River, NJ: Prentice Hall, p. 38.

Klein, D. R. 2000. Arctic grazing systems and industrial development: Can we minimize conflicts? *Polar Res.* 19:91.

Krebs, C. J., S. A. Boutin, and R. Boonstra. 2001. *Ecosystem Dynamics of the Boreal Forest: The Kluane Project.* New York: Oxford University Press.

Lavigne, D. M., C. J. Callaghan, and R. J. Smith. 1996. Sustainable utilization: The lessons of history. In *The Exploitation of Mammal Populations*, V. J. Taylor, and N. Dunstone (eds). London: Chapman and Hall, p. 250.

Leopold, A. 1933. *Game Management.* New York: Charles Scribner and Sons.

Leopold, A. 1943. Deer irruptions. Wis. Conserv. Bull. (reprinted in Wis. Conserv. Dept. Publ. 321:1–11).

Levins, R. 1969. Some demographic and genetic consequences of environmental heterogeneity for biological control. *Bull. Entomol. Soc. Amer.* 15:237.

Lima, S. L., and P. A. Zollner. 1996. Towards a behavioral ecology of ecological landscapes. *Trends Ecol. Evol.* 11:131.

Lotka, A. J. 1925. *Elements of Physical Biology.* Baltimore: Williams & Wilkins.

Ludwig, D. 2000. Limitations of economic valuation of ecosystems. *Ecosystems* 3:31.

Ludwig, D. 2001. Can we exploit sustainably? In *Conservation of Exploited Species*, J. D. Reynolds, G. M. Mace, K. H. Redford, and J. G. Robinson (eds). Cambridge: Cambridge University Press, p. 16.

Ludwig, D., R. Hilborn, and C. J. Walters. 1993. Uncertainty, resource exploitation, and conservation: Lessons from history. *Science* 260:17.

MacArthur, R. H., and E. O. Wilson. 1967. *The Theory of Island Biogeography.* Princeton: Princeton University Press.

Mahoney, S. P. M. 2004. The seven sisters: Pillars of the North American wildlife conservation model. *Bugle* 141.

Mahoney, S. P. M., and J. A. Schaefer. 2002. Long-term changes in demography and migration of Newfoundland caribou. *J. Mammal.* 83:957.

Marsh, G. P. 1864. *Man and Nature.* New York: Charles Scribner.

McKee, J K., P. W. Sciulli, D. Fooce, and T. A. Waite. 2003. Forecasting global biodiversity threats associated with human population growth. *Biol. Conserv.* 115:161.

Meine, C. 1988. *Aldo Leopold — His Life and Work.* Madison: University of Wisconsin Press.

Millennium Ecosystem Assessment. 2005. *Ecosystems and Human Well-Being: Synthesis.* Washington, DC: Island Press.

Mulder, M. B., and P. Coppolillo. 2005. *Conservation: Linking Ecology, Economics and Culture.* Princeton: Princeton University Press.

National Research Council (NRC). 1997. *Wolves, Bears, and Their Prey in Alaska: Biological and Social Challenges in Wildlife Management.* National Academy Press.

Naveh, Z., and A. Lieberman. 1984. *Landscape Ecology: Theory and Application.* New York: Springer.

Paterson, B. 2006. Ethics for wildlife conservation: Overcoming the human–nature dualism. *Bioscience* 56:144.

Pereira, H. M., and H. D. Cooper. 2006. Towards global monitoring of biodiversity change. *Trends Ecol. Evol.* 21:123.

Peters, R. H. 1983. *The Ecological Implications of Body Size.* New York: Cambridge University Press.

Pickett, S. T. A., V. T. Parker, and P. Fielder. 1992. The new paradigm in ecology: Implications for biology above the species level. In *Conservation Biology: The Theory and Practice of Nature Conservation*, P. Fielder, and S. Jain (eds). New York: Chapman and Hall, p. 65.

Pimm, S. L. 1991. *The Balance of Nature? Ecological Issues in the Conservation of Species and Communities.* Chicago: University of Chicago Press.

Pimm, S. L. 1997. The value of everything. *Nature* 387:231.

Rasker, R., M. V. Martin, and R. L. Johnson. 1992. Economics — Theory versus practice in wildlife management. *Conserv. Biol.* 6:338.

Reiger, J. F. 1975. *American Sportsmen and the Origins of Conservation.* New York: Winchester Press.

Robertson, D. P., and R. B. Hull. 2001. Beyond biology: Toward a more public ecology for conservation. *Conserv. Biol.* 15:970.

Rosenzweig, M. L., and R. H. MacArthur. 1963. Graphical representation and stability conditions of predator–prey interactions. *Am. Nat.* 97:209.

Scoones, I. 1999. New ecology and social sciences: What prospects for a fruitful engagement? *Annu. Rev. Anthro.* 28:479.

Sedjo, R. A. 1996. Toward an operational approach to public forest management. *J. Forest.* 94:24.

Sutherland, W. J. 1998. The importance of behavior in conservation biology. *Anim. Behav.* 56:801.

Sutherland, W. J., and J. A. Gill. 2001. The role of behavior in studying sustainable exploitation. In *Conservation of Exploited Species*, J. D. Reynolds, G. M. Mace, K. H. Redford, and J. G. Robinson (eds). Cambridge: Cambridge University Press, p. 16.

Taylor, V. J., and N. Dunstone (eds). 1996. *The Exploitation of Mammal Populations.* London: Chapman and Hall, p. 250.

Trefethen, J. B. 1975. *An American Crusade for Wildlife.* New York: Winchester Press.

Troll, C. 1939. Luftbildplan und ökologische Bodenforschung (Aerial photography and ecological studies of the earth). *Zeitschrift der Gesellschaft für Erdkunde*, Berlin, p. 241.

Volterra, V. 1926. Variazioni e fluttuaxioni del numero d'individui in specie animali conviventi. *Academic Lincei Series* 6:26.

Vucetich, J. A., R. O. Peterson, and T. A. Waite. 1997. Effects of social structure and prey dynamics on extinction risk in gray wolves. *Conserv. Biol.* 11:957.

Wallington, T. J., R. J. Hobbs, and S. A. Moore. 2005. Implications of a current ecological thinking for biodiversity conservation: A review of salient issues. *Ecol. Soc.* 10:15.

Webb, G. J. W. 2002. Conservation and sustainable use of wildlife — An evolving concept. *Pacific Conserv. Biol.* 8:12.

Willers, B. 1994. Sustainable development: A new world deception. *Conserv. Biol.* 8:1146.

World Resources Institute. 2000. *World Resources 2000–2001. People and Ecosystems: The Fraying Web of Life.* New York: Oxford University Press.

20 Wildlife and Ranching: From Externality to Profit Center

Barry H. Dunn

CONTENTS

Truth resides in panoramic view rather than a local view of events.

Polybius, *Greek statesman, 2nd century*

On the scale of human history, the emergence of ranching as a business is a very recent phenomenon. Raising and harvesting grazing animals for food, fiber, and by-products on large acreages of semi-arid rangeland is certainly not. But, doing it in the context of a for profit business, for the sustenance of the inhabitants of towns and cities many miles distant, began only after the U.S. Civil War. It was primarily a response to the economic drivers of the industrialization and urbanization of American society. It followed the destruction of America's indigenous herds of bison (*Bison bison*), pronghorn (*Antilocapra americana*), elk (*Cervus elaphus*), and deer (*Odocoileus* spp.) on its prairies and rangelands as a result of policy decisions concerning Native Americans, immigrant settlement, and economic exploitation. The creation of ranches with cowboys, a transportation system of cattle drives, and a central auction marketing system were all utilitarian responses to economic opportunity. By the turn of the twentieth century, the use of rangeland resources for the production of sheep and cattle had became a major force in America's food and fiber system, and grew dramatically through the first half of that century.

For all but a few visionary leaders, like Theodore Roosevelt, John Muir, Aldo Leopold, and the Kleberg family of South Texas, wildlife on America's rangelands were an after-thought to settlement and a means to short-term economic gain. After the Civil War, wildlife was viewed as an externality to the food and fiber system of American agriculture — an un-bargained-for cost or benefit. The benefits of wildlife were largely for hunting for food and as recreation for a few. The costs of the decimation of entire populations of wildlife species were ignored.

The turn of the twenty-first century has brought with it new demographic, social, and economic realities. America's population is three to four generations removed from production agriculture, and is 98% urban or suburban. Its citizenry is relatively affluent and has increasing sources of disposable income. Their image of and the value they place on wildlife and the open spaces of rangeland resources has changed dramatically. The cow–calf and stocker production of the beef industry are the only remaining use of rangelands in America's food system. Hunting and observing wildlife is now bargained for, as reflected by land prices, lease rates, and state and federal laws and regulations. Places like "The Last Great Habitat" of South Texas are no longer considered part of the great American desert, but are treasured for their ecological diversity. As a result, ranchers and land managers have new and exciting business opportunities, and the American society has a chance to redeem itself.

THEORETICAL CHANGE: THE EMERGENCE OF CAPITALISM

The theoretical debate of the eighteenth century concerning emerging economies centered on egalitarianism and the development of capitalism as a functional replacement for what in retrospect is referred to as *feudalism* (West 1975). In the eighteenth century, feudalism represented the historical context of property ownership, government, service, and commerce, which helped fuel the American and French revolutions. Much of the wealth and power in Europe during this era still resided in aristocratic families that had gained its title by heritage or marriage. The migration of Europeans to the Americas was driven not only by religious oppression, but also by the desire for economic opportunity, and away from restrictive tenant relationships of land and resource ownership and management that emerged in the waning years of feudalism. By the eighteenth and nineteenth centuries, the feudal and aristocratic system of land ownership and commerce lay in collapse, as multiple famines and disease outbreaks swept across Europe. And yet, much of the property and associated resources continued to be owned and controlled by aristocratic families. Adam Smith's theory of a capitalist economy, where property and rights to access could be earned, found a fertile proving ground with the birth of the United States. Polices relating to land and resource ownership and use that emerged in the young United States reflected a rejection of its European roots. These policies sought to provide opportunity and access as rights of citizenship. However, they also set up a riddle. For the most part, land was to be held in private ownership (Macpherson 1975), but wildlife was from the earliest days of our nation regarded as a public good. So, over 200 years later, much of the habitat of thousands of wildlife species is privately owned, while wildlife itself belongs to the public. Can publicly owned wildlife be successfully managed when their habitat is privately owned? Attempts to solve the riddle are reflected in dynamic law and policy (Anderson and Hill 1975), but are forever framed by foundational policies and the social morays they reflect.

THE POWER OF POLICY

Public policy represents the manifestation of accepted theory by a society. Both historical and present-day policy of the United States toward the wildlife resources of North America serve as excellent examples of the complex nature and power of public policy. The present-day state and nature of these resources are a reflection of past public policy. Societies exhibit their collective will through the policies that they adopt either through processes of consensus or abdication. Many, like laws or regulations, are developed through debate and discussion and are clearly stated, communicated, and need various methods of ratification. Others are unstated and reflect the will, opinion, or belief systems of the day. Others still reflect the opinions and wishes of elite minorities; either class, economic, religious, or political. They are policy because the majority abdicates their responsibility due to ignorance, ambivalence, or apathy, or as a result of various forms of force or coercion by the minority. At any one time, public policy of a society is a mixture of law, regulation, and the will, opinion, and belief systems of the majority, the minority, or both. Policy is also complex and dynamic. It exists in many different organizations and levels within a society. Policy can also appear to be at times schizophrenic, as different segments of society push conflicting agendas and programs. The future state of wildlife resources in Texas, the United States, and North America will be a direct result of the wisdom of present-day policy, in the context of their biological and ecological potentials.

FROM THEORY TO POLICY TO MANAGEMENT: WILDLIFE AS AN EXTERNALITY

While the history of extinction or near-extinction of wildlife species like the passenger pigeon (*Ectopistes migratorius*), beaver (*Castor canadensis*), and bison, during the nineteenth century has an economic aspect, the management of wildlife during our nation's first 100 plus years is a classic example of an externality: un-bargained-for costs and benefits. During that time, wildlife was alternatively viewed as a source of food or fiber, short-term economic gain, leverage point for agendas, or as competition for scarce resources. That is not to say that economic benefits were not gained by the exploitation of wildlife. Some individuals and organizations in the nineteenth century made fortunes from the harvest of wildlife, but they were un-bargained-for. Licenses were not required to hunt passenger pigeons or bison, or any species. Millions of acres of prairies in the western United States were grazed at no cost for decades by individuals who controlled the limited water resources around creeks and rivers.

A hundred years later, the costs and benefits of these myopic decisions are incalculable, and policy was the driving force. While the creation of wildlife and natural resources as an externality has policy roots in countless laws, regulations, court opinions, and unwritten views and value systems, three are offered as examples.

THE JEFFERSONIAN GRID: THE LENS WITH WHICH WE VIEW WILDLIFE AND NATURAL RESOURCES

The Land Ordinance of 1785, drafted by Thomas Jefferson, was one of the most important pieces of legislation affecting land and resource use in American history. In the original American colonies, land ownership and title had followed the ancient method of metes and bounds. This surveying system described a parcel of land in terms of the natural features and adjacent parcels that it was in relationship with, such as trees, streams, and rocks. The metes-and-bounds system led to constant conflicts between land owners, and land owners and governments, as the natural features that it was based upon changed over time. Jefferson's proposal became the most influential land use policy in the United States until the Homestead Act of 1862 (u-s-history.com 2006). It provided for a very strict grid with which land would be surveyed and on which both governments and commerce would be based. It has become known as the "Jeffersonian Grid" (Encarta 2006). Its basic unit is a rectangle, referred to as a township. A township consists of 36 "sections," each 1-square mile and consisting of 640 acres (Figure 20.1). Each section can be easily subdivided into "quarters," which consist of 160 acres, and be further subdivided into 40-acre units (Figure 20.2). This surveying methodology was applied to all land acquired by the United States in the Northwest Territory, the Louisiana Purchase, and the annexation of Texas, New Mexico, Arizona, and California. Today, it remains both the fundamental methodology relating to the commerce of land and the basis of many levels of government (Figure 20.3).

36	31	32	33	34	35	36	6
1	6	5	4	3	2	1	6
12	7	8	9	10	11	12	7
13	18	17	16	15	14	13	18
24	19	20	21	22	23	24	19
25	30	29	28	27	26	25	30
36	31	32	33	34	35	36	31
1	6	5	4	3	2	1	6

FIGURE 20.1 Diagram of the sectioning of a township in the Jeffersonian grid.

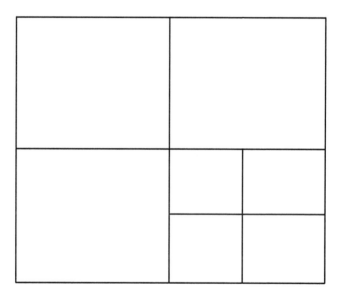

FIGURE 20.2 Diagram of the division of a section in the Jeffersonian grid into 160-acre "quarters" and further division into 40-acre "40s."

Adoption of national, state, and local policy in the late nineteenth century encouraging individual ownership and management of land and its resources reflects the rapid adoption of Adam Smith's theory of capitalism. If Jefferson, as the author of the Constitution, described and defined how citizens of the United States would be governed, it can be argued that his "Jeffersonian Grid" described and defined how its citizens would view land and natural resource management. In the agrarian communities of Jefferson's era, one family could live off the products of 40 or 80 acres, and ship surpluses to support nearby villages and towns. However, the blanket application of the "Jeffersonian Grid" through policy to the ownership and management of land in the arid and semi-arid regions west of the 100th meridian has proven to be unsustainable. The externalities are incalculable. In regions where an acre, 40 acres, 80 acres, or a quarter section is not enough land to sustain even a single cow for a year; basing land, natural resource, and business decisions on these homogenized micro units has led to many unexpected outcomes. These would

FIGURE 20.3   Example of the Jeffersonian grid as shown in farm ground.

include soil loss due to the cultivation of highly erodable land, overgrazing of grasslands, depletion of water resources, and loss of wildlife habitat due to fragmentation of land ownership and
management.

MANIFEST DESTINY: THE POWER OF A BELIEF SYSTEM

In 1845, an influential journalist named John L. O'Sullivan coined the phrase "manifest destiny"
to advocate the United State's annexation of the Republic of Texas (*The World Book Encyclopedia*
1993a). O'Sullivan continued to use the phrase in his writings, arguing that it was God's intent that the
United States should control all of North America: "it is America's manifest destiny to overspread the
continent" (Encarta 2006). While the term "manifest destiny" has been used to generically describe
a broad nineteenth century philosophy towards U.S. expansionism, its influence in American policy
and resource management was completely theoretical. While widely debated, this philosophy/theory
expressed an underpinning belief that the United States had been chosen by God, because of its virtue,
to spread its institutions across the continent and, later in the century, into regions of the Pacific.

    While the roots of manifest destiny can be traced to the Bible and the early leaders of the United
States including Thomas Jefferson and John Quincy Adams, this was not a benign philosophy idly
debated by our nation's elite. It became the theoretical underpinning of wide-ranging policy that
impacted the use and management of the natural resource wealth of the nation. Manifest destiny was
at least symbolic of nineteenth century America's belief system towards its mission. At most, it was
a fundamental expression of racism and arrogance, whose policy outcomes were the genocide of

Native Americans, the decimation of multiple wildlife populations, destruction of native ecosystems, and depletion of our nation's mineral wealth.

The long-term implications of this theoretical belief system on management of wildlife and natural resources are wide spread and long lasting. It is staggering to consider that, in the past 200 years, the United States has lost complete biomes covering millions of acres and biological diversity in many others. During the second half of the nineteenth century, many key wildlife species in the Great Plains region of the United States faced extinction or became extinct. For example, beaver, bison, elk, antelope, white-tailed deer (*Odocoileus virginianus*), whooping cranes (*Grus americana*), passenger pigeons, and many upland bird species. While the story of population decimation for each species varies, the theme is very similar; that wildlife species and their habitat were placed on the earth for the unbridled use of Americans in their quest to fulfill their destiny of controlling the North American continent and beyond. If the success of expansionism based on the philosophy of manifest destiny is measured by the achievement of the original goals of its proponents, it was very successful. If its un-bargained-for costs, its externalities, are considered, it should give pause to current leaders and policy makers in charge of policy directed at the management of wildlife and natural resources.

THE HOMESTEAD ACT OF 1862: ECONOMIC DEVELOPMENT AND EXPLOITATION

While Texas was not included, the Homestead Act of 1862 is symbolic of the settlement of the arid and semi-arid regions of the United States, and of its consequences on wildlife and natural resources. It remains the classic example of the effects of fragmentation of land and management on wildlife and natural resources. It was signed into law by President Abraham Lincoln in 1862. It had been debated for many years, but had become politically viable only after the succession from the Union by the southern states. The Homestead Act turned over 270 million acres of government-owned land in 30 states to private citizens for a nominal filing fee and the where-with-all to spend the required time on the parcel of land to "prove it up" (National Park Service 2006). Its purpose was a mixture of expansionism, economic development, and the achievement of Jefferson's dream of a democracy based on the strength of the yeoman farmer. In response, immigrants, predominantly from feudal Europe, streamed to America for opportunity and freedom. The result was 600,000 homestead claims by 1900 and over 310,000 newly established farms (National Park Service 2006). The produce of these farms became the food supply for the industrialization and urbanization of the United States.

While fraud and corporate misuse of the Homestead Act are widely reported and discussed (Nebraska Studies 2006), the nature and scope of the footprint of the homesteaders is not. There are thousands of anecdotal reports of the indiscriminate use of wildlife as a food source and the rapid decline in populations of deer, elk, pronghorn, bison, upland game birds, and waterfowl. There are also countless reports and stories of every available tree in the landscape being used for fuel or timber. The requirement of the Homestead contract to break the grasslands into farmland is a matter of fact (*The World Book Encyclopedia* 1993b), and the expansion of these acreages into millions of tilled acres of highly erodible soils is credited as the seed for the dustbowl of the 1930s. What is lacking is a comprehensive accounting of the aggregate impact of the settlement of the west on wildlife populations, ecosystem diversity, water, and fragmentation. The widespread adoption of barbed wire as a delineating fence line of ownership and control is also emblematic of the fragmentation of one of the world's largest grasslands. Its ownership and management remain fragmented 144 years after the passage of the original act. In 1906, Liberty Hyde Bailey, Dean of Agriculture of one of the most prestigious universities of its time, Cornell University, summarized it this way: "farming has not yet adapted to the natural conditions and climate and market and other environmental factors" (Bailey 1996).

It is noteworthy that Texas was the only State in the west that was not homesteaded, and that, in 2006, it is home to the "Last Great Habitat" and many privately owned ranches that are of a size and scope rare in other states. Is it coincidental? Or is it the result of different policy applied to land ownership and use, many years ago?

AN ALTERNATIVE VIEW OF RESOURCE USE

During the second half of the nineteenth century, proposals for viewing land-based resources in alternative ways were also part of American history. Perhaps the most famous were those of John Wesley Powell. In the 1870s, Major Powell was head of the Geographical and Geological Survey of the Rocky Mountain Region, which in 1879 was consolidated with other efforts to form the U.S. Geological Survey. His extensive travels in the semi-arid and arid regions of the west led him to propose a very different land use policy concerning these regions of the United States. On April 1, 1877, Powell set forth his philosophical and pragmatic plan for the development and settlement of the west in his "Report on the Lands of the Arid Region of the United States, with a More Detailed Account of the Lands of Utah" (Stegner 1992). Wallace Stegner summarized the report as "a complete revolution in the system of land surveys, land policy, land tenure, and farming methods in the West, and a denial of almost every cherished fantasy and myth associated with the westward migration and the American dream of the Garden of the World" (Stegner 1992). Powell proposed that the lands of the West be settled and developed for agricultural purposes based not on a standard "quarter section" per homestead, but on their site potential as defined by precipitation, soil, and water. His views were not widely held, and development and settlement of the West proceeded based on the policies represented by the Jeffersonian Grid, manifest destiny, and the Homestead Act of 1862 (Manning 1995).

FROM THEORY TO POLICY TO MANAGEMENT: WILDLIFE AS A PROFIT CENTER

In his classic and foundational book on capitalistic economies published in 1776, *The Wealth of Nations*, Adam Smith wrote (Smith 1937):

> The things which have the greatest value in use have frequently little or no value in exchange; and on the contrary, those which have the greatest value in exchange have frequently little or no value in use. Nothing is more useful than water: but it will purchase scarce anything; scarce anything can be had in exchange for it. A diamond, on the contrary, has scarce any value; but a very great quantity of other goods may frequently be had in exchange for it.

This has become to be known as Smith's "diamond/water paradox" (Figure 20.4). This theory describes the concept of consumer value based on scarcity and abundance. It contains the fundamental principle that perfectly explains how American society has changed its value of and attitude toward wildlife and wildlife habitat. In nineteenth-century America, wildlife was as abundant as water, and of little value. In twenty-first-century America, wildlife is like a diamond. It has little utility, but because of scarcity, it has extreme value.

THE CREATION OF DIAMONDS

The lens with which we view, exchange, and bargain for land and its related benefits like wildlife has been unchanged for over 200 years. It is still an acre, a "40," a quarter section, or a section. But, over our nation's history, policies at the national level related to wildlife and natural resources have

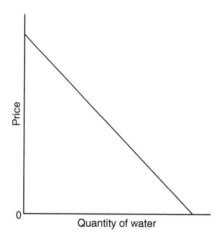

FIGURE 20.4   Adam Smith's diamond/water paradox (Adapted from Case, K. E., and R. C. Fair. 1996. *Principles of Economics*, 4th edn. Upper Saddle River, NJ: Prentice Hall).

changed dramatically. The first National Wildlife Refuge was established in 1903. The National Park System was created in 1916, and the Taylor Grazing Act was passed in 1934. The Natural Resources Conservation Service was formed in 1935. The nation's wilderness areas are 42 years old, and the Endangered Species Act is 33 years old, is up for renewal, and is highly controversial. Were these historic and watershed pieces of legislation reactionary responses to the decimation of wildlife populations and natural resources that took place during the late nineteenth and early twentieth centuries? Or perhaps they were the driving force of change, or a reflection of changing values? They were most likely reflections of all three.

During the past 30 years, populations of many wildlife species have recovered to acceptable levels. But the number of bison, bald eagles, and many other important wildlife species can still be counted in the thousands, not millions. Many waterfowl and upland birds remain relatively scarce because of an unrelenting destruction of the prairie habitat in which they nest. For example, despite policy efforts to take land out of agricultural production to improve wildlife habitat with programs like the Conservation Reserve Program (CRP), there were almost an identical number of acres harvested for crops in 2002 as in 1974 (USDA-National Agricultural Statistics Service 2004). This means that for every acre farmers and ranchers enrolled in programs like CRP to improve wildlife habitat, they converted an equal number of acres of untilled native prairie to crop production and destroyed wildlife habitat.

Access is also a part of scarcity, and access to wildlife has become complex. While the federal lands of the western United States are home to many wildlife species valued by hunters and sportsman, access to them is restricted by regulation of harvest numbers, remoteness and ruggedness of terrain, and the expense in dollars and time to travel to them. Access to many upland game birds and waterfowl is also restricted by property ownership, mostly by farmers and ranchers, whose number is shrinking. In 2002, there were approximately 2.13 million farms and ranches in the United States, nearly 500,000 fewer than in 1974 (USDA-National Agricultural Statistics Service 2004). As a result, access to private land for the purpose of hunting is controlled by dramatically fewer individuals. Also, as the business of hunting becomes a larger and larger part of ranching and farming, access for many interested individuals may actually decrease, as large acreages are removed through private and restrictive leasing. So, with an ever-increasing human population, and access to wildlife restricted by regulation, ownership, land use, and cost, consumer demand for wildlife as game has created a scarcity. Wildlife has become, as Adam Smith described, "a diamond." Evidence is reflected in the increase in rural land prices in the counties of South Texas that have growing hunting- and wildlife-centered businesses. The Texas Real Estate Center (2006) reports that average rural land prices in the ten counties that make up the Rio Grande Plains region of Texas have increased from $462 per acre in 1996 to $1200 per acre in 2005, a 159% increase in the 10-year period. They also report that average rural land prices during the same period in the nine-county South Coastal Prairie region of Texas increased from $750 per acre to $1450 per acre. These dramatic increases reflect interest in hunting, guaranteeing the purchaser access to scarce resources.

CHANGING DEFINITIONS AND CREATING OPPORTUNITIES

Ranching has historically been associated with the production of food and fiber. But ranching is being redefined as the management of financial, biological, and physical inputs inherent to arid and semi-arid rangelands (Butler 2002). These inherent inputs would include the indigenous wildlife found on a ranch. This change necessitates a fundamental shift in thought processes to include the management of wildlife resources in ranch management. Management is defined as handling with a degree of skill (Merriam-Webster's Dictionary 2001). Perhaps then, this represents an acknowledgement that in the twenty-first century, wildlife is no longer a wild resource, but is largely the result of decision making by the owners and managers of ranches. Control switches from public policy directed toward resources treated as externalities to the management of scarce resources by individuals who have vested interests in the outcome of their decision making (Rasker et al. 1992). Economic opportunity has been created and businesses follow.

Commercialization of wildlife resources is controversial (Rasker et al. 1992). In 1877, John Wesley Powell warned against it (Stegner 1992). Management regimes can range from non-consumptive uses like photography, to intense commercial hunting of native game, to the importation of non-native wildlife species for either or both consumptive and non-consumptive uses (Barber and Schulz 1997). Ethical questions arise. For example, will the enjoyment of wildlife for consumptive and nonconsumptive uses be only for the wealthy? Will the profits derived from wildlife be returned to the people, communities, and land from which they are derived, or removed to distant metropolises? In the future, wildlife may be valued for intrinsic qualities not recognized in today's marketplace. If the concept of national security is expanded to include biodiversity, then wildlife and its management may have values beyond current paradigms (Chardonnet et al. 2002). Certainly, market-driven economies are full of examples of the abuse and misuse of economic opportunity. It is the responsibility of all interested parties to engage in the development of policies and business models directed toward and concerning wildlife resources and the emerging privatization of their associated benefits (Child 1995; Sethi and Somanathan 1996).

CONCLUDING THOUGHTS

During the development of the United States, management of its wildlife resources was founded upon the emerging theory of capitalism, and framed with a series of polices that created societal paradigms in which wildlife was managed as an externality. Examples of those policies include the Jeffersonian Grid, manifest destiny, and the Homestead Act of 1862. The impact of the historical management of wildlife resources of the United States has been dramatic. In a theoretical free-market economy, private ownership and management of scarce resources allows for the benefits of that ownership and management to go to those willing to invest in them. It also places upon the owners and managers the burden and responsibility for the stewardship of those same scarce resources. The relatively recent movement to create wildlife-oriented profit centers in ranch businesses is a dramatic shift in the management of wildlife resources and will have many expected, but also many unexpected, outcomes.

The riddle remains intact 230 years after it emerged. How should publicly owned wildlife resources be managed when critical parts, if not the majority, of their habitat remains under private ownership? Changing societal attitudes toward and values of this nation's wildlife resources creates opportunity for concerned parties to learn from the past to enhance the future through enactment of thoughtfully developed and judiciously applied policies concerning management of its wildlife resources.

REFERENCES

Anderson, T. L., and P. J. Hill. 1975. The evolution of property rights and a study of the American west. *J. Law Econom.* 18:167.

Bailey, L. H. 1996. *The State and Farmer.* St Paul: University of Minnesota.

Barber, E. D., and C. E. Schulz 1997. *Wildlife, Biodiversity and Trade. Environment and Development Economics* 2. Cambridge: Cambridge University Press.

Butler, L. D. 2002. Economic survival of western ranching: Searching for answers. In *Ranching West of the 100th Meridian: Culture, Ecology, Economics*, R. L. Knight, W. C. Gilgert, and E. Marston (eds). Washington, DC: Meridian Island Press, p. 196.

Case, K. E., and R. C. Fair. 1996. *Principles of Economics*, 4th edn. Upper Saddle River, NJ: Prentice Hall.

Chardonnet, Ph., B. des Clers, J. Fischer, R. Gerhold F. Jori, and F. Lamarque. 2002. The value of wildlife. *Rev. Sci. Tech. Off. Int. Epiz.* 21:44.

Child, G. 1995. *Wildlife and People: The Zimbabwean Success.* Harare, NY: Wisdom Foundation.

Encarta, Manifest Destiny, United States History. 2006. http://encarta.msn.com/encyclopedia_761568247/Manifest_Destiny.html.

Macpherson, C. B. 1975. Capitalism and the changing concept of property. In *Feudalism, Capitalism and Beyond*, E. Kamenka, and R. S. Neale (eds). New York: St Martin's Press, p. 106.

Manning, R. 1995. *Grassland: The History, Biology, Politics, and Promise of the American Prairie.* New York: Viking.

Merriam-Webster's Dictionary, 10th edn. 2001. Springfield: Encyclopedia Britannica.

National Park Service. 2006. Homestead National Monument of America, *The Homestead Act.* http://www.nps.gov/home/historyculture/index.htm.

Nebraska Studies. 2006. *The Impact of the Homestead Act on Nebraska.* http://www.nebraskastudies.org/0500/frameset.html.

Rasker, R., M. V. Martin, and R. L. Johnson. 1992. Economics: Theory versus practice in wildlife management. *Conserv. Biol.* 6:338.

Sethi, R., and E. Somanathan. 1996. The evolution of social norms in common property resource use. *Amer. Econ. Rev.* 86:799.

Smith, A. 1937. *The Wealth of Nations, Modern Library Edition.* New York: Random House.

Stegner, W. 1992. *Beyond the Hundredth Meridian.* New York: Penguin Books.

Texas Real Estate Center. 2006. *Trends in Texas Rural Land Values — 2005.* http://recenter.tamu.edu/data/agp/rlt20.htm.

The World Book Encyclopedia. 1993a. Chicago, IL: World Book, Inc., 13:141.

The World Book Encyclopedia. 1993b. Chicago, IL: World Book, Inc., 9:304.

USDA-National Agricultural Statistics Service. 2004. United States Summary and State Data, 2002 Census of Agriculture, AC-02-CD-1, 16.

u-s-history.com. 2006. Public Land Policy Ordinance of 1785. http://www.u-s-history.com/pages/h1150.html (accessed May 2006).

West, F. J. 1975. On the ruins of feudalism-capitalism. In *Feudalism, Capitalism and Beyond*, E. Kamenka, and R. S. Neale (eds). New York: St Martin's Press.

Index

Milton Keynes UK
Ingram Content Group UK Ltd.
UKHW051945071024
449327UK00026B/2176

9 780367 388959